DISCOVERING
THE UNIVERSE

THIRD EDITION

William J. Kaufmann, III

W. H. Freeman and Company
New York

To Lawrie Kirkham, comrade and confidant

Front cover: A NASA sounding rocket, launched from White Sands Missile Range in New Mexico, carried an X-ray camera above the Earth's atmosphere to take this photograph of the Sun. The image shows a region of the Sun's atmosphere called the inner corona. Glowing gases, whose temperature is about 3 million K, trace complex and intricate patterns. Many of the brightest areas are located over sunspots, which possess strong magnetic fields. By studying the behavior of hot gases shown in this image, astronomers can learn how magnetic fields influence the solar atmosphere. (Cover photo X-ray image courtesy of Leon Golub, SAO and IBM Research)

page v: The Horsehead Nebula (AAT Board); **page vi:** The Canada-France-Hawaii Telescope on Mauna Kea (W. J. Kaufmann); **page vii:** An astronaut on the Moon (NASA); **page viii:** The Trifid Nebula (NOAO); **page ix:** The Milky Way Galaxy (NASA); **page x:** Stefan's Quintet (NOAO).

Library of Congress Cataloging-in-Publication Data
Kaufmann, William J.
Discovering the universe / William J. Kaufmann III.—3rd ed.
    p.   cm.
Includes bibliographic references and index.
ISBN 0-7167-2296-8
1. Astronomy.   2. Cosmology.   I. Title.
QB43.2.K376   1992   92-5572
520—dc20   CIP

Illustration credits are listed on page 410.

Printed in the United States of America

1 2 3 4 5 6 7 8 9 0 KP 9 9 8 7 6 5 4 3 2

Contents Overview

Contents

Preface

An adventuresome spirit pervades astronomy, a sense that things wonderful and unimagined are yet to be discovered. For millennia people have been fascinated by topics that astronomers explore today: the creation of the universe, the formation of the Earth and other planets, the motions of the stars, the structure of space and time. Armed with the powers of observation, the laws of physics, and the resourcefulness of the human mind, astronomers survey alien worlds, follow the life cycles of stars, and probe the distant reaches of the cosmos.

In *Discovering the Universe* we join astronomers both past and present in their quest for knowledge about the cosmos. We will investigate phenomena and explore realms far removed from our daily experience. Indeed, many of the objects that astronomers study are too vast, distant, or intangible to sample directly; in fact, many of the phenomena observed today occurred very long ago. When viewed in the context of the evolution of the universe, even the dimensions of space and time take on new meaning.

This book was written with the conviction that the subjects astronomers concern themselves with can—and should—be understood by just about everyone. In today's increasingly technological world, we owe it to ourselves to stay informed as best we can. In this spirit, *Discovering the Universe* offers a broad view of astronomy to readers having little or no background in science or mathematics. The narrative is largely descriptive, and a minimum of equations and formulas appear.

As its title implies, this book is as much about the nature of scientific inquiry as it is about the physical universe itself. Of all the lessons to be derived from an introductory science course, perhaps none is more important than an understanding of how scientists reason. In this book I have tried to convey how astronomers have come to know what they know. By studying the methods that astronomers have used to explore and understand the universe, we sharpen our own reasoning skills and are better able to address other scientific issues.

Organization

I wrote this text for a one-term, descriptive astronomy course. It is considerably shorter and less rigorous than its parent text, *Universe,* which offers a different selection of detail and much more expansive coverage. Having just nineteen succinct chapters, *Discovering the Universe* can be covered in a course as short as ten weeks.

The traditional Earth-outward organization of the text emphasizes how our understanding of the universe developed and invites the reader to share in the excitement of astronomical discovery. The first celestial objects we will examine are those that were observed by ancient astronomers. Moving outward from the planets to the stars and galaxies, we encounter a modern realm of observations, including those from outside the visible range and those made from space. These new observations in turn raise new questions, which draw the reader on to the limits of our universe and understanding.

The first five chapters introduce the foundations of astronomy, including descriptions of such naked-eye phenomena as eclipses and planetary motions and such basic tools as Kepler's laws and the optics of telescopes. A discussion of the formation of the solar system in Chapter 6 prepares the reader for the next four chapters, which cover the planets.

Chapter 11 introduces the Sun and sets the stage for stellar astronomy in Chapter 12. In Chapters 13 through 15, stellar evolution is described chronologically from birth to death. Molecular clouds, star clusters, nebulae, neutron stars, black holes, and various other phenomena are presented in the sequence in which they naturally occur in the life of a star, thus unifying the discussion of a wide variety of objects that astronomers find scattered about the universe.

A survey of the Milky Way introduces galactic astronomy in Chapter 16, followed by a chapter each on galaxies and quasars. Chapter 19, on cosmology, emphasizes exciting recent developments in our understanding of the physics of the early universe. An afterword addresses the question of whether intelligent life might exist elsewhere in the universe.

Changes in this edition

My first step in planning the third edition of *Discovering the Universe* was to turn to a broad spectrum of instructors who had used the second edition. The resulting comments guided my rewriting of the text. Numerous suggestions that friends and colleagues have been sending me since the publication of the second edition also served to inspire this edition. Many of the improvements made in this latest edition therefore reflect considerable classroom experience.

The most obvious change in the third edition is an enlarged page size. Larger pages and a two-column format allowed my publisher to enlarge the photographs and illustrations, resulting in a friendlier, more appealing, and more colorful book. Many of the illustrations that were black and white in the second edition have been redrawn in full color for the third edition.

To show students how scientists reason, I begin the first chapter with a description of the scientific method. To inspire students to look up at the stars, I have added a brief discussion on identifying constellations in the second chapter. The third chapter now includes an explanation of the difference between mass and weight. The latest assessment of the Hubble Space Telescope's difficulties and the funding of such projects as the Advanced X-ray Astrophysics Facility augment the fourth chapter.

The chapters on the solar system are updated to reflect recent developments, including the latest images of Venus from *Magellan* and Gaspra from *Galileo*. The important role of collisions between bodies is a unifying theme that underlies the chapters on the solar system. The chapter on the Sun now includes a discussion of the solar neutrino problem and expanded coverage of solar seismology.

New and enlarged Hertzsprung–Russell diagrams grace Chapters 12, 13, and 14, which introduce stars and stellar evolution. A carbon monoxide map of the giant molecular cloud in Orion and a comparison of visual and infrared images of the nebula M17 enhance the discussion of star formation. The latest information on the supernova SN 1987A, including a prediction that it will brighten around the year 2000, is contained in Chapter 14. A discussion of gravitational lenses in Chapter 15 is now highlighted with an image from the Hubble Space Telescope of the Einstein cross.

Collisions and mergers of galaxies are now a prominent topic in Chapter 17, which includes supercomputer simulations of galaxy–galaxy interactions as well as a remarkable radio map of the M81–M82–NGC 3077 group. The supermassive-black-hole model of the "central engine" in quasars and active galactic nuclei has been significantly enhanced with both illustrations and caveats. Observational evidence pointing to the existence of such black holes now includes the latest images from the Hubble Space Telescope of the light cusps in M32 and M87.

The final chapter on cosmology has been rewritten to reflect the growing dilemma about the age of the universe. A Hubble constant of 80 km/s/Mpc has been adopted in keeping with recent determinations. A discussion of a nonzero cosmological constant gives the student a glimpse of how cosmologists try to reconcile observations and models. With these changes, I have ventured to create a comprehensive yet entertaining text that gives a fair and accurate picture of the full scope of astronomy.

Finally, end-of-chapter material has been expanded significantly. The number of review and advanced questions has nearly doubled. The review questions focus almost exclusively on testing a student's thoroughness in reading and understanding the material in each chapter. Problems that involve calculation, reasoning, or research are now confined to the advanced-problems section. All the references at the end of each chapter have been annotated with thoughtful comments to guide the student in selecting material for further reading.

Pedagogical emphasis

I very much care that students find this book a pleasure to read and learn from. The field is dynamic, compelling, and relevant, and no book about it should be otherwise. I hope that readers will turn these pages with relish and that their interest in astronomy will grow daily.

For those taking astronomy as a first science course, I would have the experience be as rewarding as possible. The greater the ease of understanding, the greater the reward. To this end, I have tried to emphasize the central ideas around which astronomy (and indeed other sciences) revolve. Each chapter begins with a one-paragraph abstract that gives a clear idea of the chapter's contents. The chapter headings are declarative sentences that highlight main concepts, and a formal summary outlines the essential facts addressed in each chapter. Each chapter concludes with a series of questions grouped by difficulty and content into three categories: review, advanced, and discussion. Answers to questions that require computation (marked with an asterisk) appear at the end of the book. I have taken care to include questions whose answers require reasoning rather than rote memorization. Finally, the glossary provides page references, so the reader may easily refer to an appropriate place in the text.

Supplements

It is a pleasure to announce the availability of the following outstanding supplementary material for the third edition of *Discovering the Universe*:

* An *Instructor's Manual* prepared by Douglas Brown at Bellevue Community College contains chapter outlines, key points and teaching strategies, sample test questions, and solutions to computational problems in the text. Included are a section on observational activities and a resource update of *Universe in the Classroom* by Andrew Fraknoi of the Astronomical Society of the Pacific, providing an invaluable bibliography of books, articles, audiovisual materials, and software.

* A printed *Test Bank* as well as computerized versions (IBM PC and Macintosh), revised and expanded by T. Alan Clark and William J. F. Wilson at the University of Calgary, Alberta, Canada, are designed for a one-term course in basic astronomy.

* Two 45-minute videotapes that cover specific portions of the text are available to adopters. These tapes of my lectures were made during the 1992 summer meeting of the Astronomical Society of the Pacific at the University of Wisconsin. The videotape on black holes, entitled *Black Holes and Warped Spacetime*, covers most of the material in Chapter 15. The videotape entitled *Cosmology and the Creation of the Universe* was designed to be an entertaining introduction to Chapter 19.

* An attractive and useful set of 100 *Overhead Transparencies* or *Slides* of full-color line diagrams from the text is also available to adopters.

For more information and to request copies of these supplements, please contact:

Professor Services Department
W. H. Freeman and Company
4419 West 1980 South
Salt Lake City, UT 84104

Telephone: 801-973-4460
FAX: 801-977-9712

Acknowledgments

I would like to begin by thanking my many friends and colleagues who offered suggestions, comments, and corrections to the second edition. Many of the improvements in this new edition are a direct result of this unsolicited input. I am also deeply grateful to the following people, who either evaluated the second edition or carefully reviewed the manuscript for the third:

David J. Batuski   University of Maine, Orono
Charles S. Hagar   San Francisco State University

Patrick C. Hecking	Thiel College
Yong Hak Kim	Saddleback College
John H. Lacy	University of Texas, Austin
David Matzke	University of Michigan, Dearborn
Stephen W. Prata	College of Marin
Ronald Stoner	Bowling Green State University
Benjamin J. Taylor	Brigham Young University
William G. Tifft	University of Arizona
Derek Wills	University of Texas, Austin
Robert Zimmerman	University of Oregon

The reviewers of preceding editions also deserve acknow-
ledgment, as their advice has had an ongoing influence:

Wyatt W. Anderson	University of Georgia
Jay Boleman	University of Central Florida
Barbara Bowman	University of California, Berkeley
John W. Burns	Mount San Antonio College
Gladwin Comes	Broward Community College
Robert J. Dukes	College of Charleston
Alexei Filippenko	University of California, Berkeley
Roger A. Freedman	University of California, Santa Barbara
John Friedman	University of Wisconsin, Milwaukee
Douglas J. Futuyama	State University of New York, Stony Brook
Paul Helminger	University of South Alabama
Richard Henry	University of Oklahoma
Hal R. Jandorf	Moorpark College
Hollis R. Johnson	Indiana University
Klaus Keil	The University of New Mexico
Yong Hak Kim	Saddleback College
Steven L. Kipp	Mankato State University
Richard S. Marasso	Sierra College
Norman L. Markworth	Stephen F. Austin State University
Steve McMillan	Drexel University

Larry C. Oglesby	Pomona College
R. P. Olowin	St. Mary's College of California
Tobias Owen	State University of New York, Stony Brook
Hugh O. Peebles, Jr.	Lamar University
James G. Peters	San Francisco State University
Terry Rettig	University of Notre Dame
Thomas H. Robertson	Ball State University
Daniel Rothstein	Kent State University
Dale R. Snider	University of Wisconsin, Milwaukee
Michael Stewart	San Antonio College
Takamasa Takahashi	St. Norbert College
Harley Thronson, Jr.	University of Wyoming
Louis Winkler	Pennsylvania State University
David R. Wood	Wright State University

Thanks are due to many other people who have participated in the preparation of this book. Foremost among them is the development editor, Kay Ueno, who worked closely with me for many months. I also thank the acquisitions editor, Jerry Lyons, and W. H. Freeman's president, Linda Chaput, for their support and encouragement during this project.

It was once again a pleasure to work with Georgia Lee Hadler, the project editor. I also thank Sheila Anderson, the production coordinator; Nancy Singer, the designer; Mara Kasler, the illustration coordinator; and Andrew Kudlacik, who did typesetting and page make-up. I also wish to acknowledge the copy editor, Paul Monsour, and the airbrush artistry of Tomo Narashima. Finally, I thank Patrick Shriner for his assistance with the videotapes and the other supplements to this edition.

Every effort has been made to make this book error-free. Nevertheless, some errors may have crept in. I would appreciate hearing from anyone who finds an error or who wishes to comment on the text. You may write to me care of the Physics Department at SDSU. I will respond personally to all correspondence.

William J. Kaufmann, III
Department of Physics
San Diego State University

The Horsehead and Orion nebulae   *New stars are forming in the clouds of interstellar gas and dust shown in this photograph, which covers an area of the sky approximately 4° × 6°. The gases glow because of the radiation emitted by newborn, massive stars. Clouds of interstellar dust block light; they appear as dark regions silhouetted against glowing background nebulosity. The Horsehead Nebula is in the lower left; the Orion Nebula is toward the upper right. Both nebulae are about 1500 light-years from Earth. (Royal Observatory, Edinburgh)*

1

Astronomy and the Universe

Astronomy, the study of the universe, seeks to answer profound questions that people have pondered since the dawn of civilization. In this preview of the following chapters, we see that by exploring the planets, astronomers learn about the formation and evolution of the solar system. By observing stars and nebulae, they learn about the life cycles of stars. By studying galaxies, they uncover important clues about the creation of the universe. Centuries of investigating the mysteries of the universe have led us to the important realization that the physical world obeys certain rules called the laws of physics. The discovery of these rules results from systems of measurement as well as from the creativity and insight of the human mind. In addition to the deep satisfaction each new discovery brings, our understanding of the universe gives us a sweeping perspective from which we can appreciate our existence on Earth.

Figure 1-1 **The starry sky** *The star-filled sky is a beautiful and inspiring sight. This photograph, taken from northern Mexico, shows Halley's Comet and a portion of the Milky Way (upper right). To get a good view of the heavens, you must be far from any city lights. (Courtesy of D. L. Mammana)*

The splendor of the star-filled night sky is one of the central experiences of life. Gazing out into the heavens, we see thousands of stars scattered from horizon to horizon. The delicate mist of the Milky Way traces a faerie path across the starry firmament, and the entire spectacle swings slowly overhead from east to west as the night progresses. No light show, no artist's brush, no poet's words can truly capture the beauty of this breathtaking panorama (Figure 1-1).

For thousands of years people have looked up at the heavens and found themselves inspired to contemplate the nature of the universe. Like our ancestors, we find our thoughts turning to profound questions as we gaze at the stars. How was the universe created? Where did the Earth, Moon, and Sun come from? What are the planets and stars made of? And how do we fit in? What is our place and role in the cosmic scope of space and time?

To wonder about the nature of the universe is one of the most characteristic of human traits. Our curiosity, our desire to explore and discover, are unique qualities that distinguish us from lower creatures. The study of the stars transcends all boundaries of culture, geography, and politics. In a literal sense, astronomy is a universal subject—its subject is the entire universe.

1-1 Astronomers use the laws of physics to construct testable theories and models for understanding the universe

Astronomy has a rich heritage that dates back to antiquity. In many ancient societies, myths and legends dominated astronomy. The heavens were thought to be populated by demons and heroes, gods and goddesses. Astronomical phenomena were explained as the result of supernatural forces and divine intervention.

In Chapter 2 we examine some of the astronomical observations and ideas of our ancestors. We shall see that the course of civilization was greatly affected by the realization that the universe is comprehensible. This first glimpse of the power and potential of the human mind is one of the great gifts to come to us from ancient Greece. Greek astronomers learned that by observing the heavens and carefully thinking about what they saw, they could discover something about how the universe operates. For example, they measured the size of the Earth and understood and predicted eclipses.

The approach of using observation and logic to explore physical reality evolved into the **scientific method**, which has become a dominant force in science and society over the past four hundred years. In essence, the scientific method requires that our ideas about the world around us be in agreement with what we actually observe. More specifically, a scientist trying to understand some phenomenon begins by proposing a **hypothesis**, which is an idea or collection of ideas that seems to explain the phenomenon. The hypothesis should not conflict with known observations and experiments, because a discrepancy with existing facts implies that the hypothesis is wrong. The hypothesis should also be testable with new observations and experiments. This crucial aspect of the scientific method requires that the scientist carefully examine the hypothesis to see what it predicts or how its implications might be tested. Only after the hypothesis has withstood such tests, by having accurately forecast the results of new experiments and observations, does the scientist feel confident that the hypothesis is on firm ground.

It is important to realize that science does not discover absolute truth. Scientists instead describe reality in terms of **models** and **theories**, which are two commonly used names for hypotheses that have withstood observational tests. For example, the model of the atom, which scientists picture as electrons orbiting a central nucleus, is a descriptive representation of nature. Similarly, Einstein's general theory of relativity describes gravity in terms of its effects on space and time. The scientific method requires that the scientist be an open-minded person, willing to discard even the most cherished ideas if they fail to agree with observation and experiment. As new discoveries are made, old models and theories must be modified or discarded in favor of more comprehensive explanations of the world around us.

Some people think that astronomy deals only with faraway places of no possible significance to life here on Earth. This opinion is quite wrong. For example, in Chapter 3 we see that the seventeenth-century scientist Isaac Newton succeeded in describing how the planets orbit the Sun. The motions of the planets, unhampered by air resistance or friction, reveal some of the most fundamental laws of nature in their simplest form. From Newton's work we obtained our first complete, coherent description of the behavior of the physical universe. The resulting body of knowledge, which is called **Newtonian mechanics**, explicitly quantifies such concepts as force, mass, acceleration, momentum, and energy. This understanding had immediate practical application in the construction of machines, buildings, and bridges. It is no coincidence that the Industrial Revolution followed hard on the heels of these theoretical and mathematical advances inspired by astronomy.

Astronomers use Newtonian mechanics, along with other physical principles (usually called the **laws of physics**), to interpret their observations and to understand the processes that occur in the universe. Light and its relationship to matter are of particular importance to astronomers. By trying to understand how objects emit radiation and how this light interacts with matter, we acquire the skills needed to analyze and interpret the wealth of information coming to us from the stars and galaxies. Astronomers use these skills to obtain fundamental information about stars, galaxies, and the evolution of the universe.

Figure 1-2 *A telescope in space* This artist's painting shows astronauts constructing a telescope, called the Large Deployable Reflector, at an Earth-orbiting space station. Unhampered by the obscuring effects of the Earth's atmosphere, this telescope will detect visible and nonvisible radiation from astronomical objects. Astronomers who are planning this ambitious project hope that the telescope will be launched around the year 2000. (NASA)

In Chapter 4 we discuss the astronomer's most important tool, the telescope. We see that the laws of physics, particularly those involving optics and light, can be used to develop new tools and techniques for examining and exploring the universe. Until recently, everything we knew about the distant universe was based on visible light. Astronomers would peer through telescopes to observe and analyze visible starlight. By the end of the nineteenth century, however, scientists had begun discovering such nonvisible forms of light as X rays and gamma rays, radio waves and microwaves, and ultraviolet and infrared radiation.

Astronomers have recently constructed telescopes that can detect these nonvisible forms of light (Figure 1-2). Whether located on Earth or in orbit above the obscuring effects of the atmosphere, these astronomical instruments give us views of the universe vastly different from anything our eyes can see. This new information is crucial to our understanding of familiar objects like the Sun and also gives us important clues about such exotic objects as neutron stars, pulsars, quasars, and black holes.

We complete our introduction to astronomy in Chapter 5 with a discussion of light. By understanding how objects emit radiation and how light interacts with matter, we acquire the skills to analyze and interpret the wealth of information coming to us from the stars and galaxies. In later chapters we shall see how astronomers use these skills to obtain fundamental information about stars, galaxies, and the evolution of the universe.

1-2 By exploring the planets, astronomers uncover clues about the formation of the solar system

This book presents the substance of modern astronomy in three segments, corresponding to three major steps out into the universe: the planets, the stars, and the galaxies. The planets that orbit the Sun are much closer to us than the stars we see in the night sky. Planets are not self-luminous; we can see them because they reflect some of the sunlight that falls on them. In contrast, stars like the Sun generate their own light. Galaxies are enormous collections of many billions of stars.

The star we call the Sun and all the celestial bodies (including the Earth) that orbit it make up the **solar system**. In Chapter 6 through 10, we explore the solar system, beginning with Earthlike worlds and moving outward toward the frigid depths of space where comets spend most of their time.

Most of the up-to-date information we have about the planets comes from numerous Soviet and American space flights, primarily those during the 1970s and 1980s (Figure 1-3). Robot spacecraft have visited all the planets except Pluto, usually the outermost world in the solar system. Through the instruments of these unmanned probes, we have flown over Mercury's cratered surface, we have peered beneath Venus's poisonous cloud cover, we have discovered enormous canyons and extinct volcanoes on Mars. We have visited the moons of Jupiter, we have seen the rings of Saturn and Uranus, we have looked down on the active atmosphere of Neptune. Along with the moon rocks brought back by the Apollo astronauts,

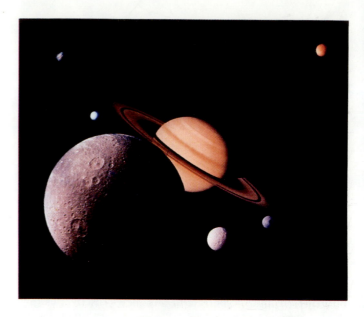

Figure 1-3 Saturn and its satellites This composite photograph shows the planet Saturn along with several of its moons. Space probes to the planets have provided us with a wealth of information about other worlds. This new knowledge gives us important insights into the formation and evolution of the solar system, as well as an appreciation of the variety of which nature is capable. (NASA)

this wealth of new and exciting information has produced a major revolution in how astronomers think about the solar system. In particular, we have come to realize that collision between objects in space is a fundamental process that has shaped many of the planets and their satellites. Craters on the Moon and on many other worlds stand in mute testimony to innumerable impacts by interplanetary rocks. The Moon itself may be the result of a catastrophic collision between the Earth and a planet-sized object shortly after the solar system was formed. As we shall see, such a collision could have torn sufficient material from the primordial Earth to create the Moon.

Throughout our journey across the solar system, we shall find that our discoveries are relevant to the quality of human life here on Earth. Until recently, our knowledge of such subjects as geology, weather, and climate was based on data from only the planet Earth. Since the advent of space exploration, however, we have had a range of other worlds to compare and contrast with our own. As a result, we have not only made important progress in understanding the creation and evolution of the Earth and the solar system, but we have also gained many valuable insights into the origins and extent of our natural resources.

1-3 By studying stars and nebulae, astronomers discover how stars are born, grow old, and eventually die

As we begin our study of the Sun and the stars in Chapters 11 and 12, we see again the surprising impact of astronomy on the course of civilization. In the 1920s and 1930s, physicists figured out how the Sun shines. At its center, thermonuclear reactions convert hydrogen into helium. This violent process releases a vast amount of energy, which eventually makes its way to the Sun's surface and escapes as light (Figure 1-4). By 1950 physicists had learned how to reproduce these thermonuclear reactions here on Earth. Hydrogen bombs operate on the same basic principles as do the thermonuclear reactions that produce energy at the Sun's center. In the decades following the invention of the hydrogen bomb, thermonuclear weapons were deployed around the world and offered the specter of devastating life on our planet. In contrast, ongoing research into peaceful applications of controlled thermonuclear reactions may result in a clean source of electrical energy early in the twenty-first century.

As we look deeper into space, we find star clusters and clouds of glowing gas, called **nebulae** (singular **nebula**),

Figure 1-5 The Orion Nebula This beautiful nebula (also called M42 or NGC 1976) is a fine example of a stellar "nursery" where stars are born. Intense radiation from the newly formed stars causes the surrounding gases to glow. Many of the stars embedded in this nebula are less than a million years old. The Orion Nebula is 1300 light-years from Earth, and the distance across the nebula is about 5 light-years. (Anglo-Australian Observatory)

scattered across the sky. These beautiful objects can tell us much about the lives of stars. Stars are born in huge clouds of interstellar gas and dust such as the Orion Nebula, shown in Figure 1-5. After many millions or billions of years, stars eventually die. Some end their lives with a spectacular detonation called a **supernova** that blows the star apart. The Crab Nebula seen in Figure 1-6 is a striking example of a supernova remnant.

As they die, stars return gas to interstellar space. This gas contains some of the original hydrogen as well as heavy elements created by thermonuclear reactions in the stars' interiors. Interstellar space thus becomes enriched with newly manufactured atoms and molecules. This process is analogous to decaying leaves and logs enriching the soil for future generations of trees. Indeed, the Sun and its planets were formed from enriched interstellar material. We therefore realize that virtually everything we touch, including the atoms in our own bodies, was created long ago in now-dead stars.

Chapters 14 and 15 discuss how dying stars can produce some of the strangest objects in the sky. The majority of dying stars become **white dwarfs**, which are very compact objects roughly the same size as the Earth. Some dead stars become **pulsars**, which emit pulses of radio waves, or **bursters**, which emit powerful bursts of X rays. Massive

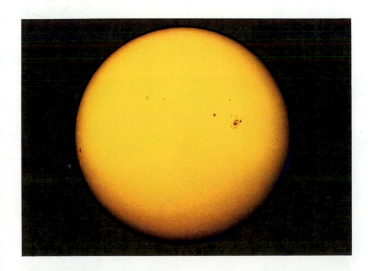

Figure 1-4 Our star—the Sun The Sun is a typical star. Its diameter is about 1.39 million kilometers (roughly a million miles), and its surface temperature is about 5500°C (10,000°F). The Sun draws its energy from thermonuclear reactions occurring at its center, where the temperature is about 15 million degrees Celsius. (NOAO)

Figure 1-6 The Crab Nebula *This nebula (also called M1 or NGC 1952) is a fine example of a supernova remnant. A dying star exploded, and this beautiful funeral shroud was caused by the gases blasted violently into space. In fact, these gases are still moving outward, at about 1000 km/s (roughly 2 million miles per hour). The Crab Nebula is 6300 light-years from Earth, and the distance across the nebula is about 6 light-years. (Lick Observatory)*

dead stars become **black holes,** surrounded by such incredibly powerful gravity that nothing—not even light—can escape from them. Many of these bizarre stellar corpses have been discovered in recent years with Earth-orbiting telescopes that detect nonvisible light, like the X rays emitted by gases falling toward a black hole.

1-4 By observing galaxies, astronomers learn about the creation and fate of the universe

Stars are not spread uniformly across the universe but are grouped together in huge assemblages called **galaxies.** Galaxies are the largest individual objects in the universe. A typical galaxy, like our own Milky Way, contains several hundred billion stars.

In Chapter 16 we tour the Milky Way Galaxy. We discover that our Galaxy has beautiful, arching spiral arms like those of M83 in Figure 1-7, which are active sites of star formation. The center of the Galaxy is emitting vast quantities of energy. Some astronomers suspect that this energy output is caused by gas falling into an enormous black hole at the galactic center.

In Chapter 17 we explore other galaxies and find that they come in a wide range of shapes and sizes. Some galaxies are quite small, containing only a few hundred million stars. Others are veritable monstrosities that devour neighboring galaxies in a process called "galactic cannibalism."

Some of the most intriguing galaxies are those that appear to be in the throes of violent convulsions. The centers of these strange galaxies, which may also harbor extremely massive black holes, are often powerful sources of X rays and radio waves.

Even more dramatic sources of energy are found still deeper in space. As described in Chapter 18, at distances

Figure 1-7 The galaxy M83 *This spectacular galaxy (also called NGC 5236) contains about 200 billion stars. The galaxy's spiral arms are outlined by many nebulae, which are sites of active star formation. This galaxy has a diameter of about 35,000 light-years and is at a distance of 12 million light-years from Earth. (Courtesy of R. J. Dufour)*

Figure 1-8 The quasar 3C48 Quasars are believed to be the most distant objects and most luminous objects that astronomers have ever seen. At first glance, a quasar is easily mistaken for a faint star. This quasar is thought to be at a distance of 4.8 billion light-years from Earth. (Palomar Observatory)

of billions of light-years from Earth, we find the mysterious **quasars**. Although quasars look like stars (Figure 1-8), they are probably the most distant and most luminous objects in the sky. A typical quasar shines with the brilliance of a hundred galaxies. Data support models suggesting that quasars draw their awesome energy from enormous black holes.

Finally, in Chapter 19, we turn to the most fundamental questions about the creation and fate of the universe. We shall see how the motions of galaxies reveal that we live in an expanding universe. Extrapolating into the past, we learn that the universe must have been born from an incredibly dense—perhaps infinitely dense—state roughly 15 billion years ago.

Most astronomers believe that the universe began with a cosmic explosion, known as the **Big Bang**, that occurred throughout all space at the beginning of time. Shortly after the Big Bang, events happened that dictated the present nature of the universe. Astronomers are making significant progress in understanding these cosmic events. Indeed, they may be about to discover the origin of some of the most basic properties of the universe. In addition, the motions of the most distant clusters of galaxies may tell us the ultimate fate of the universe: whether it will continue expanding forever or someday stop and collapse back on itself.

The foregoing discoveries and theories of today's astronomy owe a great debt to astute observers and thinkers spanning several centuries and to the first stirrings of the human imagination that led to the development of systems of measurement. Curiosity, insight, and creativity together with observation, measurement, and disciplined inquiry fostered the development of all science. Given the human need to know and understand, it seems inevitable that we would begin to unravel even the deepest mysteries of the universe. The basic tools we need for this bold endeavor in astronomy were developed long ago and are quite straightforward.

1-5 *Astronomers use angles to denote the apparent sizes and positions of objects in the sky*

Astronomers have inherited many useful concepts from antiquity. For example, ancient mathematicians invented angles and a system of angular measure that is still used to denote the positions and apparent sizes of objects in the sky.

An **angle** is the opening between two lines that meet at a point. **Angular measure** is a method of describing the size of an angle. The basic unit of angular measure is the **degree**, designated by the symbol °. A full circle is divided into 360°. A right angle measures 90°. As shown in Figure 1-9, the angle between the two "pointer stars" in the Big Dipper is about 5°.

Astronomers use angular measure to describe the apparent sizes of celestial objects. For example, imagine looking up at the full moon. The angle covered by the Moon's diameter is nearly $\frac{1}{2}°$. We therefore say that the **angular diameter**, or **angular size**, of the Moon is $\frac{1}{2}°$. Alternatively, astronomers say that the Moon "subtends" an angle of $\frac{1}{2}°$. In this context, *subtend* means "to extend across."

To talk about smaller angles, we subdivide the degree into 60 minutes of arc (abbreviated 60 arc min or 60′). A minute of arc is further subdivided into 60 seconds of arc (abbreviated 60 arc sec or 60″). Thus,

$$1° = 60 \text{ arc min} = 60'$$

$$1' = 60 \text{ arc sec } = 60''$$

Your thumbnail held at arm's length subtends an angle of 1°, and a dime viewed from a distance of 1 mile has an angular diameter of about 2 arc sec.

The *Astronomical Almanac* for 1992 states that on December 28, Venus had an angular diameter of 19.55

Figure 1-9 The Big Dipper The Big Dipper is an easily recognized grouping of seven bright stars. The angular distance between the two "pointer" stars at the front of the Big Dipper is about 5°. For comparison, the angular diameter of the Moon is about $\frac{1}{2}°$. Ten full moons could fit side by side between the two pointer stars.

arc sec, as viewed from Earth. That is a very convenient and precise statement of how big the planet appeared in Earth's sky on that date.

From everyday experience, we know that an object looks big when it is nearby but small when it is far away. The angular size of an object therefore does not necessarily tell you anything about its actual physical size. In order to convert angular size to physical size, you also need to know the distance to the object. For instance, the fact that the Moon's angular diameter is $\frac{1}{2}°$ does not tell you how big the Moon really is. But if you also happen to know the distance to the Moon, then it is possible to calculate the Moon's physical diameter.

1-6 Powers-of-ten notation is a useful shorthand system for writing numbers

Astronomy is a subject of extremes. As we examine various environments, we find an astonishing range of conditions, from the incredibly hot, dense centers of stars to the frigid, near-perfect vacuum of interstellar space. To de-

scribe such divergent conditions accurately, we need a wide range of both large and small numbers. Astronomers avoid such confusing terms as "a million billion billion" by using a standard shorthand system: All the cumbersome zeros that accompany a large number are consolidated into one term consisting of 10 followed by an **exponent**, which is written as a superscript and called the **power of ten**. The exponent indicates how many zeros you would need to write out the long form of the number. Thus,

$$10^0 = 1$$
$$10^1 = 10$$
$$10^2 = 100$$
$$10^3 = 1000$$
$$10^4 = 10,000$$

and so forth. The exponent tells you how many tens must be multiplied together to give the desired number. For example, ten thousand can be written as 10^4 ("ten to the fourth") because $10^4 = 10 \times 10 \times 10 \times 10 = 10,000$.

With this notation, numbers are written as a figure between 1 and 10 multiplied by the appropriate power of 10. The distance between the Earth and the Sun, for example, can be written as 1.5×10^8 km. Once you get used to it, you will find this notation more convenient than writing "150,000,000 kilometers" or "one hundred and fifty million kilometers."

This shorthand system can also be applied to numbers that are less than 1 by using a minus sign in front of the exponent. A negative exponent tells you that the location of the decimal point is as follows:

$$10^0 = 1$$
$$10^{-1} = 0.1$$
$$10^{-2} = 0.01$$
$$10^{-3} = 0.001$$
$$10^{-4} = 0.0001$$

and so forth. For example, the diameter of a hydrogen atom is 1.1×10^{-8} cm. That is more convenient than saying "0.000000011 centimeter" or "eleven-billionths of a centimeter."

Using the powers-of-ten notation, one can write familiar numerical terms as follows:

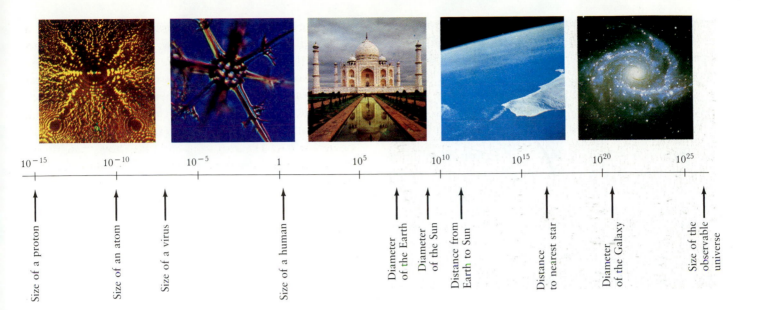

10^{-15}	10^{-10}	10^{-5}	1	10^5	10^{10}	10^{15}	10^{20}	10^{25}

Size of a proton · Size of an atom · Size of a virus · Size of a human · Diameter of the Earth · Diameter of the Sun · Distance from Earth to Sun · Distance to nearest star · Diameter of the Galaxy · Size of the observable universe

Figure 1-10 Examples of powers-of-ten notation *The scale gives the sizes of objects in meters, ranging from subatomic particles at the left to the entire observable universe on the right. The photograph at the left shows tungsten atoms, 10^{-10} meter in diameter. Second from left is a crystalline skeleton, 10^{-4} meter (0.1 millimeter) in size, of a diatom—a single-celled organism.*

At the center is the Taj Mahal, within reach of our unaided senses. On the right, looking across the Indian Ocean toward the South Pole, we see the curvature of the Earth, 10^7 meters in diameter. At the far right is a galaxy, 10^{21} meters (100,000 light-years) in diameter. (Courtesy of Scientific American Books; NASA; AAT)

$$\text{one thousand} = 10^3 = 1000$$
$$\text{one million} = 10^6 = 1,000,000$$
$$\text{one billion} = 10^9 = 1,000,000,000$$
$$\text{one trillion} = 10^{12} = 1,000,000,000,000$$

and also

$$\text{one thousandth} = 10^{-3} = 0.001$$
$$\text{one millionth} = 10^{-6} = 0.000001$$
$$\text{one billionth} = 10^{-9} = 0.000000001$$
$$\text{one trillionth} = 10^{-12} = 0.000000000001$$

Powers-of-ten notation bypasses all the awkward zeros so that a wide range of circumstances such as those shown in Figure 1-10 can be described in a convenient fashion. Figure 1-10 also shows how clearly the powers-of-ten notation expresses the scale of objects, ranging from subatomic particles like the proton to the size of the observable universe.

1-7 Astronomical distances are often measured in astronomical units, parsecs, or light-years

As we turn toward the stars in the second half of this book, we shall find that some of our traditional units of measure become cumbersome. It is fine to use kilometers or miles to measure the diameters of craters on the Moon or the heights of volcanoes on Mars. But it is as awkward to use kilometers to express distances to stars or galaxies as it would be to talk about the distance from New York to San Francisco in millimeters or inches. Astronomers have therefore devised new units of measure.

When discussing distances across the solar system, astronomers use a unit of length called the **astronomical unit** (abbreviated AU), which is the average distance between the Earth and the Sun:

$$1 \text{ AU} = 1.496 \times 10^8 \text{ km} = 93 \text{ million miles}$$

Thus, the distance between the Sun and Jupiter can be conveniently stated as 5.2 AU.

When talking about distances to the stars, astronomers choose between two different units of length. One is the **light-year** (abbreviated ly), which is the distance that light travels in one year:

$$1 \text{ ly} = 9.46 \times 10^{12} \text{ km}$$

One light-year is roughly equal to 6 trillion miles. Proxima Centauri, the nearest star other than the Sun, is 4.2 ly from Earth, for example.

The second commonly used unit of length is the **parsec** (abbreviated pc). Imagine taking a journey far into space, beyond the orbits of the outer planets. As you look back toward the Sun, the Earth's orbit subtends a smaller angle in the sky the farther you are from the Sun. The distance at which 1 AU subtends an angle of 1 arc sec, as shown in Figure 1-11, is defined as 1 parsec. The parsec turns out to be longer than the light-year. Specifically,

$$1 \text{ pc} = 3.09 \times 10^{13} \text{ km} = 3.26 \text{ ly}$$

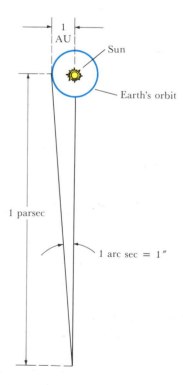

Figure 1-11 A parsec *The parsec, a unit of length commonly used by astronomers, is equal to 3.26 ly. The parsec is defined as the distance at which 1 AU perpendicular to the observer's line of sight subtends an angle of 1 arc sec.*

Thus, the distance to the nearest star can be stated as 1.3 pc as well as 4.2 ly. Whether one uses light-years or parsecs is a matter of personal taste.

For even greater distances, astronomers commonly use **kiloparsecs** and **megaparsecs** (abbreviated kpc and Mpc), in which the prefixes simply mean "thousand" and "million," respectively:

$$1 \text{ kpc} = 10^3 \text{ pc}$$
$$1 \text{ Mpc} = 10^6 \text{ pc}$$

For example, the distance from Earth to the center of our Milky Way Galaxy is about 8 kpc, and the rich cluster of galaxies in the direction of the constellation of Virgo is 20 Mpc away.

Some astronomers prefer to talk about thousands or millions of light-years rather than kiloparsecs and megaparsecs. Once again, the choice is a matter of personal taste.

1-8 Astronomy is an adventure of the human mind

An underlying theme of this book is the idea that the universe is comprehensible. Physical reality is not a hodgepodge of unrelated things behaving in arbitrary and unexplainable ways. Rather, we find strong evidence for the existence of fundamental laws of physics that govern the nature and behavior of everything in the universe. This powerful concept of universally applicable laws enables us to explore realms far removed from our earthly experience. Thus a scientist can do experiments in a laboratory to determine the properties of light or the behavior of atoms, then use this knowledge to discover the life cycles of stars and the structure of the universe.

Such discoveries have had a direct and profound influence on humanity. The past four centuries of civilization clearly show that major scientific advances sooner or later make their way into our daily lives. The world around us is filled with examples of the impact of science in technology, commerce, medicine, entertainment, and transportation. In the near future we can look forward to enjoying the benefits of space technology. Weightlessness and the near-perfect vacuum of space will enable us to manufacture a wide range of exceptional substances, from exotic alloys to ultrapure medicines (Figure 1-12).

The dreams of Jules Verne and H. G. Wells pale when compared with the reality of today. Ours is an age of

Figure 1-12 An Earth-orbiting space station A new generation of high-technology materials could easily be manufactured in space. Exotic alloys, foam metals, ultrapure semiconducting crystals, and rare vaccines are among the obvious practical applications of zero-gravity industry. This artist's conception shows the first phase of the space station currently being planned by NASA. (Courtesy of Lockheed)

exploration and discovery more profound than any since Columbus and Magellan set sail. We have walked on the Moon, dug into the Martian soil. We have discovered active volcanoes and barren ice fields on the satellites of Jupiter. We have visited the shimmering rings of Saturn. Never before has so much been revealed in so short a time. As you proceed through this book, you will come to realize that one of the great lessons of modern astronomy is the awesome power of the human mind to reach out, to explore, to observe, and to comprehend, thereby transcending the limitations of our bodies and the brevity of human life.

Summary

- The universe is comprehensible.

- Observations of the heavens have led to discovery of some of the fundamental laws of nature.

- The scientific method is a procedure involving the formulation of hypotheses. These are tested by observation or experimentation in order to build consistent models or theories that accurately describe phenomena in the universe.

- Exploration of the planets provides information about the origin and evolution of the solar system, as well as the history and resources of the Earth.

- Study of the stars and nebulae provides information about the origin and history of the Sun and the solar system.

- Observations of galaxies provide information about the origin and history of the universe.

- Astronomers use angles to denote the positions and sizes of objects in the sky.

  The size of an angle is measured in degrees, minutes of arc, and seconds of arc.

- The powers-of-ten notation system is a convenient shorthand method of writing numbers.

- Astronomers use a variety of distance units, including the parsec (pc) and the light-year (ly).

  The astronomical unit (AU) is commonly used to express distances across the solar system.

  Parsecs and light-years are used to express distances to stars and galaxies.

Review questions

The answers to all computational problems, which are preceded by an asterisk, are at the end of the book.

1 What are degrees, minutes of arc, and seconds of arc used for? What is the relationship between these units of measure?

2 With the aid of a diagram, explain what it means to say that the Moon subtends an angle of $\frac{1}{2}°$.

3 What is an exponent? How are exponents used in the powers-of-ten notation?

4 What is the advantage of the powers-of-ten notation?

5 How is an AU defined? Give an example of where it would be convenient to use this unit of measure.

6 What is the advantage to the astronomer of using the light-year as a unit of distance?

7 What is a parsec and how is it related to a kiloparsec and a megaparsec?

Advanced questions

* **8** Give the word or phrase that corresponds to the following standard abbreviations: (**a**) km, (**b**) cm, (**c**) s, (**d**) km/s, (**e**) mph, (**f**) m, (**g**) m/s, (**h**) hr, (**i**) yr, (**j**) g, (**k**) kg. Which of these are units of speed? (*Hint:* You may have to refer to dictionary. All of these abbreviations should be part of your working vocabulary.)

* **9** How many seconds of arc equal 1°?

* **10** Write the following numbers using powers-of-ten notation: (**a**) ten million, (**b**) four hundred thousand, (**c**) six one-hundredths, (**d**) seventeen billion, (**e**) your age, (**f**) the number of pages in this book.

* **11** The speed of light is 3×10^8 m/s. How long does it take for light to get from the Sun to the Earth? (*Hint:* An object traveling at velocity v for a time t covers a distance d given by $d = vt$.)

* **12** The bright star Sirius is 2.64 pc from Earth. (**a**) How long does it take for light from Sirius to reach the Earth? (**b**) What is the distance to Sirius in kilometers?

* **13** The diameter of the Sun is 1.4×10^9 m, and the distance to the nearest star, Proxima Centauri, is 4.3 ly. If the Sun were reduced to the size of a basketball (about 30

cm in diameter), at what distance would Proxima Centauri be from the Sun on this reduced scale?

* **14** How many Suns laid side by side would it take to reach from Earth to the nearest star?

Discussion questions

15 How do astronomical observations and experiments differ from those of other sciences?

16 Scientists believe that the universe is comprehensible. Discuss what this means and the thinking behind it.

For further reading

Asimov, I. *The Measure of the Universe*. Harper & Row, 1983 • This extensive journey through the universe exemplifies the usefulness of powers-of-ten notation.

Chaisson, E. *Cosmic Dawn*. Little, Brown, 1981 • This excellent book presents the scientist's worldview by discussing the origins of matter and life.

Dickinson, T. *The Universe and Beyond*. Camden House, 1986 • A beautifully illustrated, nontechnical coffee-table book that takes the reader on a tour of modern astronomy. Contrary to what the title suggests, the book stays firmly within our universe.

Gore, R. "The Once and Future Universe." *National Geographic*, June 1983 • A well-written, beautifully illustrated introduction to modern astronomy.

Graham-Smith, F., and Lovell, B. *Pathways to the Universe*. Cambridge University Press, 1988 • An introduction to astronomy by two distinguished British astronomers, consisting of 21 separate topics, from "How We Became Astronomers" to "Cosmology."

Hartmann, W., and Miller, R. *Cycles of Fire*. Workman, 1987 • An oversized, colorful album of paintings by two noted astronomical artists, accompanied by an excellent nontechnical tour of stars, nebulae, and galaxies.

Morrison, P., Morrison, P., and the Office of Charles and Ray Eames. *Powers of Ten*. Scientific American Books, 1982 • A tour of the universe where each step corresponds to a power of ten.

Seielstad, G. *Cosmic Ecology*. University of California Press, 1983 • This eloquent survey traces evolutionary processes from the Big Bang to the development of intelligence on Earth.

2

Discovering the Heavens

Ancient cultures made important astronomical observations and developed ideas that set the stage for modern astronomy. In this chapter, we learn that constellations described by the ancient Babylonians and Greeks can be used to find our way around the sky, which is conveniently described as the celestial sphere. We discover that the seasons are related to the tilt of the Earth's axis of rotation and that the axis itself is slowly changing its orientation. We also learn about the phases of the Moon and their relationship to the Moon's motion about the Earth and the Earth's motion about the Sun. We see how ancient astronomers attempted to measure the size of the Earth and the distances from Earth to the Sun and to the Moon. Eclipses of the Sun and the Moon, which are dramatic phenomena, played important roles in many of these early theories and measurements. Indeed, ancient astronomers understood eclipses well enough to predict their occurrences with some degree of reliability.

Circumpolar star trails   This long exposure is aimed at the south celestial pole and shows the apparent rotation of the sky. The foreground building is part of the Anglo-Australian Observatory, which houses one of the largest telescopes in the Southern Hemisphere. Many of the photographs in this book were taken with this telescope, whose primary mirror is 3.9 meters (12.8 feet) in diameter. During the exposure, someone carrying a flashlight walked along the dome's outside catwalk. Another flashlight made the wavy trail at ground level. (Anglo-Australian Observatory)

The beauty of the star-filled night sky or the drama of an eclipse would suffice to make astronomy fascinating. But there are practical reasons as well for an interest in the universe. The ancient Greeks knew the connection between the seasons and the relative orientation of the Sun and the Earth. Many early seafaring cultures were aware that the tides are influenced by the position of the Moon.

Ancient civilizations placed great emphasis on careful astronomical observation. Hundreds of impressive monuments that dot the British Isles, such as Stonehenge (Figure 2-1), provide evidence of this preoccupation with astronomy. Alignments of the stones point to the rising and setting locations of the Sun and Moon at various times during the year. Similar astronomically oriented monuments are found in the Americas. The Medicine Wheel in Wyoming, constructed high atop a windswept plateau by the Plains Indians, has stones and markers aligned with the rising points of several bright stars and the Sun.

Architects of the Mayan city of Chichén Itzá on the Yucatán Peninsula of Mexico built an astronomical observatory, the Caracol, nearly a thousand years ago. The Caracol has a cylindrical tower that contains windows aligned with the northernmost and southernmost rising and setting points of both the Sun and the planet Venus. A similar four-story adobe building, probably constructed during the fourteenth century, is located at the Casa Grande site in Arizona. And in the ruined city of Tiahuanaco in Bolivia, ancient engineers built the Temple of the Sun with walls aligned in the north–south or east–west directions with an accuracy better than 1°. All these struc-

tures bear witness to careful, patient astronomical observations by the people of many ancient civilizations.

2-1 Eighty-eight constellations cover the entire sky

The roots of astronomy reach back to the dawn of civilization. This ancient heritage is most apparent in the **constellations** (from the Latin word meaning "group of stars"). Perhaps it was shepherds tending their flocks at night or priests studying the starry heavens who first imagined pictures among groupings of stars.

You may already be familiar with some of these patterns in the sky such as the Big Dipper, which is actually part of a large constellation called Ursa Major (the Great Bear). Many of these constellations, such as Orion in Figure 2-2, have names from ancient myths and legends. Although some star groupings vaguely resemble the figures they are supposed to represent, most do not.

On modern star charts the entire sky is divided into 88 constellations. Some constellations cover very large areas (Ursa Major is one of the biggest), others very small areas. The constellations are used to specify certain regions of the sky. Thus, for example, we might speak of "the galaxy M31 in Andromeda" much as we would refer to "the Ural Mountains in Russia."

The Earth rotates once every 24 hours (that is why we have day and night), and so the constellations rise in the east and set in the west, as do the Sun and Moon. This daily, or **diurnal**, motion of the stars is apparent in time-

Figure 2-1 Stonehenge This astronomical monument was constructed nearly four thousand years ago on Salisbury Plain in southern England. Originally, the monument consisted of 30 blocks of gray sandstone, each standing 4 m (13.5 ft) high, set in a circle 30 m (97 ft) in diameter. These stones were topped with a continuous circle of smaller stones. Inside the circle are geometrical arrangements of other stones, most notably a horseshoe-shaped set of larger stones opening toward the northeast. (Courtesy of the British government)

Figure 2-2 Orion Orion is a prominent winter constellation. From the United States, Orion is easily seen high above the southern horizon from December through March. Because of the time exposure—four minutes on Kodak Ektachrome—the colors of the stars are very noticeable. The fanciful drawing of Orion is from an 1835 star atlas. (Courtesy of Robert Mitchell and Janus Publications)

exposure photographs such as the picture on the first page of this chapter.

People who spend time outdoors at night are familiar with the diurnal motion of the constellations. If you lack this knowledge, take the time to observe the basic facts of astronomy yourself. Go outdoors soon after dark, find a spot away from bright lights, and note the patterns of stars in the sky. A few hours later, check again. You will find that the entire pattern of stars (including the Moon, if it is visible) has shifted. New constellations will have risen above the eastern horizon, while other constellations will have disappeared below the western horizon. If you check again just before dawn, you will find, low in the western sky, the stars that were just rising when the night began.

The constellations that you can see in the sky gradually change over the course of a year. This shift occurs as the Earth orbits the Sun (Figure 2-3). The Earth takes a full year to go once around the Sun, which means that the darkened, nighttime side of the Earth is gradually turned toward different parts of the heavens. If you follow a particular star on successive evenings, you find that it rises approximately 4 minutes earlier each night.

Constellations can help you orient yourself and find your way around the sky. For instance, if you live in the Northern Hemisphere, you can use the Big Dipper in Ursa Major to find the north direction by drawing a straight line through the two stars at the front of the Big Dipper's bowl, as shown in Figure 2-4. The first moderately bright star you come to is Polaris, also called the North Star because it is located almost directly over the Earth's north pole. If you draw a line from Polaris straight down to the horizon, you have found the north direction.

By drawing a line through the two stars at the rear of the bowl of the Big Dipper you can find Leo (the Lion). As Figure 2-4 shows, that line points toward Regulus, the brightest star in the "sickle," which traces the lion's mane. By following the handle of the Big Dipper, you can locate the bright reddish star Arcturus in Boötes (the Shepherd) and the prominent bluish star Spica in Virgo (the Virgin). The saying "Follow the arc to Arcturus and speed to Spica" may help you remember these stars, which are conspicuous in the evening sky during the spring and summer months.

During the winter months in the Northern Hemisphere you can see some of the brightest stars in the sky. Many of them are in the vicinity of the "winter triangle," which connects bright stars in the constellations of Orion (the

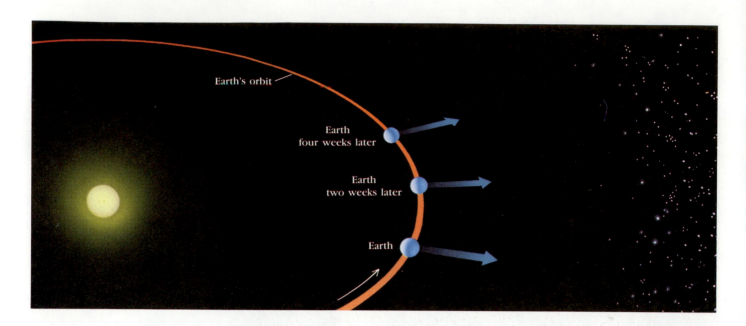

Figure 2-3 Our changing view of the night sky *As we orbit the Sun, the nighttime sky of the Earth gradually turns toward* *different parts of the heavens. Thus, the constellations that we can see change slowly from one night to the next.*

Hunter), Canis Major (the Large Hunting Dog), and Canis Minor (the Small Hunting Dog), as shown in Figure 2-5. The winter triangle is nearly overhead during the middle of winter at midnight.

A similar feature, called the "summer triangle" graces the summer sky. As shown in Figure 2-6, this triangle connects the brightest stars in Lyra (the Harp), Cygnus (the Swan), and Aquila (the Eagle). A conspicuous portion

Figure 2-4 The Big Dipper as a guide This star chart shows how the Big Dipper can be used to point out the north star as well as the brightest stars in three other constellations. The angular distance from Polaris to Spica is about 101°, so a large portion of the sky is shown.

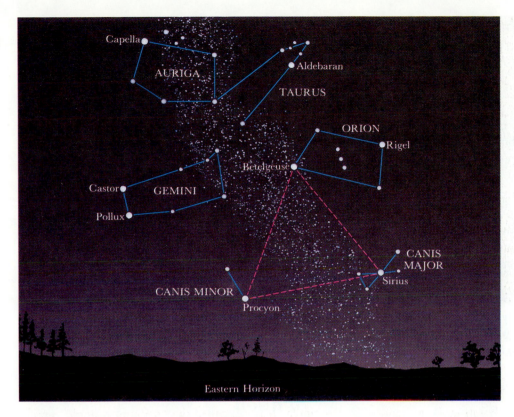

Eastern Horizon

Figure 2-5   The winter triangle
This star chart shows the eastern sky as it appears during the evening in December. Three of the brightest stars in the sky make up the winter triangle, which is about 26° on a side. In addition to the constellations involved in the triangle, Gemini (the Twins), Auriga (the Charioteer), and Taurus (the Bull) are also shown.

of the Milky Way forms a beautiful background for these constellations, which are nearly overhead during the middle of summer at midnight.

A set of star charts for the evening hours of selected months of the year is included at the end of this book. You may find stargazing a surprisingly enjoyable experience and these star charts will help you identify many well-known constellations.

Eastern Horizon

Figure 2-6   The summer triangle
This star chart shows the northeastern sky as it appears in the evening in June. The angular distance from Deneb to Altair is about 38°. In addition to the three constellations involved in the summer triangle, the faint constellations of Sagitta (the Arrow), and Delphinus (the Dolphin) are also shown.

2-2 It is often convenient to imagine that the stars are located on the celestial sphere

As you gaze at the heavens on a clear, dark night, you might think that you can see millions of stars. Actually, the unaided human eye can detect only about six thousand stars over the entire sky. At any one time, you can see roughly three thousand stars because only half of the sky is above the horizon.

Many ancient societies believed the Earth to be at the center of the universe. They also imagined that the stars were attached to the inside surface of a huge hollow sphere that encircled the Earth. This imaginary sphere, called the **celestial sphere,** is still a useful concept.

Of course, the stars are actually scattered at various distances from Earth. Many of the brightest stars you can see in the sky are 10 to 1000 light-years away. These distances are so immense, however, that all the stars appear to be equally remote, fixed to a spherical backdrop.

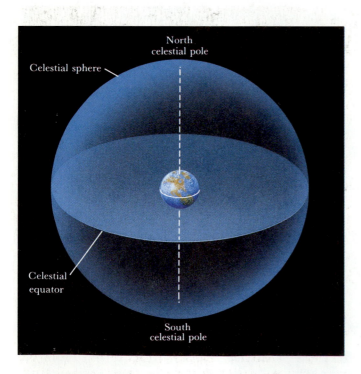

Figure 2-7 The celestial sphere *The celestial sphere is the apparent sphere of the sky. The celestial equator and pole are obtained by projecting the Earth's equator and axis of rotation out into space. The north celestial pole is therefore located directly over the Earth's north pole, while the south celestial pole is directly above the Earth's south pole.*

We can use this backdrop as a reference to specify the directions to objects in the sky.

As shown in Figure 2-7, the Earth is at the center of the celestial sphere. We can project key geographic features out into space to establish directions and bearings on the celestial sphere. If we project the Earth's equator onto the celestial sphere, we obtain the **celestial equator.** The celestial equator divides the sky into northern and southern hemispheres, just as the Earth's equator divides the Earth into two hemispheres.

We can also imagine extending the Earth's north and south poles out into space along the Earth's axis of rotation. Doing so gives us the **north celestial pole** and the **south celestial pole,** also shown in Figure 2-7. With the celestial equator and poles as reference features, astronomers denote the position of an object in the sky much in the same way that longitude and latitude are used to specify a location on Earth.

2-3 The seasons are caused by the tilt of the Earth's axis of rotation

In addition to rotating on its axis every 24 hours, the Earth revolves around the Sun every $365\frac{1}{4}$ days. The seasonal changes we experience on Earth during a year result from the way the Earth's axis of rotation is tilted with respect to the plane of the Earth's orbit around the Sun.

The Earth's axis of rotation is not perpendicular to the plane of the Earth's orbit. Instead, the Earth's axis is tilted 23.4° away from the perpendicular, as shown in Figure 2-8. The Earth maintains this tilted orientation as it orbits the Sun. Thus, during part of the year the Northern Hemisphere is tilted toward the Sun and the Southern Hemisphere is tilted away, producing summer in the north and winter in the south. Half a year later the situation is reversed, with winter in the Northern Hemisphere (now tilted away from the Sun) and summer in the Southern Hemisphere. During March and September, when spring and fall begin, both hemispheres receive roughly equal amounts of illumination from the Sun.

Imagine that you could see the stars during the day, so that you could follow the Sun's apparent motion against the background constellations. As the Earth moves along its orbit, the Sun would appear to shift its position gradually from day to day, tracing out a path in the sky called the **ecliptic** (Figure 2-9). The Sun takes one year to complete a trip around the ecliptic. Since there are about 365

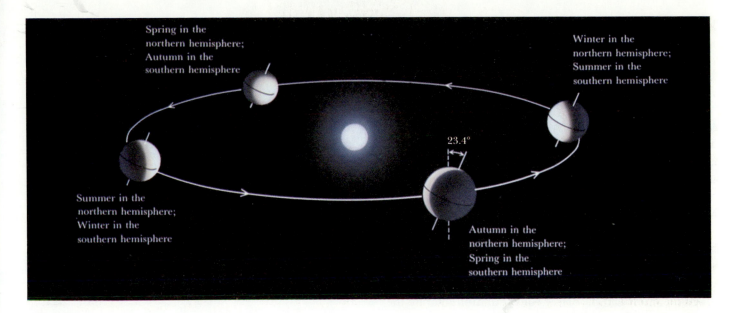

Figure 2-8 The seasons *The Earth's axis of rotation is inclined 23.4° away from the perpendicular to the plane of the Earth's orbit. The Earth maintains this orientation (with its north pole aimed at the celestial north pole near the star called Polaris)* *throughout the year as the Earth orbits the Sun. Consequently, the amount of solar illumination and the number of daylight hours at any location on Earth varies in a regular fashion throughout the year.*

days in a year and 360° in a circle, the Sun appears to move along the ecliptic at a rate of approximately 1° per day.

Because of the tilt of the Earth's axis of rotation, the ecliptic and the celestial equator are inclined to each other by 23.4°, as shown in Figure 2-9. These two circles intersect at only two points, which are exactly opposite each other on the celestial sphere. Both points are called **equinoxes** (from the Latin words meaning "equal night") because, when the Sun appears at either point, daytime and nighttime are of equal length at all locations on Earth.

The **vernal equinox,** which takes place about March 21, marks the beginning of spring in the Northern Hemisphere, as the Sun moves northward across the celestial equator. The **autumnal equinox** marks the moment when fall begins in the Northern Hemisphere (about September 22), as the Sun moves southward across the celestial equator.

We should always remember that the Northern and Southern hemispheres experience opposite seasons at any given time. For example, March 21 marks the beginning of autumn for people in Australia. The equinoxes were named during a time when virtually all astronomers lived north of the equator.

Between the vernal and autumnal equinoxes are two other significant locations along the ecliptic. The point on the ecliptic farthest north of the celestial equator is called

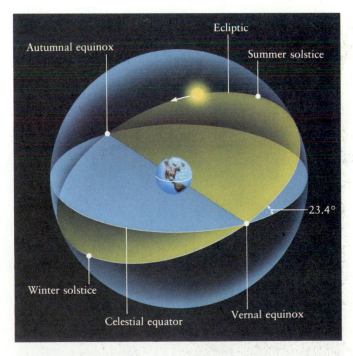

Figure 2-9 The ecliptic, equinoxes, and solstices *The ecliptic is the apparent annual path of the Sun on the celestial sphere. This path is inclined to the celestial equator by 23.4° because of the tilt of the Earth's axis of rotation. The ecliptic and the celestial equator intersect at two points called the equinoxes. The northernmost point on the ecliptic is called the summer solstice. The corresponding southernmost point is called the winter solstice.*

the **summer solstice.** It is the location of the Sun at the moment summer begins in the Northern Hemisphere, about June 21. At the beginning of winter, about December 21, the Sun is farthest south of the celestial equator at a point called the **winter solstice.**

Because the Earth is round, people located at different latitudes have different views of the celestial sphere. For instance, a person at the Earth's north pole sees the north celestial pole directly overhead. A person on the Equator sees both the celestial north and south poles on the horizon. At an intermediate latitude, such as in the United States, the celestial sphere is tilted as shown in Figure 2-10.

Seasonal changes in the Sun's daily path across the sky are diagrammed in Figure 2-10. On the first day of spring or fall (when the Sun is at one of the equinoxes), the Sun rises directly in the east and sets directly in the west. Daytime and nighttime are of equal duration. During the summer months, when the Northern Hemisphere is tilted toward the Sun, sunrise occurs in the northeast and sunset occurs in the northwest. The Sun spends more than 12 hours above the horizon in the Northern Hemisphere and passes high in the sky at noontime. Incidentally, the point in the sky directly overhead is called the **zenith,** as shown in Figure 2-10. At the summer solstice, the Sun is as far north as it gets, giving the greatest number of daylight hours to the Northern Hemisphere.

During the winter months, when the Northern Hemisphere is tilted away from the Sun, sunrise occurs in the southeast. Daylight lasts for less than 12 hours as the Sun

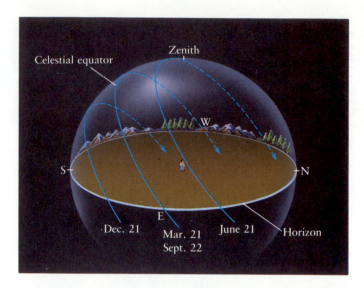

Figure 2-10 *The Sun's daily path* This sketch shows the path of the Sun on certain days of the year, as seen from middle latitudes in the Northern Hemisphere. On the first day of spring and the first day of fall, the Sun rises precisely in the east and sets precisely in the west. During summer, the Sun rises in the northeast and sets in the northwest. The maximum northerly excursion of the Sun occurs at the summer solstice. In the winter, the Sun rises in the southeast and sets in the southwest, with its maximum southerly excursion occurring at the winter solstice.

skims low over the southern horizon and sets in the southwest. Night is longest in the Northern Hemisphere when the Sun is at the winter solstice.

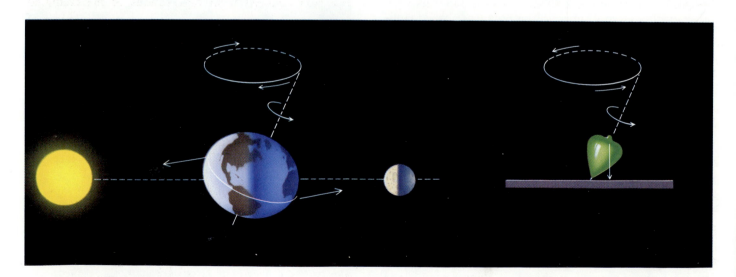

Figure 2-11 *Precession* The gravitational pulls of the Moon and the Sun on the Earth's equatorial bulge cause the Earth to precess. As the Earth precesses, its axis of rotation slowly traces out a circle in the sky. The situation is analogous to that of a spinning top. As the top spins, the Earth's gravitational pull causes the top's axis of rotation to move in a circle.

2-4 *Precession is a slow, circular motion of the Earth's axis of rotation*

Ancient astronomers realized that the Moon orbits the Earth. They knew that the Moon takes roughly four weeks to go once around the Earth. Indeed, the word *month* comes from the same Old English root as the word *moon.*

As seen from the Earth, the Moon is never far from the ecliptic. In other words, the Moon's path among the constellations is close to the Sun's path. The Moon's path remains within a band called the **zodiac** that extends about 8° on either side of the ecliptic. Thirteen constellations lie along the ecliptic, and the Moon is generally found in one of them. As the Moon moves along its orbit, it appears north of the celestial equator for about two weeks and then south of the celestial equator for about the next two weeks.

Both the Sun and the Moon exert a gravitational pull on the Earth. We shall discuss gravity in greater detail in Chapter 3. For now, it is sufficient to realize that gravity is the universal attraction of matter for other matter. The gravitational pull of the Sun and Moon affects the Earth's rotation because the Earth is not a perfect sphere. Our planet is slightly fatter, by about 43 kilometers (27 miles), across the Equator than it is from pole to pole. The Earth is therefore said to have an "equatorial bulge." The gravitational pull of the Moon and Sun on this equatorial bulge gradually changes the orientation of the Earth's axis of rotation.

Imagine a spinning toy top, as illustrated in Figure 2-11. If the top were not spinning, gravity would pull it over on its side. But when it is spinning, the combined actions of gravity and rotation cause the top's axis of rotation to trace a circle—a motion called **precession.**

As the Sun and Moon move along the zodiac, each spends half the time north of the Earth's equatorial bulge and half the time south of it. The gravitational pull of the Sun and Moon tugging on the equatorial bulge tries to "straighten up" the Earth. In other words, as sketched in Figure 2-11, the gravity of the Sun and Moon tries to pull the Earth's axis of rotation toward a position perpendicular to the plane of the ecliptic. But the Earth is spinning. As with the toy top, the combined actions of gravity and rotation cause the Earth's axis to trace out a circle in the sky while remaining tilted about 23.4° away from the perpendicular.

The Earth's rate of precession is fairly slow. It takes 26,000 years for the north celestial pole to trace out a complete circle around the sky, as shown in Figure 2-12. At the present time, the Earth's axis of rotation points within 1° of the star Polaris. In 3000 BC, it was pointing near the star Thuban in the constellation of Draco (the Dragon). In AD 14,000, the "pole star" will be Vega in Lyra (the Harp). Of course, the south celestial pole executes a similar circle in the southern sky.

As the Earth's axis of rotation precesses, the Earth's equatorial plane also moves. Because the Earth's equatorial plane defines the location of the celestial equator in the sky, the celestial equator also precesses. The intersections of the celestial equator and the ecliptic define the equinoxes, and so these key locations in the sky also shift slowly from year to year. The entire phenomenon is often called the **precession of the equinoxes.** Today the vernal equinox is located in the constellation Pisces (the Fishes). Two thousand years ago, it was in Aries (the Ram). Around the year AD 2600, the vernal equinox will move into Aquarius (the Water Bearer).

2-5 *Lunar phases are due to the Moon's orbital motion*

Ancient astronomers knew that the Moon shines by reflected sunlight. They also understood that the Moon takes roughly four weeks to orbit the Earth. This knowledge was deduced from observations of the changing **phases of the Moon,** which occur as varying amounts of the Moon's illuminated hemisphere are exposed to observers on the Earth.

Figure 2-13 is a photograph of both the Earth and Moon from space. Because the Moon orbits the Earth every 27.3 days, it takes approximately four weeks for the Moon to complete its cycle of phases. Figure 2-14 shows the Moon at several locations on its orbit (as viewed from above the Earth's north pole) along with the resulting phases.

The phase called **new moon** occurs when the unlit hemisphere of the Moon faces the Earth. We cannot see the Moon in this phase because it is in the same part of the sky as the Sun.

During the next seven days, more and more of the illuminated hemisphere of the Moon is exposed to our Earth-based view. The term *waxing* means "growing larger," and so this phase is called **waxing crescent moon.** At **first quarter moon,** the angle between the Sun, Moon,

Figure 2-12 The path of the north celestial pole As the Earth precesses, the north celestial pole slowly traces out a circle among the northern constellations. At the present time, the north celestial pole is near the moderately bright star Polaris, which serves as the "pole star."

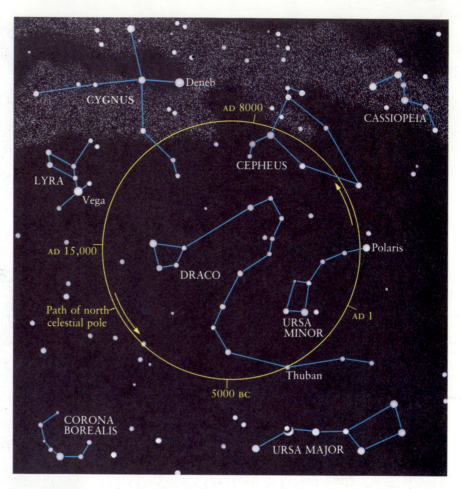

Figure 2-13 The Earth and the Moon The Moon circles the Earth every 27.3 days at an average distance of 384,400 km (238,900 mi). This picture was taken in 1977 from Voyager 1 shortly after it was launched toward Jupiter and Saturn. (NASA)

and Earth is 90°, and so we see half of the Moon's illuminated hemisphere.

During the next week, still more of the illuminated hemisphere is seen from Earth, in a phase called **waxing gibbous moon.** When the Moon stands opposite the Sun in the sky, we see the entire illuminated hemisphere, producing the phase called **full moon.** Moonrise always occurs at sunset during full moon.

Over the subsequent two weeks, we see less and less of the illuminated hemisphere as the Moon continues along its orbit. The term *waning* means "growing smaller," and so this progression produces the phases called **waning gibbous moon, last quarter moon,** and **waning crescent moon,** as diagrammed in Figure 2-14.

Because the position of the Sun in the sky determines the local time, we can correlate the Moon's phase and location with the time of day. For example, during first quarter moon, the Moon is approximately 90° east of the Sun in the sky; hence, moonrise occurs approximately at noon. Figure 2-15 is a series of photographs showing the various phases of the Moon.

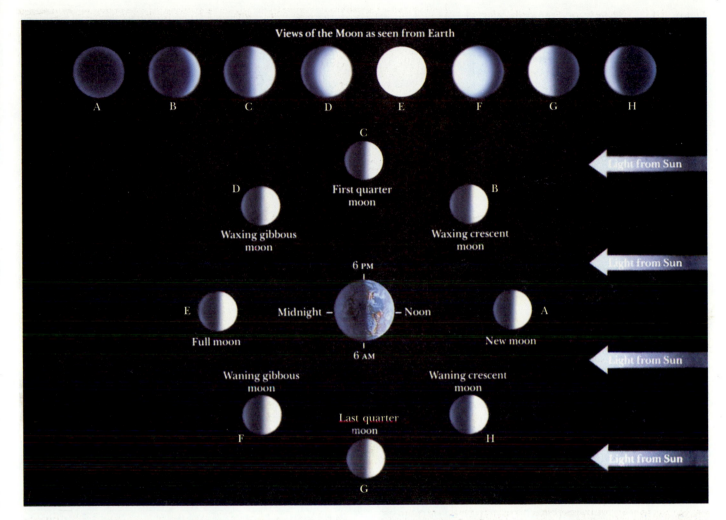

Figure 2-14 *The phases of the Moon This diagram shows the Moon at eight locations on its orbit (as viewed from above the Earth's north pole) along with the resulting lunar phases. Light from the Sun illuminates one half of the Moon, while the other half is dark. As the Moon orbits the Earth, we see varying amounts of the Moon's illuminated hemisphere. It takes 29.5 days for the Moon to go through all its phases.*

It takes about one month for the Moon to complete an orbit around the Earth. However, astronomers are careful to distinguish between two types of months, depending on whether the Moon's motion is measured relative to the stars or to the Sun. Neither interval corresponds exactly to the months of our calendar.

The **sidereal month** is the time it takes for the Moon to complete one full orbit of the Earth, measured *with respect to the stars*. This interval is the Moon's true orbital period, equal to 27.3 days.

The **synodic month** is the time it takes for the Moon to complete one cycle of phases, for example, from one new moon to the next. Consequently, the synodic month is measured *with respect to the Sun* and is equal to about 29.5 days.

The longer duration of the synodic month results from the Earth's orbiting the Sun while the Moon is orbiting the Earth. Consequently, to get from one new moon to the next, the Moon must travel more than 360° along its orbit, as shown in Figure 2-16. The synodic month is thus approximately two days longer than the sidereal month.

The Moon stays in orbit about the Earth because of the gravitational attraction between these two bodies. However, the Sun also pulls on the Moon, continually affecting the Moon's speed along its orbit. The result is that both the sidereal and the synodic months are variable. The sidereal month (average length = 27.3 days) can vary by as much as 7 hours, while the synodic month (average length = 29.5 days) can vary by as much as 12 hours.

Figure 2-15   The Moon's appearance
The Moon always keeps the same side
facing the Earth. Earth-based observers
therefore always see the same craters and
lunar mountains, regardless of the phase.
(Lick Observatory)

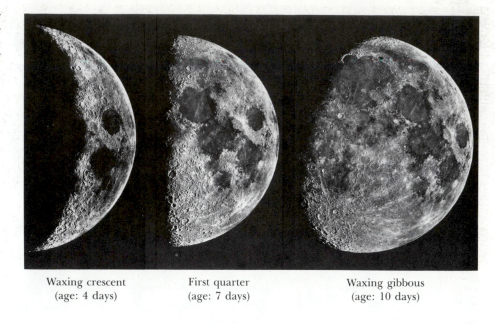

Waxing crescent
(age: 4 days)

First quarter
(age: 7 days)

Waxing gibbous
(age: 10 days)

Figure 2-16   The sidereal and synodic months
The sidereal month is the time it takes the
Moon to complete one revolution with respect
to the background stars. However, because the
Earth is constantly moving along its orbit
about the Sun, the Moon must travel through
more than 360° to get from one new moon to
the next. Thus the synodic month is slightly
longer than the sidereal month.

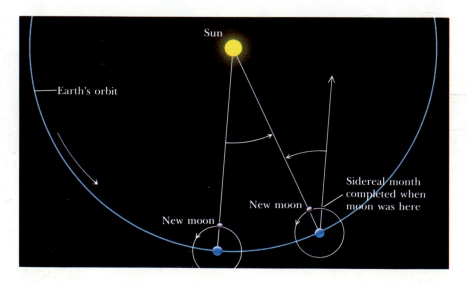

2-6   Ancient astronomers measured the size of the Earth and attempted to determine distances to the Sun and Moon

More than two thousand years ago, Greek astronomers were aware of the Earth's spherical shape. Eclipses of the Moon provided the convincing observations. During a lunar eclipse, the Moon passes through the Earth's shadow. Ancient astronomers noticed that the edge of the Earth's shadow is always circular. Because a sphere is the only shape that casts a circular shadow from any angle, the astronomers concluded that the Earth is spherical.

Around 200 BC, the Greek astronomer Eratosthenes devised a way to measure the circumference of the Earth. He was intrigued by reliable reports from the town of Syene in Egypt (the modern Aswân) that the Sun shone directly down vertical wells on the first day of summer. Eratosthenes knew that the Sun never appeared at the zenith from his home in Alexandria on the Mediterranean Sea almost due north of Syene. He measured the position

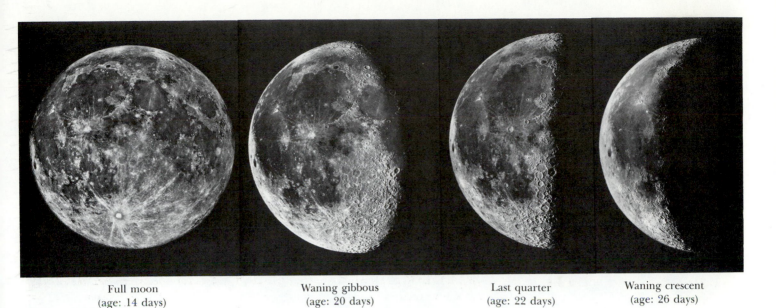

Full moon
(age: 14 days)

Waning gibbous
(age: 20 days)

Last quarter
(age: 22 days)

Waning crescent
(age: 26 days)

of the Sun at local noon on the summer solstice in Alexandria and found it to be about 7° south of the zenith, as shown in Figure 2-17. Because this angle is about $\frac{1}{50}$ of a circle, Eratosthenes concluded that the distance from Alexandria to Syene was $\frac{1}{50}$ of the Earth's circumference.

Eratosthenes was only one of several brilliant astronomers to emerge from the distinguished Alexandrian school. Around 280 BC, one of the first Alexandrian astronomers, Aristarchus of Samos, devised a method to determine the relative distances to the Sun and Moon.

Aristarchus knew that, at the moment of first or last quarter moon, the Sun, Moon, and Earth form a triangle, with a right angle at the Moon's location, as diagrammed in Figure 2-18. For reasons lost in antiquity, he also argued that the angle between the Moon and the Sun is 3° less than a right angle (that is, 87°) at first and last quarter. Using the rules of Euclidean geometry, Aristarchus concluded that the Sun is about 20 times farther from us than the Moon is. We now know that the actual figure is about 390 rather than 20. Nevertheless, it is impressive that people were logically trying to measure distances across the solar system more than two thousand years ago.

Aristarchus also used lunar eclipses in an equally bold attempt to determine the relative sizes of the Earth, Moon, and Sun. Observing how long it takes for the Moon to move through the Earth's shadow, Aristarchus

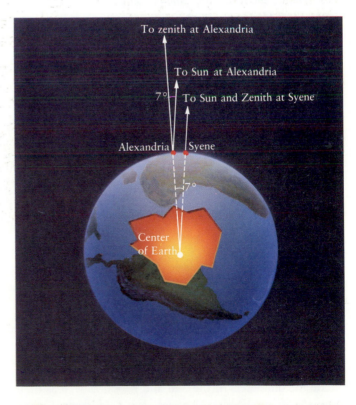

Figure 2-17 *Eratosthenes' method of determining the Earth's size Eratosthenes noticed that the Sun is about 7° south of the zenith at Alexandria when it is directly overhead at Syene. The angle is about $\frac{1}{50}$ of a circle, and so the distance between Alexandria and Syene must be about $\frac{1}{50}$ of the Earth's circumference.*

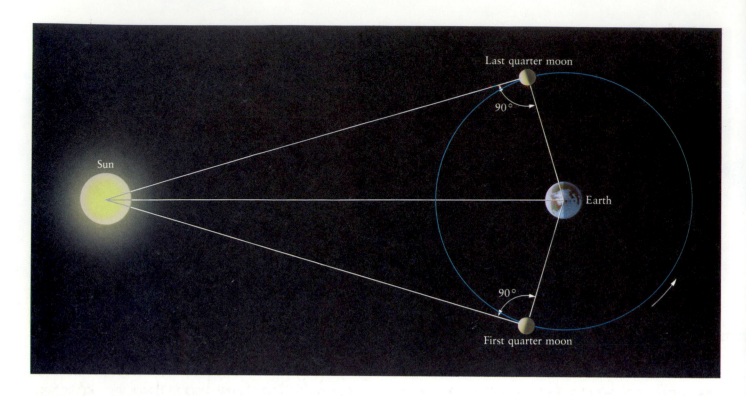

Figure 2-18 *Aristarchus's method of determining distances to the Sun and Moon Aristarchus knew that the Sun, Earth, and Moon form a right triangle at first quarter and last quarter phases. How he measured the acute angles in these right tri-* *angles is not known. Nevertheless, using Euclidean geometry, he argued that the Sun is about 20 times farther from the Earth than the Moon. The sizes of the Earth, the Moon, and the Moon's orbit are greatly exaggerated in this diagram*

estimated that the diameter of the Earth is about three times the diameter of the Moon. To determine the diameter of the Sun, he simply pointed out that the Sun and the Moon have the same angular size in the sky, and so their diameters must be in the same ratio as their distances. In other words, because Aristarchus believed that the Sun is 20 times farther from the Earth than the Moon, he concluded that the Sun must be 20 times larger than the Moon. According to modern measurements, the Sun is about 400 times larger in diameter than the Moon.

You can now appreciate the significance of Eratosthenes' measurement of the Earth's circumference. The Greeks knew that the Earth's diameter is equal to its circumference divided by the constant called π (pi). Knowing the Earth's diameter, Alexandrian astronomers could calculate the diameters of the Sun and Moon as well as their distances from Earth. Although some of these ancient measurements are far from their modern values, our ancestors' achievements stand as impressive exercises in observation and reasoning.

2-7 Eclipses occur only when the Sun and Moon are both on the line of nodes

Eclipses are among the most spectacular of nature's phenomena. In only a few minutes, the Sun can seemingly be blotted from the sky as broad daylight is transformed into an eerie twilight, or the brilliant full moon on a cloud-free night gradually darkens to a deep red.

A **lunar eclipse** occurs when the Moon passes through the Earth's shadow. This can happen only when the Sun, Earth, and Moon are in a straight line at full moon. A **solar eclipse** occurs when the Earth passes through the Moon's shadow. As seen from Earth, the Moon moves in front of the Sun. This can happen only when the Sun, Moon, and Earth are aligned at new moon.

New moon and full moon both occur at intervals of 29.5 days, but solar and lunar eclipses happen much less often. Eclipses occur infrequently because the Moon's or-

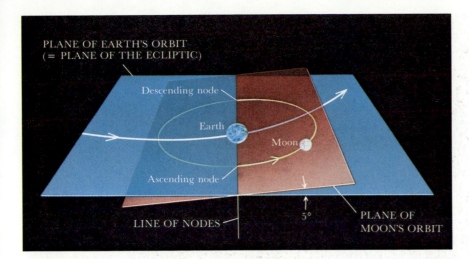

Figure 2-19 The line of nodes *The plane of the Moon's orbit is tilted slightly with respect to the plane of the Earth's orbit. These two planes intersect along a line called the line of nodes.*

bit is tilted slightly (5°) out of the plane of the Earth's orbit, as shown in Figure 2-19. Because of this tilt, new moon and full moon usually occur when the Moon is either above or below the plane of the Earth's orbit. In such positions, a perfect alignment between the Sun, Moon, and Earth is not possible and an eclipse cannot occur.

The plane of the Earth's orbit and the plane of the Moon's orbit intersect along a line called the **line of nodes,** which passes through the Earth and is pointed in a particular direction in space, as shown in Figure 2-20. Eclipses can occur only when both the Sun and Moon are on or very near the line of nodes, because only then do the Sun, Earth, and Moon lie along a straight line.

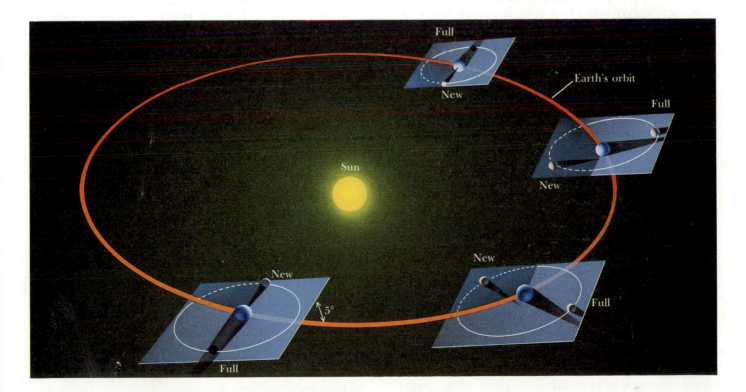

Figure 2-20 Conditions for eclipses *A solar eclipse occurs only if the Moon is very near the line of nodes at new moon. A lunar eclipse occurs only if the Moon is very near the line of nodes at full moon. When new moon or full moon phases occur away from the line of nodes, no eclipse is seen because the Moon and the Earth do not pass through each other's shadows.*

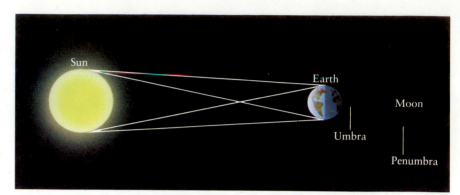

Figure 2-21 *The geometry of a lunar eclipse* *People on the nighttime side of the Earth see a lunar eclipse when the Moon moves through the Earth's shadow. The umbra is the darkest part of the shadow. In the penumbra, only part of the Sun is covered by the Earth.*

Knowing the orientation of the line of nodes is clearly important to anyone who wants to predict eclipses. However, predicting eclipses is complicated by the fact that the direction of the line of nodes gradually changes. The constant gravitational pull of the Sun on the Moon causes the orientation of the Moon's orbit to gradually shift in space. The resulting slow, westward movement of the line of nodes is one of several details that astronomers must include in their calculations for times of upcoming eclipses.

At least two but no more than five solar eclipses occur each year. (The last year in which five solar eclipses occurred was 1935.) Lunar eclipses occur just about as frequently as solar eclipses, but the maximum number of eclipses (both solar and lunar) possible in a year is seven.

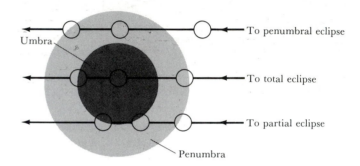

Figure 2-22 *Various lunar eclipses* *This diagram shows the Earth's umbra and penumbra at the distance of the Moon's orbit. Different kinds of lunar eclipses are seen, depending on the Moon's path through the Earth's shadow.*

2-8 Lunar and solar eclipses can be either partial or total, depending on the alignment of the Sun, Earth, and Moon

During the few hours of a lunar eclipse, you can see the Earth's shadow move across the Moon. The Earth's shadow has two distinct parts, as diagrammed in Figure 2-21. The **umbra** is the darkest part of the shadow, from which no portion of the Sun's surface can be seen. In the **penumbra**, only part of the Sun's surface is blocked out.

Three kinds of lunar eclipses can occur, depending on exactly how the Moon travels through the Earth's shadow. First, the Moon might pass through only the Earth's penumbra, creating a **penumbral eclipse**. During this type of eclipse, none of the lunar surface is completely shaded by the Earth; as seen from the Moon, only part of the Sun is covered by the Earth. At mid-eclipse, the Moon merely looks a little dimmer than usual from Earth and there is

Table 2-1   Lunar eclipses, 1992–1997

Date	Percentage eclipsed (100 = total)	Duration of totality
1992 June 15	69	——
1992 December 10	100	1 hr 14 min
1993 June 4	100	1 38
1993 November 29	100	0 50
1994 May 25	28	——
1995 April 15	12	——
1996 April 4	100	1 24
1996 September 27	100	1 12
1997 March 24	93	——
1997 September 16	100	1 06

Figure 2-23 A total eclipse of the Moon This photograph was taken by an amateur astronomer during the lunar eclipse of September 6, 1979. Notice the distinctly reddish color of the Moon. (Courtesy of M. Harms)

through the Earth's atmosphere is deflected into the Earth's umbra. Most of this deflected light is red, and so the darkened Moon glows faintly in reddish hues during totality, as shown in Figure 2-23.

As with the Earth's shadow, the darkest part of the Moon's shadow is also called the umbra. You must be within the Moon's umbra in order to see a **total solar eclipse,** for that is the only region from which the Moon completely covers the Sun. Because the Sun and the Moon have nearly the same angular diameter as seen from Earth—about $\frac{1}{2}°$— the Moon "fits" over the Sun during a total solar eclipse. This is important to astronomers because the hot gases called the **solar corona** that surround the Sun can be photographed and studied in detail during the few precious moments when the eclipse is total (Figure 2-24).

The Moon's umbra is surrounded by a region of partial shadow called the penumbra. During a solar eclipse, the Moon's penumbra extends over a large portion of the Earth's surface. When the observer is within the Moon's penumbra, the Sun's surface appears only partly covered by the Moon. This is called a partial eclipse of the Sun.

no "bite" taken out of the Moon by the Earth's umbra. Since the Moon still looks full, it is easy to miss a penumbral eclipse.

Most people notice a lunar eclipse only if the Moon passes into the Earth's umbra. During the umbral phase of such an eclipse, a bite seems to be taken out of the Moon. If the Moon's orbit is oriented so that only part of the lunar surface passes through the umbra, then we see a **partial eclipse.** When the Moon travels completely into the umbra, as sketched in Figure 2-22, we see a **total eclipse** of the Moon. The maximum duration of totality occurs when the Moon travels directly through the center of the umbra. The Moon's speed through the Earth's shadow is roughly 1 kilometer per second (2300 miles per hour), which means that totality can last for as much as 1 hour 42 minutes.

As an example of the frequency of lunar eclipses, Table 2-1 lists all the eclipses, both total and partial, from 1992 through 1997. Penumbral eclipses are not included in this listing.

Even during a total eclipse, the Moon does not completely disappear. A small amount of sunlight passing

Figure 2-24 A total eclipse of the Sun During a total solar eclipse, the Moon completely covers the Sun's disk, and the solar corona can be seen. This halo of hot gases extends for thousands upon thousands of kilometers into space. This photograph was taken by an amateur astronomer during the solar eclipse of July 11, 1991. (Courtesy of W. Dellinges)

Figure 2-25 The geometry of a total solar eclipse During a total solar eclipse, the tip of the Moon's umbra traces an eclipse path across the Earth's surface. People inside the eclipse path see a total solar eclipse, whereas people inside the penumbra see only a partial eclipse.

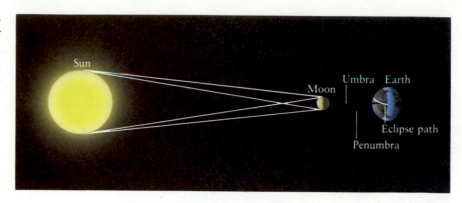

Only the tip of the Moon's umbra reaches the Earth's surface, as shown in Figure 2-25. As the Earth turns, the tip of the umbra traces an **eclipse path** across its surface. Only those people inside this path are treated to the spectacle of a total solar eclipse. (Figure 2-26 shows the dark spot produced by the Moon's umbra on the Earth's surface during a total solar eclipse.)

During a total eclipse, the Earth's rotation and the orbital motion of the Moon cause the umbra to race along the eclipse path at speeds in excess of 1700 kilometers per hour (1050 miles per hour). People along the eclipse path therefore observe totality for only a few moments. Totality never lasts for more than $7\frac{1}{2}$ minutes at any one location on the eclipse path. In a typical total solar eclipse, various factors such as the Earth–Moon distance limit the duration of totality to much less than the maximum $7\frac{1}{2}$ minutes.

The Moon's orbit around the Earth is not quite a perfect circle. The distance between the Earth and the Moon, which averages 384,400 kilometers (238,900 miles), varies by a few percent as the Moon goes around the Earth. The width of the eclipse path depends primarily on the Earth–Moon distance during an eclipse. The eclipse path is widest—up to 270 kilometers (170 miles)—if the Moon happens to be at the point in its orbit nearest the Earth. Usually it is much narrower. In fact, sometimes the Moon's umbra does not even reach down to the Earth's surface. If the alignment for a solar eclipse occurs when the Moon is farthest from the Earth, then the Moon's umbra falls short of the Earth, and no one sees a truly total eclipse. From the Earth's surface, the Moon appears too small to cover the Sun completely, and a thin ring of light is seen around the edge of the Moon at mid-eclipse. An eclipse of this type is called an **annular eclipse** (Figure 2-27). The length of the Moon's umbra is nearly 5000 kilometers (3100 miles) shorter than the average distance between the Moon and the Earth's surface. Thus, the

Figure 2-26 The Moon's shadow on the Earth This photograph was taken from an Earth-orbiting satellite during the total eclipse of March 7, 1970. The Moon's umbra appears as a dark spot on the eastern coast of the United States. (NASA)

Figure 2-27 *An annular eclipse of the Sun* This composite of five exposures taken at sunrise in Costa Rica shows the progress of an annular eclipse of the Sun that occurred on December 24, 1974. Note that at mid-eclipse the limb of the Sun is visible around the Moon. (Courtesy of D. di Cicco

Moon's shadow often fails to reach the Earth, making annular eclipses more common than total eclipses.

To illustrate the frequency of solar eclipses, Table 2-2 lists all the total, partial, and annular solar eclipses from 1992 through 1997.

2-9 Ancient astronomers achieved limited ability to predict eclipses

A total solar eclipse is a dramatic event. The sky begins to darken, the air temperature falls, and the winds increase as the Moon's umbra races toward you. All nature responds: Birds go to roost, flowers close their petals, and crickets begin to chirp as if evening had arrived. As totality approaches, the landscape is bathed in shimmering bands of light and dark as the last few rays of sunlight

Table 2-2 Solar eclipses, 1992–1997

Date	Area	Type	Notes
1992 January 4–5	Central Pacific	Annular	
1992 June 30	South Atlantic	Total	Max. length 5 min 20 sec
1992 December 24	North Pacific	Partial	84% eclipsed
1993 May 21	Canada, Arctic, Northern Europe	Partial	74% eclipsed
1993 November 13	Antarctic	Partial	93% eclipsed
1994 May 10	Pacific, Mexico, United States, Canada	Annular	
1994 November 3	Peru, Brazil, South Atlantic	Total	Max. length 4 min 23 sec
1995 April 29	South Pacific, Peru, Brazil, South Atlantic	Annular	
1995 October 24	Iran, India, East Indies, Pacific	Total	Max. length 2 min 5 sec
1996 April 17	South Pacific, Antarctic	Partial	88% eclipsed
1996 October 12	Europe, North Atlantic, Arctic	Partial	76% eclipsed
1997 March 9	China, Russia, Arctic	Total	Max. length 2 min 50 sec
1997 September 2	Australia, New Zealand, Antarctic	Partial	90% eclipsed

peek out from behind the edge of the Moon. And finally the corona blazes forth in a star-studded daytime sky. It is an awesome sight.

In ancient times, the ability to predict eclipses must have seemed very desirable. Archaeological evidence suggests that astronomers in many civilizations struggled to predict eclipses, with varying degrees of success. The number and placement of certain holes in the ground around Stonehenge suggest some ability to predict eclipses more than four thousand years ago. One of three priceless manuscripts to survive the devastating Spanish conquests shows that Mayan astronomers in Mexico and Guatemala had a fairly reliable method for predicting eclipses. And there are many apocryphal stories such as the one about the great Greek astronomer Thales of Miletus, who is said to have predicted the eclipse of 585 BC, which occurred during the middle of a war. The sight was so awesome and unexpected that the soldiers put down their arms and declared peace.

Rather than actual predictions, it seems that ancient astronomers actually produced eclipse "warnings" with varying degrees of reliability. Working with historical records, these astronomers generally searched for cycles and regularities from which to anticipate future eclipses. One of the most famous eclipse cycles, used by the Babylonians and others, is the **saros**, an interval of 18 years 11.3 days in which eclipses repeat themselves.

Ancient astronomers, particularly those of Greece, gave humanity a new and powerful way of thinking about the world. They gave the first clear demonstration that the tools of logic, reason, and mathematics can be used to discover and understand the workings of the universe. This general approach to reality underlies all modern science. Several ancient Greek astronomers even went so far as to suggest that the motions of the planets in the sky could be explained if all the planets, including Earth, orbit the Sun. As we shall see in the next chapter, these gifted thinkers were nearly two thousand years ahead of their time.

Summary

- It is convenient to imagine the stars fixed to the celestial sphere with the Earth at its center.

  The surface of the celestial sphere is divided into 88 regions called constellations.

- The celestial sphere appears to rotate around the Earth once in each day and night; in fact, of course, it is the Earth that is rotating.

  The poles and equator of the celestial sphere are determined by extending the axis of rotation and the equatorial plane of the Earth to the celestial sphere.

- Earth's axis of rotation is tilted at an angle of 23.4° from the perpendicular to the plane of the Earth's orbit.

  The seasons are caused by this tilt; one hemisphere or the other is tipped toward the Sun at certain times of the year.

- Equinoxes and solstices are significant points along the Earth's orbit, determined by the relationship between the Sun's path on the celestial sphere (the ecliptic) and the celestial equator.

  The Earth's axis of rotation moves slowly in a phenomenon called precession.

  Precession is caused by the gravitational pull of the Sun and Moon on the Earth's equatorial bulge.

- The phases of the Moon are caused by the relative positions of the Earth, Moon, and Sun.

  The Moon completes one orbit around the Earth with respect to the stars in a sidereal month averaging 27.3 days.

  The Moon completes one cycle of phases in a synodic month averaging 29.5 days.

- Ancient astronomers made great progress in determining the sizes and relative distances of the Earth, Moon, and Sun.

Around 280 BC, Aristarchus attempted to measure the distances from the Earth to the Moon and to the Sun. He also estimated the relative sizes of the Earth, Moon, and Sun.

Around 200 BC, Eratosthenes measured the size of the Earth by comparing the position of the Sun at the same moment at two different locations.

- A lunar eclipse occurs when the Moon moves through the Earth's shadow. This happens when the Sun and Moon are both on the line of nodes at full moon. The line of nodes is the line where the planes of the Earth's orbit and the Moon's orbit intersect.

- A solar eclipse occurs when the Earth passes through the Moon's shadow. This happens when the Sun and Moon are both on the line of nodes at new moon.

- The shadow of an object has two parts: the umbra, where the light source is completely blocked, and the penumbra, where the light source is only partially obscured.

  Depending on the relative positions of the Sun, Moon, and Earth, lunar eclipses may be penumbral, partial, or total, and solar eclipses may be partial, annular, or total.

Review questions

1 How are constellations useful to astronomers?

2 What is the celestial sphere and why is this ancient concept still useful today?

3 What is the celestial equator and how is it related to the Earth's equator? How are the north and south celestial poles related to the Earth's axis of rotation?

4 At what location on Earth is the north celestial pole on the horizon?

5 With the aid of a diagram, explain why the tilt of the Earth's axis relative to the Earth's orbit causes the seasons as we orbit the Sun.

6 What are the vernal and autumnal equinoxes? What are the summer and winter solstices? How are these four points related to the ecliptic and the celestial equator?

7 What is precession and how does it affect our view of the heavens?

8 Why does the Moon exhibit phases?

9 What is the difference between a sidereal month and a synodic month? Which is longer? Why?

10 How did Eratosthenes determine the size of the Earth?

11 How did Aristarchus try to estimate the distance from the Earth to the Sun and Moon?

12 What is the line of nodes and how is it related to the occurrence of solar and lunar eclipses?

13 What is a penumbral eclipse of the Moon? Why do you suppose that it is easy to overlook such an eclipse?

14 How is an annular eclipse of the Sun different from a total eclipse of the Sun? What causes this difference?

Advanced questions

* 15 Where do you have to be on the Earth in order to see the Sun at the zenith? If you stay at that location for a full year, on how many days will the Sun pass through the zenith?

* 16 Where do you have to be on Earth in order to see the south celestial pole directly overhead? What is the maximum possible elevation of the Sun above the horizon at that location? On what date is this maximum elevation observed?

* 17 At what point on the horizon does the vernal equinox rise?

18 Is it possible for a total eclipse of the Sun to be followed three months later by a lunar eclipse? Why?

19 Can one ever observe an annular eclipse of the Moon? Why?

20 Consult a star map of the southern sky and determine which, if any, bright southern stars could some day become south celestial pole stars.

21 During a lunar eclipse, does the Moon enter the Earth's shadow from the east or from the west? Explain why.

Discussion questions

22 Examine a list of the 88 constellations. Are there any that obviously date from modern times? Where are they located? Why do you suppose they do not have archaic names?

23 Describe the seasons if the Earth's axis of rotation were tilted 0° and 90° to its orbital plane.

24 Describe the cycle of lunar phases that would be observed if the Moon moved about the Earth in an orbit perpendicular to the plane of the Earth's orbit. Is it possible for solar and lunar eclipses to occur under these circumstances?

For further reading

Allen, D., and Allen, C. *Eclipse*. Allen & Unwin, 1987 • A well-written history of our understanding of eclipses, including the mythology they engendered and the science behind them.

Allen, R. *Star Names: Their Lore and Meaning*. 1899. Dover reprint, 1963 • A classic book detailing the origins of constellations and star names.

Berry, R. *Discover the Stars*. Harmony Books, 1987 • An excellent manual featuring clear, beautiful star charts showing the whole sky along with enlarged views with more detail.

Chartrand, M. *Skyguide*. Western Publications, 1982 • This pocket-sized book gives a concise introduction to observational astronomy with superb illustrations and excellent star charts.

Cornell, J. *The First Stargazers*. Scribner, 1981 • This introduction to "archeoastronomy," the archeological study of ancient astronomy, includes excellent sections on Stonehenge, the Americas, China, and Egypt.

Gallant, R. *The Constellations: How They Came to Be*. Four Winds Press, 1979 • The stories of 50 constellations are accompanied by star charts and useful information.

Krupp, E. *Echoes of the Ancient Skies*. Harper & Row, 1983 • A superb introduction to archeoastronomy. Krupp skillfully brings together the mythology, rituals, and monuments of many cultures, demonstrating both similarities and differences.

Menzel, D., and Pasachoff, J. "Solar Eclipse: Nature's Superspectacular." *National Geographic*, August 1970 • A beautifully illustrated article about observing a total solar eclipse.

Ridpath, I., and Tirion, W. *Universe Guide to Stars and Planets*. Universe Books, 1985 • All 88 constellations are covered in exquisitely drawn star charts along with sky maps for different seasons.

Sagan, C. "The Shores of the Cosmic Ocean." In Sagan, C., *Cosmos*. Random House, 1980 • This chapter is an excellent recounting of Eratosthenes' experiment to measure the size of the Earth.

Whitney, C. *Whitney's Star Finder*. Knopf, 1981 • A useful guide for the novice who wants to identify stars and constellations.

Williamson, R. *Living the Sky: The Cosmos of the American Indian*. Houghton-Mifflin, 1984 • A fascinating tour of the astronomical thought and monuments of native American Indian tribes.

3

Gravitation and the Motions of the Planets

Ancient astronomers believed that the heavens rotated around a stationary Earth. During the Renaissance, several brilliant thinkers spearheaded a revolution that dethroned Earth from its central location. First came Copernicus, who argued that the planets go around the Sun. Later Kepler demonstrated that planetary orbits are in fact ellipses. Observations by Galileo, the first person to turn a telescope toward the skies, supported the idea that the Earth goes around the Sun. Discoveries by Kepler and Galileo set the stage for Isaac Newton, who formulated basic laws of physics. Newton's efforts led him to a precise, mathematical description of the force of gravity, which holds the planets in their orbits about the Sun. Then, in the early twentieth century, Albert Einstein proposed the radically new idea that gravity affects the curvature of space and the flow of time. The relationships between gravity, space, and time are central to all our modern ideas about the structure of the universe.

Apollo 11 *leaving the Moon* *A lunar module returns from the Moon after completing a successful manned lunar mission. This photograph was taken from the command module* Columbia, *in which the astronauts returned to Earth. All of the orbital maneuvers to the Moon and back were based on Newtonian mechanics and Newton's law of gravity. These same principles are used by astronomers to understand a wide range of phenomena, from the motions of double stars to the rotation of the Galaxy. (NASA)*

It is not obvious that the Earth moves around the Sun. Indeed, our daily experience strongly suggests that the opposite is true. The daily rising and setting of the Sun, Moon, and stars could lead us to believe that the entire cosmos revolves about us, with the Earth the center of the universe. That was just what most observers did believe for thousands of years.

3-1 *Ancient astronomers invented geocentric cosmology to explain planetary motions*

The ancient Greeks conceived of certain principles that even today guide modern scientists. For instance, around 550 BC Pythagoras and his followers put forth the idea that natural phenomena could be described with mathematics. About 200 years later, Aristotle asserted that the universe is governed by physical laws. These two concepts found their highest expression in the works of great astronomers and physicists who struggled to explain planetary motion.

Ancient Greek astronomers were among the first to leave a written record of their attempts to explain the motion of the planets. Most Greeks assumed that the Sun, the Moon, the stars, and five planets revolve about the Earth, and thus their view of the universe is said to be "geocentric." A theory of the universe is called a **cosmology**, and thus a Greek thinker such as Pythagoras or Aristotle gave credence to a **geocentric cosmology**.

The Greeks and other ancient cultures knew of five planets: Mercury, Venus, Mars, Jupiter, and Saturn. These planets are quite obvious in the sky because from night to night they slowly shift their positions with respect to the background of the "fixed" stars in the constellations. In fact, the word *planet* comes from a Greek term meaning "wanderer." Furthermore, some of these planets are very bright in the night sky. For example, at its maximum brilliance Venus is 16 times brighter than the brightest star.

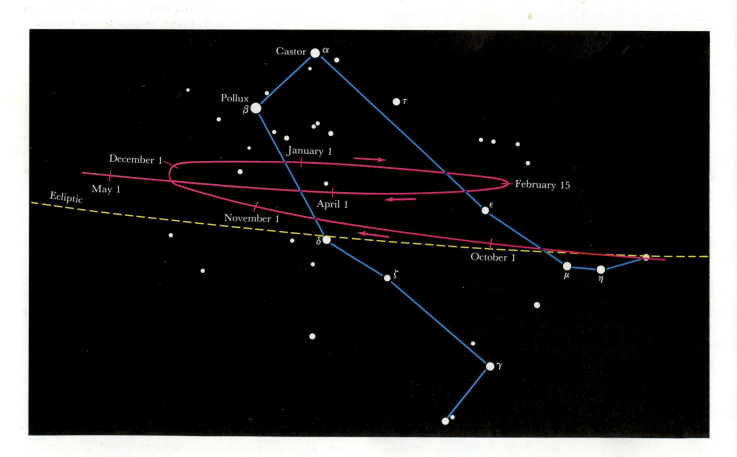

Figure 3-1 The path of Mars in 1992–1993 *From the fall of 1992 through the spring of 1993, Mars moved across the con-* *stellation of Gemini. From November 30 through February 14, Mars's motion was retrograde.*

Observations of the positions of the planets against the stars from night to night make it clear that the planets do not move at uniform rates across the constellations. Explaining the motions of the five planets was one of the main challenges facing the astronomers of antiquity. It was not easy to develop a comprehensive geocentric theory of the universe.

As seen from Earth, the planets move primarily across the constellations of the zodiac. As mentioned in Chapter 2, these constellations circle the sky in a band centered on the ecliptic. If you follow a planet as it travels across the zodiac from night to night, you find that the planet usually moves slowly eastward against the background stars. This eastward movement is called **direct motion.** Occasionally, however, the planet seems to stop and then back up for several weeks or months. This occasional westward movement is called **retrograde motion.** These motions are much slower than the daily rotation of the sky caused by the Earth's rotation. Both direct and retrograde motion are best detected by mapping the position of a planet against the background stars from night to night over a long period of observation. Figure 3-1 shows direct and retrograde paths of Mars from October 1992 through May 1993.

Greek astronomers devised various theories to account for retrograde motion and the loops that the planets trace out against the background stars. One of the most successful ideas was investigated by Hipparchus and elaborated by the last of the great Greek astronomers, Ptolemy, who lived in Alexandria during the second century AD. The basic concept is sketched in Figure 3-2. Each planet is assumed to move in a small circle called an **epicycle,** which in turn moves in a larger circle called a **deferent,** which is approximately centered on the Earth. As viewed from Earth, the epicycle moves eastward along the deferent, and both circles rotate in the same direction (counterclockwise in Figure 3-2).

Most of the time, the motion of the planet on its epicycle adds to the eastward motion of the epicycle on the deferent. Thus the planet is seen to be in direct (eastward) motion against the background stars throughout most of the year. However, when the planet is on the part of its epicycle nearest the Earth, its motion along the epicycle subtracts from the motion of the epicycle along the deferent. The planet thus appears to slow and then halt its usual eastward movement among the constellations, even seeming to go backward for a few weeks or months. This concept of epicycles and deferents provides a general explanation of the retrograde loops that the planets execute.

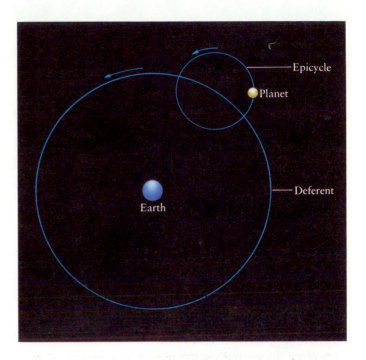

Figure 3-2 A geocentric explanation of planetary motion Each planet revolves about an epicycle, which in turn revolves about a deferent centered approximately on the Earth. As seen from Earth, the speed of the planet on the epicycle alternately adds to or subtracts from the speed of the epicycle on the deferent, thus producing alternating periods of direct and retrograde motion.

Using the wealth of astronomical data in the library at Alexandria, including records of planetary positions covering hundreds of years, Ptolemy deduced the sizes of the epicycles and deferents and the rates of revolution needed to produce the recorded paths of the planets. After years of arduous work, Ptolemy assembled his calculations in 13 volumes collectively called the *Almagest,* in which the positions and paths of the Sun, Moon, and planets were described with unprecedented accuracy. In fact, the *Almagest* was so successful that it became the astronomer's bible, and for over a thousand years Ptolemy's cosmology endured as a useful description of the workings of the heavens.

Eventually, however, things began to go awry. Tiny errors and inaccuracies that were unnoticeable in Ptolemy's day compounded and multiplied over the years, especially with regard to precession. Fifteenth-century astronomers made some cosmetic adjustments to the Ptolemaic system. However, the system became less and less satisfactory as more complicated and arbitrary details were added to keep it consistent with the observed motions of the planets.

3-2 Nicolaus Copernicus devised the first comprehensive heliocentric cosmology

Imagine driving on a freeway at high speed. As you pass a slowly moving car, it appears to move backward even though it is traveling in the same direction as your car. This sort of observation inspired the ancient Greek astronomer Aristarchus to suggest a more straightforward explanation of retrograde motion—one in which all the planets, including the Earth, revolve about the Sun. The retrograde motion of Mars, for example, occurs when the Earth overtakes and passes Mars, as shown in Figure 3-3. The occasional retrograde movement of a planet is the result of our changing viewpoint—an idea that is beautifully simple compared to an Earth-centered system with all its "circles upon circles."

Aristarchus had demonstrated that the Sun is bigger than the Earth, which made it sensible to consider the possibility that the smaller Earth might be orbiting the larger Sun. In Aristarchus's day, however, it seemed inconceivable that something as huge as the Earth could be in motion. Almost two thousand years elapsed before someone had the insight and determination to work out the details of a **heliocentric** (Sun-centered) **cosmology**. That person was a sixteenth-century Polish lawyer, physician, mathematician, economist, monk, and artist named Nicolaus Copernicus. Copernicus's astronomical studies detailed the advantages of a straightforward Sun-centered cosmology over the cumbersome Earth-centered theory. This work instilled a revolutionary concept that pervades modern science: Simplicity is a hallmark of correctness. Thus, nowadays, when a theory becomes unduly elaborate and complicated, scientists begin to suspect that the theory is probably wrong.

Copernicus realized that, with a heliocentric perspective, he could determine which planets are closer to the Sun than the Earth is and which are farther away. Because Mercury and Venus are always observed fairly near the Sun, Copernicus concluded that their orbits must be smaller than the Earth's. The other visible planets—Mars, Jupiter, and Saturn—can be seen in the middle of the night, when the Sun is far below the horizon, which can occur only if the Earth comes between the Sun and a planet. Copernicus concluded that the orbits of Mars, Jupiter, and Saturn are larger than the Earth's orbit.

It is often useful to specify various points on a planet's orbit, as in Figure 3-4. These points help us identify certain geometrical arrangements, or **configurations**, between the Earth, another planet, and the Sun. For example, when Mercury or Venus is between the Earth and the Sun, we say the planet is at **inferior conjunction;** when they are on the opposite side of the Sun, they are at **superior conjunction.**

The angle between the Sun and a planet as viewed from the Earth is called the planet's **elongation.** At **greatest eastern elongation,** Mercury or Venus is as far east of the Sun as it can be. At such times, the planet appears above the western horizon after sunset and is often called an "evening star." Similarly, at **greatest western elongation,** Mercury or Venus is as far west of the Sun as it can be and rises before the Sun, gracing the eastern predawn sky as a "morning star."

When one of the outer planets is behind the Sun, we say that the planet is at **conjunction.** When it is exactly opposite the Sun in the sky, we say that it is at **opposition.** It is not difficult to determine when a planet happens to be located at one of the key positions in Figure 3-4. For example, when Mars is at opposition, it appears high in the sky at midnight.

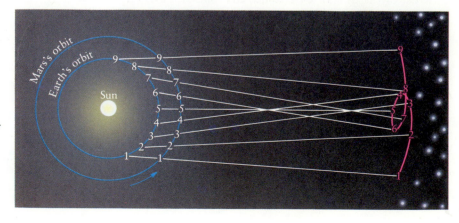

Figure 3-3 A heliocentric explanation of planetary motion *The Earth travels around the Sun more rapidly than does Mars. Consequently, as the Earth overtakes and passes this slower-moving planet, Mars appears to move backward for a few months.*

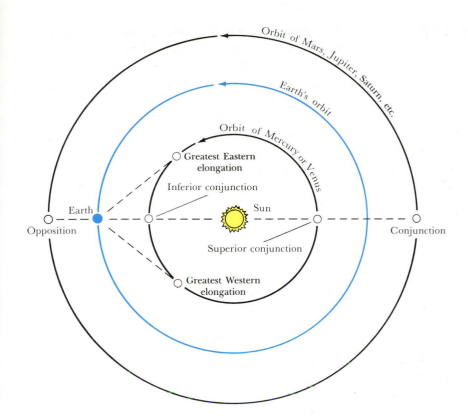

Figure 3-4 *Planetary configurations* *It can be useful to specify key points along a planet's orbit as shown in this diagram. These points identify specific geometric arrangements between the Earth, another planet, and the Sun.*

Although it is easy to follow a planet as it moves from one configuration to another, these observations in themselves do not immediately provide relevant data about the planet's actual orbit around the Sun. Copernicus realized that the Earth, from which we make the observations, is also moving. He was therefore careful to distinguish between two characteristic time intervals, or **periods**, of each planet. The **synodic period** is the time that elapses between two successive identical configurations (as seen from the Earth)—from one opposition to the next, for example, or from one conjunction to the next. The **sidereal period** is the true orbital period of a planet, the time it takes the planet to complete one orbit of the Sun.

The synodic period of a planet can be determined by observing the sky, but the sidereal period must be calculated. Copernicus performed such calculations and obtained the results shown in Table 3-1. Knowing the sidereal periods of the planets, Copernicus was able to devise a straightforward geometric method of determining the distances of the planets from the Sun. His answers turned out to be remarkably close to the modern values, as shown in Table 3-2. From these two tables it is apparent that the farther a planet is from the Sun, the longer the planet takes to travel around its orbit.

Copernicus compiled his ideas and calculations in a book entitled *De revolutionibus orbium coelestium* ("On

Table 3-1 The synodic and sidereal periods of the planets

Planet	Synodic period	Sidereal period
Mercury	116 days	88 days
Venus	584 days	225 days
Earth	——	1.0 year
Mars	780 days	1.9 years
Jupiter	399 days	11.9 years
Saturn	378 days	29.5 years

Table 3-2 Average distances of the planets from the Sun (in astronomical units)

Planet	Copernicus	Modern
Mercury	0.38	0.39
Venus	0.72	0.72
Earth	1.00	1.00
Mars	1.52	1.52
Jupiter	5.22	5.20
Saturn	9.07	9.54

the Revolutions of the Celestial Spheres") that was published in 1543, just in time to reach Copernicus on his deathbed. Although he assumed that the Earth travels around the Sun along a circular path, he found that perfectly circular orbits cannot accurately describe the paths of the other planets. Copernicus had to add an epicycle to each planet to account for the slight variation in speed along its orbit. Thus, according to Copernicus, each planet revolves around a small epicycle, which in turn orbits the Sun along a circular path.

Placing the Sun at the center of the universe was a revolutionary proposal and one Renaissance astronomer, Tycho Brahe, tried to test Copernicus's ideas with detailed observations of the sky. Brahe spent his lifetime making accurate observations of the positions of the stars and planets, achieving an unprecedented level of precision. Brahe realized that these observations might reveal the motion of the Earth around the Sun.

We know that, when we walk from one place to another, nearby objects appear to shift their positions against the background of more distant objects. If Copernicus was correct, nearby stars should shift slightly against the background stars as the Earth orbits the Sun. In spite of Brahe's accurate observations, he could not detect any shifting of star positions. He therefore concluded that Copernicus was wrong. Actually, the stars are so far away that Brahe's naked-eye observations could not detect the tiny shifting of star positions that has now been confirmed with telescopic observations.

In spite of Tycho Brahe's failure to confirm Copernicus's heliocentric cosmology, his attempt to test the theory was quite revolutionary. At a time when heretics were burned at the stake for questioning dogma, Brahe was arguing that objective observations could be used to determine the validity or falsehood of an idea.

Brahe's astronomical records were destined to play an important role in the development of a heliocentric cosmology. Upon his death in 1601, many of his charts and books fell into the hands of his gifted assistant, Johannes Kepler.

3-3 Johannes Kepler proposed elliptical paths of the planets about the Sun

Until Johannes Kepler's time, at the beginning of the seventeenth century, astronomers had assumed that heavenly objects move in circles. The circle was considered the most perfect and the most harmonious of all geometric shapes. People of that time believed that a perfect God, who resides in heaven with all the celestial bodies, would use only perfect circles to control the motions of the planets. Kepler spent several years trying in vain to make circular motion fit Tycho Brahe's observations. Kepler's first major contribution to astronomy was the discovery that the planetary orbits are not circles.

Kepler turned from circles to ovals, but he found they did not accurately describe the orbits of the planets about the Sun. Then he began working with a slightly different curve called an **ellipse**.

An ellipse can be drawn with a loop of string, two thumbtacks, and a pencil, as shown in Figure 3-5. Each thumbtack is at a **focus** (plural **foci**). The longest diameter across an ellipse, called the **major axis**, passes through both foci. Half of that distance is called the **semimajor axis**, whose length is usually designated by the letter a.

To Kepler's delight, the ellipse turned out to be the curve he had been searching for. He published this discovery in 1609 in a book known today as *New Astronomy*. This important discovery is now called **Kepler's first law** and is stated as follows:

The orbit of a planet about the Sun is an ellipse with the Sun at one focus.

Kepler realized that planets do not move at uniform speeds along their orbits. A planet moves most rapidly

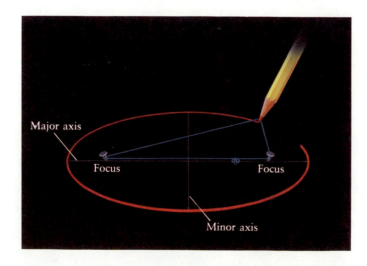

Figure 3-5 The construction of an ellipse *An ellipse can be drawn with a pencil, a loop of string, and two thumbtacks, as shown in this diagram. If the string is kept taut, the pencil traces out an ellipse. The two thumbtacks are located at the two foci of the ellipse.*

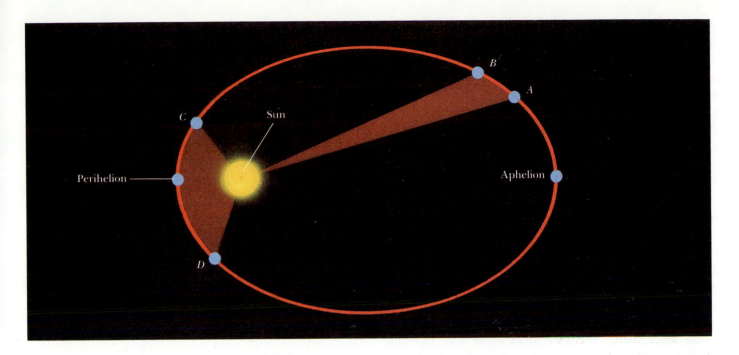

Figure 3-6 *Kepler's first and second laws* *According to Kepler's first two laws, every planet travels around the Sun along an elliptical orbit with the Sun at one focus in such a way that the line joining the planet and the Sun sweeps out equal areas in equal intervals of time.*

when it is nearest the Sun, at a point on its orbit called the **perihelion.** A planet moves most slowly when it is farthest from the Sun, at a point called the **aphelion.**

After much trial and error, Kepler discovered a way to describe how fast a planet moves along its orbit. This discovery, now called the **law of equal areas** or **Kepler's second law,** is illustrated in Figure 3-6. Suppose that it takes 30 days for a planet to go from point *A* to point *B*. During that time, the line joining the Sun and the planet sweeps out a nearly triangular area. Kepler discovered that the line joining the Sun and the planet sweeps out an equal area during any other 30-day interval. In other words, if the planet also takes a month to go from point *C* to point *D*, then the two shaded segments in Figure 3-6 are equal in area. Kepler's second law, also published in *New Astronomy*, can thus be stated:

A line joining a planet and the Sun sweeps out equal areas in equal intervals of time.

One of Kepler's later discoveries, published in 1619, stands out because of its impact on future developments

in astronomy. Now called the **harmonic law** or **Kepler's third law,** it states a relationship between the sidereal period of a planet and the length of the semimajor axis of the planet's orbit:

The squares of the sidereal periods of the planets are proportional to the cubes of the semimajor axes of their orbits.

If a planet's sidereal period P is measured in years and the length of the semimajor axis a of its orbit is measured in astronomical units, then Kepler's third law is simply stated as

$$P^2 = a^3$$

The length of the semimajor axis can be regarded as the average distance between a planet and the Sun. Using data from Tables 3-1 and 3-2, we can demonstrate Kepler's third law as shown in Table 3-3. This relationship can also be displayed on a graph as in Figure 3-7.

It is testimony to Kepler's genius that his three laws are rigorously obeyed in any situation where two objects

Table 3-3 A demonstration of Kepler's third law

Planet	Sidereal period P (years)	Semimajor axis a (AU)	P^2	a^3
Mercury	0.24	0.39	0.06	0.06
Venus	0.61	0.72	0.37	0.37
Earth	1.00	1.00	1.00	1.00
Mars	1.88	1.52	3.53	3.51
Jupiter	11.86	5.20	140.7	140.6
Saturn	29.46	9.54	867.9	868.3

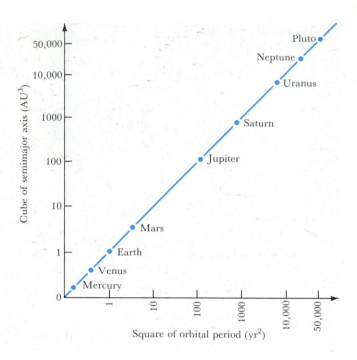

Figure 3-7 Kepler's third law *On this graph, the squares of the periods of the planets (P^2) are plotted against the cubes of the semimajor axes (a^3) of their orbits. The fact that the points fall along a straight line is verification of Kepler's discovery that $P^2 = a^3$.*

orbit each other under the influence of their mutual gravitational attraction. Kepler's laws are obeyed not only by planets circling the Sun but also by artificial satellites orbiting the Earth and by two stars revolving about each other in a double star system. Throughout this book, we shall see that Kepler's laws have a wide range of practical applications.

3-4 Galileo's discoveries with a telescope strongly supported a heliocentric cosmology

While Kepler was making rapid progress in central Europe, an Italian physicist was making equally dramatic observational discoveries. Galileo Galilei did not invent the telescope, but he was the first person to point one of the new devices toward the sky and publish his observations. He saw things that no one had ever dreamed of. He saw mountains on the Moon and sunspots on the Sun. He also discovered that Venus exhibits phases (Figure 3-8).

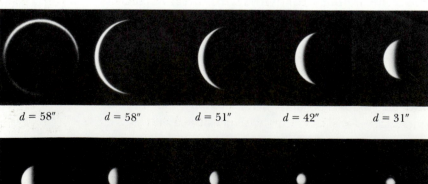

$d = 58''$	$d = 58''$	$d = 51''$	$d = 42''$	$d = 31''$
$d = 24''$	$d = 18''$	$d = 15''$	$d = 12''$	$d = 10''$

Figure 3-8 The phases of Venus *This series of photographs shows how the appearance of Venus changes as it moves along its orbit. The number below each view is the angular diameter (d) of the planet in seconds of arc. (New Mexico State University Observatory)*

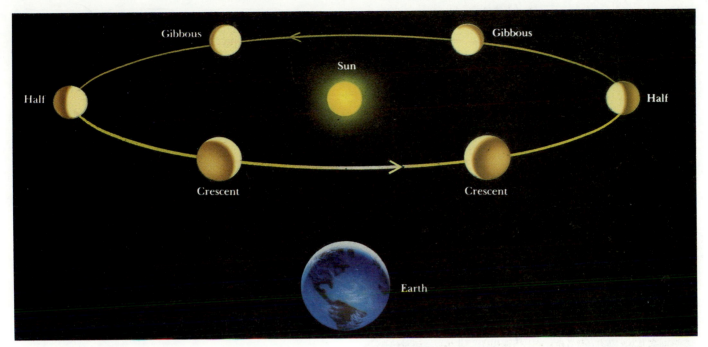

Figure 3-9 *The changing appearance of Venus* *The phases of Venus are correlated with the planet's angular size and its angular distance from the Sun, as sketched in this diagram. Galileo's* *observations of this correlation clearly supported the idea that Venus orbits the Sun.*

After only a few months of observation, Galileo noticed that the apparent size of Venus as seen through his telescope was related to the planet's phase. The planet appears smallest at gibbous phase and largest at crescent phase. There is also a correlation between the phases of Venus and the planet's angular distance from the Sun. These relationships clearly supported the conclusion that Venus goes around the Sun (Figure 3-9).

In 1610 Galileo also discovered four moons orbiting Jupiter (Figure 3-10). He realized that they are orbiting Jupiter because they move back and forth from one side of the planet to the other. Confirming observations made in 1620 are shown in Figure 3-11. Astronomers soon demonstrated that these four moons obey Kepler's third law: The square of a moon's orbital period about Jupiter is proportional to the cube of its average distance from the planet. In honor of their discoverer, these four moons are today called the Galilean satellites.

These telescopic observations constituted the first fresh influx of fundamentally new astronomical data in almost two thousand years. In contradiction to prevailing opinions, these discoveries strongly suggested a heliocentric view of the universe. The Roman Catholic church attacked Galileo's ideas because they were not reconcilable with certain passages in the Bible or with the writings of Aristotle and Plato. Nevertheless, there was no turning back.

Figure 3-10 *Jupiter and its largest moons* *This photograph, taken by an amateur astronomer with a small telescope, shows the four Galilean satellites alongside an overexposed image of Jupiter. Each satellite is bright enough to be seen with the unaided eye were it not overwhelmed by the glare of Jupiter. (Courtesy of C. Holmes)*

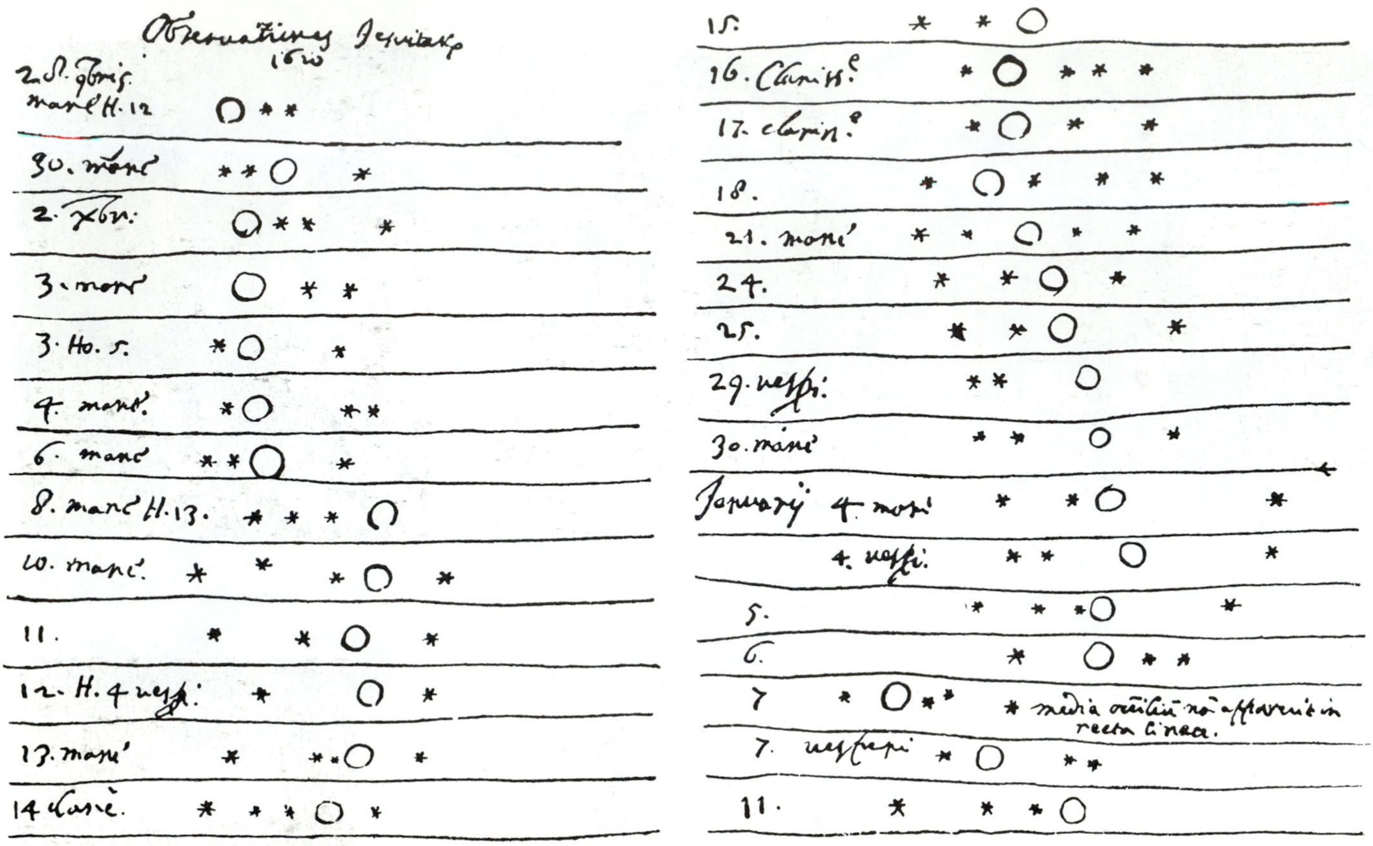

Figure 3-11 Early observations of Jupiter's moons In 1610 Galileo discovered four "stars" that move back and forth across Jupiter from one night to the next. He concluded that these are four moons that orbit Jupiter much as our Moon orbits the Earth. This drawing shows observations made by Jesuits in 1620. (Yerkes Observatory)

Galileo was condemned to spend his latter years under house arrest "for vehement suspicion of heresy," but his revolutionary ideas soon inspired a sickly English boy born on Christmas Day of 1642, less than a year after Galileo died. The boy's name was Isaac Newton.

3-5 Isaac Newton formulated a description of gravity that accounts for Kepler's laws and explains the motions of the planets

Until the mid-seventeenth century, virtually all mathematical astronomy had been entirely empirical, characterized by trial and error. From Ptolemy to Kepler, essentially the same approach had been used. Astronomers would work directly from data and observations, adjusting ideas and calculations until they finally came out with the right answers.

Isaac Newton introduced a new approach. He made three assumptions, now called **Newton's laws**, about the nature of reality. These are quite general statements that apply to all forces and bodies. Newton then showed that Kepler's three laws follow logically from these laws of motion and from a formula for the force of gravity that Newton derived. Using this formula, Newton accurately described the observed orbits of the Moon, comets, and other objects in the solar system.

Newton's first law, also called the law of inertia, states that:

A body remains at rest or moves in a straight line at a constant speed unless acted upon by an outside force.

Copernicus
(1473–1543)

Kepler
(1571–1630)

Galileo
(1564–1642)

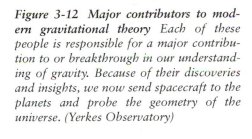

Figure 3-12 Major contributors to modern gravitational theory Each of these people is responsible for a major contribution to or breakthrough in our understanding of gravity. Because of their discoveries and insights, we now send spacecraft to the planets and probe the geometry of the universe. (Yerkes Observatory)

Newton
(1642–1727)

Einstein
(1879–1955)

At first, this law might seem to conflict with your everyday experience. For example, if you give a chair a shove, it does not continue at a constant speed forever, but rather comes to rest after gliding only a short distance across the floor. From Newton's viewpoint, however, an "outside force" does indeed act on the moving chair: friction between the chair's legs and the floor. If there were no friction, the chair would continue in a straight path at a constant speed.

Newton's first law tells us that there must be an outside force acting on the planets, or else they would leave their curved orbits and move away from the Sun along straight-line paths at constant speeds. Since this does not happen,

Newton concluded that the continuous action of this force confines the planets to their elliptical orbits.

Isaac Newton did not invent the idea of gravity (Figure 3-12). An educated seventeenth-century person had a vague appreciation of the fact that some force pulls things down to the ground. It was Newton, however, who gave us a precise description of the action of gravity. Using his first law, Newton mathematically proved that the force acting on each of the planets is directed toward the Sun. This discovery led him to suspect that the force pulling a falling apple straight down to the ground is the same as the force on the planets that is always aimed straight at the Sun.

Newton's second assumption does two things; it defines the concept of a **force,** and it describes how a force changes the motion of an object. To appreciate these concepts, we must first understand quantities that describe motion, such as speed, velocity, and acceleration.

Imagine an object in space. Push on the object and it will begin to move. At any moment, you can specify the object's motion by giving both its speed and direction. Speed and the direction of motion together constitute the object's velocity. If you continue to push on the object, its speed will increase—in other words, it will accelerate.

Acceleration is the rate at which velocity changes. Since velocity involves both speed and direction, acceleration can result from changes in either. Also note that, contrary to popular use of the term, acceleration is not restricted to increases in speed alone. A slowing down, a speeding up, or a change in direction all involve accelerations.

A planet revolving about the Sun along a perfectly circular orbit is an example of acceleration that involves change of direction only. As the planet moves along its orbit, its speed remains constant. But the planet's direction of motion is continually changing, and so the planet is continually being accelerated.

Newton's second law says that:

The acceleration of an object is proportional to the force acting on the object.

In other words, the harder you push on an object, the greater is the resulting acceleration. This law can be succinctly stated as an equation. If a force F acts on an object, the object will experience an acceleration a such that

$$F = ma$$

where m is the mass of the object.

The **mass** of an object is a measure of the total amount of material in the object, usually expressed in grams or kilograms. For example, the mass of the Sun is 2×10^{30} kg. The mass of a hydrogen atom is 1.7×10^{-24} g. The mass of the author of this book is 79 kg. The Sun, a hydrogen atom, and the author have these masses regardless of where they happen to be in the universe.

It is important not to confuse the concept of mass with that of weight. **Weight** is the force with which an object presses down on the ground (due to gravity's pull), and force is usually expressed in pounds, newtons, or dynes. For example, the force with which the author presses down on the ground is 174 pounds.

Note that the author weights 174 pounds only when he is on the Earth. He would weigh less on the Moon and more on Jupiter. Floating in space, he would have no weight at all; he would be "weightless." Nevertheless, under all these circumstances, he would always have exactly the same mass. Thus we see that mass is an inherent property of matter unaffected by details of the environment. Whenever we describe the properties of planets, stars, or galaxies, we speak of their masses, never of their weights.

Newton's final assumption, called Newton's third law, is the famous statement about action and reaction:

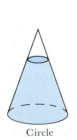
Circle

Ellipse

Parabola

Hyperbola

*Figure 3-13 **Conic sections** A conic section is any one of a family of curves obtained by slicing a cone with a plane, as shown in this diagram. The orbit of one body about another can be any one of these curves: a circle, an ellipse, a parabola, or a hyperbola.*

Whenever one body exerts a force on a second body, the second body exerts an equal and opposite force on the first body.

For example, if you weigh 165 pounds, you are pressing down on the floor with a force of 165 pounds. Newton's third law tells us that the floor is also pushing up against your feet with an equal force of 165 pounds. (If it were not, you would fall through the floor.) In the same way, Newton realized that, because the Sun is exerting a force on each planet to keep it in orbit, each planet must also be exerting an equal and opposite force on the Sun.

Using his own three laws and Kepler's three laws, Newton succeeded in formulating a general statement describing the nature of the force called **gravity** that keeps the planets in their orbits. Newton's **universal law of gravitation** states:

Two bodies attract each other with a force that is directly proportional to the product of their masses and inversely proportional to the square of the distance between them.

In other words, if two objects have masses m_1 and m_2 and are separated by a distance r, then the gravitational force F between these two masses is

Figure 3-14 Halley's Comet Halley's Comet orbits the Sun with an average period of about 76 years. During the twentieth century, the comet passed near the Sun twice—once in 1910 and again in 1986. This photograph shows how the comet looked in 1986. (Lumicon)

$$F = G\,\frac{m_1 m_2}{r^2}$$

In this formula, G is a number called the **universal constant of gravitation,** whose value has been determined from laboratory experiments.

Using his law of gravity, Newton found that he could mathematically prove the validity of Kepler's three laws. For example, whereas Kepler discovered by trial and error that $P^2 = a^3$, Newton demonstrated mathematically that this equation follows logically from his law of gravity.

Newton also discovered new features of orbits around the Sun. For example, his equations soon led him to conclude that the orbit of an object around the Sun could be any one of a family of curves called conic sections. A **conic section** is any curve that you get by cutting a cone with a plane, as shown in Figure 3-13. You can get circles and ellipses by slicing all the way through the cone. By slicing the cone at a steep angle, you can get two "open" curves called **parabolas** and **hyperbolas.** Comets hurtling

toward the Sun from the depths of space sometimes have hyperbolic orbits.

Newton's ideas and methods turned out to be incredibly successful in describing a wide range of situations. The orbits of the planets and their satellites could now be calculated with unprecedented precision. In addition, Newton's laws and mathematical techniques could be used to predict new phenomena. For example, one of Newton's friends, Edmund Halley, was intrigued by historical records of a comet that was sighted about every 76 years. Using Newton's methods, Halley worked out the details of the comet's orbit and predicted its return in 1758. It was first sighted on Christmas night of that year, and to this day the comet bears Halley's name (Figure 3-14).

Perhaps the most dramatic success of Newton's ideas was their role in the discovery of the eighth planet from the Sun. The seventh planet, Uranus, had been discovered by William Herschel in 1781 during a telescopic survey of the sky. Fifty years later, however, it was clear that Uranus was not following its predicted orbit. Two mathematicians, John Couch Adams in England and U. J. Leverrier

Figure 3-15 Uranus and Neptune *The discovery of Neptune was a major triumph for Newtonian mechanics. The existence of Neptune (shown here with one moon) was deduced from deviations in the predicted orbit of Uranus (shown with three moons).*

This discovery was a major triumph for Newtonian mechanics. Both planets have more satellites than appear in these photographs. (Lick Observatory)

in France, independently calculated that the deviations of Uranus from its orbit could be explained by the gravitational pull of a yet unknown, more distant planet. They each predicted that the planet would be found at a certain location in the constellation of Aquarius. A brief telescopic search on September 23, 1846, revealed Neptune less than 1° from the calculated position. Although sighted with a telescope (Figure 3-15), Neptune was really discovered with pencil and paper.

Over the years, Newton's ideas were successfully used to predict and explain of many phenomena, and Newtonian mechanics became the cornerstone of modern physical science. Even today, as we send astronauts to the Moon and probes to the outer planets, Newton's equations are used to calculate the orbits and trajectories of these spacecraft.

There was one instance, however, in which Newtonian mechanics was not quite in agreement with observations. During the mid-1800s, Leverrier pointed out that Mercury was not following its predicted orbit. As the planet moves along its elliptical orbit, the orbit itself slowly rotates (or precesses) as shown in Figure 3-16. Because most of Mercury's precession is caused by the gravitational pull of the other planets, it is largely explained with New-

tonian mechanics. There is, however, an unexplained excess rotation of Mercury's major axis amounting to only 43 arc sec per century. Although this motion is very tiny, all attempts to account for it met with failure. Nevertheless, Newton's laws proved eminently successful in all other situations.

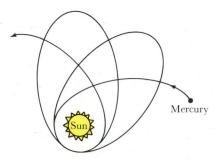

Figure 3-16 The advance of Mercury's perihelion *As Mercury moves along its orbit, the orbit rotates. Consequently, Mercury traces out a rosette figure about the Sun, greatly exaggerated in this diagram. The excess rotation of Mercury's orbit (beyond that predicted by Newtonian mechanics) amounts to only 43 arc sec per century.*

It is a testament to Isaac Newton's genius that his three laws were precisely the three basic ideas needed for a full understanding of the motions of the planets. Newton brought a new dimension of elegance and sophistication to our understanding of the workings of the universe. That is where things probably would have remained had it not been for the insight, vision, and genius of Albert Einstein.

3-6 Albert Einstein's theory states that gravity affects the shape of space and the flow of time

In the cosmologies of many ancient civilizations, the Earth and its inhabitants occupied a special place at the center of the universe. Then Copernicus's cosmology made the Earth just one of a number of planets orbiting the Sun. We have seen that this new approach proved quite fruitful. Within less than a century, Kepler's accurate description of planetary orbits led directly to Newton's universal law of gravitation. Astronomers then began to explore the possibility that even the Sun might occupy no special location in the universe. Today we know that our Sun is merely one of many stars scattered throughout the Milky Way Galaxy, which is one of many galaxies in the universe.

Albert Einstein introduced an even more powerful extension of Copernicus's conviction that we do not occupy a special place in the universe. Einstein believed that the fundamental laws of the universe should not depend on a person's location or motion. In other words, the laws of physics should be the same, whether we happen to be sitting on the Earth or moving through space at 95% of the speed of light. Inspired by this viewpoint, Einstein endeavored to reformulate the laws of physics so that all observers would have equal status.

In 1916 Albert Einstein formulated a new theory of gravity, called the **general theory of relativity,** that is independent of one's location or motion. To achieve this goal, Einstein had to abandon the notion that rulers have fixed lengths and clocks always tick at an unchanging rate. The basic idea in general relativity is that the gravity of an object alters the properties of space and time around the object.

Einstein's description of gravity is radically different from Newton's. According to Newtonian mechanics, space is perfectly flat and extends infinitely far in all directions. Similarly, Newtonian clocks monotonously tick at an unchanging rate, never speeding up or slowing down. In this rigid, unalterable framework of space and time, gravity is described as a "force" that acts at a distance. The planets literally pull on each other across empty space with a strength described by Newton's universal law of gravitation.

The general theory of relativity does not treat gravity as a force at all. Instead, gravity causes space to become curved and time to slow down. Far from any sources of gravity, space is flat and clocks tick at their normal rate. But near a source of gravity, space becomes curved and clocks tick more slowly than normal. The stronger the source of gravity, the greater are its effects on space and time. In the vicinity of a massive object like the Sun, space becomes curved as shown in Figure 3-17. A planet or

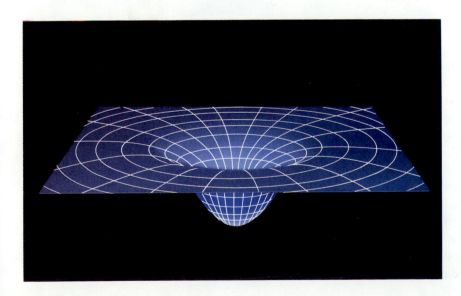

Figure 3-17 The gravitational curvature of space *According to Einstein's general theory of relativity, space becomes curved near a massive object such as the Sun. This diagram shows a two-dimensional analogy of the shape of space around a massive object.*

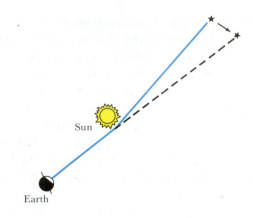

Figure 3-18 The gravitational deflection of light Light rays passing near the Sun are deflected from their straight-line paths. By relativistic standards, the Sun's gravity is very weak. The maximum angle of deflection is only 1.75 arc sec for a light ray grazing the Sun's surface.

spacecraft passing near this massive object is deflected from a straight-line path because space itself is curved.

One of the first things Einstein did with his new theory was to calculate the orbits of the planets. With only one minor exception, general relativity theory gave almost exactly the same answers as Newtonian theory. Mercury is the only planet to pass close enough to the Sun for the curvature of space to give rise to motions that differ noticeably from those predicted by Newtonian mechanics. Indeed, throughout most of the solar system the curvature of space is so slight that Einstein's equations are essentially equivalent to Newtonian calculations. However, Einstein's theory did predict that Mercury's orbit should be a slowly rotating ellipse, thereby explaining a phenomenon that had frustrated astronomers for half a century.

To test the validity of his theory Einstein predicted new phenomena. He calculated that light rays passing near the surface of the Sun should appear to be deflected from their straight-line paths because the space through which they are moving is curved. In other words, gravity should bend light rays, an effect not predicted by Newtonian mechanics because light has no mass.

Figure 3-18 shows a beam of light from a star passing by the Sun and continuing down to the Earth. Because the light ray is bent, the star appears shifted from its usual location. At most, the apparent position of a star whose light rays graze the Sun's surface is shifted by 1.75 arc sec.

This prediction was first tested during a total solar

eclipse in 1919. During the precious moments of totality, when the Moon blocked out the blinding solar disk, astronomers succeeded in photographing the stars around the Sun. Careful measurements revealed that the stars were shifted from their usual positions by an amount consistent with Einstein's theory. General relativity had passed another important test.

Einstein made another prediction. Because gravity causes time to slow down, a clock on the first floor of a building should tick slightly more slowly than a clock on the roof, as shown in Figure 3-19. This prediction was tested in 1959 in a four-story building at Harvard University using a radioactive substance as an accurate timepiece. Again Einstein was proven correct.

During the past several decades, these and similar tests have confirmed general relativity over and over again. This remarkable theory now stands as our most precise and complete description of gravity.

It is important to emphasize that Einstein did not prove Newton wrong. Rather, Einstein demonstrated that Newtonian gravitation is accurate only when applied to weak

Figure 3-19 The gravitational slowing of time A clock on the ground floor of a building is closer to the Earth than a clock at a higher elevation. Therefore, according to general relativity, the clock on the ground floor should tick more slowly than the clock on the roof.

gravity. If extremely powerful gravitational fields (such as those of neutron stars and black holes) are being studied, only a relativistic calculation will yield correct analyses.

Einstein's general theory of relativity evoked a sensation when it was first proposed. Newtonian gravitation had been overthrown by a radically different approach that worked better. After the initial excitement, however, interest in general relativity rapidly waned. No one could imagine places where the curvature of space might have a major effect. Relativistic theory offered only a little more accuracy, as in the case of Mercury's orbit. The complex mathematics of relativity seemed burdensome, and Newton's simpler approach gave reasonable precision in almost all conceivable circumstances.

Although Newtonian mechanics is still widely used, in recent years there has been a reawakening of interest in general relativity, primarily because of dramatic advances in our understanding of the evolution of stars. As we shall see in later chapters, we now have a reasonably complete picture of how stars are born, what happens to them as they mature, and where they go when they die. In particular, it appears that the most massive dying stars are doomed to collapse completely upon themselves, thereby producing some of the most bizarre objects in the universe—black holes. Gravity around one of these massive stellar corpses is so strong that it punches a hole in the fabric of space. Because of this intense gravitational effect, black holes can be described and discussed only in terms of general relativity.

As we set our sights on understanding the universe as a whole, we must once again turn to general relativity. All the matter in all the stars and galaxies is responsible for the overall curvature or shape of space. From the viewpoint of general relativity, it is therefore reasonable to ask about the actual shape of the universe. We shall see that the answer suggests the ultimate fate of the cosmos.

Summary

- Ancient astronomers believed that the Earth is at the center of the universe. They invented a complex system of epicycles and deferents to explain the direct and retrograde motions of the planets.

- A heliocentric (Sun-centered) theory simplifies the general explanation of planetary motions.

   In a heliocentric system, the Earth is one of several planets that orbit the Sun.

   The sidereal period of a planet, which is measured with respect to the stars, is the planet's true orbital period. Its synodic period is measured with respect to the Sun as seen from the moving Earth (for example, from one opposition to the next).

- Ellipses describe the paths of the planets around the Sun much more accurately than do circles. Kepler's three laws give important details about elliptical orbits.

- The invention of the telescope led to new discoveries (phases of Venus, moons of Jupiter, and so forth) that supported a heliocentric view of the universe.

- Newton based his explanation of the universe on three assumptions, or laws of motion. These laws and his universal law of gravitation can be used to deduce Kepler's laws and extremely accurate descriptions of planetary motions.

   The mass of an object is a measure of the amount of matter in the object; its weight is a measure of the force with which the gravity of some other object pulls on it.

   In general, the path of one object about another, such as that of a comet about the Sun, is one of the curves called conic sections: a circle, an ellipse, a parabola, or a hyperbola.

- Although Newtonian mechanics accurately describes and predicts numerous phenomena, Einstein's general theory of relativity is more accurate where extremely intense gravitational fields are involved.

   The general theory of relativity explains that gravity causes space to be curved and time to slow down.

Review questions

1 How did ancient astronomers differentiate between a star and a planet?

2 What is an epicycle, and how is it used in Ptolemy's explanation of the retrograde motions of the planets?

3 How did Copernicus explain the retrograde motions of the planets?

4 Which planets can never be seen at opposition? Which planets can never be seen at inferior conjunction?

5 At what configuration (superior conjunction, greatest eastern elongation, etc.) would it be best to observe Mercury or Venus with an Earth-based telescope? At what configuration would it be best to observe Mars, Jupiter, or Saturn? Explain your answers.

6 What is the difference between the synodic and sidereal periods of a planet?

7 In your own words, state Kepler's three laws of planetary motion. Why are these laws important?

8 In what ways did the astronomical observations of Galileo support a heliocentric cosmology?

9 How did Newton's approach to understanding planetary motions differ from that of his predecessors?

10 What is the difference between mass and weight?

11 What are conic sections and in what way are they related to the orbits of objects in the solar system?

12 Why is the discovery of Neptune a major confirmation of Newton's universal law of gravitation?

13 Why is Einstein's general theory of relativity a better description of gravity than Newton's universal law of gravitation?

Advanced questions

14 Is it possible for an object in the solar system to have a synodic period of exactly one year? Explain your answer.

* 15 A line joining the Sun and an asteroid was found to sweep out 5.2 square astronomical units of space in 1990. How much area was swept out in 1991? In five years?

* 16 A comet moves in a highly elongated orbit about the Sun with a period of 1000 years. What is the length of the semimajor axis of the comet's orbit? What is the farthest the comet can get from the Sun?

* 17 The orbit of a spacecraft about the Sun has a perihelion distance of 0.5 AU and an aphelion distance of 3.5 AU. What is the spacecraft's orbital period?

18 Look up the dates of various greatest eastern and western elongations for Mercury in a year of your choice. Does it take longer to go from eastern to western elongation or vice versa? Why do you suppose this is the case?

* 19 Suppose that the Earth were moved to a distance of 10 AU from the Sun. How much stronger or weaker would the Sun's gravitational pull be on the Earth?

20 Look up orbital data for the four largest moons of Jupiter. Demonstrate that these data obey Kepler's third law.

Discussion questions

21 Which planet would you expect to exhibit the greatest variation in apparent brightness as seen from Earth? Explain your answer.

22 Use two thumbtacks, a loop of string, and a pencil to draw several ellipses. Describe how the shape of the ellipse varies as you change the distance between the thumbtacks.

For further reading

Drake, S. "Newton's Apple and Galileo's Dialogue." *Scientific American,* August 1980 • This excellent article shows how Newton's thinking was influenced by the work of Galileo.

Gardner, M. *The Relativity Explosion.* Vintage, 1976 • An excellent book about Einstein's ideas and relativity theory.

Gingerich, O. "The Galileo Affair." *Scientific American,* August 1982 • This article gives many insights into the life and times of Galileo with special emphasis on his relationship with the Church.

Kaufmann, W. *Relativity and Cosmology.* Harper & Row, 1977 • This slim book, which emphasizes relativity, describes the evolution of our understanding of gravitation.

Wilson, C. "How Did Kepler Discover His First Two Laws?" *Scientific American,* March 1972 • An excellent treatment of the steps that Kepler took toward his great discoveries.

The summit of Mauna Kea on Hawaii *Astronomers prefer to build observatories on isolated mountaintops far from city lights, where the air is dry, stable, and cloud-free. The summit of Mauna Kea, shown here during winter, offers the best observing site in the world. The dome in the foreground houses the 10-m Keck reflector, the largest telescope in the world. To the right of the Keck Observatory is NASA's Infrared Telescope Facility. The domes on the ridge in the background enclose (from left to right) the 3.6-m Canada-France-Hawaii telescope, the 2.2-m University of Hawaii telescope, and the U.K. Royal Observatory 3.8-m infrared telescope. (California Association for Research in Astronomy)*

4

Light, Optics, and Telescopes

The telescope is the astronomer's most important tool because it gives big, bright, sharp images of distant astronomical objects . Refracting telescopes, which use large lenses to collect incoming starlight, were popular in the nineteenth century, but modern astronomers strongly prefer reflecting telescopes, which gather light with large concave mirrors. Astronomers attach a variety of equipment to telescopes with which to record and analyze incoming starlight. The Earth's atmosphere is transparent to both visible light and radio waves. Ground-based radio telescopes can therefore be used to view the universe at wavelengths much longer than those of visible light. At most other wavelengths, the Earth's atmosphere is opaque, and so observations must be performed from space. The wealth of information coming from both ground-based and Earth-orbiting observatories is giving us deep insight into the nature of the universe.

The telescope is the single most important tool of astronomy. Using a telescope, we can see extremely faint objects in space far more clearly than we can with the naked eye. Telescopes have played a major role in revealing the universe since Galileo first used one and saw craters on the Moon four centuries ago.

Traditionally telescopes have been used to detect visible light. Light from a distant object is brought, by either lenses or mirrors, to a focus, where the resulting image is viewed or photographed. Recently, however, astronomers have built telescopes that detect nonvisible forms of light such as X rays and radio waves. To appreciate these developments and to understand how telescopes work, we must first learn something about the basic properties of light.

4-1 Light is electromagnetic radiation and is characterized by its wavelength

Galileo and Newton made important contributions to our modern understanding of light as well as to our theories of gravity and mechanics. In the early 1600s, Galileo made one of the first attempts to measure the speed of light. At night he stood on one hilltop while an assistant stood on another hilltop at a known distance, each holding a shuttered lantern. First Galileo opened the shutter of his lantern. As soon as the assistant saw the flash of light, he opened his own lantern. Using the rate of his pulse as a clock, Galileo attempted to measure the time that elapsed between opening his shutter and seeing the light from the assistant's lantern. By knowing this time and the distance between hilltops, he could then compute the speed at which light traveled to the distant hilltop and back again. He soon concluded that light travels so rapidly that slow human reactions and crude clocks, like his pulse, make it impossible to measure its speed.

The first successful measurement of the speed of light was made in 1675 by Olaus Roemer, a Danish astronomer, who carefully timed eclipses of Jupiter's satellites. Roemer discovered that the moment at which a satellite enters Jupiter's shadow seems to depend on the distance between the Earth and Jupiter. When the Earth–Jupiter distance is short (around the time of opposition), eclipses are observed slightly earlier than when Jupiter and the Earth are widely separated. Roemer correctly interpreted this phenomenon as the result of the time it takes light to travel across space. The greater the distance to Jupiter, the longer we must wait for the image of an eclipse to reach our eyes. From his timing measurements, Roemer concluded that it takes 16.5 minutes for light to traverse the diameter of the Earth's orbit (2 AU). Because no one in Roemer's day knew the length of the astronomical unit, he could not actually compute the speed of light in kilometers per second.

The first accurate laboratory experiments to measure the speed of light were performed in the mid-1800s, most of them using elaborate optical equipment. From these experiments, we now know that the speed of light in a vacuum is about 3×10^8 meters per second (186,000 miles

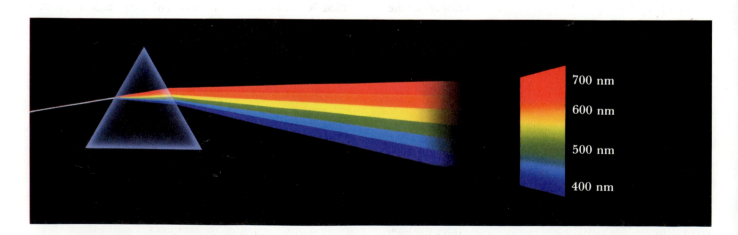

Figure 4-1 A prism and a spectrum *When a beam of white light passes through a glass prism, the light is broken into a rainbow-colored band called a spectrum. The numbers on the right side of the spectrum indicate wavelengths, as described in the text.*

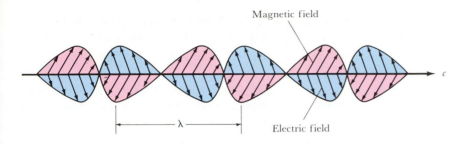

Magnetic field

λ

Electric field

c

Figure 4-2   Electromagnetic radiation   All forms of light consist of oscillating electric and magnetic fields that move through empty space at a speed of 3×10^8 m/s. The distance between two successive crests is called the wavelength of the light, which is usually designated by the lowercase Greek letter λ (lambda).

per second). Incidentally, standard abbreviations for units of speed, like m/s for meters per second, or km/hr for kilometers per hour, will be used throughout the rest of this book.

A pioneering breakthrough in understanding light came from a simple experiment performed by Isaac Newton in the late 1600s. He passed a beam of sunlight through a glass prism that spread the light out into the colors of the rainbow, as shown in Figure 4-1. This rainbow, called a **spectrum,** suggested to Newton that white light is actually a mixture of all colors. In Newton's day, most people erroneously believed that the colors were somehow added to the light by the glass. Newton disproved this idea by passing the spectrum through a second prism inverted with respect to the first. Since only white light emerged from this second prism, he had shown that the second prism had reassembled the colors of the rainbow to give back the original beam of sunlight.

From his many experiments in optics, Newton argued that light is composed of indetectably tiny particles of energy. In fact, as we shall see in the next chapter, the modern theory of quantum mechanics, developed in the early 1900s, also takes this viewpoint. In the mid-1600s, however, a rival explanation was proposed by the Dutch astronomer Christian Huygens, who suggested that light travels in the form of waves rather than particles. Scientists now realize that light possesses both particlelike and wavelike characteristics.

The English physicist Thomas Young confirmed the wave nature of light in 1801. Young demonstrated that the shadows of objects in light of a single color are not crisp and sharp. Instead, the boundary between illuminated and shaded areas is overlaid with patterns of closely spaced dark and light bands. These patterns are similar to the patterns produced by water waves passing the edge of a reef or barrier in the ocean. Young showed that a wavelike description of light could explain the results of his experiment, but Newton's particle theory could not.

Analyses of similar experiments soon produced overwhelming evidence for the wavelike behavior of light.

Further insight into the wave character of light came from calculations by the Scottish physicist James Clerk Maxwell in the 1860s. Maxwell succeeded in describing all the basic properties of electricity and magnetism in four equations. By combining these equations, Maxwell demonstrated that electrical and magnetic effects should travel through space in the form of waves. Furthermore, his calculations proved that these waves should travel with a speed of about 3×10^8 m/s. Maxwell's suggestion that these waves do exist and are observed as light was soon confirmed by a variety of experiments.

Light consists of vibrating electric and magnetic fields, as shown in Figure 4-2. Because of its electric and magnetic properties, light is called **electromagnetic radiation.** The distance between two successive wave crests is called the **wavelength** of the light.

The wavelength of visible light is extremely small, less than a thousandth of a millimeter. To express these tiny distances conveniently, scientists use a unit of length called the nanometer (abbreviated nm), where 1 nm = 10^{-9} m. Experiments demonstrated that visible light has wavelengths covering the range from about 400 nm for violet light to about 700 nm for red light. Intermediate colors of the rainbow fall between these wavelengths (see Figure 4-1).

Visible light includes a specific, narrow range of wavelengths. But Maxwell's equations placed no restrictions on the wavelengths that electromagnetic radiation can have. It was therefore possible that electromagnetic waves existed with wavelengths both longer and shorter than the 400–700 nm range of visible light. Researchers began to look for invisible forms of light, forms to which the cells of the human retina do not respond.

The British astronomer William Herschel discovered radiation just beyond the red end of the visible spectrum. In 1888 the German physicist Heinrich Hertz succeeded

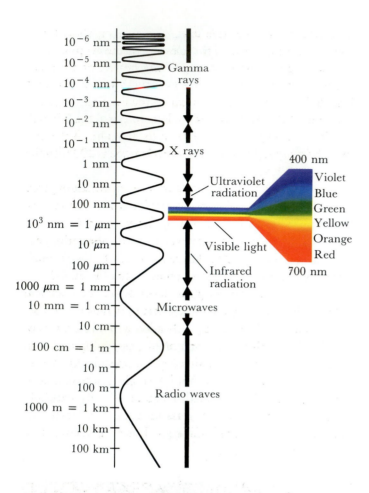

Figure 4-3 The electromagnetic spectrum *The full array of all types of electromagnetic radiation is called the electromagnetic spectrum. It extends from the shortest-wavelength gamma rays to the longest-wavelength radio waves. Visible light forms only a tiny portion of the full electromagnetic spectrum.*

in producing light having wavelengths longer than about 10 cm, now known as **radio waves**. In 1895 Wilhelm Röntgen invented a machine that produced light with a wavelength shorter than 10 nm, now known as **X rays**. Modern versions of Röntgen's machine are today found in medical and dental offices.

Over the years radiation in many other ranges of wavelength has been discovered. Visible light is only a tiny fraction of the full extent of possible wavelengths, collectively called the **electromagnetic spectrum**. As shown in Figure 4-3, the electromagnetic spectrum stretches from the longest-wavelength radio waves to the shortest-wavelength **gamma rays**. For example, at wavelengths slightly longer than visible light, **infrared radiation** covers the range from about 700 nanometers to 1 millimeter (abbreviated 1 mm). From roughly 1 millimeter to 10 cen-

timeters is the range of **microwaves**, beyond which is the domain of radio waves. Infrared and microwave wavelengths are sometimes measured in micrometers (μm), where 1 μm = 1000 nm.

At wavelengths shorter than those of visible light, **ultraviolet radiation** extends from about 400 down to 10 nanometers. Wavelengths between about 10 and 0.01 nanometers are X rays, beyond which is the domain of gamma rays. It should be noted that these arbitrary divisions are simply rough boundaries that allow us to identify certain broad sections of the electromagnetic spectrum.

Although these various types of electromagnetic radiation share many basic properties (for example, they all travel at the speed of light), they interact very differently with matter. Your body is transparent to X rays but not to visible light; your eyes respond to visible light but not to gamma rays; your radio detects radio waves but not ultraviolet light. Consequently, astronomers use fundamentally different kinds of telescopes to detect radiation in these various ranges of wavelength. For example, a radio telescope that detects radio waves from space is quite different from either an X-ray telescope or an ordinary optical telescope. Because optical telescopes are the most common and familiar astronomical tool, we shall discuss them in detail before turning to the exotic instruments that reveal the nonvisible sky.

4-2 A refracting telescope uses a lens to concentrate incoming starlight at a focus

Although light travels at about 3×10^8 m/s in a vacuum, it moves more slowly through a dense substance such as glass. The abrupt slowing of light entering a piece of glass is analogous to the motion of a person walking from a boardwalk onto a sandy beach: Her pace suddenly slows as she steps from the smooth pavement into the sand. And just as the person stepping back onto the boardwalk easily resumes her original pace, light exiting a piece of glass resumes its original speed.

In addition to undergoing an abrupt change in speed, a light ray is bent as it passes from one transparent medium into another. To describe this phenomenon, called **refraction**, imagine a light ray entering a piece of glass at an oblique angle, as shown in Figure 4-4. Furthermore, imagine drawing a perpendicular to the surface of the glass at the point where the light strikes the glass. As a light ray goes from the air into the glass, the light ray is bent

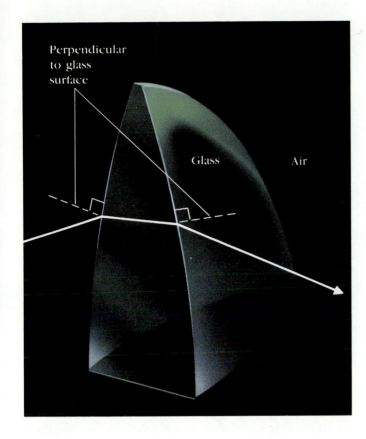

telescope is the **objective lens.** The smaller, short-focal-length lens at the rear of the telescope is the **eyepiece lens.**

The **magnification,** or **magnifying power,** of a refracting telescope is equal to the focal length of the objective lens divided by the focal length of the eyepiece lens. For example, if the objective of a telescope has a focal length of 100 cm and the eyepiece has a focal length of 0.5 cm, then the magnifying power of the telescope is 200 (usually written as 200×).

If you build a telescope using only the instructions given so far, you will probably be disappointed with the results. You will see stars surrounded by fuzzy, rainbow-colored halos. This optical defect, called **chromatic aberration,** exists because a lens bends different colors of light through different angles, just as a prism does (recall Figure 4-1).

The speed of light in glass depends on the chemical composition of the glass—a fact that opticians use to correct for chromatic aberration. By adding small amounts of various chemicals to a vat of molten glass, an optician can manufacture different kinds of glass. Specifically, a thin lens can be mounted just behind the main objective lens of a telescope as diagrammed in Figure 4-7. By carefully choosing different kinds of glass for these two lenses, the optician can ensure that different colors of light come to a focus at the same point.

Figure 4-4 Refraction A light ray entering a piece of glass is bent toward the perpendicular. As the ray leaves the piece of glass, it is bent away from the perpendicular.

toward the perpendicular direction. Upon emerging from the other side of a piece of glass, light resumes its original high speed, and the light ray is bent away from the perpendicular direction. The exact amount of refraction depends upon the angle at which the light ray strikes the glass and on the speed of light in the glass.

Because of the refracting property of glass, a convex lens—one that is fatter in the middle than at the edges—causes incoming light rays to converge to a point called the **focus,** as shown in Figure 4-5. If the light source is extremely far away, then the incoming light rays are parallel, and they come to a focus at a specific distance from the lens, called the **focal length** of the lens.

Stars and planets are so far away that light rays from these objects are essentially parallel. Consequently, a lens always focuses light from an astronomical object as shown in Figure 4-5. An image of the object is formed at the focus, and a second lens can be used to magnify and examine this image. Such an arrangement of two lenses is called a **refracting telescope,** or **refractor** (Figure 4-6). The large-diameter, long-focal-length lens at the front of the

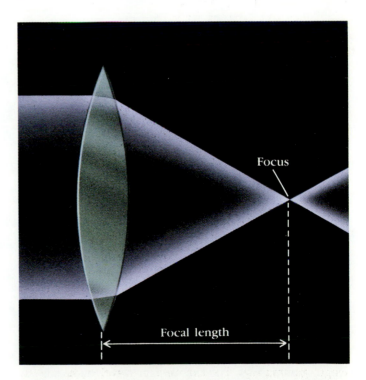

Figure 4-5 A convex lens A convex lens causes parallel light rays to converge to a focus. The distance from the lens to the focus is the focal length of the lens.

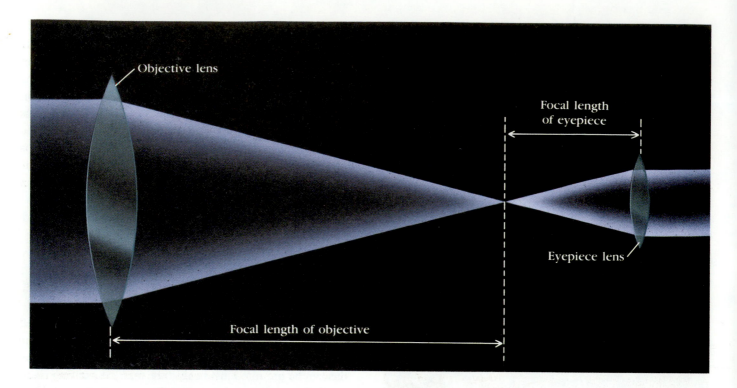

Figure 4-6 A refracting telescope *A refracting telescope consists of a large, long-focal-length objective lens and a small, short-focal-length eyepiece lens. The eyepiece lens magnifies the image formed at the focus of the objective lens.*

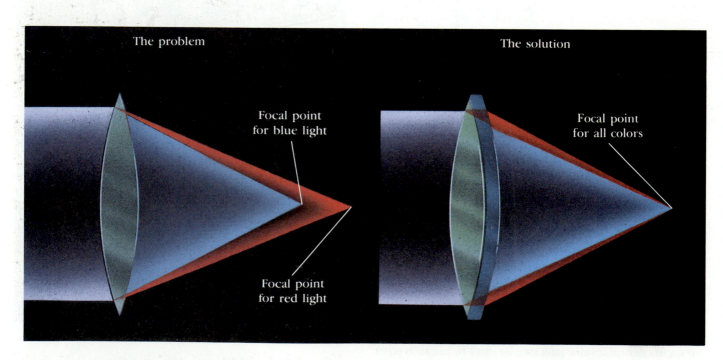

Figure 4-7 Chromatic aberration *A single lens suffers from a defect called chromatic aberration in which different colors of light have slightly different focal lengths. This problem is corrected by adding a second lens made of a different kind of glass from that of the first lens.*

which frequently form when molten glass is poured into a mold. Consequently, the glass for the lens is extremely expensive. Second, glass is opaque to certain kinds of light. Even visible light is dimmed substantially in passing through the thick slab of glass at the front of a refractor, and ultraviolet radiation is largely absorbed by the glass lens. Third, it is impossible to completely eliminate chromatic aberration in a refracting telescope. Fourth, it is difficult to support such a heavy lens without blocking the path of light into the telescope. All of these problems can be avoided by using mirrors instead of lenses.

4-3 A reflecting telescope uses a mirror to concentrate incoming starlight at a focus

Reflection can be described very simply. To understand reflection, imagine drawing a perpendicular to a mirror's surface at the point where a light ray strikes the mirror, as shown in Figure 4-9. The angle between the arriving (incident) light ray and the perpendicular is always equal to

Figure 4-8 A large refracting telescope *This giant refracting telescope, built in the late 1800s, is housed at Yerkes Observatory near Chicago. The objective lens is 102 cm (40 in.) in diameter, and the telescope tube is 19.5 m (63.5 ft) long. (Yerkes Observatory)*

Chromatic aberration is the most severe of a host of optical problems that must be solved in designing a high-quality refracting telescope. Nineteenth-century master opticians devoted their lives to overcoming these problems, and several magnificent refractors were constructed in the late 1800s. The largest refracting telescope, completed in 1897, is located at the Yerkes Observatory near Chicago (Figure 4-8) and has an objective lens 102 cm (40 in.) in diameter. The second largest refracting telescope is located at Lick Observatory near San Jose, California. This refractor has an objective lens whose diameter is 91 cm (36 in.). All refractors have extremely long focal lengths. For example, the Yerkes refractor has a focal length of 19.35 m (63.5 ft).

Few major new refracting telescopes have been constructed in the twentieth century. There are many reasons for the modern astronomer's lack of interest in this type of telescope. First, because faint light must pass readily through the objective lens, the glass from which the lens is made must be totally free of defects such as bubbles,

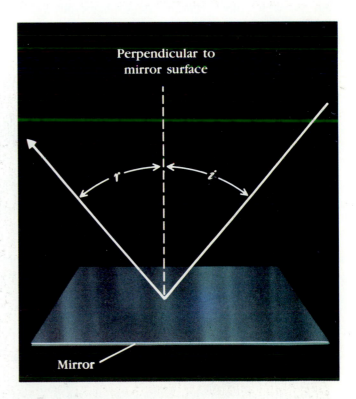

Figure 4-9 Reflection *The angle at which a beam of light strikes a mirror (the angle of incidence i) is always equal to the angle at which the beam is reflected from the mirror (the angle of reflection r).*

the angle between the reflected ray and the perpendicular. Knowing this, Isaac Newton realized that a concave mirror will cause parallel light rays to converge to a focus, as shown in Figure 4-10. The distance between the reflecting surface and the focus is called the focal length of the mirror.

An image of a distant object is formed at the focus of a concave mirror. In order to view the image, Newton placed a small, flat mirror at a 45° angle in front of the focal point, as sketched in Figure 4-11a. This secondary mirror deflects the light rays to one side of the **reflecting telescope, or reflector,** where the astronomer can place an eyepiece lens to magnify the image. A telescope having this optical design is appropriately called a **Newtonian reflector.** The magnifying power of such a reflecting telescope is calculated in the same way as for a refractor: The focal length of the primary mirror is divided by the focal length of the eyepiece.

Modifications of Newton's original design are commonly used in professional observatories. The primary mirrors of many major reflectors are so large that the astronomer can actually sit at the undeflected focal point, directly in front of the primary mirror. This arrangement is called a **prime focus** (Figure 4-11b). The "observing cage," in which the astronomer rides, blocks only a small fraction of the incoming starlight.

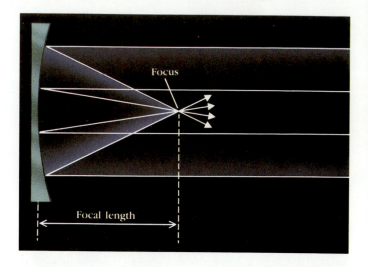

Figure 4-10 A concave mirror *A concave mirror causes parallel light rays to converge to a focus. The distance between the mirror and the focus is the focal length of the mirror.*

Another popular design, called a **Cassegrain focus,** has the advantage of placing the focal point at a convenient, accessible location. A hole is drilled directly through the center of the primary mirror. A convex secondary mirror

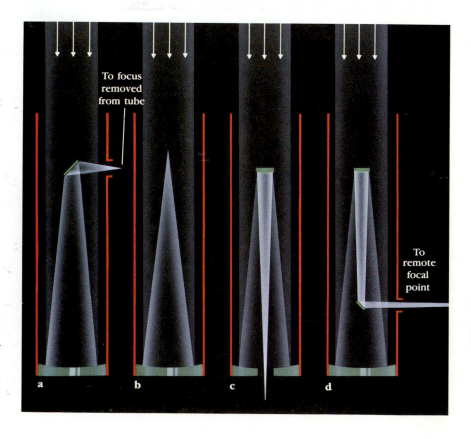

Figure 4-11 Reflecting telescopes *Four of the most popular optical designs for reflecting telescopes:* (**a**) *Newtonian focus,* (**b**) *prime focus,* (**c**) *Cassegrain focus, and* (**d**) *coudé focus.*

Furthermore, reflection is not affected by the wavelength of the light. Therefore, the image formed by a mirror does not suffer from chromatic aberration. Finally, the mirror can be fully supported by braces on its back, so that a large and heavy mirror can be mounted without much danger of breakage or distortion of its shape.

There are 16 reflectors around the world with primary mirrors measuring at least 3 meters in diameter. One of the largest, a 6-meter telescope, is located in the Caucasus Mountains in Russia. The famous 200-inch (5-meter) telescope at the Palomar Observatory is near San Diego, California. In the early 1970s, a matching pair of telescopes were built at Kitt Peak in Arizona and at Cerro Tololo in Chile. Both have mirrors 4 meters (13.1 feet) in diameter. These two telescopes (Figure 4-12) allow astronomers to observe the entire sky with essentially the same instrument.

Astronomers prefer large telescopes because of their ability to produce bright, sharp images. A large mirror intercepts and focuses more starlight than does a small mirror (Figure 4-13). A large mirror therefore produces brighter images and detects fainter stars than does a small mirror. The **light-gathering power** of a telescope is directly related to the area of the telescope's primary mirror. For example, the 200-inch mirror at Palomar Observatory

Figure 4-12   The 4-m telescope at Cerro Tololo   This telescope is located on a mountaintop near Santiago, Chile. Its twin is at the Kitt Peak Observatory in Arizona. Both telescopes have been in operation since the early 1970s. (NOAO)

placed in front of the original focal point is used to reflect the light rays back through the hole (Figure 4-11c).

Alternatively, a series of mirrors can be used to channel the light rays away from the telescope to a remote focal point. Heavy optical equipment that could not be mounted directly on the telescope is located at the resulting **coudé focus** (Figure 4-11d).

To make a reflector, an optician grinds and polishes a large slab of glass into the appropriate concave shape. The glass is then coated with silver or aluminum or a similar highly reflective substance. Defects inside the glass such as bubbles or flecks of dirt do not detract from the telescope's effectiveness, as they would in the objective lens of a refracting telescope.

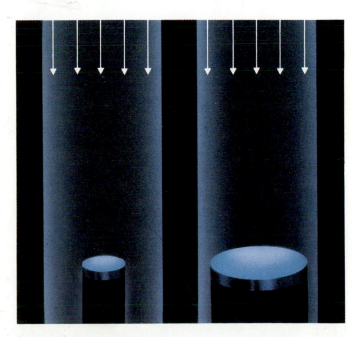

Figure 4-13   Light-gathering power   Because a large mirror intercepts more starlight than does a small mirror, a large mirror produce a brighter image.

Figure 4-14 The Multiple-Mirror Telescope This aerial photograph shows the six 1.8-m (5.9-ft) mirrors that together constitute the first multiple-mirror telescope. The total area of the six mirrors is equal to one 4.5-m mirror. (Multiple-Mirror Telescope)

has four times the area of the 100-inch mirror at Mt. Wilson Observatory. Therefore, the Palomar telescope has four times the light-gathering power of the Mt. Wilson telescope.

A large telescope also increases the sharpness of the image and the degree of detail that can be seen. This property is called **angular resolution.** Poor angular resolution causes star images to be fuzzy and blurred together. A telescope with good angular resolution produces star images that are sharp and crisp.

The angular resolution of a telescope is measured as the angle between two adjacent stars whose images can just barely be distinguished under ideal observing conditions. The smaller the angle, the sharper the image. Large modern telescopes—such as those at Palomar, Kitt Peak, and Cerro Tololo—are calculated to have angular resolutions better than 0.1 arc sec. In practice, however, this exceptional angular resolution is never achieved. Turbulence and impurities in the air cause star images to jiggle around, or twinkle. Even when photographed through the largest telescopes, a star looks like a tiny circular blob rather than a pinpoint of light.

The angular diameter of a star's image, called the **seeing disk,** is a realistic measure of the best possible resolution. The size of the seeing disk varies from one observatory site to another. At Palomar and Kitt Peak, the seeing disk is roughly 1 arc sec. The best conditions in the world (with a seeing disk of 0.2 arc sec) have been reported at the observatory on top of Mauna Kea, the tallest volcano on the island of Hawaii.

Significant engineering problems are associated with building large reflectors. Very large mirrors (more than about 4 meters in diameter) are slabs of glass so heavy that the mirror's shape changes slightly as the telescope is turned toward different parts of the sky. The mirror actually sags under its own weight, thereby detracting from the sharpness and quality of the resulting image. New techniques for building thin, lightweight mirrors should alleviate this problem.

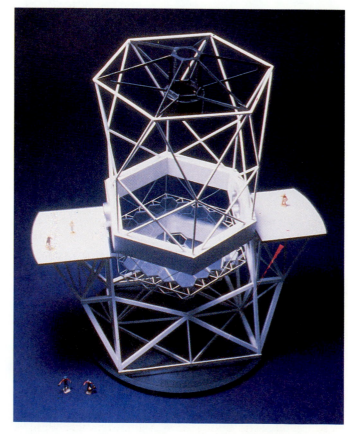

Figure 4-15 The 10-m Keck telescope This model shows the design of the giant Keck telescope at the summit of Mauna Kea on the island of Hawaii. Thirty-six hexagonal mirrors, each measuring 1.8 m (5.9 ft) across, have the effect of one mirror 10 m in diameter. (California Institute of Technology)

Another approach to improving the image is to mount several smaller mirrors together and aim them at the same focal point. The Multiple-Mirror Telescope atop Mt. Hopkins in Arizona has six mirrors, each measuring 1.8 m (6 ft) in diameter, mounted together as shown in Figure 4-14. The total light-gathering power of this arrangement is equivalent to one 4.5-m mirror. The design has proven so successful that astronomers around the world are building even larger multiple-mirror telescopes.

The first of these giant multiple-mirror instruments is the recently completed 10-meter Keck telescope on the summit of Mauna Kea in Hawaii. Thirty-six hexagonal mirrors are mounted side by side to give a primary that is 10 m (400 in.) in diameter, as shown in Figure 4-15.

An optical distortion called **spherical aberration** must be minimized when building reflecting telescopes. At issue is the precise shape of a mirror's concave surface. A spherical surface is easy to grind and polish, but different parts of a spherical mirror have slightly different focal lengths (Figure 4-16), which results in a fuzzy image. The easiest way to eliminate spherical aberration is to polish the mirror's surface to a parabolic shape. Because a parabola reflects parallel light rays to a common focus, many reflecting telescopes have parabolic mirrors.

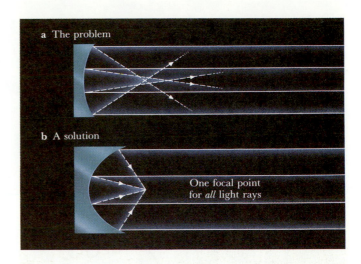

Figure 4-16 Spherical aberration *Different parts of a spherically concave mirror reflect light to slightly different focal points. This difficulty can be corrected by using a parabolic mirror.*

4-4 Electronic devices are often used to record the image at a telescope's focus

The invention of photography during the nineteenth century was a boon for astronomy. By taking a long exposure with a camera mounted at the focus of a telescope, an astronomer could record extremely faint features that could not be seen by looking through the telescope. Today astronomical photography continues to reveal details in galaxies, star clusters, and nebulae.

Astronomers have long realized, however, that a photographic plate is not a very efficient light detector. Only 5% of the light striking a photographic plate succeeds in triggering the chemical reaction in the photographic emulsion that is needed to produce an image. Thus, roughly 95% of the light falling onto a photographic plate is wasted.

The most sensitive and efficient light detector currently available to astronomers is the newly invented **charge-coupled device (CCD)**. A CCD is a thin silicon wafer roughly the size of a postage stamp (Figure 4-17). The

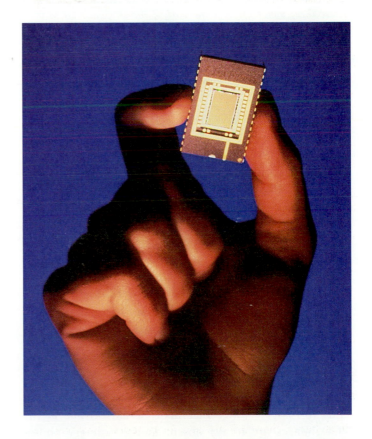

Figure 4-17 A charge-coupled device (CCD) *This tiny silicon rectangle contains 163,840 light-sensitive electric circuits that store images. At the end of each exposure, additional circuits in the silicon chip control the transfer and readout of the data to a waiting computer. (Smithsonian Institution Astrophysical Observatory)*

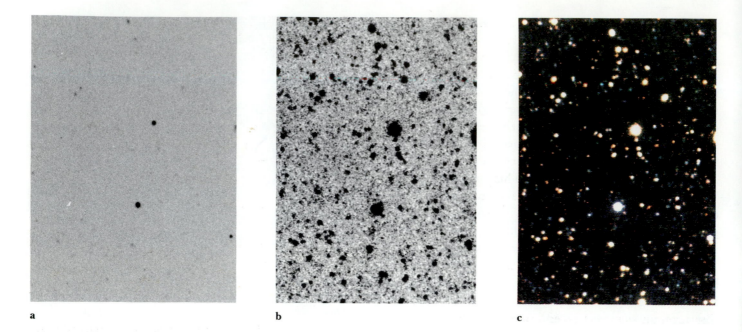

a   b   c

Figure 4-18   Ordinary photography versus a CCD image
These three views of the same part of the sky, each taken with
the 4-m telescope shown in Figure 4-12, compare CCDs to ordinary photographic plates. (a) A negative print (black stars and
white sky) of a 45-minute exposure on a photographic plate. (b)
The sum of fifteen 500-second CCD images. Notice that many

faint stars and galaxies virtually invisible in the ordinary photograph can be clearly seen in this CCD image. (c) This color view
was produced by combining a series of CCD images taken
through colored filters. The total exposure time was 6 hours.
(Courtesy of P. Seitzer, NOAO)

wafer is divided into an array of small, light-sensitive
squares called picture elements or, more commonly, **pixels.** For example, one of the latest CCDs has over four
million pixels arranged in 2048 rows by 2048 columns.
When an image from a telescope is focused on the CCD,
an electric charge builds up in each pixel in proportion to
the intensity of the light falling on that pixel. When the
exposure is finished, the amount of the charge on each
pixel is read into a computer. From the computer, the
image can be transferred onto ordinary photographic film
or to a television monitor. Over certain wavelength ranges, nearly 75% of the light falling on a CCD can be
recorded.

Figure 4-18 shows one photograph and two CCD images of the same region of the sky, both taken with the
same telescope. Notice that many details visible in the
CCD images are totally absent in the ordinary photograph. In fact, the CCD pictures in Figure 4-18 show
some of the faintest stars and galaxies ever recorded. Because of their extraordinary sensitivity and their use in
conjunction with computers, CCDs are playing an increasingly important role in astronomy.

4-5   A radio telescope uses a large concave dish to reflect radio waves to a focus

Until recently all information that astronomers could
gather about the universe was based on ordinary visible
light. With the discovery of nonvisible electromagnetic
radiation, scientists began to wonder if objects in the universe might also emit radio waves, X rays, and infrared
and ultraviolet radiation.

The first evidence of nonvisible radiation from outer
space came from the work of a young radio engineer,
Karl Jansky of Bell Telephone Laboratories. Using long
antennas, Jansky was investigating the sources of radio
static that affects short-wavelength radiotelephone
communication. In 1932, he realized that a certain kind
of radio noise is strongest when the constellation of
Sagittarius is high in the sky. The center of our Galaxy
is located in the direction of Sagittarius, and Jansky
concluded that he was detecting radio waves from elsewhere in the Galaxy.

Astronomers were not quick to pursue this line of research. Only one person, Grote Reber (an electronics engineer living in Illinois), pursued the matter. In 1936 Reber built the first radio telescope in his backyard to map radio emission from the Milky Way. His design was modeled after an ordinary reflecting telescope, with a concave "dish" (reflecting antenna) measuring 9.1 m in diameter. The radio receiver at the focal point of the metal dish was tuned to a wavelength of 1.85 m.

By 1944, when Reber completed his map of the Milky Way, astronomers had begun to take notice of these developments. Shortly after World War II, radio telescopes began to spring up around the world. Radio observatories are today as common as major optical observatories.

Like Reber's prototype, the standard radio telescope has a large concave dish (Figure 4-19). A small antenna tuned to the desired wavelength is located at the focus. The incoming signal is relayed to amplifiers and recording instruments, which are typically located in a room at the base of the telescope's pier.

At first astronomers were not enthusiastic about detecting radio noise from space, in part because of the poor angular resolution of early radio telescopes. The angular resolution of any telescope worsens as the wavelength increases. In other words, the longer the wavelength, the fuzzier the picture. Because radio radiation has very long wavelengths, astronomers thought that radio telescopes could produce only blurry, indistinct views.

Very large radio telescopes can produce somewhat sharper radio images, because the bigger the dish, the better the angular resolution. For this reason, most modern radio telescopes have dishes more than 30 m in diameter. Nevertheless, even the largest radio dish in existence cannot come close to the resolution of the best optical telescopes.

A very clever technique was devised to circumvent this limitation and produce high-resolution radio images. Unlike ordinary light, radio signals can be carried over electrical wires, which means that two radio telescopes separated by many kilometers can be hooked together. This technique is called **interferometry**, because the incoming radio signals are made to "interfere," or blend together, so that the combined signal is sharp and clear. The result is impressive: The effective angular resolution is equivalent to that of one gigantic dish with a diameter equal to the distance between the two telescopes.

Interferometry techniques were exploited for the first time in the late 1940s and gave astronomers their first detailed views of radio objects in the sky. More recently radio telescopes separated by thousands of kilometers have

Figure 4-19 A radio telescope *The dish of this radio telescope is 45.2 m (148 ft) in diameter. It is one of several large instruments at the National Radio Astronomy Observatory near Green Bank, West Virginia. (NRAO)*

been linked together to produce images that are much sharper and crisper than those from optical telescopes. This technique is called **very-long-baseline interferometry (VLBI)**. The best angular resolution would be obtained by two telescopes on opposite sides of the Earth. In that case, features as small as 0.00001 arc sec could be distinguished at radio wavelengths, which is 100,000 times better than the sharpest pictures from ordinary optical telescopes.

One of the finest systems of radio telescopes began operating in 1980 in the desert near Socorro, New Mexico. Called the Very Large Array (VLA), it consists of 27 concave dishes, each 26 m (85 ft) in diameter. The 27 telescopes are arranged along the arms of a gigantic Y covering an area 27 km (17 mi) in diameter. Only a portion of the VLA is shown in Figure 4-20. This system produces radio views of the sky with resolution comparable to that of the very best optical telescopes.

Radio astronomers often use "false color" to display their radio views of astronomical objects. An example is shown in Figure 4-21*b*. The most intense radio emission is shown in red, the least intense in blue. Intermediate colors of the rainbow represent intermediate levels of radio intensity. Black indicates that there is no detectable radio radiation. Astronomers working at other nonvisible wavelength ranges also frequently use false-color techniques to display views obtained from their instruments.

Figure 4-20   The Very Large Array (VLA)
*The 27 radio telescopes of the VLA system are arranged along the arms of a **Y** in central New Mexico. The north arm of the array is 19 km long; the southwest and southeast arms are each 21 km long. (NRAO)*

4-6 Telescopes in orbit around the Earth detect radiation that does not penetrate the atmosphere

As the success of radio astronomy began to mount, astronomers started exploring the possibility of making observations at other nonvisible wavelengths. Unfortunately, the Earth's atmosphere is opaque to many wavelengths. Very little radiation other than visible light and radio waves manages to penetrate the air we breathe.

Because water vapor is the main absorber of infrared radiation from space, locating infrared observatories at sites of low humidity can overcome much of the atmosphere's hindrance. For example, the summit of Mauna Kea on Hawaii is exceptionally dry, and infrared observations are the primary function of NASA's 3-meter telescope there.

Another possibility is to take a telescope up in an airplane. That is the basic idea behind the Kuiper Airborne Observatory (KAO), shown in Figure 4-22. The airplane carries a 1-m reflecting telescope to an altitude of 12 km (40,000 ft), placing the observatory above 99% of the atmospheric water vapor.

a

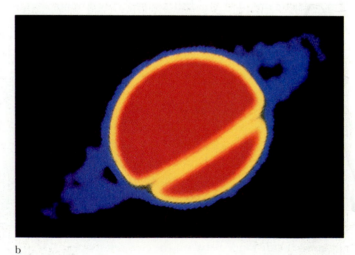

b

Figure 4-21   Optical and radio views of Saturn   (a) This picture was taken by a camera on board a spacecraft as it approached Saturn. Sunlight reflected from the planet's cloudtops *and rings is responsible for this view. (b) This false-color picture, taken by the VLA, shows radio emission from Saturn at a wavelength of 2 cm. (NASA; NRAO)*

Figure 4-22   The Kuiper Airborne Observatory *This C-141 jet airplane carries a 1-m reflecting telescope specifically designed for infrared observations. The observing door (through which the telescope is aimed) can be seen on the fuselage just in front of the wing. (NASA)*

A telescope in Earth orbit offers the best arrangement. In 1983 the Infrared Astronomical Satellite (IRAS) was launched into a 900-km-high polar orbit (Figure 4-23). During its ten-month mission, this 60-cm, cryogenically cooled reflector revealed the full richness and variety of the infrared sky. For the first time, astronomers saw dust bands in our solar system, dust disks around nearby stars, and distant galaxies that emit most of their radiation at infrared wavelengths.

In 1993 the European Space Agency (ESA) will launch an infrared telescope that will spend several years making observations at infrared wavelengths that never penetrate the Earth's atmosphere. Budget cuts have frustrated NASA plans to develop and launch SIRTF (the Space Infrared Telescope Facility), which would survey the infrared sky with unprecedented resolution.

The best ultraviolet observations are also made from space. During the early 1970s, both Apollo and Skylab

Figure 4-23   The Infrared Astronomical Satellite (IRAS) *This satellite contains a small reflecting telescope that gave astronomers their first in-depth look at the infrared sky. Launched in 1983, IRAS contributed to research topics running the gamut from asteroids to the large-scale distribution of matter in the universe. (NASA)*

Figure 4-24 Orion as seen in ultraviolet, infrared, and visible wavelengths *An ultraviolet view* (**a**) *of the constellation of Orion was obtained during a brief rocket flight on December 5, 1975. The 100-second exposure captured wavelengths ranging between 125 and 200 nm. The "false-color" view* (**b**) *from the Infrared Astronomical Satellite uses color to display specific ranges of infrared wavelengths: Red indicates long-wavelength radiation; green, intermediate-wavelength radiation; and blue, short-wavelength radiation. For comparison, an ordinary optical photograph* (**c**) *and a star chart* (**d**) *are included. (Courtesy of George R. Carruthers, NRL; NASA; R. C. Mitchell, Central Washington University)*

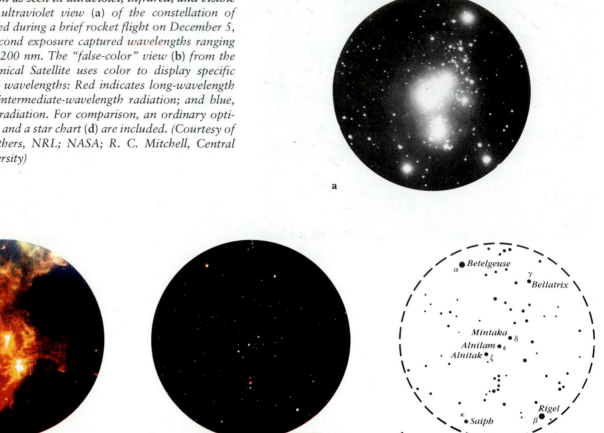

Figure 4-25 The International Ultraviolet Explorer (IUE) Since 1978, this 671-kg satellite has produced observations in the far-ultraviolet. The dark blue panels at the midsection of the satellite are solar cell arrays that provide electric power for the radio transmitters and other electronic equipment. (NASA)

astronauts carried small telescopes above the Earth's atmosphere to give us some of our first views of the ultraviolet sky. Small rockets have also been used to place ultraviolet cameras briefly above the Earth's atmosphere. A typical view is shown in Figure 4-24, along with a corresponding infrared view from IRAS, a view in visible light, and a star chart.

Some of the finest ultraviolet astronomy has been accomplished by the International Ultraviolet Explorer (IUE), which was launched in 1978. The satellite (Figure 4-25) is built around a Cassegrain telescope with a 45-cm (18-in.) mirror and a total focal length of 6.74 m (22 ft). Observations cover the range from 116 to 320 nm.

For decades astronomers have dreamed of having a major observatory in space. Although satellites like IRAS and IUE gave us excellent views at selected wavelength regions, astronomers were enthusiastic about the prospect of one large telescope that could be operated over a wide range of wavelengths—from the infrared through the visible range and out into the far-ultraviolet. This was the mission of the Hubble Space Telescope (HST), which was carried aloft by the Space Shuttle in 1990 (Figure 4-26).

Soon after HST was placed in orbit, astronomers discovered that the telescope's 2.4-m primary mirror suffers from spherical aberration. The mirror should have been able to concentrate 70% of a star's light into an image 0.1 arc sec in diameter. Instead, the figure is only 20%; the remaining 80% is smeared out over an area 1 arc sec in diameter. A star's image therefore consists of a central spot of modest brightness surrounded by a hazy glow.

One way astronomers can cope with this problem is to use only the 20% of incoming light that is properly focused and, with computer processing, discard the remaining poorly focused 80%. An example of this procedure is shown in Figure 4-27. Although this technique produces sharp, crisp images, it is practical only on brighter objects with which astronomers can afford to waste light. Unfortunately, many of the observing projects scheduled on the HST involve extremely dim galaxies and nebulae. These observations have been postponed until a shuttle mission in 1993 installs new optics to compensate for the spherical aberration.

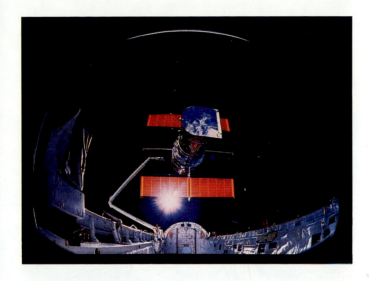

Figure 4-26 The Hubble Space Telescope (HST) *This photograph of HST hovering above the Space Shuttle's cargo bay was taken as the telescope was being deployed. During its 15-year lifetime, HST will study the heavens at wavelengths from the infrared through the ultraviolet. (NASA)*

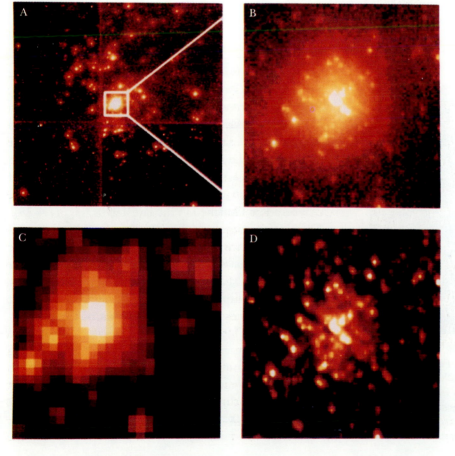

Figure 4-27 Computer enhancement of HST images *These four images show a star cluster in a nearby galaxy called the Large Magellanic Cloud. (a) The raw picture from the HST shows the cluster and its surroundings. (b) Note the hazy glow that envelops the stars in this enlargement of the cluster. (c) This ground-based photograph is typical of the best image that can be obtained with a large telescope on the Earth's surface. (d) This computer-enhanced version of panel (b) demonstrates that computer processing can overcome HST's optical defects. (Space Telescope Science Institute, NASA)*

Because neither X rays nor gamma rays penetrate the Earth's atmosphere, observations at these extremely short wavelengths also must be done from space. Astronomers got their first look at the X-ray sky with brief rocket flights during the late 1940s. Several small satellites launched during the early 1970s viewed the entire X-ray and gamma-ray sky, revealing hundreds of previously unknown sources, including several black hole candidates.

Although heroic in their day, these preliminary efforts pale in comparison to the detailed views and results from three huge satellites launched between 1977 and 1979. Called High Energy Astrophysical Observatories (HEAO), these satellites each carried an array of X-ray and gamma-ray detectors. Thousands of sources were discovered all across the sky. The second satellite in this series was especially successful in producing high-quality X-ray images of a wide range of exotic objects. This satellite, called the Einstein Observatory, was launched near the hundredth anniversary of Albert Einstein's birth. X-ray views from this observatory appear throughout this book.

The successor to the Einstein Observatory is ROSAT, launched into a nearly circular Earth-orbit in 1990. This satellite carries several X-ray detecting instruments, including an X-ray telescope and a wide-field X-ray camera. Scientists are hopeful that ROSAT, which grew from a German venture into an international project, will pro-

Figure 4-29 The Compton Gamma-Ray Observatory This photograph of the Compton Observatory hovering about the Space Shuttle's cargo bay was taken as the spacecraft was being deployed in 1991. Previously known as the Gamma Ray Observatory, it was renamed in honor of Arthur Holly Compton, an American physicist who made important discoveries about gamma rays. The best views of the high-energy gamma-ray sky come from this 15-ton observatory, which carries four gamma-ray detectors. (NASA)

Figure 4-28 The Advanced X-ray Astrophysics Facility (AXAF) If funding is approved by the U.S. Congress, this X-ray telescope will be placed into Earth orbit in the late 1990s. Twelve cylindrical mirrors—the largest being about 1 m long and 1.25 m in diameter—will focus incoming X rays to produce images and observations of a wide variety of objects. (Courtesy of TRW)

vide a wealth of information about the X-ray sky during the next few years.

ROSAT is an important evolutionary step on the way to building AXAF, the Advanced X-Ray Astrophysics Facility, which is in the planning stages at NASA. Like the Hubble Space Telescope, AXAF will be long-lived, adaptable, and controllable by astronomers on the ground. If this project does not fall prey to a NASA budget cut, AXAF could be launched in the late-1990s (Figure 4-28).

The electromagnetic radiation with the shortest wavelengths and the most energy are gamma rays. In 1991 the Compton Gamma-Ray Observatory was carried aloft by the Space Shuttle (Figure 4-29). This orbiting observatory carries four instruments that are performing a variety of observations, giving us tantalizing views of the gamma-ray sky.

The advantages and benefits of these Earth-orbiting observatories cannot be overemphasized. We are no longer limited to the narrow ranges of wavelengths that manage to leak through our shimmering, hazy atmosphere (Figure 4-30). For the first time we are really seeing the universe.

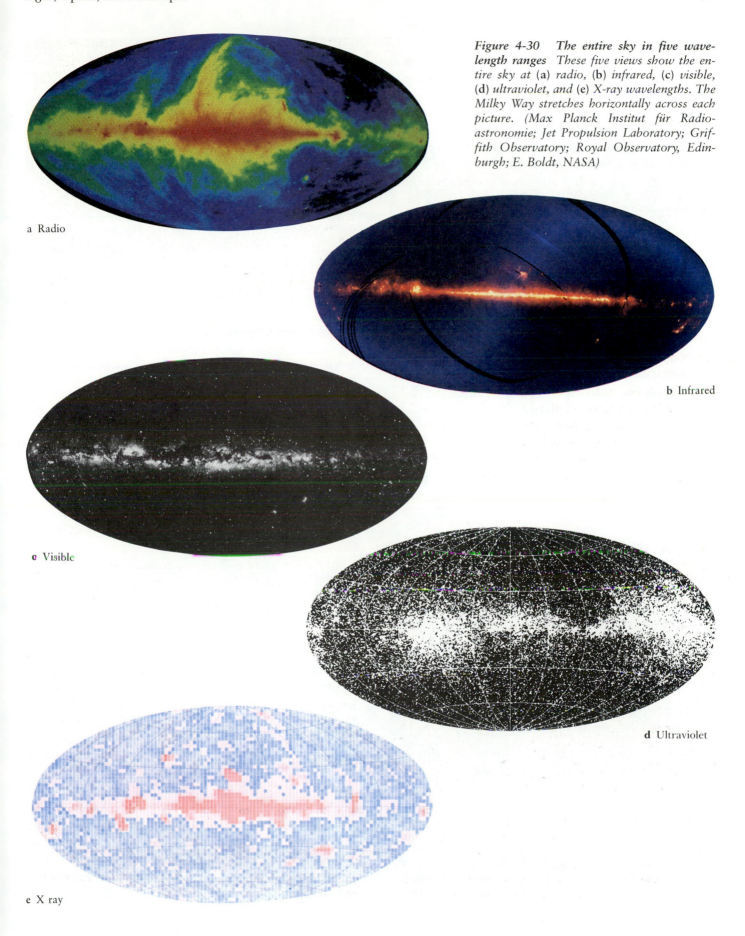

a Radio

b Infrared

c Visible

d Ultraviolet

e X ray

Figure 4-30 **The entire sky in five wavelength ranges** *These five views show the entire sky at* (**a**) *radio,* (**b**) *infrared,* (**c**) *visible,* (**d**) *ultraviolet, and* (**e**) *X-ray wavelengths. The Milky Way stretches horizontally across each picture. (Max Planck Institut für Radioastronomie; Jet Propulsion Laboratory; Griffith Observatory; Royal Observatory, Edinburgh; E. Boldt, NASA)*

Summary

- Electromagnetic radiation consists of vibrating electric and magnetic fields that carry energy through space at the speed of light (3×10^8 m/s).

- Visible light, radio waves, microwaves, infrared and ultraviolet radiation, X rays, and gamma rays are all forms of electromagnetic radiation.

  Visible light forms only a small portion of the electromagnetic spectrum.

- The wavelength of light is associated with its color; wavelengths of visible light range from about 400 nm for violet light to 700 nm for red light.

  Infrared radiation, microwaves, and radio waves have wavelengths larger than those of visible light; ultraviolet radiation, X rays, and gamma rays have wavelengths that are shorter.

- Refracting telescopes, or refractors, produce images by bending light rays as they pass through glass lenses.

  Chromatic aberration is an optical defect whereby light of different wavelengths fails to come to a common focus.

  Limitations of glass purity, opacity to certain wavelengths, and structural difficulties have made it inadvisable to build extremely large refractors.

- Reflecting telescopes, or reflectors, produce images by reflecting light rays from concave mirrors to a focus point.

  Reflectors are not subject to many of the problems that limit the usefulness of refractors.

- Charge-coupled devices (CCDs) are often used at a telescope's focus to record faint images.

- Radio telescopes have large reflecting antennas (dishes) that are used to focus radio waves.

  Very sharp radio images are produced with combinations or arrays of radio telescopes linked together in a technique called interferometry.

- The Earth's atmosphere is transparent to visible light and radio waves arriving from space, but it absorbs much of the radiation at other wavelengths.

For observations at other wavelengths, astronomers depend upon telescopes carried above the atmosphere by high-altitude airplanes, rockets, or satellites.

Satellite-based observatories are giving us a wealth of new information about the universe, permitting coordinated observation of the sky at all wavelengths.

Review questions

1 Give everyday examples of the phenomena of refraction and reflection.

2 With the aid of a diagram, describe a refracting telescope.

3 What is chromatic aberration, and how can it be corrected?

4 With the aid of a diagram, describe a reflecting telescope. Describe four different ways in which an astronomer can have access to the focal point.

5 Explain some of the advantages of reflecting telescopes over refracting telescopes.

6 What is spherical aberration, and how can it be corrected?

7 What is meant by the angular resolution of a telescope?

8 What limits the angular resolution of the 5-m telescope at Palomar?

9 Why can radio astronomers observe at any time during the day, whereas optical astronomers are mostly limited to nighttime observing?

10 Why must X-ray telescopes be placed above the Earth's atmosphere?

11 Why will very large telescopes of the future make use of multiple mirrors?

Advanced questions

12 Quite often advertisements appear for telescopes that extol their magnifying abilities. Is this a good criterion for evaluating telescopes? Explain your answer.

∗ **13** The observing cage in which an astronomer sits at the prime focus of the 5-m telescope on Palomar Mountain is about 1 m in diameter. What fraction of the incoming starlight is blocked by the cage? (*Hint:* The area of a circle of radius r is πr^2, where $\pi \approx 3.14$).

∗ **14** Compare the light-gathering power of the Palomar 5-m telescope with that of the fully dark-adapted human eye, which has a pupil diameter of about 5 mm.

15 Show by means of a diagram why the image formed by a simple refracting telescope is "upside down."

∗ **16** Suppose your Newtonian reflector has a mirror with a diameter of 20 cm and a focal length of 2 m. What magnification do you get with eyepieces whose focal lengths are (**a**) 9 mm, (**b**) 20 mm, and (**c**) 55 mm?

17 At the time this book went to press, funding for AXAF was in jeopardy. Consult recent issues of such magazines as *Sky & Telescope* and *Science News* to learn about the current status of this project. What is the latest timetable for the construction and launching of the spacecraft?

18 At the time this book went to press, NASA was planning a shuttle mission to the Hubble Space Telescope in 1994. Consult recent issues of such magazines as *Sky & Telescope* and *Science News* to learn about the current status of this mission. When is the scheduled launch date? Have mechanical difficulties with HST's solar arrays and gyroscopes overshadowed the telescope's optical problems? Which pieces of optical equipment will be replaced, and how will they compensate for the spherical aberration of the primary mirror?

Discussion questions

19 Discuss the advantages and disadvantages of using a small telescope in Earth orbit versus a large telescope on a mountaintop.

20 If you were in charge of selecting a site for a new observatory, which factors would you consider important?

For further reading

Bok, B. "The Promise of the Space Telescope." *Mercury*, May/June 1983 • An excellent summary of the questions and issues to be examined by the Hubble Space Telescope.

Burns, J., and others. "Observatories on the Moon." *Scientific American*, March 1990 • This article explores the exciting possibility of establishing major astronomical observatories on the Moon.

Chaisson, E., and Villard, R. "Hubble Space Telescope: The Mission." *Sky & Telescope*, April 1990 • This article, written by two astronomers at the Space Telescope Science Institute prior to HST's launch, describes the capabilities of the telescope and gives an excellent overview of the main observing projects.

Davies, J. *Satellite Astronomy*. Ellis Horwood/John Wiley, 1988 • A comprehensive introduction to astronomy done from satellites.

Fienberg, R. "Hubble's Agony and Ecstasy." *Sky & Telescope*, January 1991 • This article describes HST's optical and mechanical problems and some of the first observations.

Janesick, J., and Blouke, M. "Sky on a Chip: The Fabulous CCD." *Sky & Telescope*, September 1987 • This superb article describes how CCDs work and why they have become an indispensable asset at well-equipped observatories.

Krisciunas, K. *Astronomical Centers of the World*. Cambridge University Press, 1988 • This book tells the story of the development of major centers of astronomy, including the observatory complexes on Mauna Kea and Kitt Peak.

Learner, R. *Astronomy through the Telescope*. Van Nostrand Reinhold, 1981 • A beautifully illustrated history of telescopes.

Powell, C. "Mirroring the Cosmos." *Scientific American*, November 1991 • This article discusses the latest telescope designs with special emphasis on "adaptive optics" that can cancel out distorting effects of the Earth's atmosphere.

Robinson, L. "The Frigid World of IRAS." *Sky & Telescope*, January 1984 • An excellent summary of the historic Infrared Astronomical Satellite mission.

Schorn, R. "Astronomy in the Next Decade." *Sky & Telescope*, April 1982 • This brief article discusses the planning, politics, and financing of astronomy in the United States.

Tucker, W., and Tucker, K. *The Cosmic Inquirers*. Harvard University Press, 1986 • This delightful book takes the reader behind the scenes in the design, construction, and operation of five major astronomical instruments, including the Very Large Array, the Infrared Astronomical Satellite, and the Hubble Space Telescope.

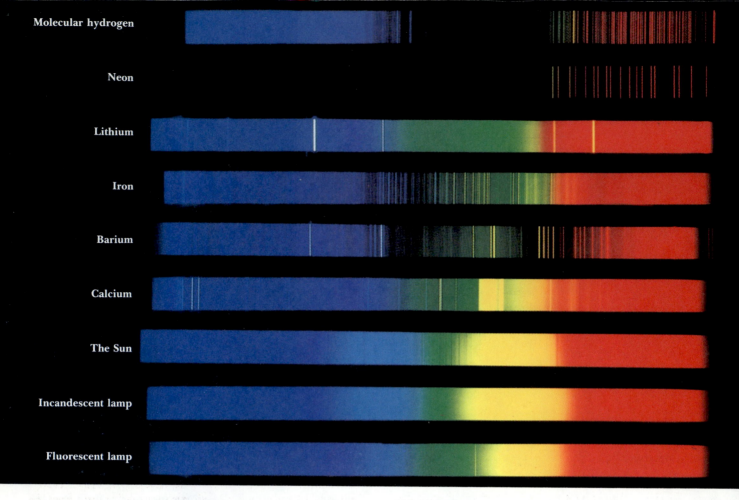

Molecular hydrogen
Neon
Lithium
Iron
Barium
Calcium
The Sun
Incandescent lamp
Fluorescent lamp

Spectra and spectral lines   When an electric spark passes through a gas, atoms of the gas emit radiation at certain wavelengths, and so a spectrum of that gas contains bright spectral lines. This photograph displays the spectra of various elements. Each element has a unique pattern of spectral lines. Spectra of the Sun and two common household lamps are also shown. The spectrum of an incandescent light bulb is quite like the continuous spectrum of a blackbody. (Courtesy of Bausch & Lomb)

5

The Nature of Light and Matter

Astronomy is built on an understanding of light, and every scientific discovery about light has led to important discoveries about the universe. In this chapter we find that light is a form of energy possessing both wavelike and particlelike properties. Because the light emitted by an object depends upon the object's temperature, we are able to determine the surface temperatures of the stars. With an understanding of the structure of atoms

comes an understanding of why each chemical element emits and absorbs light at specific wavelengths. Astronomers use this knowledge to determine the composition of stars and galaxies. Even the motion of a light source affects its wavelength, permitting us to deduce its speed of approach or retreat. These are but a few of the reasons that understanding light is a prerequisite to understanding the universe.

Most of our knowledge about the universe comes from studying light arriving at the Earth from distant objects. When we speak of "light" from stars and galaxies, we mean the full electromagnetic spectrum (see Figure 4-3). Indeed, most scientists use the terms *light* and *electromagnetic radiation* interchangeably. The human eye responds only to a narrow band of wavelengths that we call "visible light." As we saw in the previous chapter, astronomers use various telescopes to detect gamma rays, X rays, ultraviolet and infrared radiation, microwaves, and radio waves emitted by astronomical objects.

Recall from Chapter 4 that electromagnetic radiation consists of oscillating electric and magnetic fields (see Figure 4-2) that move through empty space at a speed of nearly 300,000 km/s (186,000 mi/s). The distance between adjacent crests of an electromagnetic wave is called the wavelength.

All objects emit electromagnetic radiation. Stars like the Sun emit radiation primarily at visible wavelengths (from 400 nm to 700 nm). During the late 1800s, physicists discovered how the intensity and wavelengths of light emitted by an object depend on the object's temperature. Before we learn about these important discoveries, we need to know how scientists measure temperature.

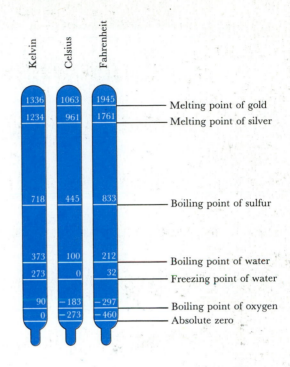

Figure 5-1 *Temperature scales* *Three temperature scales are in common use: Kelvin, Celsius, and Fahrenheit. Scientists usually prefer the Kelvin scale, which is measured upward from the coldest possible temperature (0 K), called absolute zero.*

5-1 *Temperatures are usually given on the Celsius or the Kelvin temperature scale*

There are three temperature scales in common use today (Figure 5-1). Temperatures are expressed throughout most of the world in degrees Celsius (°C). The **Celsius temperature scale** is based on the behavior of water, which freezes at 0°C and boils at 100°C at sea level. This scale, once known as the centigrade scale, was renamed in honor of the Swedish astronomer Anders Celsius, who proposed it in 1742.

For many purposes, scientists prefer to express temperatures in a unit called the **kelvin** (K), named after the British physicist William Thomson (Lord Kelvin), who made many important contributions to our knowledge about heat and temperature. On the Kelvin temperature scale, water freezes at 273 K and boils at 373 K. Because water must be heated through a change of either 100 K or 100°C to go from the freezing point to the boiling point, the size of a kelvin is the same as the size of a degree Celsius. A temperature expressed in kelvin is always equal to the temperature in degrees Celsius *plus* 273. Scientists

usually prefer the Kelvin scale because of its physical interpretation of the meaning of temperature.

All substances are made of tiny **atoms** (typical atomic diameters are about 10^{-10} m = 0.1 nm) that are constantly in motion. The temperature of a substance is directly related to the average speed of its atoms. If something is hot, its atoms move at high speeds. If a substance is cold, its atoms move much more slowly.

The coldest possible temperature is the temperature at which atoms move as slowly as possible (they can never quite stop completely). The minimum possible temperature, called **absolute zero,** is the starting point for the Kelvin scale. Absolute zero is 0 K, or −273°C.

In the United States, many people still use the Fahrenheit scale, expressing temperatures in **degrees Fahrenheit** (°F). When the German physicist Gabriel Fahrenheit introduced this scale in the early 1700s, he intended 0°F to represent the coldest temperature then achievable (with a mixture of ice and saltwater) and 100°F to represent the

Table 5-1 Various temperatures on different scales

	Kelvin (K)	Celsius (°C)	Fahrenheit (°F)
Absolute zero	0	−273	−460
Liquid helium boils	4	−269	−452
Liquid oxygen boils	90	−183	−297
Water freezes	273	0	32
"Room temperature"	295	22	72
Water boils	373	100	212
Sulfur boils	718	445	833
Iron melts	1808	1535	2795
Iron boils	3273	3000	5432
Carbon boils	5100	4827	8721

temperature of a healthy human body. On the Fahrenheit scale, water freezes at 32°F and boils at 212°F. Table 5-1 shows various temperatures expressed on the Kelvin, Celsius, and Fahrenheit scales.

Many scientists who study the solar system express temperatures (such as the temperatures on the surfaces of the planets) using the Celsius scale. Temperatures relating to stars, galaxies, and other astronomical phenomena, however, are usually given in kelvin.

5-2 Any object emits electromagnetic radiation with intensity and wavelengths related to its temperature

The total amount of energy radiated by an object depends upon its temperature. The hotter the object, the more energy it emits as electromagnetic radiation. The dominant wavelength of the emitted radiation also depends upon the temperature. A cool object emits most of the energy at long wavelengths, whereas a hotter object emits most of the energy at shorter wavelengths.

These basic phenomena are familiar to anyone who has watched a welder or blacksmith heat a bar of iron (as

shown in Figure 5-2). As it starts to heat up, the bar begins to glow with a deep red color. As the temperature rises, the bar begins to give off a brighter, reddish-orange light. At still higher temperatures, it shines with a brilliant, yellowish-white light. If the bar could be prevented from melting and vaporizing, at very high temperatures it would emit a dazzling, blue-white light.

In 1879 the Austrian physicist Josef Stefan summarized the results of his experiments on this phenomenon by stating that *an object emits energy at a rate proportional to the fourth power of the object's temperature (measured in kelvin)*. In other words, if you double the temperature of an object (for example, from 500 to 1000 K), the energy emitted from the object's surface each second increases by a factor of $2^4 = 16$. If you triple the temperature (for instance, from 500 to 1500 K), the rate at which energy is emitted increases by a factor of $3^4 = 81$.

Five years after Stefan announced this law, another Austrian physicist, Ludwig Boltzmann, showed how it could be derived mathematically from basic assumptions about atoms and molecules. The law today is commonly known as the **Stefan–Boltzmann law.**

The energy emitted from each square meter of an object's surface each second is called the **energy flux** (*F*). In this context, flux means "rate of flow," and thus *F* is a measure of how much energy is flowing out of the object. Using this term, we can write the Stefan–Boltzmann law as

$$F = \sigma T^4$$

where *T* is the temperature of the object in kelvin and σ (sigma) is a number called the Stefan–Boltzmann constant.

Boltzmann showed that this law is best obeyed by an idealized object called a **blackbody.** By definition, a blackbody does not reflect any light; it instead absorbs all radiation falling on it. Since it does not reflect any light, the radiation that a blackbody does emit is entirely the result of its temperature.

Ordinary objects are not perfect blackbodies—we can see them because they reflect light. For example, we can see planets in the sky because they reflect sunlight. The amount of energy that comes from such an object is different from the amount calculated from the Stefan–Boltzmann law. Stars, however, do behave like blackbodies. Stars produce their own light; they do not reflect light from other stars. Virtually all the light that comes

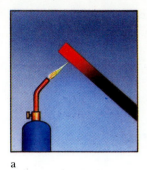

a   b   c

Figure 5-2 Heating a bar of iron *This sequence of drawings shows the changing appearance of a bar of iron as it is heated. As the temperature increases, the amount of energy radiated by the bar increases. The color of the bar also changes because, as the temperature goes up, the dominant wavelength of light emitted by the bar decreases. These effects are described by the Stefan–Boltzmann law and Wien's law.*

from a star is emitted by its hot gases. Astronomers can therefore use the Stefan–Boltzmann law to relate a star's brightness to its surface temperature.

Any object emits radiation over a wide range of wavelengths, but there is always a particular wavelength (λ_{max}) at which the emission of energy is strongest. This dominant wavelength gives a glowing hot object its characteristic color.

In 1893 the German physicist Wilhelm Wien discovered that *the dominant wavelength (λ_{max}) of radiation emitted by a blackbody is inversely proportional to its temperature.* In other words, the hotter an object, the shorter the dominant wavelength of the electromagnetic radiation it emits. This relationship is today called Wien's law.

Wien's law can also be stated as a simple equation. If λ_{max} is measured in meters, then

$$\lambda_{max} = \frac{2.9 \times 10^{-3}}{T}$$

where T is the temperature of the blackbody measured in kelvin.

From the Stefan–Boltzmann law, we see that any object with a temperature above absolute zero (0 K) emits some electromagnetic radiation. From Wien's law, we find that a very cold object with a temperature of only a few kelvin emits primarily microwaves. An object at "room temperature" (about 295 K) emits primarily infrared radiation. One with a temperature of a few thousand kelvin emits mostly visible light. Something with a temperature of a few million kelvin emits most of its radiation in the X-ray wavelengths.

As an example, consider energy coming from the Sun. The Sun emits energy over a wide range of wavelengths, but the maximum intensity of sunlight is at a wavelength of roughly 500 nm = 5×10^{-7} m. From Wien's law, we find the Sun's surface temperature ($T_\odot$) is

$$T_\odot = \frac{2.9 \times 10^{-3}}{5 \times 10^{-7}} = 5800 \text{ K}$$

Wien's law is very useful in computing the surface temperatures of stars because it does not require knowledge of the star's size or brightness; all we need to know is the dominant wavelength of the star's electromagnetic radiation.

5-3 Studies of blackbody radiation led to the discovery that light has particlelike properties

The Stefan–Boltzmann law and Wien's law describe only two basic properties of **blackbody radiation,** the electromagnetic radiation emitted by a hypothetical blackbody. A more complete picture is given by **blackbody curves,** such as those in Figure 5-3. These curves show the relationship between the wavelength and the intensity of light emitted by a blackbody at a given temperature.

The total area under a blackbody curve is proportional to the energy emitted; the wavelength corresponding to the peak of the curve is the dominant wavelength λ_{max}. Note that the blackbody curves clearly illustrate both of the laws we have discussed: A cool object has a low curve that peaks at a long wavelength, and a hotter object has a much higher curve that peaks at a shorter wavelength.

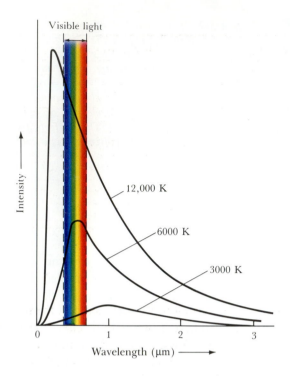

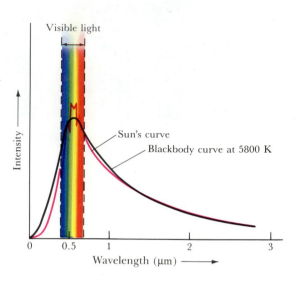

Figure 5-3 Blackbody curves *Three representative blackbody curves are shown here. Each curve shows the intensity of radiation at every wavelength emitted by a blackbody at a particular temperature. On this graph, wavelength is measured in micrometers (μm) where 1 μm = 1000 nm. Note the range of wavelengths of visible light.*

Figure 5-4   The Sun as a blackbody *This graph compares the intensity of sunlight over a wide range of wavelengths with the intensity of radiation from a blackbody at a temperature of 5800 K. Measurements of the Sun's intensity were made above the Earth's atmosphere. The Sun mimics a blackbody remarkably well.*

Figure 5-4 shows how the intensity of sunlight varies with wavelength. Note that the peak of the curve is at a wavelength of about 0.5 μm = 500 nm, which is nearly in the middle of the wavelength range for visible light. The blackbody curve for a temperature of 5800 K is also plotted in Figure 5-4. Note how closely the observed intensity curve for the Sun matches the blackbody curve. The close correlation between the observed intensity curves for most stars and the idealized blackbody curves justifies using the laws of blackbody radiation.

By the end of the nineteenth century, physicists realized that they had reached an impasse in their understanding of light. All attempts to explain the characteristic shapes of blackbody curves had failed. A breakthrough came in 1900, when the German physicist Max Planck discovered that he could derive a mathematical formula for the blackbody curves if he assumed that electromagnetic radiation was emitted in separate packets of energy. This important assumption was verified in 1905 by Albert Einstein, who described light as consisting of particlelike packets called **photons.** *The energy carried by a photon of light is inversely proportional to its wavelength.* In other

words, long-wavelength photons (such as radio waves and microwaves) carry little energy, whereas short-wavelength photons (like X rays and gamma rays) carry much more energy. This relationship between the energy of a photon and its wavelength is called **Planck's law.**

The energy of a photon is usually expressed in **electron volts (eV).** An electron volt is a tiny amount of energy, especially compared to the common unit of kilowatt-hours in which household electric energy is usually measured (a 1000-watt electric appliance consumes 1 kilowatt-hour of energy in 1 hour). Visible photons each carry between 2 and 3 eV. It takes 2.25×10^{25} eV to equal 1 kilowatt-hour. Table 5-2 gives the energy carried by photons of various wavelengths.

Together, Planck's law and Wien's law relate the temperature of an object to the energy of the photons that it emits, as in Table 5-2. A cool object emits primarily long-wavelength photons that carry little energy, and a hot object gives off mostly short-wavelength photons that carry much more energy. In later chapters, we shall find these ideas very useful in understanding how stars of various temperatures interact with gas and dust in space.

Table 5-2 Some properties of electromagnetic radiation

	Wavelength (cm)	Photon energy (eV)	Blackbody temperature (K)
Radio	$>10^{-1}$	$<10^{-5}$	<0.03
Microwave	10^{-1} to 10^{-4}	10^{-5} to 10^{-2}	0.03 to 30
Infrared	10^{-4} to 7×10^{-7}	0.01 to 2	30 to 4100
Visible	7×10^{-7} to 4×10^{-7}	2 to 3	4100 to 7300
Ultraviolet	4×10^{-7} to 10^{-9}	3 to 10^3	7300 to 3×10^6
X ray	10^{-9} to 10^{-11}	10^3 to 10^5	3×10^6 to 3×10^8
Gamma ray	$<10^{-11}$	$>10^5$	$>3 \times 10^8$

Note: The symbol > means "greater than"; the symbol < means "less than."

5-4 Each chemical element produces its own unique set of spectral lines

In 1814, the German master optician Joseph von Fraunhofer repeated Newton's classic experiment of shining a beam of sunlight through a prism (recall Figure 4-1), but he subjected the resulting rainbow-colored spectrum to intense magnification. Fraunhofer discovered that the solar spectrum contains hundreds of fine dark lines, which became known as **spectral lines.** Fraunhofer counted over 600 such lines, and today we know more than 20,000. Hundreds of spectral lines are visible in the photograph of the Sun's spectrum shown in Figure 5-5.

Half a century later, chemists discovered that they could produce spectral lines in laboratory experiments. Certain chemicals are easy to identify by the distinctive colors

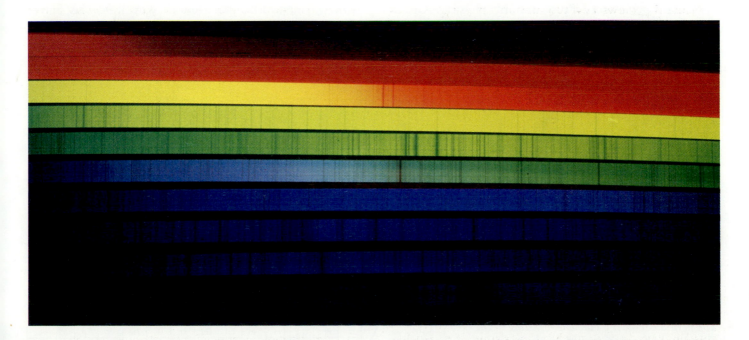

Figure 5-5 The Sun's spectrum Numerous spectral lines are seen in this photograph of the Sun's spectrum. The spectrum is so long that it had to be cut into convenient segments to fit on this page. (NOAO)

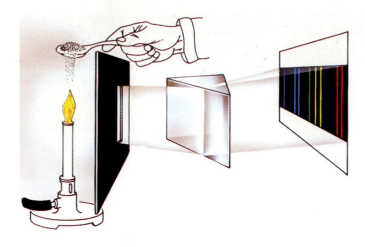

Figure 5-6 The Kirchhoff–Bunsen experiment In the mid-1850s, Kirchhoff and Bunsen discovered that when a chemical substance is heated and vaporized, the resulting spectrum exhibits a series of bright spectral lines. In addition, they found that each chemical element produces its own characteristic pattern of spectral lines.

emitted when they are sprinkled into a flame. About 1857 the German chemist Robert Bunsen invented a special gas burner that produces a clean, colorless flame. This Bunsen burner became very useful in analyzing substances because its flame had no color of its own to confuse with the color produced by a substance sprinkled into it.

Bunsen's colleague, Prussian-born physicist Gustav Kirchhoff, suggested that light from the colored flames could best be studied by passing it through a prism (Figure 5-6). They promptly discovered that the spectrum from a flame consists of a pattern of thin, bright spectral lines against a dark background. They next found that *each chemical element produces its own characteristic pattern of spectral lines.* Thus was born in 1859 the technique of **spectral analysis,** the identification of chemical substances by their unique patterns of spectral lines.

A chemical **element** is a fundamental substance because it cannot be broken down into more basic chemicals. By the mid-1800s, chemists had conclusively identified such familiar elements as hydrogen, oxygen, carbon, iron, gold, silver, and so forth. Spectral analysis promptly led to the discovery of additional elements, many of which are quite rare.

After Bunsen and Kirchhoff had recorded the prominent spectral lines of all the known elements, they soon began to discover other spectral lines in mineral samples. In 1860, for instance, they found a new line in the blue portion of the spectrum of mineral water. After chemically isolating the previously unknown element responsible for the line, they named it cesium (from the Latin *caesius,* "gray-blue"). The next year a new spectral line in the red portion of the spectrum of a mineral sample led them to the discovery of the element rubidium (from *rubidus,* "red").

During a solar eclipse in 1868, astronomers found a new spectral line in the light coming from the upper surface of the Sun while the main body of the Sun was hidden by the Moon. This line was attributed to a new element that was named helium (from the Greek *helios,* "Sun"). Helium was not actually discovered on the Earth until 1895, when it was identified in gases obtained from a uranium mineral.

Figure 5-7 The periodic table of the elements The periodic table is a convenient listing of the elements arranged according to their weights and chemical properties.

1 H																	2 He
3 Li	4 Be											5 B	6 C	7 N	8 O	9 F	10 Ne
11 Na	12 Mg											13 Al	14 Si	15 P	16 S	17 Cl	18 A
19 K	20 Ca	21 Sc	22 Ti	23 V	24 Cr	25 Mn	26 Fe	27 Co	28 Ni	29 Cu	30 Zn	31 Ga	32 Ge	33 As	34 Se	35 Br	36 Kr
37 Rb	38 Sr	39 Y	40 Zr	41 Nb	42 Mo	43 Tc	44 Ru	45 Rh	46 Pd	47 Ag	48 Cd	49 In	50 Sn	51 Sb	52 Te	53 I	54 Xe
55 Cs	56 Ba	57 La	72 Hf	73 Ta	74 W	75 Re	76 Os	77 Ir	78 Pt	79 Au	80 Hg	81 Tl	82 Pb	83 Bi	84 Po	85 At	86 Rn
87 Fr	88 Ra	89 Ac	104	105	106												

58 Ce	59 Pr	60 Nd	61 Pm	62 Sm	63 Eu	64 Gd	65 Tb	66 Dy	67 Ho	68 Er	69 Tm	70 Yb	71 Lu
90 Th	91 Pa	92 U	93 Np	94 Pu	95 Am	96 Cm	97 Bk	98 Cf	99 Es	100 Fm	101 Md	102 No	103 Lr

Figure 5-8 Iron in the Sun's atmosphere *The upper spectrum is a portion of the Sun's spectrum from 420 to 430 nm. Numerous dark spectral lines are visible. The lower spectrum is a corresponding portion of the spectrum of vaporized iron. Several bright spectral lines can be seen against the black back-* *ground. The fact that the iron lines coincide with some of the solar lines proves that there is some iron (albeit a very tiny amount) in the Sun's atmosphere. (The Observatories of the Carnegie Institution of Washington)*

A list of the chemical elements is most conveniently displayed in the form of a **periodic table** (Figure 5-7). Each element is assigned a unique **atomic number,** and the elements are arranged in the periodic table in that sequence. With a few exceptions, this sequence also corresponds to increasing average mass of the atoms of the elements. Thus hydrogen (the symbol H) with atomic number 1 is the lightest element. Iron (Fe) has atomic number 26 and is a relatively heavy element.

All the elements in a single vertical column of the periodic table have similar chemical properties. For example, the elements in the far right column are all gases at Earth-surface temperature and pressure, and they all tend to be very reluctant to react chemically with other elements.

In addition to nearly a hundred naturally occurring elements, Figure 5-7 lists several artificially produced elements. All of them are heavier than uranium (the symbol U) and are highly radioactive, which means that they change into lighter elements shortly after being created in the laboratory.

5-5 A spectrograph is an optical device that records a spectrum

The discovery that each chemical element produces its own unique pattern of spectral lines gave scientists the ability to determine the chemical composition of a remote astronomical object by identifying the spectral lines in its spectrum. For example, Figure 5-8 shows a portion of the Sun's spectrum along with the spectrum of iron. No other chemical can mimic iron's particular pattern of spectral lines at these wavelengths. It is iron's own distinctive "fingerprint." Since the spectral lines of iron appear in the Sun's spectrum, we can safely conclude that the Sun's atmosphere contains some vaporized iron.

Bunsen and Kirchhoff collaborated in the design and construction of the first **spectroscope,** a device consisting of a prism and several lenses by which a spectrum could be magnified and examined. After photography was invented, scientists preferred to produce a permanent photographic record of spectra. A device for photographing a spectrum is called a **spectrograph.** These instruments have become one of the astronomer's most important tools, perhaps second only to the telescope itself.

In its basic form, a spectrograph consists of a slit, two lenses, and a prism arranged to focus the spectrum of an astronomical object onto a small photographic plate, as shown in Figure 5-9. This optical device typically is mounted at the focal point of a telescope, and the image

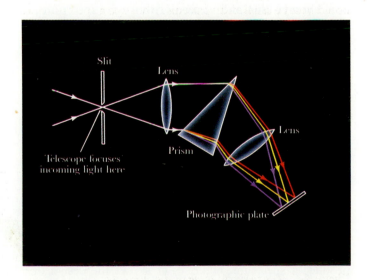

Figure 5-9 A prism spectrograph *This optical device uses a prism to break up the light of an object into a spectrum. Lenses focus that spectrum onto a photographic plate.*

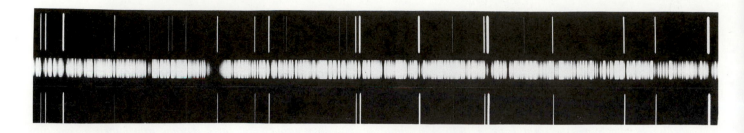

Figure 5-10 A spectrogram *The photographic record of a spectrum is called a spectrogram. This spectrogram shows the spectrum of a star. Numerous dark lines can be seen against the brighter background of the spectrum. Above and below the star's spectrum are bright lines produced by an iron arc. These bright lines are the comparison spectrum. (Palomar Observatory)*

of the object to be examined is focused on the slit. After the spectrum of a star or galaxy has been photographed, the exposed portion of the photographic plate is covered and light from a known source (usually an electric spark that vaporizes a small amount of iron) is focused on the slit. This exposes a "comparison spectrum" above and below the spectrum of the object being examined, as in Figure 5-10. The wavelengths of the spectral lines in the comparison spectrum are already known from laboratory experiments. These lines can therefore be used to identify and measure the wavelengths of the lines in the spectrum of the star or galaxy under study.

There are drawbacks to this old-fashioned spectrograph. A prism does not disperse the colors of the rainbow evenly: Blue and violet portions of a spectrum are spread out more than the red portion. In addition, because the blue and violet wavelengths must pass through more glass than the red wavelengths (examine Figure 4-1), light is absorbed unevenly across the spectrum. Indeed, a glass prism is opaque to near-ultraviolet wavelengths.

A better device for breaking starlight into the colors of the rainbow is a **grating**, a piece of glass on which thousands of closely spaced lines are cut. Some of the finest gratings have as many as 10,000 lines per centimeter, which are usually cut by drawing a diamond back and forth across the piece of glass. The spacing of the lines must be very regular. Light rays reflected from different parts of the grating interfere with each other so as to produce a spectrum. Figure 5-11 shows the design of a modern grating spectrograph.

In recent years the television and electronics industries have produced a variety of light-sensitive devices that are

significantly better than photographic film for recording spectra. For instance, many observatories now record spectra with a charge-coupled device, or CCD (see Figure 4-17). As described in Chapter 4, a CCD is a silicon chip approximately the size of a postage stamp divided into

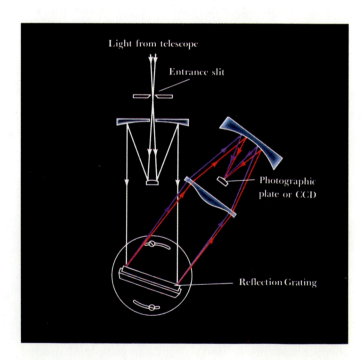

Figure 5-11 A grating spectrograph *This optical device uses a grating to break up the light of an object into a spectrum. An arrangement of lenses and mirrors focuses that spectrum onto a photographic plate or CCD.*

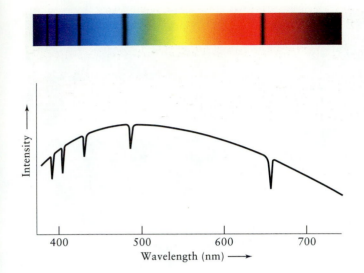

Figure 5-12 *The spectrum of hydrogen gas* *When photographic film is placed at the focus of a spectrograph, a rainbow-colored spectrum is obtained. When a CCD is placed at the focus of a spectrograph, a graph of intensity versus wavelength is produced by computer controlled equipment attached to the CCD. Note that the dark spectral lines appear as dips in the intensity-versus-wavelength curve.*

thousands of tiny squares. When light falls on one of these tiny squares (a pixel), an electric charge builds up in proportion to the amount of light. When the exposure is finished, electronic equipment measures how much charge has accumulated in each pixel. The final result is a graph on which light intensity is plotted against wavelength. Dark lines in the rainbow-colored spectrum appear as depressions or valleys, while bright lines in the spectrum appear as peaks. For example, Figure 5-12 shows both a picture and a plot of a spectrum in which five dark spectral lines appear. These particular spectral lines are caused by hydrogen.

5-6 Spectral lines are bright or dark depending on conditions in the spectrum's source

During his pioneering experiments with spectra, Kirchhoff noticed that sometimes he would see dark spectral lines, called absorption lines, among the colors of the rainbow. But a different experiment might give bright

spectral lines, called emission lines, against an otherwise dark background. By the early 1860s, Kirchhoff's studies had progressed sufficiently for him to formulate three statements that describe the conditions under which various types of spectra are observed. These statements, called Kirchhoff's laws, are as follows:

Law 1 A hot object or a hot, dense gas produces a **continuous spectrum**—a complete rainbow of colors without any spectral lines.

Law 2 A hot, rarefied gas produces an **emission line spectrum**—a series of bright spectral lines against a dark background.

Law 3 A cool gas in front of a continuous source of light produces an **absorption line spectrum**—a series of dark spectral lines among the colors of the rainbow.

Figure 5-13 illustrates the formation of absorption and emission lines. Note that the bright lines in the emission spectrum of a particular gas occur at exactly the same wavelengths as the dark lines in the absorption spectrum of that gas. Also note the importance of the relative temperatures of the gas cloud and its background. Absorption lines are seen if the background is hotter than the gas. Emission lines are seen if the background is cooler.

At the time of these discoveries, scientists knew that all matter is composed of atoms. An atom is the smallest particle of a chemical element that still has the properties of that element. Kirchhoff's laws tell us that the atoms of a gas somehow extract light of very specific wavelengths from the white light that passes through the gas. Thus, dark absorption lines are created in the continuous spectrum of the white-light source. The atoms then radiate light of precisely these same wavelengths in all directions, so that an observer at an oblique angle (without the white-light source in the background) detects bright emission lines.

Why do the atoms of a gas absorb light only at particular wavelengths? Why do they then emit light only at these same wavelengths? Traditional theories of electromagnetism could not answer these questions. The answers came early in the twentieth century with the development of quantum mechanics and nuclear physics.

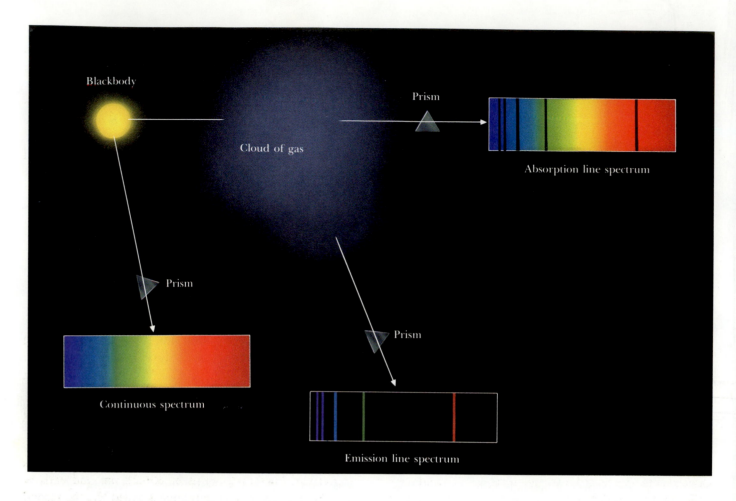

Figure 5-13 *Continuous, absorption line, and emission line spectra This schematic diagram summarizes how different types of spectra are produced. A hot, glowing object emits a continuous spectrum. If this source of light is viewed through a* cool gas, dark absorption lines appear in the resulting spectrum. When the gas is viewed against a cold, dark background, bright emission lines are seen in the spectrum.

5-7 An atom consists of a small, dense nucleus surrounded by electrons

The first important clue about the internal structure of atoms came from an experiment conducted in 1910 by Ernest Rutherford, a gifted chemist and physicist from New Zealand. Rutherford and his colleagues at the University of Manchester in England were investigating the recently discovered phenomenon of radioactivity. Certain radioactive elements such as uranium and radium were known to emit massive particles with considerable speed. In one series of experiments, Rutherford and his associates directed a beam of these particles against a thin

sheet of metal. Almost all the particles passed through the metal sheet with little or no deflection. To the surprise of the experimenters, however, an occasional particle bounced back from the metal sheet as though it had struck something very dense.

Rutherford was quick to realize the implications of this experiment. Most of the mass of an atom is concentrated in a compact, massive lump of matter that occupies only a small part of the atom's volume. The majority of the particles pass freely through the nearly empty space that makes up most of the atom, but a few particles happen to strike the dense mass at the center and bounce back.

Rutherford proposed a new model for the structure of an atom. According to this model, a massive, positively charged **nucleus** at the center of the atom is orbited by

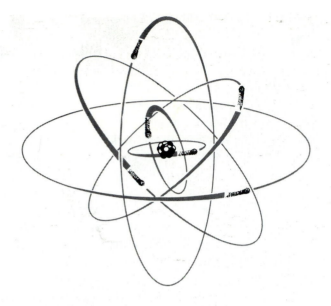

Figure 5-14 *Rutherford's model of the atom* Electrons orbit *the atom's nucleus, which contains most of the atom's mass. The nucleus contains two types of particles: protons and neutrons.*

tiny, negatively charged electrons (Figure 5-14). Rutherford concluded that at least 99.98% of the mass of an atom is concentrated in a nucleus whose diameter is only about one ten-thousandth the diameter of the atom.

We know today that the nucleus of an atom contains two types of particles: protons and neutrons. A proton has almost the same mass as a neutron, and each is about 2000 times more massive than an electron. A proton has a positive electric charge and a neutron has none. The electrical attraction between the positively charged protons and the negatively charged electrons holds an atom together.

Normally, the number of electrons orbiting an atom is equal to the number of protons in the nucleus, thus making the atom electrically neutral. Furthermore, the number of protons in an atom's nucleus equals the atomic number for that particular element. Thus, a hydrogen nucleus has 1 proton, a helium nucleus has 2,

and so forth, up to uranium with 92 protons in its nucleus.

The number of protons in an atom's nucleus determines what element that atom is. The number of neutrons in the nuclei of the atoms of the same element may, however, vary. For example, oxygen, the eighth element on the periodic table, with an atomic number of 8, has exactly eight protons, but it may have eight, nine, or ten neutrons. These three slightly different kinds of oxygen are called **isotopes.** The isotope with eight neutrons is by far the most abundant variety.

5-8 Spectral lines are produced when an electron jumps from one energy level to another within an atom

The challenge of reconciling Rutherford's atomic model with the observations of spectral analysis was undertaken by the young Danish physicist Niels Bohr, who joined Rutherford's group at Manchester in 1911.

Bohr began by trying to understand the structure of hydrogen, the simplest and lightest of the elements. A hydrogen atom consists of a single electron and a single proton. Hydrogen has a simple spectrum consisting of a pattern of lines that begins at 656.28 nm and ends at 364.56 nm. The first spectral line is called H_α, the second H_β, the third H_γ, and so forth, ending with H_∞ at 364.56 nm. The closer you get to 364.56 nm, the more spectral lines you see.

The regularity in this spectral pattern had been described mathematically in 1885 by Johann Jakob Balmer, an elderly German schoolteacher. Balmer used trial and error to discover a formula for calculating the wavelengths of the hydrogen lines. Since his formula was successful, the spectral lines of hydrogen at visible wavelengths are today called **Balmer lines,** and the entire pattern from H_α to H_∞ is called the **Balmer series.** The spectrum of the star shown in Figure 5-15 exhibits more than two dozen Balmer lines from H_{13} through H_{40}.

Figure 5-15 *Balmer lines in the spectrum of a star* This portion of the spectrum of a star *called HD 193182 shows nearly two dozen Balmer lines. The series converges at 364.56 nm, just to the left of H_{40}. This star's spectrum also contains the first 12 Balmer lines (H_α through H_{12}), but they are not visible in this particular spectrogram. (The Observatories of the Carnegie Institution of Washington)*

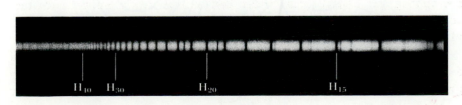

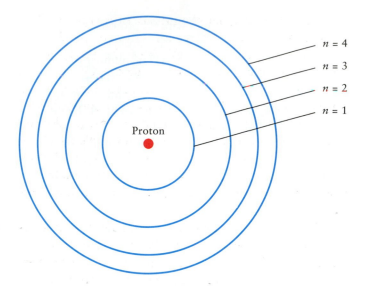

Figure 5-16 The Bohr model of the hydrogen atom *According to Bohr's model of the atom, an electron circles the nucleus only in allowed orbits* n = 1, 2, 3, *and so on. The first four Bohr orbits are shown here.*

Bohr's goal was to mathematically derive Balmer's formula from basic laws of physics. He began by assuming that the electron in a hydrogen atom orbits the nucleus only in certain specific orbits. As shown in Figure 5-16, it is customary to label these orbits $n = 1$, $n = 2$, $n = 3$, and so on. They are called the Bohr orbits.

Bohr argued that for an electron to jump from one orbit to another, the hydrogen atom must gain or lose a specific amount of energy. An electron jump from an inner orbit to an outer orbit would require energy; an electron jump from an outer orbit to an inner one would release energy.

The energy gained or released by the atom when the electron changes orbits is the difference in energy between these two orbits. According to Planck and Einstein, the packet of energy gained or released is a photon whose energy is inversely proportional to its wavelength. Using these ideas, Bohr mathematically derived the formula that Balmer had discovered by trial and error. Furthermore, Bohr's discovery elucidated the meaning of the Balmer series: All the Balmer lines are produced by electron transitions between the second Bohr orbit ($n = 2$) and higher orbits ($n = 3, 4, 5$, and so forth). As an example, Figure 5-17 shows the electron transition that gives rise to the H_α spectral line, which has a wavelength of 656.28 nm.

In addition to giving the wavelengths of the Balmer series, Bohr's formula correctly predicts the wavelengths of other series of spectral lines that occur at nonvisible wavelengths. For example, electron transitions between the lowest Bohr orbit ($n = 1$) and all higher orbits ($n = 2, 3, 4, \ldots$) also produce spectral lines. These transitions produce the **Lyman series,** which is entirely in the ultraviolet. At infrared wavelengths is the **Paschen series,** which arises out of transitions to and from the third Bohr orbit ($n = 3$). Additional series exist at still longer wavelengths.

Bohr's ideas help explain Kirchhoff's laws. Each spectral line corresponds to one specific transition between the orbits of the atoms of a particular element. An absorption line is created when an electron jumps from an inner orbit to an outer orbit, extracting the required photon from an outside source of energy such as the continuous spectrum of a hot, glowing object. An emission line is produced

Figure 5-17 The absorption and emission of an H_α photon *This schematic diagram, drawn according to the Bohr model of the atom, shows what happens when a hydrogen atom absorbs or emits a photon whose wavelength is 656.28 nm.* **(a)** *The photon is absorbed by the atom as the electron jumps from orbit* n = 2 *up to orbit* n = 3. **(b)** *The photon is emitted by the atom as the electron falls from orbit* n = 3 *down to orbit* n = 2.

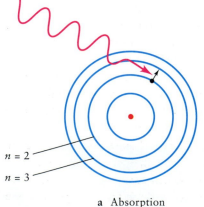

a Absorption   b Emission

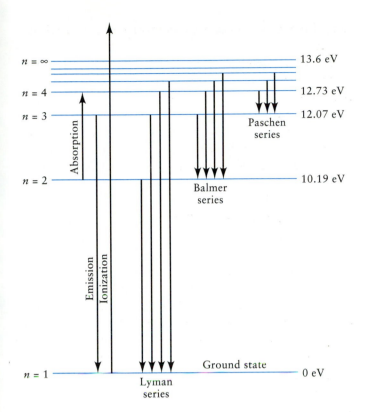

Figure 5-18 *The energy level diagram of hydrogen The structure of the hydrogen atom is conveniently displayed in a diagram showing the energy levels above the ground state. A variety of electron jumps, or transitions, are shown—including those that produce some of the most prominent lines in the hydrogen spectrum.*

when the electron falls back down to a lower orbit and gives up a photon.

Today's view of the atom owes much to the Bohr model, but it is different in certain ways. The modern picture is based on **quantum mechanics,** a branch of physics dealing with photons and subatomic particles that was developed during the 1920s. As a result of this work, physicists no longer imagine that electrons move in specific orbits about the nucleus. Instead, electrons are now said to occupy certain allowed **energy levels** in the atom.

An extremely useful way of displaying the structure of an atom is with an **energy level diagram,** such as that shown in Figure 5-18 for hydrogen. The lowest energy level, called the **ground state,** corresponds to the $n = 1$ Bohr orbit. An electron can jump from the ground state up to the $n = 2$ level only if the atom absorbs a Lyman-alpha photon of wavelength 121.6 nm. As noted earlier, the energy of a photon is usually expressed in electron

volts. The Lyman-alpha photon has an energy of 10.19 eV, and so the energy level $n = 2$ is shown on the diagram as having an energy 10.19 eV above the energy of the ground state, usually assigned a value of 0 eV. Similarly, the $n = 3$ level is 12.07 eV above the ground state, and so forth up to the $n = \infty$ level at 13.6 eV. If the atom absorbs a photon of an energy greater than 13.6 eV, an electron from the ground state will be knocked completely out of the atom. This process, in which high-energy photons knock electrons out of atoms, is called **ionization.** In general, an atom that has been stripped of one or more electrons is called an **ion.**

With the work of people like Planck, Einstein, Rutherford, and Bohr, the interchange between astronomy and physics came full circle. Modern physics was born when Newton had set out to understand the motions of the planets. Two and a half centuries later, physicists in their laboratories discovered the basic properties of light and the structures of atoms. As we shall see in later chapters, the fruits of their labors have important astronomical applications.

5-9 The wavelength of a spectral line is affected by the relative motion between the source and the observer

Christian Doppler, a professor of mathematics in Prague, pointed out in 1842 that wavelength is affected by motion. As shown in Figure 5-19, light waves from an approaching light source are compressed. The circles represent waves emitted from consecutive positions as the source moves along. Because each successive wave is emitted from a position slightly closer to you, you see a shorter wavelength than you would if the source were stationary. All the spectral lines in the spectrum of an approaching source are shifted toward the short-wavelength (blue) end of the spectrum. This phenomenon is called a **blueshift.**

Conversely, light waves from a receding source are stretched out. You see a longer wavelength than you would if the source were stationary. All the spectral lines in the spectrum of a receding source are shifted toward the longer-wavelength (red) end of the spectrum, producing a **redshift.** In general, the effect of relative motion on wavelength is called the **Doppler effect.**

Suppose that λ_0 is the wavelength of one particular spectral line from a source that is not moving. It is the

Figure 5-19 The Doppler effect This diagram shows why wavelength is affected by motion between the light source and an observer. A source of light is moving toward the left. The four circles (numbered 1 through 4) indicate the location of light waves that were emitted by the moving source when it was at points S_1 through S_4, respectively. Note that the waves are compressed in front of the source but stretched out behind it. Consequently, wavelengths appear shortened (blueshifted) if the source is moving toward the observer and lengthened (redshifted) if the source is moving away from the observer. Motion perpendicular to an observer's line of sight does not affect wavelength.

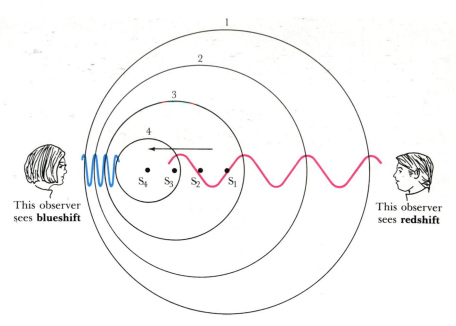

This observer sees **blueshift**

This observer sees **redshift**

wavelength that you might look up in a reference book or determine in a laboratory experiment for this spectral line. If the source is moving, this particular spectral line is shifted to a different wavelength λ. The size of the wavelength shift is usually written as $\Delta\lambda$, where $\Delta\lambda = \lambda - \lambda_0$. Thus $\Delta\lambda$ is the difference between the wavelength listed in reference books and the wavelength that you actually observe in the spectrum of a star or galaxy.

Christian Doppler proved that the wavelength shift $\Delta\lambda$ is governed by the simple equation

$$\frac{\Delta\lambda}{\lambda_0} = \frac{v}{c}$$

where v is the speed of the source measured along the line of sight between the source and the observer. As usual, c is the speed of light (3×10^8 m/s).

For example, the spectral lines of hydrogen appear in the spectrum of the bright star Vega as shown in Figure 5-20. The prominent hydrogen line H_α has a normal wavelength of 656.285 nm, but in Vega's spectrum this line is located at 656.255 nm. The wavelength shift is calculated thus:

$$\Delta\lambda = \lambda - \lambda_0$$
$$= 656.255 \text{ nm} - 656.285 \text{ nm}$$
$$= -0.030 \text{ nm}$$

and so the star is approaching us with a speed of

$$v = c\left(\frac{\Delta\lambda}{\lambda_0}\right)$$
$$= 3 \times 10^5 \text{ km/s}\left(\frac{-0.030}{656.285}\right)$$
$$= -14 \text{ km/s}$$

The minus sign indicates that the star is moving toward us.

A speed determined in this fashion is called a **radial velocity,** because v is the component of the star's motion parallel to our line of sight, or along the "radius" drawn from Earth to the star. Of course, a sizable fraction of a star's motion may be perpendicular to our line of sight. This transverse movement, called **proper motion,** does not affect wavelengths.

The Doppler effect is a powerful tool that astronomers use in a wide range of situations to discover the motions of objects in space. For instance, careful spectroscopic studies of the Sun's surface reveal the speed of hot gases as they rise and fall. Similarly, Doppler shift measurements of stars in double-star systems give crucial data about the speeds of the stars along their orbits. Indeed, by measuring the redshifts of distant galaxies, we can determine the rate at which the entire universe is expanding. In later chapters we shall refer to the Doppler effect whenever we need to convert an observed wavelength shift to a speed.

Figure 5-20 The spectrum of Vega *Several Balmer lines are seen in this photograph of the spectrum of Vega, the brightest star in Lyra (the Harp). All the spectral lines are shifted very* *slightly toward the blue side of the spectrum, indicating that Vega is approaching us. (NOAO)*

Summary

- Electromagnetic radiation (including visible light) has wavelike properties and travels through empty space at the constant speed $c = 3 \times 10^8$ m/s.

- A blackbody is a hypothetical object that is a perfect absorber of electromagnetic radiation at all wavelengths. Since a blackbody does not reflect any light from outside sources, the radiation that it does emit depends only on its temperature.

  Stars closely approximate the behavior of blackbodies.

- The Stefan–Boltzmann law relates the temperature of a blackbody to the rate at which it radiates energy.

  The energy flux from a blackbody is proportional to the fourth power of its temperature (in kelvin).

- Wien's law relates the temperature of an object to the dominant wavelength at which it radiates energy.

  The dominant wavelength of radiation emitted by a blackbody is inversely proportional to its temperature (in kelvin).

- The intensities of radiation emitted at various wavelengths by a blackbody at a given temperature are shown by a blackbody curve.

  An explanation of blackbody curves led to the discovery that light (electromagnetic radiation) has particlelike properties; particles of light are called photons.

- Planck's law relates the energy of a photon to its wavelength.

  The energy of a photon is inversely proportional to its wavelength.

- Kirchhoff's three laws of spectral analysis describe the conditions under which absorption lines, emission lines, and a continuous spectrum can be observed.

- Spectroscopy, the study of spectra, provides information about the chemical composition of remote astronomical objects.

  Spectral lines serve as distinctive "fingerprints" for detecting elements and chemical compounds in a light source.

- An atom consists of a small, dense nucleus (composed of protons and neutrons) surrounded by electrons that occupy only certain allowed energy levels.

- The spectral lines of a particular element correspond to the various electron transitions between allowed energy levels in the atoms of the element.

  When an electron jumps from one energy level to another, a photon of the appropriate energy (and hence a specific wavelength) is absorbed or emitted by the atom.

- The spectrum of hydrogen at visible wavelengths consists of the Balmer series, which arises from electron transitions between the second energy level of the hydrogen atom and higher levels.

- The process by which an atom absorbs an energetic photon and loses an electron entirely is called ionization.

- The spectral lines of an approaching light source are shifted toward short wavelengths (a blueshift); the spectral lines of a receding light source are shifted toward long wavelengths (a redshift).

    The equation describing the Doppler effect states that the size of a wavelength shift is proportional to the radial velocity between the light source and the observer.

Review questions

1 What is a blackbody? What does it mean to say that a star behaves like a blackbody? If stars behave like blackbodies, why are they not black?

2 What is the Stefan–Boltzmann law? Why do you suppose that astronomers are interested in it?

3 What is Wien's law? How could you use it to determine the temperature of a star's surface?

4 Using Wien's law and the Stefan–Boltzmann law, explain the changes in color and brightness that are observed as the temperature of a hot, glowing object increases.

5 Describe the experimental evidence that supported the Bohr model of the atom.

6 Explain how the spectrum of hydrogen is related to the structure of the hydrogen atom.

7 Why do different elements have different patterns of lines in their spectra?

8 What is the Doppler effect, and why is it important to astronomers?

9 Explain why the Doppler effect tells us only about the motion directly along the line of sight between a light source and an observer.

Advanced questions

* 10 Approximately how many times around the world could a beam of light travel in one second?

* 11 The bright star Regulus in the constellation of Leo (the Lion) has a surface temperature of 12,200 K. What is the dominant wavelength (λ_{max}) of the light it emits?

* 12 The bright star Procyon in the constellation of Canis Minor, the Little Dog, emits the greatest intensity of radiation at a wavelength (λ_{max}) of 445 nm. What is the surface temperature of the star?

* 13 Imagine a star the same size as the Sun but with a surface temperature twice that of the Sun. At what wavelength would that star emit most of its radiation? How many times brighter than the Sun would that star be? (*Hint:* The Sun emits the greatest intensity of radiation in the middle of the visible spectrum, at a wavelength of about 500 nm.)

* 14 Imagine a star the same size as the Sun with a surface temperature of 2900 K. Suppose both the Sun and this star were located at the same distance from you. Which would be brighter? By how much? (*Hint:* The Sun's surface temperature is 5800 K.)

* 15 The wavelength of H_β in the spectrum of the star Megrez in the Big Dipper is 486.112 nm. Laboratory measurements demonstrate that the normal wavelength of this spectral line is 486.133 nm. Is the star coming toward us or moving away from us? At what speed?

* 16 In the spectrum of the bright star Rigel, H_α has a wavelength of 656.331 nm. Is the star coming toward us or moving away from us? How fast?

* 17 Imagine driving down a street toward a traffic light. How fast would you have to go so that the red light would appear green?

Discussion questions

18 Compare identifying chemicals by spectral line patterns with identifying people by line patterns in their fingerprints.

19 Suppose you look up at the night sky and observe some of the brightest stars with your naked eye. Is there any way of telling which stars are hot and which are cool? Explain.

For further reading

Bova, B. *The Beauty of Light*. Wiley, 1988 • A fascinating introduction to all aspects of light. Includes sections on the eye, color, and the technology of producing and detecting light.

Cline, B. *Men Who Made a New Physics*. Signet, 1965 • This book describes the history of the discoveries that led to our modern understanding of light and atoms.

Feynman, R. *QED*. Princeton University Press, 1985 • Using many carefully thought-out analogies, this slim volume discusses how photons and particles interact. The author won a Nobel Prize for his work in this field.

Gingerich, O. "Unlocking the Chemical Secrets of the Cosmos." *Sky & Telescope*, July 1981 • This brief article offers some fascinating insights into the work of Kirchhoff and Bunsen.

Gribbin, J. *In Search of Schroedinger's Cat*. Bantam, 1984 • This excellent introduction to quantum mechanics provides both background and historical material along with a fine review of current thinking.

Sobel, M. *Light*. University of Chicago Press, 1987 • An excellent nontechnical introduction to all aspects of light.

Snow, C. *The Physicists*. Macmillan, 1981 • This book describes many of the people who contributed to a modern understanding of light and matter.

van Heel, A., and Velzel, C. *What Is Light?* McGraw-Hill, 1968 • This book surveys our modern understanding of the phenomenon of light.

Weinberg, S. *Subatomic Particles*. Scientific American Books, 1983 • This book beautifully describes the various developments, both theoretical and experimental, leading up to our modern understanding of light and matter. The author won a Nobel Prize for his work in this field.

6

Our Solar System

With telescopes and space probes, astronomers have amassed a rich body of information about our solar system. A survey of the major physical characteristics of the planets reveals two distinct classes. Small, rocky planets like the Earth are found near the Sun, while huge, gaseous planets, like Jupiter, are located far from the Sun. This dichotomy is a direct result of conditions under which the planets formed. Our Moon and six other giant satellites are large enough to be considered planets in their own right. Numerous smaller satellites, asteroids, and comets also populate the solar system. The planets are composed of substances left over from generations of stars that lived out their lives and shed their matter into space long before the Sun was born. Our solar system was probably created from a huge cloud of interstellar gas and dust that contracted under the force of gravity. Most of the matter fell toward the center of that contracting cloud to form the Sun; outlying material accumulated in clumps to form the planets. As the planets began to develop, they must have been hit by numerous chunks of rock and ice that orbited the protosun 4.5 billion years ago. Our Moon may have formed from material torn from the Earth during one such collision with a planet-sized rock.

Nebulae in Sagittarius *Planets are probably forming along with new stars in these nebulae in Sagittarius. The type of planet that forms at a particular distance from a star depends on such conditions as the temperature there and the substances—rock fragments, ice crystals, and gases—at that distance. (Royal Observatory, Edinburgh)*

Throughout human history, people have wondered about the nature and origin of the Sun, Moon, stars, and planets. Unlike our ancestors, however, we possess a wealth of information about the universe. Especially within the past few decades, telescopic observations and interplanetary spacecraft have given us vast quantities of data from which to formulate and test theories. We see that the Sun, planets, moons, asteroids, comets, and meteoroids make up our tiny niche in the universe. Many of these objects are exceedingly ancient and contain records of the cosmic events that created our solar system.

In addition to gleaning information from the objects that orbit the Sun, we can observe the ongoing formation of stars elsewhere in our Galaxy. Stars and planetary systems are now being formed in many beautiful nebulae scattered across the heavens. A general understanding of the creation of stars coupled with knowledge of the Sun and its satellites gives us a reasonably comprehensive picture of how the solar system was created. Many details still need to be worked out, but the overall scenario seems remarkably sound. For the first time, we can truly appreciate what is unique and what is commonplace about our world. We have begun to fathom our connection with the rest of the cosmos and our place in the universe.

6-1 The planets are classified as terrestrial or Jovian by their physical attributes

A brief overview of the solar system distinguishes two classes of planets. First, notice the striking dichotomy in the orbits of the planets, as shown in Figure 6-1. The orbits of the four inner planets (Mercury, Venus, Earth, and Mars) are crowded close to the Sun. In contrast, the orbits of the next four (Jupiter, Saturn, Uranus, and Neptune) are widely spaced at greater distances from the Sun.

As you might expect, a planet's surface temperature is related to its distance from the Sun. The four inner planets are quite warm. For example, noontime temperatures on Mercury climb to 600 K (621°F). At noon in the middle of summer on Mars, it is sometimes as warm as 300 K (81°F). Of course, the outer planets, which receive much less solar radiation, are much cooler. Typical temperatures range from about 150 K (−189°F) in Jupiter's cloudtops to 63 K (−346°F) on Neptune. Temperature plays a major role in determining whether substances exist as solids, liquids, or gases, thus profoundly affecting the appearance of the planets.

Most of the planets' orbits are nearly circular. As discussed in Chapter 3, Kepler discovered that these orbits are actually ellipses. Astronomers denote the elongation of an ellipse by its **eccentricity**, which is a measure of how far the foci are from the center of the ellipse. The "flatter" an ellipse is, the greater its eccentricity. The eccentricity of a circle is zero. Most planets have orbital eccentricities that are very close to zero. The exceptions are Mercury and Pluto. In fact, for a portion of its orbit, Pluto is nearer the Sun than its neighbor, Neptune.

The planetary orbits all lie in nearly the same plane. In other words, the orbits of the planets are inclined at only small angles to the plane of the ecliptic. Again, Pluto is a notable exception. The plane of Pluto's orbit is tilted at 17° to the plane of the Earth's orbit (Table 6-1).

As we compare the physical properties of the planets, we again find that they fall naturally into two groups—the four inner planets and the four outer planets—with Pluto once more an exception. Some of the most important properties of the planets are their diameters, masses, and densities.

The four inner planets are quite small. The Earth, with a diameter of 12,756 km (7926 mi) is the largest. In sharp contrast, the four outer planets are much larger. First place goes to Jupiter, whose equatorial diameter is 142,984 km (88,846 mi). Pluto, again the anomaly, is even smaller than the inner planets, despite its position as the outermost planet. Its diameter is only about 2300 km (1430 mi), roughly two-thirds the size of our Moon. Figure 6-2 shows the Sun and the planets drawn to the same scale. The diameters of the planets are given in Table 6-2.

Determining the mass of a planet is quite straightforward if that planet has a satellite. The satellite's orbit obeys Newtonian mechanics, which relate the orbit to the planet's mass. Astronomers can observe the satellite and measure its orbital period and semimajor axis. With this information, they can calculate the planet's mass.

If the planet does not have a satellite, astronomers must rely on observations of a comet or spacecraft as it passes near the planet. The planet's gravity (which is directly related to its mass) deflects the path of the comet or spacecraft. By measuring the size of this deflection and using Newtonian mechanics, astronomers can determine the planet's mass. The four inner planets have low masses, the next four planets have substantially larger masses. Again, Jupiter is first, with a mass almost 318 times greater than the mass of the Earth. The masses of the planets are given in Table 6-2.

Average density (mass divided by volume) is a physical property that can often be used to deduce important information about the composition of an object. Scientists commonly express average density in kilograms per cubic meter. The four inner planets have large average den-

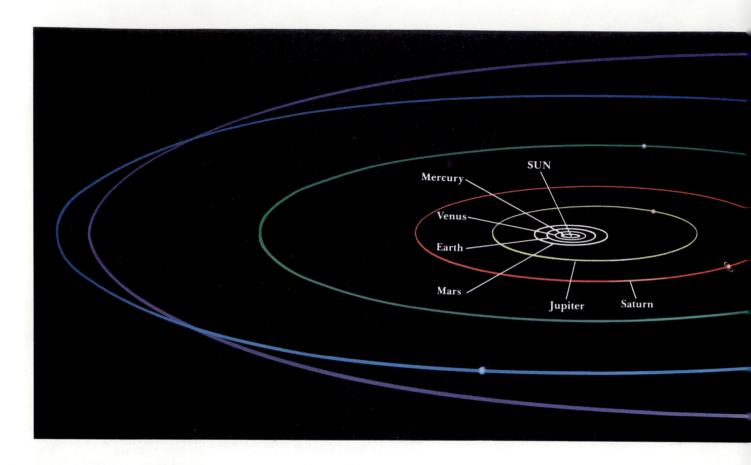

Figure 6-1 The solar system *This scale drawing shows the distribution of planetary orbits around the Sun. Four inner planets are crowded close to the Sun; five outer planets orbit at much greater distances from the Sun.*

Table 6-1 Orbital characteristics of the planets

| | Mean distance from Sun | | Orbital period | | Inclination to ecliptic |
	(AU)	(10⁶ km)	(yrs)	Eccentricity	(degrees)
Mercury	0.39	58	0.24	0.206	7.0
Venus	0.72	108	0.62	0.007	3.4
Earth	1.00	150	1.00	0.017	0.0
Mars	1.52	228	1.88	0.093	1.8
Jupiter	5.20	778	11.86	0.048	1.3
Saturn	9.54	1427	29.46	0.056	2.5
Uranus	19.19	2871	84.01	0.046	0.8
Neptune	30.06	4497	164.79	0.010	1.8
Pluto	39.53	5914	248.54	0.248	17.1

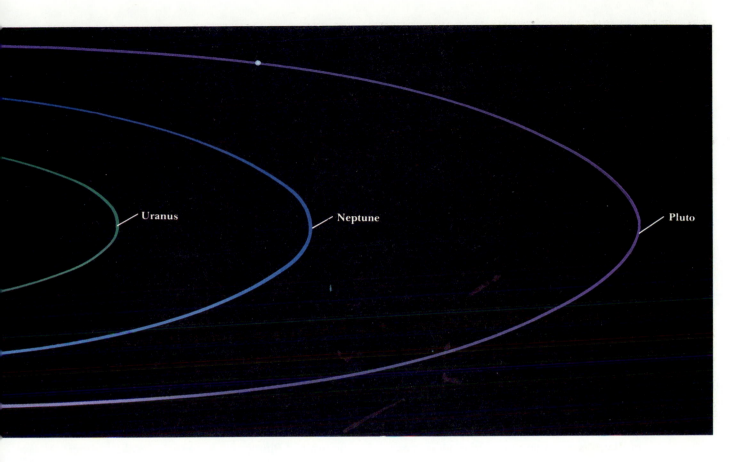

Uranus   Neptune   Pluto

sities (see Table 6-2). The average density of the Earth is about 5500 kg/m³. For comparison, the average density of a typical rock is only about 3000 kg/m³, and the average density of water is 1000 kg/m³. The Earth must therefore contain a large amount of material that is denser than rock. This information provides our first clue that some of the smaller planets, like the Earth, have iron cores.

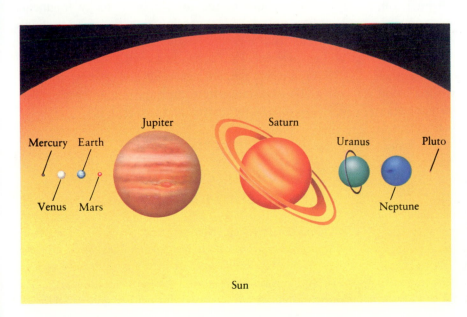

Mercury Earth   Jupiter   Saturn   Uranus   Pluto

Venus   Mars   Neptune

Sun

Figure 6-2   The Sun and the planets   *This drawing shows the nine planets in front of the disk of the Sun, with all ten bodies drawn to the same scale. The four planets that have orbits nearest the Sun (Mercury, Venus, Earth, and Mars) are small and made of rock. The next four planets (Jupiter, Saturn, Uranus, and Neptune) are large and composed primarily of gas.*

Table 6-2 Physical characteristics of the planets

	Diameter		Mass		Average density (kg/m³)
	(km)	(Earth = 1)	(kg)	(Earth = 1)	
Mercury	4,878	0.38	3.3×10^{23}	0.06	5430
Venus	12,102	0.95	4.9×10^{24}	0.81	5250
Earth	12,756	1.00	6.0×10^{24}	1.00	5520
Mars	6,786	0.53	6.4×10^{23}	0.11	3950
Jupiter	142,984	11.21	1.9×10^{27}	317.94	1330
Saturn	120,536	9.45	5.7×10^{26}	95.18	690
Uranus	51,118	4.01	8.7×10^{25}	14.53	1290
Neptune	49,528	3.88	1.0×10^{26}	17.14	1640
Pluto	2,300	0.18	1.3×10^{22}	0.002	2030

In sharp contrast, the outer planets have very low densities. Indeed, Saturn is less dense than water. These low densities suggest that the giant outer planets are composed primarily of such light elements as hydrogen and helium. Again, Pluto is an exception. Although it is even smaller than the dense inner planets, its average density seems to be much like those of the giant outer planets. Table 6-2 lists the average densities of the planets.

These differences in size, mass, and density lead us to consider the four inner and four outer planets as two distinct groups. The four inner planets are called **terrestrial planets** because they resemble the Earth (in Latin, *terra*). They are small and dense, with craters, canyons, and volcanoes being common on their hard, rocky surfaces (Figure 6-3). The four outer planets are called **Jovian planets** because they resemble Jupiter (*Jove* was another name for the Roman god Jupiter). Vast, swirling cloud formations dominate the appearance of these enormous gaseous spheres (Figure 6-4). The solid cores of these planets are probably not much bigger than the Earth, buried beneath atmospheres that are tens of thousands of kilometers thick.

Although it is officially a planet, Pluto clearly is an oddity. Its physical properties are not typical of either the terrestrial or the Jovian planets. Pluto's density is 2030 kg/m³, which suggests that it is probably a mixture of ice and rock. Some astronomers speculate that Pluto is one of perhaps a thousand tiny, frozen worlds at the outskirts of the solar system.

In addition to the nine planets, many smaller objects orbit the Sun. Between the orbits of Mars and Jupiter are thousands of small rocks (typical diameters about 40 km) called **asteroids.** The largest asteroid, Ceres, has a diameter of about 940 km. Quite far from the Sun, well beyond the orbit of Pluto, are chunks of ice called **comets.** Many comets have highly elongated orbits that occasionally bring them close to the Sun. When this happens, the Sun's radiation vaporizes some of the comet's ices, thereby producing a long, flowing tail (Figure 6-5).

Astronomers believe that asteroids and comets are debris left over from the formation of the solar system. In the inner regions of the solar system, rocky fragments have been able to endure continuous exposure to the Sun's heat. Far from the Sun, chunks of ice have survived for billions of years. Thus debris in the solar system naturally divides into two families (asteroids and comets) arranged according to distance from the Sun, just as the two categories of planets (terrestrial and Jovian) are.

6-2 Seven large satellites resemble terrestrial planets

All the planets except Mercury and Venus are orbited by moons, which are properly termed **satellites.** Sixty satellites are known (Jupiter, Saturn, and Uranus each have at least 15), and certainly dozens of others await discovery. The known satellites fall into two distinct categories. There are seven giant satellites, roughly comparable to Mercury in

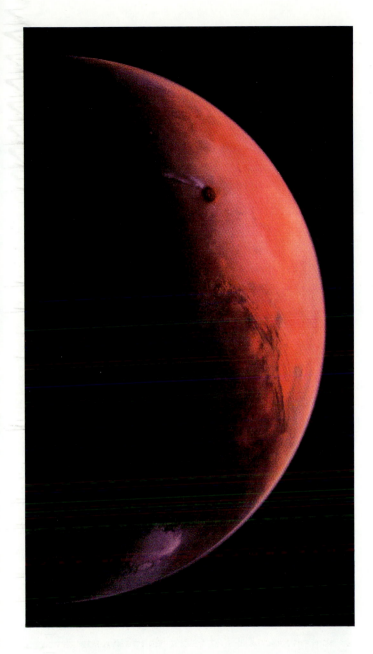

Figure 6-3 (above) A terrestrial planet Mars is a typical terrestrial planet. Its diameter is only 6786 km, and it has an average density of 3950 kg/m³, indicating that it is composed of rock. Volcanoes, canyons, and craters can be seen in this photograph taken by the Viking 2 spacecraft in 1976. (NASA)

Figure 6-4 (top right) A Jovian planet Jupiter is the largest of the Jovian planets. Its equatorial diameter is 142,984 km. Its average density is only 1330 kg/m³, indicating that the planet is composed primarily of light elements. Two of Jupiter's moons, Io and Europa, are seen in this view taken by the Voyager 1 spacecraft in 1979. Each of these satellites is approximately the same size as Earth's moon. (NASA)

Figure 6-5 A comet The solid part of a comet is a chunk of ice roughly 10 km in diameter. When a comet passes near the Sun, solar radiation vaporizes some of the comet's ices, and the resulting gases form a tail millions of kilometers long. This photograph shows a comet that was seen in January 1974. (NASA)

Table 6-3 The seven giant satellites

Satellite name	Parent planet	Diameter (km)	Average density (kg/m³)
Moon	Earth	3476	3340
Io	Jupiter	3630	3570
Europa	Jupiter	3138	2970
Ganymede	Jupiter	5262	1940
Callisto	Jupiter	4800	1860
Titan	Saturn	5150	1880
Triton	Neptune	2700	2070

size, which are listed in Table 6-3. All the other satellites are much smaller, with diameters of less than 2000 km.

The interplanetary missions of *Voyager 1* and *Voyager 2* have revealed many fascinating characteristics of the giant satellites (Figure 6-6). For example, Jupiter's satellite Io is one of the most geologically active worlds in the solar sys-tem, with numerous volcanoes belching sulfur-rich com-pounds. Saturn's largest satellite, Titan, has an atmos-phere nearly twice as dense as Earth's. Figure 6-7 shows all seven giant satellites along with Mercury, all to the same scale.

Some scientists like to classify the seven giant satellites as planets, even though they do not orbit the Sun inde-pendently. Their hard surfaces (of either rock or ice) would place them in the terrestrial category. This ex-panded definition of a terrestrial planet gives us ten Earth-like worlds to compare with our own planet.

6-3 The relative abundance of the elements is the result of cosmic processes

Some elements are very common, others are quite rare. Hydrogen is by far the most abundant substance, making up nearly three-quarters of the mass of all the stars and galaxies in the universe.

Figure 6-6 A giant satellite This view, taken from Voyager 2 *in July 1979, shows Callisto, the second largest moon of Jupiter. Its diameter is almost exactly the same as that of Mercury. Num-erous craters pockmark Callisto's icebound surface. (NASA)*

Figure 6-7 The smaller terrestrial worlds Mercury, our Moon, and the largest satellites of Jupiter, Saturn, and Neptune are shown here to the same scale. Each of these worlds has its own unique characteristics, and all are large enough to be clas-sified as terrestrial planets. Only Titan possesses a substantial atmosphere. (NASA)

Figure 6-8 A mass-loss star This star is shedding material. The nebulosity around the star is caused by starlight reflected from dust grains. These grains may have condensed from the material cast off by the star. (Anglo-Australian Observatory)

Helium is the second most abundant element. Together hydrogen and helium account for 98% of the mass of all the material in the universe—leaving only 2% for all the other elements combined.

There is a good reason for this overwhelming abundance of hydrogen and helium. Most astronomers believe that the universe began roughly 20 billion years ago with a violent event called the Big Bang. Only the lightest elements—hydrogen, helium, and a tiny amount of lithium—emerged from the enormously high temperatures following this cosmic event. All the other elements were manufactured deep inside stars at later times. In Chapters 13 and 14, we shall explore the processes by which elements are created at a star's center. For the moment, let us simply note that if it were not for the nuclear reactions inside stars, there would be no heavy elements in the universe today.

Near the ends of their lives, stars cast much of their matter out into space. This process can be a comparatively gentle one in which a star's outer layers are gradually expelled. Figure 6-8 shows a star that is losing material in this fashion. Alternatively, a star may end its life with a spectacular detonation called a supernova explosion, which blows the star apart. Either way, the interstellar gases in the galaxy become enriched with heavy elements dredged up from the dying star's interior, where they were created. New stars that form out of this enriched material thus have an ample supply of heavy elements from which to develop a system of planets, satellites, asteroids, and comets.

Our solar system is made of matter created in stars that existed billions of years ago. The Sun is a fairly young star, only 5 billion years old. All the heavy elements in our solar system were created and cast off by ancient stars during the first 10 billion years of our Galaxy's existence. We are literally made of star dust (Figure 6-9).

Stars create the different elements in different amounts. For example, the elements carbon, oxygen, silicon, and

Figure 6-9 A dusty region of star formation These young stars in the constellation of Orion are still surrounded by much of the gas and dust from which they formed. The bluish, wispy nebulosity is caused by starlight reflecting off abundant interstellar dust grains. These grains are made of heavy elements produced by earlier generations of stars. (Anglo-Australian Observatory)

Table 6-4 Abundances of the most common elements

Atomic number	Element	Symbol	Relative abundance
1	Hydrogen	H	1×10^{12}
2	Helium	He	6×10^{10}
6	Carbon	C	4×10^{8}
7	Nitrogen	N	9×10^{7}
8	Oxygen	O	7×10^{8}
10	Neon	Ne	7×10^{7}
12	Magnesium	Mg	4×10^{7}
14	Silicon	Si	5×10^{7}
16	Sulfur	S	2×10^{7}
26	Iron	Fe	3×10^{7}

iron are readily produced in a star's interior, whereas gold is created only under special circumstances. Thus, gold is rare in the solar system but carbon is not.

A convenient way to express the abundance of the various elements is to say how many atoms of a particular element are found for every trillion (that is, 10^{12}) hydrogen atoms in space. For example, for every trillion hydrogen atoms there are about 60 billion (6×10^{10}) helium atoms. From chemical analysis of Earth rocks, Moon rocks, and meteorites, scientists have been able to determine the relative abundances of the elements in our part of the Galaxy. The most abundant elements are listed in Table 6-4.

In addition to these ten very common elements, five others are moderately abundant: sodium, aluminum, argon, calcium, and nickel. These elements have abundances between 10^{6} and 10^{7} relative to the standard trillion hydrogen atoms. All other elements are much rarer. For instance, for every trillion hydrogen atoms in the solar system, there are only six atoms of gold.

6-4 The planets were formed by the accumulation of material in the solar nebula during the birth of the Sun

Hydrogen and helium are the most common elements, but they are not plentiful on the inner four terrestrial planets. Warmth from the nearby Sun causes temperatures on the four inner planets to be comparatively high. The higher the temperature of a gas, the greater the speed of its atoms. The lightweight atoms of hydrogen and helium move swiftly enough to escape from these small planets.

A different situation prevails on the four Jovian planets. Far from the Sun, temperatures are low and the atoms of hydrogen and helium move slowly. The relatively strong gravity of the massive Jovian planets easily prevents the lightest gases from escaping into space. In fact, the Jovian planets are composed primarily of hydrogen and helium.

Temperature must also have been an important factor in determining conditions inside the vast cloud of gas and dust, called the **solar nebula,** out of which the solar system formed. It is thus extremely useful to categorize the abundant elements and their common compounds according to their behavior at various temperatures.

First, hydrogen and helium are gaseous except at extremely low temperatures and extraordinarily high pressures. Second, rock-forming compounds of iron and silicon are solids except at temperatures exceeding 1000 K. Finally, there is an intermediate class of common chemicals such as water, carbon dioxide, methane, and ammonia. At low temperatures (typically below 200 to 300 K), these substances become ices when they solidify. At somewhat higher temperatures, they can exist as liquids or gases. In Table 6-5, the common planet-forming substances are listed according to these three broad classifications.

Before the birth of the Sun, atoms in the solar nebula were so widely spaced that no substance could exist as a liquid. Matter in this vast cloud existed either as a gas or as tiny grains of dust and ice. Astronomers find it useful to speak of **condensation temperature** to specify whether a substance is a solid or a gas under these conditions of extremely low pressure. Above its condensation temperature, a substance is a gas; below its condensation temperature, the substance solidifies into tiny specks of dust or snowflakes.

Rock-forming substances have very high condensation temperatures, typically in the range of 1300 to 1600 K. The ices have condensation temperatures in the range of 100 to 300 K. The condensation temperatures of hydro-

Table 6-5 Common planet-forming substances

Gas	Ice	Rock
Hydrogen (H)	Water (H_2O)	Iron (Fe)
Helium (He)	Methane (CH_4)	Iron sulfide (FeS)
Neon (Ne)	Ammonia (NH_3)	Olivine (($Mg,Fe)_2SiO_4$)
	Carbon dioxide (CO_2)	Pyroxene ($CaMgSi_2O_6$)

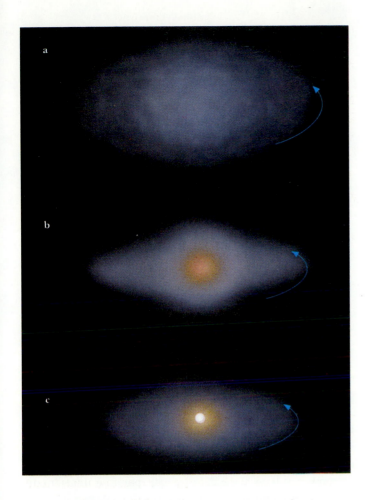

matter called the **protosun.** As the nebula continued to contract, atoms in the cloud collided with one another with increasing speed and frequency, thereby causing temperatures deep inside the solar nebula to climb.

The solar nebula also must have had an overall slight amount of rotation, or **angular momentum** as it is properly called. Otherwise, everything would have fallen straight into the protosun, leaving nothing behind to form the planets. Mathematical studies show that the combined effects of gravity and angular momentum can transform a shapeless cloud into a rotating flattened disk that is warm at the center and cold at its edges, as sketched in Figure 6-10. The transformation of the solar nebula into a flattened disk explains why the orbits of the planets are all nearly in the same plane. Furthermore, astronomers find disks of material surrounding other stars, like the one shown in Figure 6-11. Planets may still be forming in this disk of debris left over from the birth of that star.

Figure 6-10   The birth of the solar system   This sequence of drawings shows three stages in the formation of the solar system. In (a) a slowly rotating cloud of interstellar gas and dust begins to contract because of its own gravity. A central condensation forms in (b) as the cloud flattens and rotates faster. In (c) a flattened disk of gas and dust surrounds the young Sun, which has begun to shine.

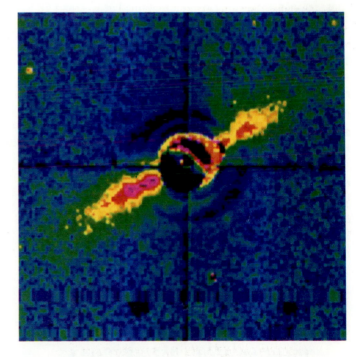

Figure 6-11   A circumstellar disk of matter   This computer-enhanced photograph shows a disk of material orbiting the star called β Pictoris. This star is located behind a small circular mask at the center of the picture that was needed to block the star's light, which otherwise would have overwhelmed light from the disk. The disk, seen nearly edge on, is believed to be very young, possibly no more than a few hundred million years old. (University of Arizona and JPL)

gen and helium are so near absolute zero that they always existed as gases during the creation of the solar system.

Initially, before the Sun was formed, the solar nebula must have been quite cold. Temperatures throughout the cloud were probably less than 50 K, which is below the condensation temperatures of all common substances except hydrogen and helium. Snowflakes and ice-coated dust grains were scattered abundantly across the solar nebula, which had a diameter of at least 100 AU and a mass roughly two or three times the mass of the Sun.

The gravitational pull of the particles on one another caused them to begin a general drift toward the center of the solar nebula. Density and pressure at the center of the solar nebula thus began to increase, producing a concentration of

Figure 6-12 Temperature distribution in the solar nebula This graph shows how the temperature probably varied across the solar nebula as the planets were forming. For the inner planets, temperatures ranged from roughly 1200 K at Mercury to about 500 K at the orbit of Mars. Beyond Jupiter, temperatures were everywhere less than 200 K. (Based on calculations by J. S. Lewis)

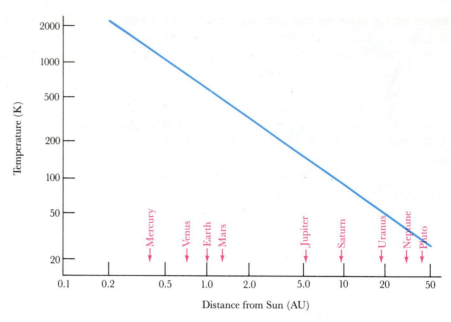

Temperatures around the newly created protosun soon climbed to 2000 K. Meanwhile, temperatures in the outermost regions of the solar nebula remained at less than 50 K. Figure 6-12 shows a probable temperature distribution throughout the solar nebula at this preliminary stage in the formation of the solar system. All the common icy substances in the inner regions of the solar nebula were vaporized by the high temperatures. Only the rocky substances remained solid. That is why the four inner planets are composed primarily of dense, rocky material. In contrast, snowflakes and ice-coated dust grains were able to survive in the cooler, outer portions of the solar nebula. As a result, the Jovian planets are made of low-density substances, including abundant quantities of methane, ammonia, and water. Indeed, many of the satellites of the Jovian planets are either partially or almost entirely composed of ices.

6-5 The terrestrial planets formed out of rocky material, whereas the Jovian planets also incorporated vast amounts of ices and gases

The formation of the four inner planets was dominated by the fusing together of solid, rocky particles. Initially neighboring dust grains and pebbles in the solar nebula collided and stuck together. Then, over a period of a few million years, these accumulations of dust and pebbles coalesced into objects called **planetesimals,** with diameters of about 100 km.

During the next stage, the gravitational attraction between the planetesimals caused them to collide and coalesce into still larger objects called **protoplanets.** This accumulation of material is called **accretion.** In recent years, astronomers have used detailed computer simulations to improve our understanding of accretion. A computer is programmed to simulate a large number of planetesimals circling a hypothetical newborn sun along orbits dictated by Newtonian mechanics. Such studies show that accretion may continue for roughly 100 million years and typically form about half a dozen planets.

A particularly successful computer simulation is summarized in Figure 6-13. The calculations begin with 100 planetesimals, each having a mass of 1.2×10^{23} kg. This procedure ensures that the total mass (1.2×10^{25} kg) equals the mass of the four terrestrial planets (Mercury through Mars) plus Earth's Moon. The initial orbits of these planetesimals are inclined to each other by angles less than 5° to simulate a thin layer of asteroidlike objects orbiting the protosun.

In this simulation, after an elapsed time of 30 million years, the 100 original planetesimals have coalesced into 22 protoplanets. After 79 million years, 11 larger protoplanets remain. Nearly another 100 million years elapse before the total number of growing protoplanets is reduced to 6. Figure 6-13c shows 4 planets following nearly circular orbits after a total elapsed time of 441 million years. This exercise does not exactly reproduce the inner solar system, because the fourth planet from the Sun ends up being the most massive. In fact, Earth is nine times more massive than Mars. Nevertheless, agreement with most characteristics of the inner solar system is very striking.

The material from which the inner protoplanets accreted was rich in mineral-forming elements having high condensation temperatures. Iron, silicon, magnesium, and sulfur were particularly abundant, followed closely by aluminum, calcium, and nickel. Energy released by the violent impact of planetesimals with the growing protoplanets, as well as by the decay of radioactive elements, melted all this rocky material. The terrestrial planets thus began their existence as spheres of molten rock.

Many details on the formation of the planets are topics of current research and debate among scientists. For instance, there are two competing theories to explain how planets like the Earth came to have dense, iron-rich cores surrounded by mantles of less-dense rock. According to one theory, called the **homogeneous accretion theory**, the inner planets grew by the accretion of planetesimals of the same general composition—a mixture of rock and iron. Initially, therefore, the inner planets were quite homogeneous. As the decay of short-lived radioactive elements melted this matter, gravity caused the denser iron-rich minerals to sink to the centers of the planets, while the less-dense silicon-rich minerals floated to their surfaces.

An alternative explanation, called the **heterogeneous accretion theory**, argues that planets formed as material condensed out of the solar nebula. Because iron and iron oxides were among the first substances to condense, the iron-rich cores of the terrestrial planets formed quite early. Then, as the solar nebula continued to cool, silicon-rich minerals condensed and were accreted onto the iron-rich planetary cores.

Neither theory may be entirely correct; an accurate account of the birth of the inner planets may involve ideas from both.

The formation of the outer planets probably also began with the accretion of planetesimals, resulting in a rocky core for each of the Jovian planets that served as a "seed" around which the rest of the planet grew. Calculations suggest that, for about a million years, each of these planetary cores accumulated a coating of additional rock and gas in a leisurely fashion until a critical point when the masses of the core and the envelope were equal. From that moment on, the accumulation of gas increased dramatically, with the envelope pulling in all the gas it could get. This runaway growth of the protoplanet continued until all the available gas was used up. The final result was a huge planet with an enormously thick, hydrogen-rich atmosphere surrounding an Earth-sized core of rocky material. This scenario occurred at four different distances from the Sun, thus creating the four Jovian planets. Figure 6-14 summarizes this story of the formation of the solar system.

During the millions of years while the planets were forming, temperatures and pressures at the center of the contracting protosun continued to climb. Finally, temperatures at the center of the protosun reached eight million kelvin, hot enough to ignite thermonuclear reactions, and the Sun was

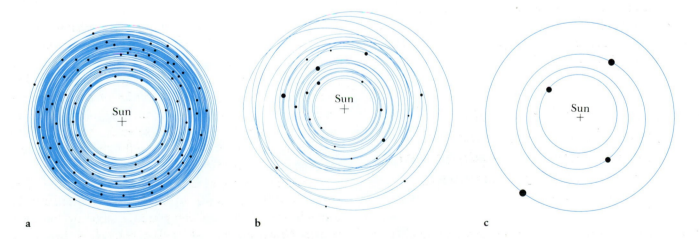

a   b   c

Figure 6-13 *Accretion of the terrestrial planets   These three drawings show the results of a computer simulation of the formation of the inner planets. (a) The simulation begins with 100 planetesimals. (b) After 30 million years, these planetesimals have coalesced into 22 protoplanets. (c) This final view is for an elapsed time of 441 million years, but the formation of the inner planets is essentially complete after only 150 million years. (Adapted from a computer simulation by George W. Wetherill)*

Figure 6-14 Formation of the solar system *This series of sketches shows major stages in the birth of the solar system spanning 100 million years. Terrestrial planets accrete from rocky material in the warm, central regions of the solar nebula. Meanwhile, the huge, gaseous Jovian planets form in the cold outer regions.* (a) *The solar nebula in its initial stages.* (b) *The early solar system after 50 million years.* (c) *Planetary formation (nearly complete) after 100 million years.*

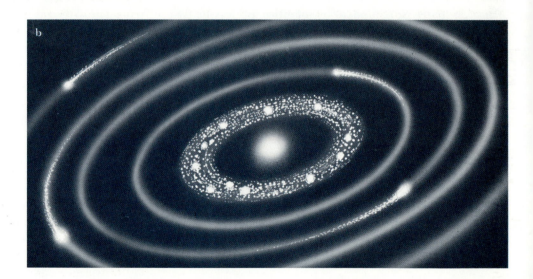

born. Sunlike stars take approximately 100 million years to form from a prestellar nebula, which means that the Sun must have become a full-fledged star at roughly the same time the accretion of the inner protoplanets was complete.

The preceding description presents conclusions drawn from many different kinds of evidence, and some of the details of this story will probably be revised as more evidence becomes available from spacecraft and orbiting observatories. However, most of the new discoveries of the past few decades have confirmed the broad outline of this history. Astronomers are thus fairly confident that the solar system did indeed form in the fashion outlined here.

A newborn star adjusts fairly violently to the onset of thermonuclear reactions at its core, and often the star's tenuous outermost layers are vigorously expelled into space (Figure 6-15). This brief burst of mass loss is observed in many young stars across the sky and is called a **T Tauri wind**, after the star in Taurus (the Bull), where it was first identified. The T Tauri wind that heralded the birth of the Sun swept the solar system clean of excess gases, thereby preventing further accretion. Many small rocks were left behind to pelt the planets over the next half billion years. For all practical purposes, however, the formation of the solar system had been completed by the time the T Tauri wind began to blow.

The Sun today continues to lose matter gradually in a mild fashion. This ongoing, gentle mass loss, called the **solar wind**, consists of high-speed protons and electrons leaking away from the Sun's outer layers. In later chapters, we shall see how each of the planets carves out its own distinctive cavity in this solar wind.

Figure 6-15 Newly formed stars *A full-fledged star is born when thermonuclear reactions ignite at its center. This ignition is often accompanied by an outpouring of particles and radiation from the star's surface that sweeps the surrounding space clean of dust and gas from which the star had formed. This photograph shows dusty material being blown away from newly created stars at the center of the so-called Trifid Nebula in the constellation of Sagittarius. (Anglo-Australian Observatory)*

6-6 Our Moon may have formed from material torn from the Earth during a collision with a protoplanet

Collisions between objects dominated the early history of the solar system. As we have seen, the planets were formed from smaller objects colliding and fusing together to build larger objects. Some of these collisions must have been quite spectacular. Astronomers have recently come to appreciate how important such collisions were in shaping and modifying the solar system.

The Moon's surface bears the scars of numerous impacts dating mostly from the final stages of the formation of the solar system (Figure 6-16). These scars, called **craters**, are also seen on other planets and satellites that do not have appreciable atmospheres or geological activity to erase these features. Indeed, the Moon's airless environment has preserved important information about the early history of the solar system. You can see a closeup view of a heavily cratered region of the lunar surface in Figure 6-17.

Radioactive dating of moon rocks brought back by the Apollo astronauts indicates that the rate of impacts declined dramatically about 3.5 billion years ago. Since that time, impact cratering has proceeded at a very low rate that continues to the present. In other words, most of the craters on the Moon and planets were formed during the first billion years of the solar system's history as the young planets swept up rocky debris left over from the solar nebula.

One of the important goals of the manned lunar landings of the late 1960s and early 1970s was to obtain clues about the Moon's formation. Prior to the Apollo program, there were three main theories about the origin of the Moon. One theory said that the Moon was pulled out from a rapidly rotating protoearth, perhaps while it was still entirely molten. A second theory suggested that the Moon was formed elsewhere in the solar system and then captured into orbit about the Earth by the force of gravity. A third theory proposed that many small rocks, captured

Figure 6-16   Our Moon   This photograph, taken by astronauts in 1972, shows thousands of craters, most of them produced by impacts of rocky debris left over from the formation of the solar system. Age-dating of moon rocks brought back by the astronauts indicate that the Moon is about 4.5 billion years old. Most of the lunar craters were formed when the Moon was less than a billion years old. (NASA)

by the gravity of the young Earth, eventually collided and fused together to form the Moon.

Chemical analyses of moon rocks brought back by the Apollo astronauts did not support any of these three theories. As we shall see in Chapter 9, moon rocks resemble material from deep within the Earth. In addition, moon rocks are bone dry, whereas Earth rocks contain small amounts of volatile substances like water. In other words, moon rocks resemble deep-seated Earth rocks that have been baked at a high temperature, driving off substances that evaporate or vaporize easily. None of the three pre-Apollo theories seemed to account adequately for these important characteristics of moon rocks.

In the mid-1980s, scientists devised a new theory inspired by the fact that violent collisions between sizable protoplanets must have occurred during the final stages of the formation of the solar system. This new hypothesis, called the **collisional ejection theory**, posits that a large asteroidlike object struck the Earth 4.5 billion years ago, shortly after our planet had formed. The impacting object may have been as large as Mars. A supercomputer simulation of this cataclysm is shown in Figure 6-18. Energy released during the collision

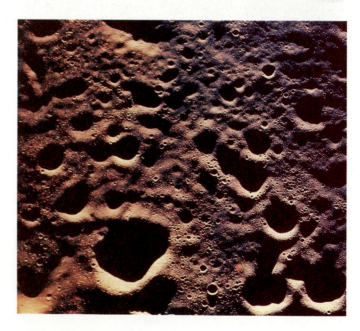

Figure 6-17   Lunar craters   This closeup photograph of the Moon's surface was taken from lunar orbit by astronauts in 1969. The area shown measures about 30 km on a side. The airless environment of the Moon has preserved numerous craters that were formed 3.5 to 4.5 billion years ago. (NASA)

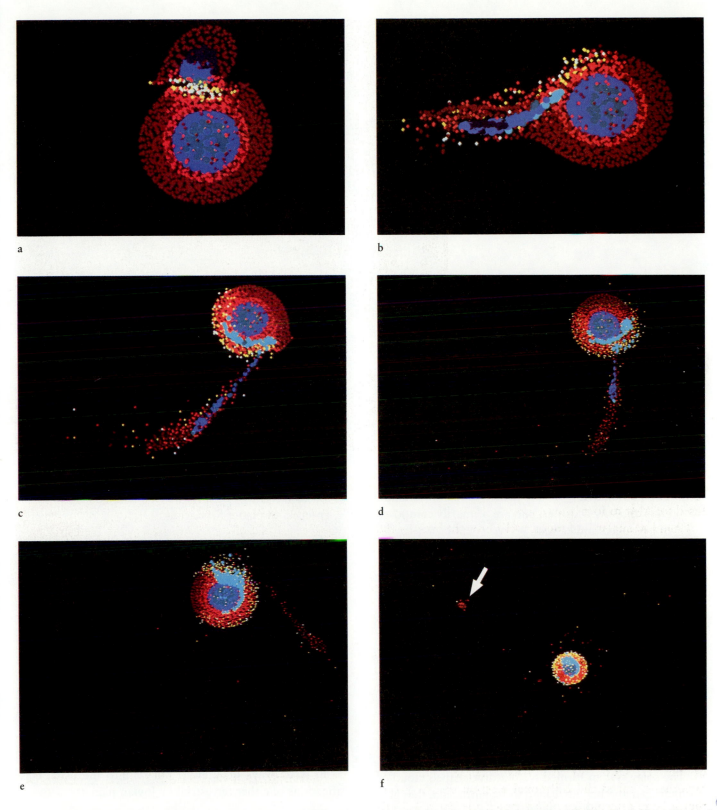

Figure 6-18 The Moon's creation *This simulation, performed on a supercomputer, shows the creation of the Moon from material ejected by the impact of a large asteroid on the Earth. To follow the ejected material as it moves away from the Earth, successive views show increasingly larger volumes of space. A huge plume of vaporized rock is expelled by the impact. Most of this material falls back to the Earth. The surviving ejected rocky matter coalesces to form the Moon (indicated by arrow) during its first orbit of the Earth. (Courtesy of W. Benz)*

produces a huge plume of vaporized rock that squirts out from the point of impact. After the ejected material cools, the Earth is orbited by debris that soon coalesces to form the Moon.

The collisional ejection theory agrees with many of the known facts about the Moon. For example, the vaporization of ejected rock during impact would have driven off volatile elements and water, leaving material that strongly resembles moon rocks. Indeed, the impact of an asteroid large enough to create the Moon could also have tipped the Earth's rotation axis relative to its orbit, just as we observe it today.

Astronomers and geologists have only recently come to appreciate the important role of collisions between objects in shaping the planets. As we explore the solar system in the next four chapters, we shall see that craters cover the surfaces of many worlds. These craters attest to a barrage that dominated the first billion years of the solar system. Of course, many impacts must have also scarred the Earth, but most of these craters have long since been erased by rain, wind, and snow. Nevertheless, we shall see that a large asteroid impact 63 million years ago may have killed off the dinosaurs and thus changed the course of life on the Earth.

Summary

- The four inner planets of the solar system share many characteristics and are distinctly different from the four giant outer planets.

  The four inner (terrestrial) planets are relatively small, have high average densities, and are composed primarily of rock.

  The giant outer (Jovian) planets have large diameters and low densities, and are primarily composed of hydrogen and helium.

- Hydrogen and helium, the two lightest elements, were formed shortly after the creation of the universe. The heavier elements were produced much later in the centers of stars and cast into space when the stars died.

  By mass, 98% of the matter in the universe is hydrogen and helium.

- The basic planet-forming substances can be classified as gas, ice, and rock, depending on their condensation temperatures. The terrestrial planets are composed primarily of rock, whereas the Jovian planets are composed largely of gas.

- The solar system formed from a disk-shaped cloud of hydrogen and helium, called the solar nebula, that also contained ice and dust particles.

  The four inner planets formed through the accretion of dust particles into planetesimals and then into larger protoplanets.

The Jovian planets probably formed through the runaway accretion of gas onto rocky protoplanetary cores.

- The Sun formed at the center of the solar nebula. After about 100 million years, temperatures at the protosun's center were high enough to ignite thermonuclear reactions.

- When the protosun became a star, excess gas and dust were vigorously blown away from the Sun, ending the process of planet formation.

- For nearly a billion years after the Sun was formed, impacts of asteroidlike objects on the young planets dominated the early history of the solar system.

  Our Moon may have been created by the shattering impact of a Mars-sized object on the Earth 4.5 billion years ago.

Review questions

1 What are the characteristics of a terrestrial planet?

2 What are the characteristics of a Jovian planet?

3 In what ways does Pluto not fit into the usual classifications of either terrestrial or Jovian planets?

4 Why is it reasonable to classify certain satellites as terrestrial planets? Which satellites are these?

5 What is meant by the "average density" of a planet? What does the average density of a planet tell us?

6 What are the ten most abundant elements in the universe? List them in order of decreasing abundance.

7 If hydrogen and helium account for 98% of the mass of all the material in the universe, why aren't the Earth and Moon composed primarily of these two gases?

8 Why are water (H_2O), methane (CH_4), and ammonia (NH_3) comparatively abundant substances?

9 Briefly outline the currently accepted theory of the formation of the solar system. How did the creation of the terrestrial planets differ from that of the Jovian planets?

10 Why is it reasonable to suppose that the solar system was formed during a relatively short interval?

11 Why are craters common on the Moon but rare on the Earth?

12 How was the Moon probably formed?

Advanced questions

13 Why do you suppose there are no Jovian planets near the Sun?

14 Would you expect very old stars to possess planetary systems? If so, what types of planets would they have? Explain.

*15 Suppose there were a planet having roughly the same mass as the Earth but located 50 AU from the Sun. What do you think this planet would be made of? On the basis of this speculation, assume a reasonable density for this planet and calculate its diameter. How many times bigger (or smaller) than the Earth would it be? (*Hint:* The volume of a sphere of radius r is given by $\frac{4}{3}\pi r^3$.)

16 Why is it reasonable to suppose that planet-sized objects collided near the end of the planet-forming era 3.5 billion years ago?

Discussion questions

17 Propose an explanation for the fact that Jupiter is orbited by satellites roughly the size of Mercury or our Moon.

18 How does studying other planets and moons help us better understand the Earth?

19 Suppose a planetary system is now forming around some protostar in the sky. In what ways might this system turn out to be similar or different from our own solar system?

20 Speculate about a connection between the Moon's creation and the Earth's seasons.

For further reading

Beatty, J. "The Making of a Better Moon." *Sky & Telescope*, December 1986 • This brief but excellent article summarizes the collisional ejection hypothesis and includes a "report card" that evaluates the various theories of the Moon's origin.

Beatty, J., and Chaikin, A., eds. *The New Solar System.* 3rd ed. Sky Publishing and Cambridge University Press, 1990 • This excellent introduction to the solar system consists of 20 lavishly illustrated chapters, each written by noted scientists.

Briggs, G., and Taylor, F. *The Cambridge Photographic Atlas of the Planets.* Cambridge University Press, 1982 • A clear, non-technical survey of the solar system that includes a superb collection of spacecraft photographs from various missions.

Cameron, A. G. W. "The Origin and Evolution of the Solar System." *Scientific American*, September 1975 • This article describes conditions in the solar nebula and the accretion of the planets.

Chapman, C. *Planets of Rock and Ice.* Scribner's, 1982 • The author eloquently conveys the excitement of planetary exploration while taking the reader on a nontechnical tour of the solar system.

Fisher, D. *The Birth of the Earth.* Columbia University Press, 1987 • This book gives a nontechnical survey of the origins of the universe, the solar system, and the Earth.

Frazier, K. *Solar System.* Time-Life Books, 1985 • This lavishly illustrated book gives an overview of our understanding of the solar system.

Lewis, J. "The Chemistry of the Solar System." *Scientific American*, March 1974 • This well-written survey of the solar system compares the chemical composition of the planets.

Moore, P., and Hunt, G. *Atlas of the Solar System.* Rand McNally, 1983 • This excellent reference includes maps, photographs, and tables summarizing information about the Sun and planets.

Morrison, D., and Owen, T. *The Planetary System.* Addison Wesley, 1988 • This superb textbook covers the full scope of planetary science.

Murray, B., ed. *The Planets*. W. H. Freeman and Company, 1983 • This book is a collection of ten articles about the solar system originally published in *Scientific American* magazine.

Preiss, B., ed. *The Planets*. Bantam, 1985 • This book is an intriguing collection of essays and science fiction stories about the solar system, written by well-known scientists and science fiction authors.

Reeves, H. "The Origin of the Solar System." *Mercury*, March/April 1977 • This excellent article by a noted French astronomer describes many intriguing details of the formation of the solar system.

Smoluchowski, R. *The Solar System*. Scientific American Books, 1983 • This book presents a brief, nicely illustrated survey of the solar system.

Wetherill, G. "The Formation of the Earth from Planetesimals." *Scientific American*, June 1981 • This article describes the accretion of the terrestrial planets from the colliding planetesimals that originally orbited the Sun.

Venus

Earth

Mars

The three largest terrestrial planets   Earth and Venus have nearly the same mass, size, and surface gravity. However, Venus's surface is perpetually shrouded by a dense cloud cover containing poisonous, corrosive gases. Mars is only about half the size and has only one-tenth the mass of Earth. The thin, dry Martian atmosphere does not protect the planet from the Sun's ultraviolet radiation. Both the Venusian and Martian environments are hostile to life forms that thrive on Earth. (NASA; ESA)

7

The Major Terrestrial Planets

In this chapter, we explore the three largest terrestrial planets: Earth, Venus, and Mars. On Earth we find an atmosphere profoundly affected by the presence of life. Mars's atmosphere is very sparse, but on Venus we find oppressive, hot, poisonous gases. We also discover that the Earth's interior is hot and partly molten because of decaying radioactive elements. The transfer of this heat outward from the Earth's interior to its crust profoundly affects our planet's surface features. Earth's crust is divided into huge plates that jostle each other, creating mountain ranges and earthquakes. On Mars such plate movement ceased long ago as this small planet rapidly cooled. Although Venus does have two continents, its crust is too thin and too weak to sustain large crustal plates that could spawn Earthlike surface features. Finally, we see that electric currents in the molten portions of the Earth's interior produce a magnetic field that shields our planet from the solar wind, whereas Venus and Mars have no such protection.

Of all the planets that orbit the Sun, we are most familiar with the Earth. We walk on its surface, we drink its water, and we breathe its air. Naturally, we know more about the Earth than about any other object in the universe. Sometimes, however, this very familiarity makes it difficult for us to view the Earth in its place as a member of the solar system. Explorations of our nearest planetary neighbors, Venus and Mars, have led to enlightening comparisons. What we have learned about both planets gives us a perspective on how various conditions affect the evolution of Earthlike worlds. Venus shows us what the Earth might have been like if it had been nearer the Sun, and Mars shows us what Earth might have been like if it had been less massive.

7-1 Earth's atmosphere is a mixture of nitrogen and oxygen, whereas the atmospheres of Venus and Mars are nearly pure carbon dioxide

Imagine an alien spacecraft approaching the inner solar system. Its occupants might first come upon Mars with its thin atmosphere and barren desert landscapes. Closer to the Sun, they would find Venus with its shroud of corrosive clouds hiding a forbiddingly hot surface. Between these two relatively unpromising planets, they would see Earth with its ever-changing ballet of delicate white cloud formations contrasting against the darker browns and blues of the continents and oceans (Figure 7-1). Perhaps it is only our own bias that leads us to guess that the aliens would focus on the Earth as the most interesting and inviting of the terrestrial planets.

At first, alien visitors might be puzzled to find that Earth's atmosphere is very different from that of its neighbors. The atmospheres of both Venus and Mars consist almost entirely of carbon dioxide, whereas only 0.03% of the Earth's atmosphere is carbon dioxide. Earth's atmosphere is predominantly a 4-to-1 mixture of nitrogen and oxygen, but these two gases are found only in very small amounts on Venus and Mars, as Table 7-1 shows. The data for Venus and Mars in the table come from measurements by Soviet and American spacecraft that landed on these planets in the 1970s.

Why do Earth's planetary neighbors possess atmospheres of nearly pure carbon dioxide when Earth's own atmosphere is composed almost entirely of nitrogen and

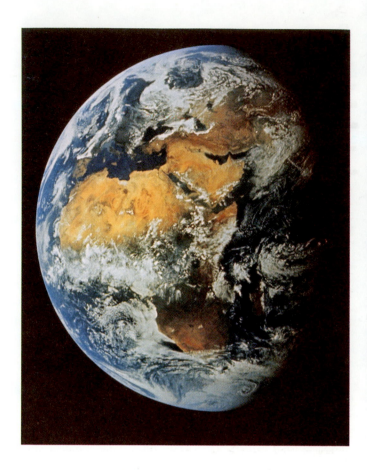

Figure 7-1 The Earth Astronauts often report that Earth is the most invitingly beautiful object visible from their spacecraft. Our blue-and-white world is the largest of the terrestrial planets. This photograph was taken in 1969 by Apollo 11 astronauts on their way to the first manned landing on Moon. Most of Africa and portions of Europe can be seen. (NASA)

oxygen? On Earth a vast amount of carbon dioxide is dissolved in the oceans and chemically bound in rocks like limestone and marble. Furthermore, from study of the Earth's geological records, we know that living organisms have been active here for at least the past three billion years. We also know that such biological processes as photosynthesis are largely responsible for the present chemical composition of Earth's atmosphere. As far as we know, large amounts of oxygen in a planetary atmosphere can arise only as a result of biological activity.

Nearing Earth's surface, aliens would find that the temperature of our atmosphere varies with altitude in a complicated way, as shown in Figure 7-2. A graph like this can be used to deduce fundamental information about a planet's atmosphere because the temperature variations indicate how sunlight is absorbed at different altitudes.

Seventy-five percent of the mass of Earth's atmosphere lies below an altitude of 11 km (roughly 7 mi, or 36,000 ft) in a layer called the **troposphere.** All Earth's weather—clouds, rain, sleet, and snow—occurs in this lowest layer. Commercial jets generally fly at the top of the troposphere to minimize buffeting and jostling.

Above the troposphere is a region called the **stratosphere,** which covers a range in altitude from 11 to 50 km (7 to 30 mi) above the Earth's surface. Ozone (O_3) molecules in the stratosphere efficiently absorb solar ultraviolet rays, thus heating the air in this layer. Certain chemicals that have been released into the air, such as the chlorofluorocarbons found in aerosol cans and air conditioners, are destroying the ozone layer. If it were not absorbed by the ozone in the stratosphere, ultraviolet radiation from the Sun would beat down on the Earth's surface. Because ultraviolet radiation breaks apart most of the delicate molecules that form living tissue, loss of the ozone layer would have a devastating effect on plants and animals.

Above the stratosphere, atmospheric temperature declines with increasing altitude in the **mesosphere,** reaching a minimum temperature at an altitude of about 80 km (50 mi). This minimum marks the bottom of the **thermosphere,** above which the Sun's ultraviolet light strips atoms of one or more electrons. Ionized atoms and molecules of oxygen and nitrogen are the most prevalent elements. The stripped electrons easily reflect radio waves over a wide range of frequencies, which is why you can tune in a distant radio station: Its transmissions bounce off this radio-reflective layer.

At sea level on Earth, the air weighs down with a pressure of 14.7 pounds per square inch. By definition, this pressure is called *one atmosphere* (1 atm). Because we are familiar with this pressure, atmospheric pressure on other planets is usefully expressed in atmospheres also. For example, the atmospheric pressure on the surface of Venus is

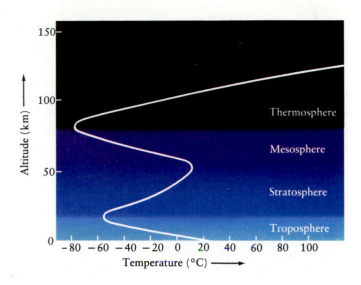

Figure 7-2 *Temperature profile of Earth's atmosphere* The atmospheric temperature varies with altitude because of the way sunlight and infrared radiation emitted by the Earth's surface interact with atoms and ions at different elevations.

90 atm, whereas the sparse Martian atmosphere exerts a pressure of only 0.01 atm. The atmospheric pressure above any planet decreases with increasing altitude, because the higher you go, the less the atmosphere above you weighs.

7-2 The surface of Venus is hidden beneath a very thick, highly reflective cloud cover

At first glance, Venus looks like Earth's twin. The two planets have almost the same mass, the same diameter, the same average density, and the same surface gravity. But Venus is closer to the Sun than the Earth is, and so it is exposed to more intense sunlight, transforming this potentially Earthlike planet into an extremely hostile world.

Since Venus is closer to the Sun than the Earth is, we always find Venus near the Sun in the sky. When visible, Venus can be seen either above the western horizon after sunset (when it is called an "evening star") or before sunrise above the eastern horizon (when it is called a "morning star").

Venus is easy to identify because it is often one of the brightest objects in the night sky. Venus's cloud cover reflects 76% of the sunlight that falls on it. From the Earth, Venus can appear 16 times brighter than the

Table 7-1   The chemical composition of three planetary atmospheres

	Venus	Earth	Mars
Nitrogen	4%	77%	3%
Oxygen	Almost zero	21%	Almost zero
Carbon dioxide	96%	Almost zero	95%
Other gases	Almost zero	2%	2%

brightest star. Indeed, only the Sun and Moon outshine Venus at its greatest brilliancy.

Earth-based telescopic views of Venus reveal a thick, nearly featureless, unbroken layer of clouds. Figure 7-3 shows a closeup view of the clouds as seen from a spacecraft orbiting the planet. This highly reflective cloud cover is responsible for our long-standing ignorance of the planet's surface.

Spacecraft have descended into Venus's clouds and provided detailed information about the Venusian atmosphere. During the 1960s, while the Americans were busy landing astronauts on the Moon, Soviet scientists concentrated on building spacecraft that could survive a descent into the Venusian cloud cover. The task proved to be more frustrating than anyone had expected because every

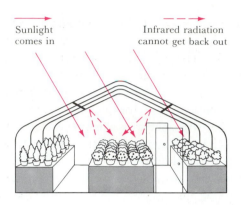

Figure 7-4 The greenhouse effect *Incoming sunlight easily penetrates the windows of a greenhouse and is absorbed by objects inside. These objects reradiate this energy at infrared wavelengths to which the glass windows are opaque. The trapped infrared radiation is absorbed by the air and objects, causing the temperature inside the greenhouse to rise.*

spacecraft sent to Venus would stop transmitting data before reaching the ground. Finally, in 1970, a Soviet spacecraft managed to transmit data for a few seconds directly from the Venusian surface. Subsequent Soviet missions during the early 1970s measured a surface temperature of 750 K (900°F) and a pressure of 90 atm, which is equal to a crushing pressure of $\frac{2}{3}$ ton per square inch.

After some initial puzzlement, astronomers quickly realized there is a straightforward explanation for Venus's high surface temperature. Perhaps you have had the experience of parking your car in the sunshine on a warm summer day. You roll up the windows, lock the car, and go on an errand for a few hours. When you return, you are annoyed to discover that the interior of your automobile has become stiflingly hot, typically at least 20°C warmer than the outside air temperature.

What happened to make your car so warm? First, energy in the form of sunlight entered the car through the windows. The solar radiation was absorbed by the dashboard, the steering wheel, and the upholstery, causing their temperature to rise about 20°C. At this elevated temperature, the seats in your car reradiate the energy primarily at infrared wavelengths. Your car windows are opaque to these wavelengths. This energy is therefore trapped inside your car and absorbed by the air and interior surfaces. As more sunlight coming through the windows becomes trapped, the temperature continues to rise.

A similar phenomenon occurs in Venus's atmosphere and to a lesser extent in Earth's. It is commonly called the

Figure 7-3 Venus *Venus's thick cloud cover efficiently traps heat from the Sun, resulting in a surface temperature (750 K, or 900°F) even hotter than that on Mercury. Unlike Earth's clouds, which are made of water droplets, Venus's clouds are very dry and contain droplets of concentrated sulfuric acid. This photograph was taken in 1979 by an American spacecraft in orbit about Venus. (NASA)*

greenhouse effect (Figure 7-4). Sunlight enters the thick Venusian clouds, where it is absorbed and reradiated at infrared wavelengths that cannot get back out. This trapped radiation produces a high surface temperature.

All the Soviet and American atmospheric probes have carried instruments that measure atmospheric pressure and temperature as they descend to the Venusian surface. The results of all the missions are summarized in Figures 7-5 and 7-6. Graphs like these are important, because they show how pressure and temperature are related to the altitude above a planet's surface—fundamental information from which the detailed structure of a planet's atmosphere can be deduced. The pressure and temperature profiles of the Venusian atmosphere are quite simple. As we saw in Figure 7-2, the relationship between temperature and altitude in the Earth's atmosphere is much more complicated.

Soviet spacecraft discovered that Venus's clouds are confined to a 22-km-thick layer located between 48 and 70 km above the planet's surface. Below that, between the altitudes of 30 and 50 km, is a 20-km-thick haze layer. Beneath the haze layer, the Venusian atmosphere is remarkably clear, all the way down to the ground.

Because Venus and Earth are so similar in size and mass, astronomers had assumed that Venus's clouds were made of water vapor, like Earth's. This assumption was proven wrong in the 1960s, when scientists had great difficulty detecting any water in the Venusian atmosphere. The latest measurements demonstrate that Venus is extremely dry, with water making up far less than 1% of the clouds. If the clouds are dry, what are they made of?

Clues to the complex chemistry of the Venusian clouds came in bits and pieces in the 1960s, when Earth-based observations and measurements from spacecraft demonstrated that the clouds efficiently absorb radiation at specific infrared and microwave wavelengths. Scientists soon proposed a surprising explanation for this absorption: droplets of sulfuric acid.

When sulfuric acid (H_2SO_4) is mixed with water (H_2O), the ions HSO_4^- and H_3O^+ are produced. These ions absorb infrared and microwave radiation at precisely the wavelengths Venus's clouds absorb it. Sulfuric acid is partly responsible for depleting the Venusian atmosphere of water by converting it into H_3O^+. The latest Soviet and American probes leave no doubt that the Venusian clouds are in fact made of droplets of concentrated sulfuric acid.

Venus has a slightly yellowish hue, and data from spacecraft have indicated why: The upper clouds contain substantial amounts of sulfur dust. Over the temperature range of these upper clouds, sulfur is distinctly yellow or yellowish-orange. At lower elevations, large concentrations of sulfur compounds (especially SO_2, OCS, and H_2S) were found along with droplets of sulfuric acid. Because of the tremendous atmospheric pressure, the droplets do

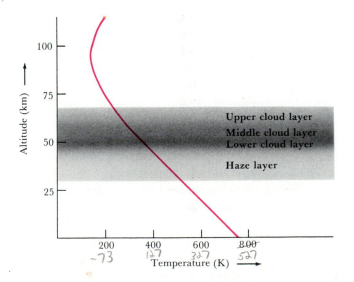

Figure 7-5 Temperature in the Venusian atmosphere *The temperature in Venus's atmosphere rises smoothly from a minimum of about 170 K (about −150°F) at an altitude of 100 km to a maximum of nearly 750 K (about 900°F) on the ground.*

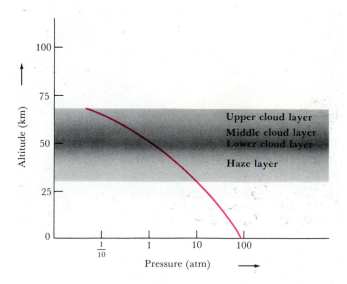

Figure 7-6 Pressure in the Venusian atmosphere *The pressure at the Venusian surface is a crushing 90 atm (1300 pounds per square inch). Above the surface, atmospheric pressure decreases smoothly with increasing altitude.*

not fall as a rain; instead, they are suspended in the clouds like an aerosol.

The sulfuric acid in Venus's atmosphere causes a number of chemical reactions. Reactions with fluorides and chlorides in surface rocks give rise to hydrofluoric acid (HF) and hydrochloric acid (HCl). Further reactions produce fluorosulfuric acid (HSO_3F), one of the most corrosive substances known, capable of dissolving lead, tin, and most rocks. The Venusian clouds are a cauldron of chemical reactions hostile to metals and other solid materials. No wonder Soviet and American spacecraft survive for only a few minutes after landing on Venus.

7-3 The Martian atmosphere is thin and dry

Mars is the only planet whose surface features can be seen through Earth-based telescopes (Figure 7-7). Even with the modest telescopes of the seventeenth century, astronomers made discoveries that hinted that Mars might be very Earthlike. For example, in 1666 the Italian astronomer Giovanni Domenico Cassini spent many nights carefully observing the motions of surface features on Mars and discovered that the planet's rotation period is 24 hours 37 minutes. Thus a day on Mars

is only slightly longer than a day on Earth. A century later, the German-born English astronomer William Herschel determined the inclination of Mars's axis of rotation. Just as Earth's equatorial plane is tilted 23.4° from the plane of its orbit, Mars's equator makes an angle of 25.2° with its orbit. As a result, Mars experiences seasons much as Earth does.

Cassini also discovered that Mars has polar caps that are large during the Martian winter but shrink with the coming of summer. Early telescopic observers also reported seasonal variations in color that seemed to indicate vegetation on the Martian surface, and some reported geometric patterns that might represent a planetwide system of artificial canals. Scientists speculated about the possibility of life on Mars, and science-fiction writers wove popular stories about invasions of Earth by hostile Martians.

Ironically, the first actual invasion came in 1976, when automated spacecraft from the Earth landed on the Martian surface. These probes sent back pictures and data indicating that Mars is a barren, desolate world (Figure 7-8). The possibility of some form of Martian life has not been completely ruled out, but it now seems quite likely that Mars is as sterile and lifeless as the Moon.

The two American spacecraft that journeyed to Mars in 1976, *Viking 1* and *Viking 2*, made numerous measurements of conditions in the Martian atmosphere during their descents. Figure 7-9 shows the recorded atmospheric temperature plotted against altitude above the Martian surface. Also shown, for comparison, is the atmospheric temperature above the Earth's surface. Note that the Martian atmosphere does not have the pronounced swings in temperature that characterize the Earth's atmosphere.

Atmospheric temperature on Earth exhibits a maximum at an altitude of 50 km because our ozone layer absorbs ultraviolet radiation. The absence of a similar maximum in the Martian atmosphere tells us that Mars has no ozone layer. The absence of an ozone layer means that the Sun's ultraviolet radiation strikes the Martian surface directly. The sterilizing effect of ultraviolet light may be an important factor in inhibiting the appearance of life on Mars.

The meteorological instruments on both Viking landers promptly confirmed that the atmospheric pressure on the surface of Mars is about 0.007 atm, as had been expected from previous flyby missions. Because the atmospheric pressure at sea level on Earth is 1 atm, the density of the Martian atmosphere is less than 1% that of the Earth's atmosphere.

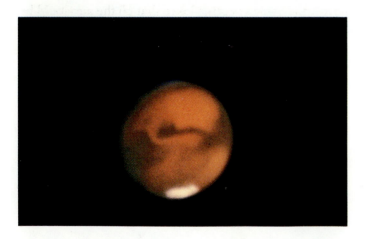

Figure 7-7 Mars viewed from the Earth *This high-quality, Earth-based photograph of Mars was taken in 1971, when the Earth–Mars distance was only 56 million kilometers (35 million miles). At that time Mars presented a disk nearly 25 arc sec in angular diameter. Circumstances as favorable as these will not be repeated for the rest of the century. This photograph accurately portrays what you might typically see through a moderate-sized telescope under excellent observing conditions. Notice the prominent southern polar cap. (Courtesy of S. M. Larson)*

Figure 7-8 Mars   *Numerous flat-bottomed craters are seen in this mosaic of about a hundred images taken by a spacecraft orbiting the planet. North is at the top. The large, heavily cratered circular area in the upper half of this view is known as Arabia. The dark area to the right of Arabia, called Syrtis Major, can be seen from Earth under favorable observing conditions. Carbon dioxide snow covers the floors of craters near the Martian south pole. (NASA, USGS)*

Direct chemical analysis of the Martian atmosphere by the Vikings' instruments reported a carbon dioxide abundance of 95%, with nitrogen at 2.7% and argon at 1.6%. The remaining fraction is mostly oxygen and carbon monoxide, with only a small amount of water vapor. If all the water vapor could somehow be squeezed out of the Martian atmosphere, it would not fill one of the five Great Lakes in North America.

In most of the pictures sent back from Mars, the sky has a distinctly pinkish-orange tint (Figure 7-10). This coloration is thought to be caused by extremely fine-grained dust suspended in the Martian atmosphere. Indeed, Earth-based observers have often reported seeing planetwide dust storms that occasionally obscure the Martian surface features for several weeks at a time. Although the Martian atmosphere is very thin, its winds are sometimes strong enough to raise large amounts of fine dust particles high into the atmosphere.

After the Viking landers had been on the Martian surface for a few weeks, their data clearly showed that the atmospheric pressure at both landing sites was dropping steadily. Mars seemed to be rapidly losing its atmosphere, and some scientists joked that all the air would be gone in a few months. A straightforward explanation was, however, readily available: Winter was coming to the

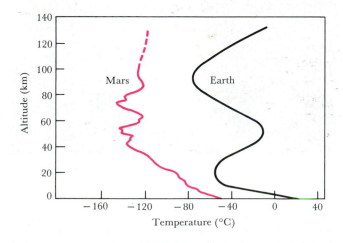

Figure 7-9 Temperature of the Martian atmosphere *Unlike Earth, Mars lacks an ozone layer in its atmosphere. Consequently, the temperature profile of the Martian atmosphere does not show a maximum at 50 km altitude as does Earth's.*

Figure 7-10 A panorama of the Viking 2 landing site *This mosaic of three views shows about 180° of the Viking 2 landing site. Northwest is at the left, southeast on the right. The flat,* *featureless horizon is approximately 3 km (2 mi) away from the spacecraft. (NASA)*

Figure 7-11 The south polar region of Mars *During the Martian winter, the temperature drops so low that carbon dioxide freezes out of the Martian atmosphere. A thin coating of carbon dioxide frost covers a cratered region around Mars's south pole. Seasonal freezing and evaporation of carbon dioxide at both poles cause planetwide variations in atmospheric pressure. (NASA)*

southern hemisphere. At the Martian south pole, it was so cold that large amounts of carbon dioxide were solidifying out of the atmosphere, covering the ground with dry-ice snow (Figure 7-11). The condensation of dry-ice snow was removing gas from the Martian atmosphere, thereby causing the atmospheric pressure to decline.

Several months later, when spring came to the southern hemisphere, the dry-ice snow rapidly evaporated and the atmospheric pressure returned to prewinter levels. With the arrival of winter in the northern hemisphere, atmospheric pressure decreased again as dry-ice snow blanketed the northern latitudes. Thus it is clear that the Martian surface experiences extreme seasonal variations in temperature and atmospheric pressure.

Aside from an occasional dust storm, the weather on Mars would seem boring to someone accustomed to the temperate zones of the Earth. Atmospheric pressure varies with the seasons in a regular, predictable fashion. The atmospheric temperature varies with the time of day, also in a monotonously repetitive way. Actually, the daily temperature variation on Mars is quite similar to the temperature changes observable in a desert on Earth (Figure 7-12). The warmest time of day is about two hours after local noon, and the coldest temperatures are recorded just before sunrise. The thin, dry Martian atmosphere is not capable of retaining much of the day's heat, however, and so the range of the daily temperature swings on Mars is larger than that on Earth.

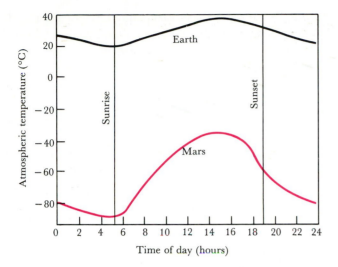

Figure 7-12 Daily temperature variation *The lower curve shows the daily temperature variation at the Viking 1 site. The upper curve shows the daily temperature variation at a desert site in California. The daily temperature range on Mars is about three times greater than that on Earth because the thin, dry Martian air does not retain heat as well as the Earth's atmosphere does.*

7-4 The Earth has oceans and its crust consists of large plates whose motions produce mountain ranges, volcanoes, and earthquakes

Were aliens to land on Earth, they would find it radically different from either of its neighbors in yet another way: The Earth is very wet. Nearly 71% of the Earth's surface is covered with water. An alien space probe to Earth might send back thousands of photographs such as the one shown in Figure 7-13, representing a random closeup view of Earth's surface. In contrast, Venus and Mars are extremely arid. No place on Earth is as dry as the surfaces of Venus and Mars. By Venusian or Martian standards, our Sahara Desert is a veritable swamp.

One of the most important geological discoveries of the twentieth century is the realization that the surface of our planet is active and constantly changing. We have learned that the Earth's crust is divided into huge **plates** that constantly jostle each other, producing earthquakes, volcanoes, and oceanic trenches.

The idea of huge, moving plates might occur to anyone who carefully examines a map of the Earth. South Amer-

ica, for example, would fit snugly against Africa were it not for the Atlantic Ocean. Indeed, the fit between land masses on either side of the Atlantic Ocean is remarkable (Figure 7-14). This observation inspired people such as Alfred Wegener to propose the idea of "continental drift," suggesting that the continents on either side of the Atlantic Ocean have simply drifted apart. After much research, Wegener published in 1915 the theory that there had originally been a single, gigantic supercontinent he called Pangaea, which began to break up and drift apart some 200 million years ago. Pangaea must have first split into two smaller supercontinents that Wegener called Laurasia and Gondwanaland. Gondwanaland later split into Africa and South America, while Laurasia divided to become North America and Eurasia.

Initially most geologists greeted Wegener's ideas with ridicule and scorn. Although it was generally accepted that the continents "float" on denser material beneath them, few geologists could accept the idea that entire continents could move around the Earth at speeds that must be as great as several centimeters per year. The "continental drifters" could not explain what forces could be shoving the massive continents around.

Then, in the mid-1950s, scientists began discovering long mountain ranges on the ocean floors, such as the Mid-Atlantic Ridge (Figure 7-15), which stretches all the way from Iceland to Antarctica. During the 1960s, careful examination of the ocean floor revealed that molten rock

Figure 7-13 A typical closeup view of Earth's surface *More than two-thirds of Earth's surface is covered with water. In contrast, there is no liquid water on Venus or Mars.*

from the Earth's interior is oozing upward along the Mid-Atlantic Ridge, which is therefore a long chain of underwater volcanoes. This upwelling of new material is forcing the ocean floor to spread. The eastern floor of the Atlantic Ocean is moving eastward, the western floor is moving westward. This **seafloor spreading** is in fact pushing South America and Africa apart at a speed of roughly 3 cm per year.

Seafloor spreading provided the mechanism that had been missing from Wegener's theory of continental drift. With some embarrassment, geologists began to review the old theories around 1964 and found a great deal of new evidence to support them. This time, however, the emphasis was upon the motion of large plates of the crust (not simply the continents). The modern theory of crustal motion came to be known as **plate tectonics.** Geologists today realize that seismic activity indicates where plates are either colliding or separating. The boundaries between

Figure 7-14 (above) Comparing the continents Africa, Europe, Greenland, and North and South America fit together as though they were once joined. The fit is especially convincing if the edges of the continental shelves (rather than today's shorelines) are used. (Adapted from P. M. Hurley)

Figure 7-15 (right) The Mid-Atlantic Ridge This artist's rendition shows the floor of the North Atlantic Ocean. The unusual mountain range in the middle of the ocean floor, called the Mid-Atlantic Ridge, is caused by lava seeping up from the Earth's interior along a rift that extends from Iceland to Antarctica. (Courtesy of Marie Tharp and Bruce Heezen)

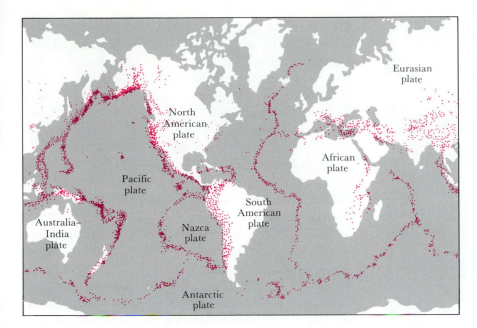

Figure 7-16 *The major plates* *Boundaries of major plates are scenes of violent seismic and geologic activity. Most earthquakes occur where plates separate or collide. Plate boundaries are therefore easily identified by plotting the locations of earthquakes on a map.*

plates are clear when the locations of earthquakes are plotted on a map (Figure 7-16).

Geologists believe that convective currents beneath the Earth's crust are responsible for the movement of the plates. **Convection** is the transport of heat energy that takes place as warm material rises while cooler material sinks. Although rock beneath the Earth's surface is not liquid, it is hot enough to permit an oozing, plastic flow throughout a soft region called the **asthenosphere.** As sketched in Figure 7-17,

hot material (magma) seeps upward along **oceanic rifts** where plates are separating. Cool material sinks back down along **subduction zones** where plates are colliding. Above the plastic asthenosphere is a low-density, rigid layer called the **lithosphere,** divided into plates that simply ride on the convection currents of the asthenosphere.

The boundaries between plates are the sites of some of the most impressive geological activity on our planet. Great mountain ranges, such as those along the western

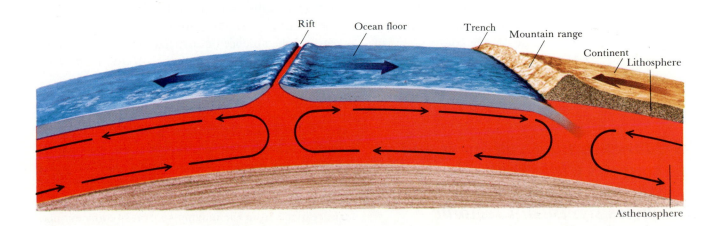

Figure 7-17 *The mechanism of plate tectonics* *Convection currents in the soft upper layer of the Earth's interior (called the asthenosphere) are responsible for pushing around rigid, low-density plates that constitute the Earth's crust (called the litho-* *sphere). New crust is formed in oceanic rifts, where lava oozes upward between separating plates. Mountain ranges and deep oceanic trenches are formed where plates collide.*

Figure 7-18   The separation of two plates   *The plates that carry Egypt and Saudi Arabia are moving apart, leaving the trench that contains the Red Sea. In this south-facing view taken by astronauts in 1966, Saudi Arabia is on the left and Egypt (with the Nile River) on the right. (NASA)*

Figure 7-19   The collision of two plates   *The plates that carry India and China are colliding and the Himalaya Mountains have been thrust upward as a result. In this photograph taken by astronauts in 1968, India is on the left, Tibet on the right, and Mount Everest is one of the snow-covered peaks near the center. (NASA)*

coasts of North and South America, are thrust up by ongoing collisions with the plates of the ocean floor. Subduction zones, where old crust is pushed back down into the Earth, are typically the locations of deep oceanic trenches, such as those off the coasts of Japan and Chile. Figures 7-18 and 7-19 show two well-known geographic features that result from tectonic activity at plate boundaries.

7-5   Mars has huge volcanoes and canyons while Venus has two continents, but neither planet shows significant tectonic activity

The convection process associated with seafloor spreading is the primary way in which heat is transported outward from the Earth's interior to the crust. Here again our planet differs from its neighbors. Venus and Mars transport heat outward to their surfaces predominantly by a mechanism called **hot-spot volcanism.** Hot areas deep inside Venus and Mars squirt molten lava up through the crust. In the absence of any tectonic activity to move the crust around, millions of years of eruptions eventually build enormous volcanoes.

Hot-spot volcanism also occurs on Earth. The Hawaiian Islands, which are in the middle of the Pacific plate, are a fine example. There is a hot spot in the Earth's interior beneath Hawaii that continuously pumps lava up through the crust. However, the Pacific plate is moving northwest at the rate of several centimeters per year. New volcanoes are created over the hot spot as older volcanoes move away from the magma source, become extinct, and eventually erode and disappear beneath the ocean. In fact, the Hawaiian Islands are the most recent additions to a long chain of extinct volcanoes that stretches all the way back toward the Kamchatka Peninsula, northeast of Japan.

Figure 7-20 Olympus Mons This computer-enhanced view looks eastward across the largest volcano on Mars. Olympus Mons is nearly three times as tall as Mount Everest. Its cliff-ringed base is about 600 km (370 mi) in diameter. (NASA)

Martian volcanoes were first photographed by a spacecraft that was placed into orbit about Mars in 1971. The largest Martian volcano, Olympus Mons, rises 24 km (15 mi) above the surrounding plains—nearly three times as high as Mount Everest (Figure 7-20). The highest volcano on Earth is Mauna Loa in the Hawaiian Islands, whose summit is only 8 km above the ocean floor. The huge size of Olympus Mons strongly suggests a lack of plate tectonics on Mars. A hot spot in Mars's interior kept pumping lava upward through the same vent for millions of years, producing one giant volcano rather than a chain of smaller volcanoes.

The base of Olympus Mons is ringed with cliffs and covers an area as big as the state of Missouri. At the volcano's summit are several overlapping volcanic craters, forming a **caldera** large enough to contain the state of Rhode Island (Figure 7-21).

Olympus Mons is one of several very large volcanoes centered just north of the Martian equator and clustered together on a huge bulge, called the Tharsis Rise, that covers an area 2500 km in diameter. Ground levels over this vast dome-shaped region are typically 5 to 6 km higher than the average ground level for the rest of the planet. Another major, though less dramatic grouping of volcanoes is located nearly on the opposite side of the planet. None of these volcanoes seems to be active today.

For reasons we do not yet understand, most of the volcanoes on Mars are in the northern hemisphere, whereas most of the craters are found in the southern hemisphere. Between the two hemispheres is a vast canyon called Valles Marineris, running roughly parallel to the Martian equator (Figure 7-22).

Valles Marineris stretches 4000 km, beginning with heavily fractured terrain in the west and ending with ancient cratered terrain in the east. If this canyon were located on Earth, it could stretch all the way from New York to Los Angeles. Many geologists suspect that Valles Marineris is a fracture in the Martian crust caused by internal stresses, not unlike the rift valleys on Earth that result from plate tectonics. One such valley is the Red Sea (recall Figure 7-18), which is 3000 km long.

Figure 7-21 The Olympus caldera This view of the summit of Olympus Mons is based on a mosaic of six pictures taken by one of the Viking spacecraft. The caldera consists of overlapping, collapsed volcanic craters and measures roughly 70 km across. The volcano itself is wreathed in midmorning clouds that formed from ice (H_2O) brought upslope by cool air currents. The cloudtops are about 8 km below the volcano's peak. (NASA)

Figure 7-22 A segment of Valles Marineris
This mosaic of Viking orbiter photographs shows a segment of Valles Marineris. The canyon is about 100 km wide in this region. The entire canyon system is about 4000 km long and up to 8 km deep. (NASA)

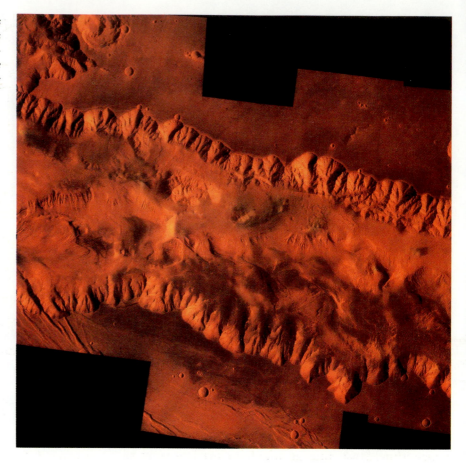

It seems that plate tectonics did not significantly affect the surface of Mars, probably because the planet cooled rapidly during its early history. A small planet loses its internal heat much more rapidly than a large one, and Mars has only half the diameter and one-tenth the mass of Earth. Plate tectonics ceased early in Mars's history as its rapidly cooling lithosphere became thick and sluggish.

Although Venus is perpetually shrouded in clouds, radar has enabled scientists to determine the planet's surface features. During the 1970s and 1980s, astronomers used large radio telescopes to transmit powerful bursts of microwaves toward Venus. This radiation easily penetrates the Venusian clouds, and images of surface features were constructed from an analysis of the reflected signal.

By far the best images of the Venusian surface have come from the highly successful *Magellan* spacecraft that arrived at Venus in 1990. Now in orbit about the planet, *Magellan* is using sophisticated radar techniques in a five-year mission to map the Venusian surface with unprecedented resolution. Figure 7-23 shows a global view of Venus constructed from *Magellan* data.

Venus is remarkably flat. Over 80% of the planet's surface is covered with volcanic plains that are the result of numerous lava flows. Indeed, the Venusian surface is dominated by volcanic features on a scale that is unparalleled on Earth.

Two large highlands, or "continents," rise well above the generally level surface of Venus. In the northern hemisphere is Ishtar Terra, named after the Babylonian goddess of love. Ishtar is approximately the same size as Australia and consists of a high plateau ringed by high mountains. The highest mountain is Maxwell Montes, whose summit rises to an altitude of 11 km. For comparison, Mount Everest on Earth rises 9 km above sea level.

The second major Venusian "continent," Aphrodite Terra (named after the Greek equivalent of Venus), lies just south of the equator. Aphrodite is slightly bigger than Ishtar and has about one-half the area of Africa.

Magellan has given us superb pictures of Venusian craters, which are relatively few (Figure 7-24). All told, Venus probably has only a thousand craters larger than a few kilometers in diameter. That is many more than have been found on Earth, but only a small fraction of the number on the Moon or Mars (recall Figures 6-16 and 7-8). Planetary geologists who study images of the planets' surfaces know that cratering rates were quite high during the early history of the solar system, when considerable interplanetary debris still orbited the Sun. Since that time, impact rates have declined in a fairly well-defined way, so that the age of a planet's surface can be inferred from the number of craters it sports. Old craters are obliterated as vulcanism or tectonics renews a planet's surface. Consequently, the more craters a planet has, the older is its surface. The average age of the Venusian surface seems to be about 400 million years. This is about twice the age of the Earth's surface, but much younger than the surface of the Moon or Mars, each of which is billions of years old.

Roughly a fifth of Venus's surface shows clear signs of tectonic activity, consisting of folded and faulted ridges that resemble one kind of mountain building on Earth. Nevertheless, Venus has neither the large tectonic plates nor the associated lithospheric recycling that dominates the Earth's surface. Deformations of Venus's surface show limited horizontal displacement, which suggests that its lithosphere is thinner and weaker than Earth's. This absence of Earthlike tectonics may be the result of Venus's high surface temperature, which slowed the rate at which Venus's crust could

Figure 7-23 A global view of Venus's surface A computer was used to map numerous Magellan images onto a simulated globe. Color is used to enhance small-scale structures. Extensive lava flows and lava plains cover about 80% of Venus's relatively flat surface. (NASA)

Figure 7-24 A Venusian crater and volcano Computer processing of Magellan images was used to construct this perspective view of the Venusian surface. The crater in the foreground is 48 km (30 mi) in diameter. The volcano near the horizon is 3 km (1.9 mi) high. Note the cracks and folds in the Venusian surface toward the right side of the image. (NASA)

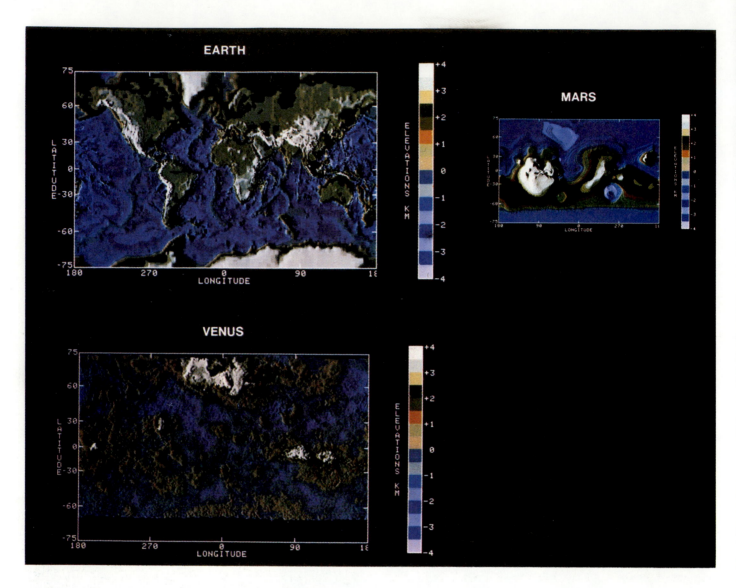

Figure 7-25 *The surfaces of Earth, Venus, and Mars* These maps of Earth, Venus, and Mars are reproduced to the same scale. Each map shows surface elevation (Earth's oceans are empty) on a color-coded scale. Elevations up to 4 km above each planet's average radius (sea level for Earth) appear in shades of tan, green, and white. Elevations down to 4 km below the planet's average radius appear in various shades of blue. Venus's two continents are prominent, as are Earth's seven. The large protrusion on Mars is the Tharsis Rise, a major volcanic region that includes Olympus Mons. (S. P. Meszaros; NASA)

cool, thereby causing its lithosphere to be thin and weak. It is therefore quite likely that Venus offers us a glimpse into the Earth's ancient history—perhaps 3 to 4 billion years ago—when Earth's newly solidified crust was neither thick enough nor strong enough to support the tectonic activity we see today.

A color-coded map of the Venusian surface is shown in Figure 7-25, with comparable maps of Earth and Mars.

Neither Venus nor Mars has long mountain chains like the Mid-Atlantic Ridge, which are indicative of active plate tectonics. Venus's lithosphere is too thin to support significant plate tectonics, while Mars's is too thick. Between these two extremes, Earth's lithosphere today has the proper thickness and structural strength to sustain the tectonic activity that gives our planet its unique surface features.

7-6 Interesting theories try to explain the absence of water on both Venus and Mars

Volcanoes provide clues about the gases that were probably spewed into the original atmospheres of Venus, Earth, and Mars. For example, during the Mount St. Helens eruption of 1980 (Figure 7-26), geologists monitored substantial amounts of sulfuric acid and other sulfur compounds emitted by the volcano. As we have seen, these compounds are common in the Venusian atmosphere.

There is compelling circumstantial evidence that significant volcanic activity is occurring on Venus today. Spacecraft orbiting the planet have detected radio bursts thought to be strokes of lightning, which usually accompany the plumes of erupting volcanoes. Spacecraft have also detected occasional instances of unexpectedly high levels of sulfur dioxide and sulfuric acid in Venus's upper atmosphere. Because these chemicals are highly reactive, they must be constantly replenished, probably by volcanic eruptions. Scientist are hopeful that *Magellan* will provide us with pictures of an erupting volcano before the spacecraft's mission officially ends in 1995.

Water vapor and carbon dioxide are among the most abundant gases in volcanic vapors. The Earth has abundant water in its oceans and atmosphere, but very little carbon dioxide in its atmosphere. In contrast, Venus's atmosphere contains much carbon dioxide but very little water. If the atmospheres of both the Earth and Venus developed mainly from the same outgassing process, what became of Earth's carbon dioxide and what happened to all the water on Venus?

The carbon dioxide on the Earth is dissolved in the oceans and chemically bound into carbonate rocks such as limestone and marble that formed in those oceans. Were the Earth to become as hot as Venus, so much carbon dioxide would be boiled out of the oceans and baked out of the crust that our planet would soon develop a thick, oppressive carbon dioxide atmosphere much like that of Venus.

Some scientists have recently proposed an interesting theory of Venus's early history that accounts for its lack of water. Venus and Earth are so similar in size and mass that it is reasonable to suppose that Venusian volcanoes outgassed an amount of water vapor roughly comparable to the total content of Earth's oceans. Although some of this water on Venus might originally have collected in oceans, heat from the Sun soon vaporized the liquid to create a thick cover of water vapor clouds. Calculations demonstrate that this water vapor would have added 300 atm of pressure to the existing 90 atm of carbon dioxide. Thus the early Venusian atmosphere would have weighed down on the planet's surface with a pressure of three tons per square inch.

This thick, humid atmosphere efficiently trapped heat from the Sun, creating a greenhouse effect far more extreme than the greenhouse effect that operates today on Venus. Calculations demonstrate that the ground temperature would have increased to 1800 K (2700°F), which is hot enough to melt rock. Indeed, the Venusian surface was probably molten down to a depth of 450 km

Figure 7-26 Eruption of Mount St. Helens in May 1980 A preponderance of data suggests that the terrestrial planets obtained their atmospheres from volcanic outgassing. Geologists study this outgassing process in eruptions of volcanoes on Earth. This spectacular recent eruption in the western United States was particularly well studied. (USGS)

a

b

Figure 7-27 A Venusian landscape (a) *This color photograph from Venera 13 shows that rocks on the Venusian surface appear orange because the thick, cloudy atmosphere absorbs the blue component of sunlight.* (b) *Computer processing used to remove the effects of orange illumination reveals the true grayish color of the rocks. In this panorama, the rocky plates covering the ground may be fractured segments of a thin layer of lava, or they may be crusty layers of sediment that have been cemented together by chemical and wind erosion. (Courtesy of C. M. Pieters and the U.S.S.R. Academy of Sciences)*

(280 mi). Soon, however, the Sun's ultraviolet radiation striking water molecules in Venus's atmosphere broke them into hydrogen and oxygen atoms. The light hydrogen atoms escaped into space. Oxygen, which is one of the most chemically active elements, readily combined with other substances in Venus's atmosphere. Thus Venus has been left with almost no water and retains the carbon dioxide atmosphere we find today. Hostile as it may seem, the modern environment on Venus is probably quite mild compared to that of earlier times.

A few photographs of the arid Venusian surface have come from Soviet spacecraft that landed on the planet. A panoramic view taken in 1981 is shown in Figure 7-27. Soviet scientists believe that this region was covered with a thin layer of lava that fractured upon cooling to create the rounded, interlocking shapes seen in the photograph. This hypothesis agrees with information obtained from the spacecraft's instruments, which indicates that the soil composition is similar to lava rocks called basalt that are common on Earth and the Moon.

Basaltic lava rock is also common on the Martian surface (Figure 7-28). Chemical analysis of the Martian soil showed a very high iron content. Iron and silicon comprise about two-thirds of a typical Martian rock's content. Surprisingly high concentrations of sulfur were also found—apparently it is 100 times more abundant in Martian rocks than in Earth rocks.

Each Viking lander had a scoop at the end of a mechanical arm for obtaining rock samples for analysis. Bits of rock were observed to cling to a magnet mounted on the scoop, thus confirming a high iron content in the soil. The familiar reddish color of the Martian surface may simply be due to an abundance of rust (iron oxide).

As soon as measurements by spacecraft confirmed a low atmospheric pressure on Mars, scientists realized that the Martian environment must be extremely dry. Water is liquid only over a certain range of temperature and pressure. If the atmospheric pressure above a body of water is very low, molecules escape from the liquid's surface, causing the water to vaporize. Any liquid water on Mars would furiously boil and rapidly evaporate into the thin Martian air. Thus, there are no oceans on Mars.

Valles Marineris was not formed by water erosion, but Mars-orbiting spacecraft have produced many photo-

Figure 7-28 Digging in the Martian soil *Viking's mechanical arm with its small scoop protrudes from the right side of this view of the Chryse plains. Several small trenches dug by the scoop in the Martian regolith appear near the left side of the picture. (NASA)*

graphs that show dried-up riverbeds and erosional features that look like the results of flash floods in the deserts of Arizona and New Mexico (Figure 7-29). These features were totally unexpected because liquid water cannot exist today on Mars.

If water once flowed on Mars, where might that water be today? Scientists immediately looked to the Martian polar caps as a possible frozen reservoir. The winter temperature is so low ($-140°C$, or $-220°F$) that carbon dioxide freezes. It was not clear, therefore, how much of the polar

Figure 7-29 Ancient river channels on Mars *Many dried riverbeds cut across this cratered terrain, located south of the Martian equator. The existence of riverbeds on Mars suggests that the planet's atmosphere was thicker and its climate more Earth-like long ago, so that liquid water could flow across its surface. Any water exposed to the sparse Martian atmosphere today would rapidly boil away. (NASA, USGS)*

Figure 7-30 The northern polar cap of Mars This mosaic of images shows the northern polar cap of Mars in midsummer. Most of the carbon dioxide frost that covers northern latitudes during the winter has evaporated. The residual polar cap shown here is mostly water ice. The reddish streaks consist of dust, and dark bluish areas are sand dunes forming a huge "collar" that surrounds the polar cap. (NASA, USGS)

caps consists of frozen carbon dioxide (dry ice) and how much is water ice.

In 1972 a spacecraft sent back photographs from orbit about Mars as summer came to Mars's northern hemisphere. During the spring, the polar cap receded rapidly, strongly suggesting that a thin layer of carbon dioxide frost was quickly evaporating in the sunlight. However, with the arrival of summer, the rate of recession abruptly slowed, suggesting that a thicker layer of water ice had been exposed. Scientists concluded that the residual polar cap that survive through the Martian summers contain a large quantity of frozen water (Figure 7-30). Calculating the volume of water ice in the residual cap is difficult, however, because we do not know how thick the layer of ice is.

Another important clue about water on Mars came from the Viking orbiters in 1976. Many close-up views show flash-flood erosion features where the water appears to have emerged from collapsed, jumbled terrain. These photographs confirmed earlier suspicions that frozen water might form a layer of **permafrost** under the

Martian surface, similar to that beneath the tundra in far northern regions on Earth. Apparently heat from volcanic activity occasionally melted this subsurface ice. The ground then collapsed as millions of tons of rock pushed the water to the surface, producing a brief flash flood as the water quickly boiled away.

As mentioned earlier, water vapor and carbon dioxide are two of the most common gases in volcanic vapors. Mars's surface gravity, although only one-third as strong as Earth's, is nevertheless capable of keeping these gases from escaping into space. Although ultraviolet light from the Sun can dissociate water into hydrogen and oxygen, which are light enough to escape into space, detailed calculations show that Mars could have lost only a small fraction of its water in this fashion. It is therefore reasonable to suppose that much of the carbon dioxide and water vapor outgassed during Mars's early history still remains on the planet. Some is in the polar caps; some is in subsurface ice and permafrost. Some water and carbon dioxide may also be chemically bound in the rocks and sand that cover the Martian surface.

7-7 Earth's interior consists of a crust, a mantle, and an iron-rich core

The densities of rocks you find on the ground are typically 3000 kg/m^3, but the average density of the Earth as a whole is 5520 kg/m^3. Thus the rocks of the Earth's crust are not representative of our planet's interior, which must be composed of a substance much denser than the crust.

Iron is a good candidate for this substance because it is the most abundant of the heavier elements (recall Table 6-4), and its presence in the Earth's interior is strongly suggested by the existence of the Earth's magnetic field. Furthermore, iron is common in meteoroids that strike the Earth, suggesting that it was common in the planetesimals from which the Earth formed.

Geologists strongly suspect that the Earth was entirely molten soon after its formation about 4.5 billion years ago. Energy released by the violent impact of numerous meteoroids and asteroids and by the decay of radioactive isotopes melted the solid material collected from the earlier planetesimals. Gravity caused the abundant, dense iron to sink toward the Earth's center, forcing less dense material to the surface. This process, called **chemical differentiation**, produced a layered structure within the

Earth: a central **core** primarily composed of iron, surrounded by a **mantle** of dense, iron-rich minerals, which in turn is surrounded by a thin crust of relatively light silicon-rich minerals.

The Earth's interior is as difficult to examine as the most distant galaxies in space. The deepest wells go down only a few kilometers, barely penetrating the surface of our planet. Geologists have, however, deduced basic properties of the Earth's interior by studying earthquakes.

Earthquakes produce several different kinds of **seismic waves,** which travel around or through the Earth in different ways and at different speeds. Geologists use sensitive **seismographs** to detect and record these vibratory motions. Seismic waves are bent as they travel through the Earth because of the varying density and composition of the Earth's interior. By studying the deflection of these waves, geologists have been able to determine properties of the Earth's interior.

By 1906 analysis of earthquake recordings led to the discovery that the Earth's iron core is molten and has a diameter of about 7000 km (4300 mi). For comparison, the overall diameter of our planet is 12,756 km (7926 mi). More careful measurements in the 1930s revealed that inside the molten core is a solid iron core with a diameter of about 2500 km (1550 mi).

The interior of our planet therefore has a curious structure: a liquid core sandwiched between a solid inner core and a solid mantle. To understand why this is so, we must examine the temperature and pressure inside the Earth and their effects on the melting point of rock.

Both temperature and pressure increase with increasing depth below the Earth's surface. The temperature of the Earth's interior rises steadily from about 20°C on the surface to nearly 5000°C at the center (Figure 7-31).

The Earth's crust is only about 30 km thick. It is composed of rocks whose melting points are far greater than typical temperatures in the crust. Hence the crust is solid.

The Earth's mantle, which extends to a depth of about 2900 km (1800 mi), is largely composed of minerals rich in iron (called *ferrum* in Latin) and magnesium. On the Earth's surface, specimens of these ferromagnesian minerals have melting points slightly over 1000°C. However, the melting point of a substance depends on the pressure to which it is subjected: The higher the pressure, the higher the melting point. As shown in Figure 7-31, the melting point of the mantle's minerals is everywhere higher than the actual temperature, and so the mantle is solid.

At the boundary between the mantle and the outer core, there is a change in chemical composition from ferromagnesian minerals to almost pure iron with a small admixture of nickel. This iron–nickel material has a lower melting point than ferromagnesian minerals, and so the melting-point curve on Figure 7-31 dips below the temperature curve as it crosses from the mantle to the outer core. The melting-point curve remains below the temperature curve down to a depth of about 5100 km. Hence, from depths of about 2900 to 5100 km the core is liquid.

At depths greater than about 5100 km, the pressure is more than three million atmospheres. This pressure is so great that the melting point of the iron–nickel mixture exceeds the actual temperature (see Figure 7-31). As a result, the Earth's inner core is solid.

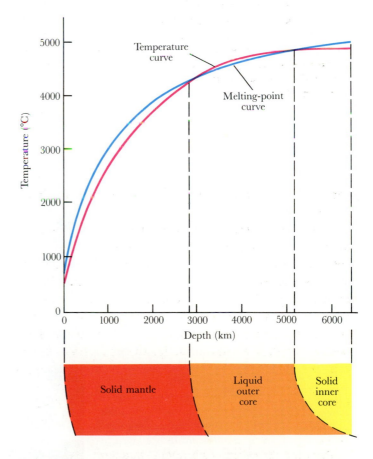

Figure 7-31 The temperature and melting point of rock inside the Earth *The temperature (red curve) rises steadily from the Earth's surface to its center. By plotting the melting point of rock (blue curve) on this graph, we can deduce which portions of the Earth's interior are solid or liquid. (Adapted from F. Press and R. Siever)*

132 Chapter 7

7-8 The Earth's magnetosphere shields us from the solar wind, but neither Venus nor Mars has a magnetic field

As almost anyone with a compass knows, the Earth has a magnetic field. Magnetism arises whenever electrically charged particles are in motion. Many geologists suspect that the Earth's magnetic field is caused by electric currents flowing in the liquid portions of the Earth's iron core. As our planet rotates, these currents produce a magnetic field in the same way that a loop of wire carrying an electric current generates a magnetic field. The Earth rotates fast enough to produce a magnetic field that dominates space for tens of thousands of kilometers and dramatically affects Earth's interaction with the solar wind.

As mentioned briefly at the end of the previous chapter, the solar wind is a relatively constant flow of charged particles (mostly electrons and protons) away from the outer layers of the Sun's upper atmosphere. The speed of these particles is faster than the speed of sound, and so we say that the solar wind is "supersonic." If a planet possesses a magnetic field, this field repels and deflects the impinging particles, thereby forming an elongated "cavity" in the solar wind. This cavity is called the planet's **magnetosphere**.

Figure 7-32 is a scale drawing of the Earth's magnetosphere, which shields us from the supersonic particles that would otherwise strike the upper atmosphere. When this supersonic flow encounters Earth's magnetic field, a bow-shaped **shock wave** is formed where particles of the solar wind are abruptly slowed to subsonic speeds. Most of these particles are deflected around the Earth's magnetic field, just as water is deflected to either side of the bow of a ship.

Close to the Earth, our planet's magnetic field is strong enough to trap charged particles that manage to leak through to the inner regions of our magnetosphere. These particles are trapped in two huge, doughnut-shaped rings called the **Van Allen radiation belts**. These belts were discovered in 1958 during the flight of the first successful Earth-orbiting satellite by the United States. They are named after physicist James Van Allen, who insisted that the satellite carry a Geiger counter to detect charged particles. The inner Van Allen belt contains mostly protons and extends over altitudes of about 2000 to 5000 km. The outer Van Allen belt contains mostly electrons and is about 6000 km thick, centered at an altitude of about 16,000 km above the Earth's surface.

Occasionally a violent event on the Sun's surface called a **solar flare** sends a burst of protons and electrons toward the Earth. Many of these particles penetrate the magnetosphere and overload the Van Allen belts. The excess particles move

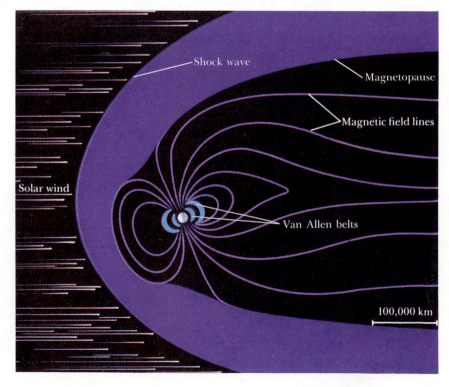

Figure 7-32 Earth's magnetosphere Earth's magnetic field carves out a cavity in the solar wind. A shock wave marks the boundary where the supersonic solar wind is abruptly slowed to subsonic speeds. Most of the particles of the solar wind are deflected around the Earth in a turbulent region colored purple in this drawing. Because of the strength of Earth's magnetic field, our planet is able to trap charged particles in two huge, doughnut-shaped rings called the Van Allen radiation belts.

Figure 7-33 The northern lights (aurora borealis) *Aurorae can be produced by a deluge of protons and electrons from a solar flare striking atoms in the Earth's upper atmosphere. Aurorae typically occur at altitudes between 100 and 140 km (about 60 to 90 mi) above the Earth's surface. (Courtesy of S.-I. Akasofu, Geophysical Institute, University of Alaska)*

along the Earth's magnetic field and rain down on the upper atmosphere near the Earth's north and south magnetic poles. As these particles collide with gases in the upper atmosphere, atoms of oxygen and nitrogen fluoresce like the gases in a fluorescent tube. The result is a beautiful, shimmering display called the **northern lights** (*aurora borealis*) or **southern lights** (*aurora australis*), depending on the hemisphere from which the phenomenon is observed (Figure 7-33).

Numerous American and Soviet space flights have failed to detect any magnetic field around Venus. This absence of a magnetic field might seem surprising since Earth and Venus are so similar in mass, size, and average density. Indeed, it is reasonable to assume that Venus has an interior structure quite similar to Earth's, with a substantial iron core, some of which is probably molten.

It seems that Venus lacks a magnetic field because of the planet's slow rotation. By sending pulses of radar waves toward Venus and analyzing the reflected signals, scientists have determined that Venus rotates backward at a very slow rate. In other words, the Sun rises in the west and sets in the east on Venus. A "day" on Venus (that is, the time from one sunrise to the next) lasts for 116.8 Earth days. Although Venus may have a sizable liquid core, the planet's leisurely rotation rate is apparently too slow to generate a planetwide magnetic field.

With virtually no magnetic field, Venus is incapable of producing a magnetosphere to protect itself from the solar wind. The solar wind therefore impinges directly on Venus's upper atmosphere, where many of the atoms become stripped of one or more electrons. An atom that has been stripped of one or more electrons is called an ion. The interaction between the ions in Venus's upper atmosphere and the supersonic charged particles of the solar wind produces a shock wave, where the supersonic flow abruptly becomes subsonic (Figure 7-34).

Although Mars rotates at about the same rate as Earth, Soviet spacecraft to Mars have detected only an extremely weak magnetic field there. The near absence of a Martian magnetic field may mean Mars lacks a molten iron core. Both Venus and Earth have average densities greater than 5000 kg/m^3, indicating the presence of substantial iron cores. But Mars's average density (3950 kg/m^3) is only slightly greater than that of ordinary crustal rocks. Thus Mars may not possess an Earthlike iron-rich core. Indeed, Mars may not be as chemically differentiated as Venus or Earth. Mars's iron may be more uniformly distributed throughout the planet, as evidenced by the high percentage of iron in rocks on the Martian surface.

Mars's magnetic field is so weak that it is largely ineffectual in warding off the solar wind. In fact, the solar wind may actually impinge directly on the outermost layers of the Martian atmosphere, just as it does on Venus's. Unfortunately, our knowledge of this aspect of the Martian environment is woefully incomplete. Another possibility is that Mars's weak field just manages to carve out a magnetosphere barely enclosing the planet's atmosphere.

Figure 7-34 Venus's interaction with the solar wind Because Venus has no magnetic field, the solar wind strikes the uppermost layers of the planet's atmosphere. Most of the particles of the solar wind are deflected around Venus in a turbulent region colored purple in this scale drawing.

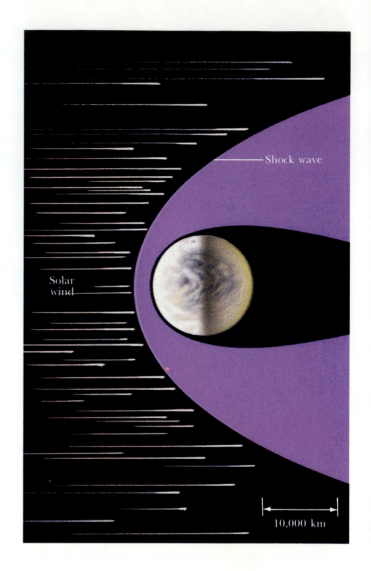

It will be up to future missions to search for a Martian magnetosphere and monitor "Marsquakes" with seismographs. Such missions could continue the search for life on Mars that began in 1976 with the Viking landers. Although the regions examined by the Viking landers seem quite sterile, some scientists point out that the polar regions of Mars may have conditions more suitable for life. These areas might be explored by a rover vehicle that could send back data and pictures from a wide range of sites. We would also learn a lot if a spacecraft were to scoop up Martian rocks and return them to Earth for laboratory analysis. The former Soviet Union was particularly successful with missions of this type to the Moon.

The United States has only one mission to Mars during the rest of the twentieth century: a small satellite called the *Mars Observer* that was launched in late 1992. Scientists in the United States and the former Soviet Union ultimately hope to establish a manned base on Mars early in the twenty-first century, perhaps in a joint venture. A courageous mission of this magnitude would be one of the greatest adventures in all human history.

Summary

- The Earth's atmosphere is primarily nitrogen and oxygen, whereas the atmospheres of Venus and Mars are almost pure carbon dioxide.

- Earth's atmosphere can be divided into layers. All weather occurs in the lowest layer, the troposphere, which extends up to an altitude of 11 km.

- Venus is similar to the Earth in size, mass, average density, and surface gravity, but it is covered by nearly featureless, unbroken clouds.

- The Venusian clouds are confined to a thick layer well above the ground. The clouds consist of droplets of concentrated sulfuric acid, with substantial amounts of yellowish sulfur dust. Active volcanoes on Venus may be a continual source of this material.

- On the surface of Venus, the pressure is 90 atm and the temperature is 750 K. The high temperature is caused by the greenhouse effect: Carbon dioxide in the atmosphere prevents infrared radiation from escaping into space.

- Mars is smaller than either Earth or Venus; a day on Mars is slightly longer than a day on Earth. A day on Venus is nearly 117 Earth days long.

- The atmospheric pressure on Mars is about $\frac{1}{100}$ that of the Earth's atmosphere and shows seasonal variations.

- Dried riverbeds and flash-flood features on the Martian surface indicate that water once flowed on Mars.

- Liquid water would quickly boil away in Mars's thin atmosphere, but the polar caps do contain a considerable amount of frozen water. A layer of permafrost may exist beneath the Martian surface.

- Study of seismic waves (vibrations produced by earthquakes) shows that the Earth has a small solid inner core surrounded by a liquid outer core; the outer core is surrounded by a dense mantle, which in turn is surrounded by a thin, low-density crust.

    The Earth's inner and outer cores are primarily composed of iron; the mantle is composed of iron–magnesium minerals; the crust is largely composed of silicon-rich minerals.

- The Earth's crust and the upper part of its mantle are divided into huge plates. Movements of these plates (the process called plate tectonics) are driven by convective currents in the mantle.

- Plate tectonics is responsible for most of the major features of the Earth's surface, including mountain ranges, volcanoes, and the shapes of the continents and oceans. The surfaces of Venus and Mars show little evidence of the motion of large crustal plates that played a major role in shaping the Earth's surface.

    The surface of Venus is quite flat, mostly covered with lava plains; there are two major continents and some large volcanoes.

    Venus's lithosphere seems to be too thin and too weak to support Earthlike tectonic activity.

- The northern hemisphere of Mars has numerous extinct volcanoes whereas the southern hemisphere has numerous flat-bottomed craters; a huge 4000-km-long canyon stretches along the equator.

- The Earth's magnetic field surrounds our planet with a magnetosphere that shields us from the solar wind; charged particles from the solar wind are trapped in two huge doughnut-shaped rings called the Van Allen radiation belts.

- Venus has no detectable magnetic field or magnetosphere; Mars has only a very weak magnetic field.

Review questions

1 Describe three ways in which the Earth is different from either Venus or Mars.

2 Why is the Earth's surface not riddled with craters like the Moon?

3 Describe the process of plate tectonics. Give specific examples of geographic features created by plate tectonics.

4 Describe the Earth's interior. What causes the Earth's inner core to be solid, whereas its outer core is molten?

5 What gives rise to the Earth's magnetic field?

6 In earlier astronomy books, Venus was often referred to as Earth's twin. What physical properties do the two planets have in common? In what ways are the two planets dissimilar?

7 What techniques have astronomers used to examine and map the surface of Venus? What kinds of surface features have they found?

8 Why is it reasonable to suppose that Venus's interior is similar to Earth's? Why doesn't Venus have a planetwide magnetic field as Earth does?

9 Why is Mars red?

10 Suppose you were in a spacecraft in orbit about Mars. What kinds of surface features would you see? What do these surface features tell you about plate tectonics on Mars?

11 Compare Olympus Mons with the Hawaiian Islands. In what way are they different manifestations of the same physical process?

12 Why does Earth have an abundance of water whereas Venus and Mars are very dry?

13 Describe the Earth's magnetosphere. What are the Van Allen radiation belts?

14 What are the northern lights?

Advanced questions

15 With carbon dioxide just about as abundant in the Martian atmosphere as in the Venusian atmosphere, why do you suppose there is little or no greenhouse effect on Mars?

*16 What fractions of the Earth's total volume are occupied by the core, the mantle, and the crust?

*17 As mentioned in the text, Africa and South America are separating at a rate of about 3 cm per year. Assuming that this rate has been constant, calculate when these two continents must have been in contact.

18 Suppose a planet's atmosphere were opaque to visible light but transparent to infrared radiation. How would this affect the planet's surface temperature? Contrast and compare the effect of the atmosphere of this hypothetical planet with the greenhouse effect caused by Venus's atmosphere.

19 *Magellan*'s exploration of Venus is scheduled to continue through June 1995. Read up on this mission in magazines like *Sky & Telescope* and *Science News*. Does the spacecraft continue to function properly? What new discoveries have been made since this textbook was published? Has *Magellan* observed any erupting volcanoes?

20 How might Venus's cloud cover change if all of Venus's volcanic activity suddenly stopped? How might these changes affect the overall Venusian environment?

21 Is it reasonable to suppose that the polar regions of Mars might harbor life forms? Explain your answer.

Discussion questions

22 The human population on Earth is currently doubling about every 30 years. Describe the various pressures placed on the Earth by uncontrolled human population growth. Can such growth continue indefinitely? If not, what natural and human controls might arise to curb this growth? It has been suggested that overpopulation problems could be solved by colonizing the Moon or Mars. Do you think this is a reasonable solution? Explain your answer.

23 If you were designing a space vehicle to land on Venus, what special features would be necessary? In what ways would this mission and landing craft differ from a spacecraft designed for a similar mission to Mars?

24 Imagine that you are an astronaut living at a base on Mars. Describe what your day might be like, what you would see, what the weather would be like, what your space suit would be like, and so on. Suppose you and your colleagues have a motorized vehicle for exploring the planet. Where would you like to go?

For further reading

Burchfiel, B. C. "The Continental Crust." *Scientific American*, September 1983 • The article describes how the Earth's surface is constantly being reworked by cycles of tectonics, volcanism, erosion, and sedimentation.

Burgess, E. *To the Red Planet*. Columbia University Press, 1978 • This book offers a fine summary of the Viking missions to Mars.

Burnham, R. "Venus: Planet of Fire." *Astronomy*, September 1991 • This well-written summary of results from the *Magellan* mission includes numerous high-resolution images of the Venusian surface.

Cattermole, P., and Moore, P. *The Story of the Earth*. Cambridge University Press, 1985 • This well-illustrated book gives a fine introduction to geology and the chronology of our planet.

Cooper, H. *The Search for Life on Mars*. Holt, Rinehart & Winston 1980 • This book by a noted journalist profiles the Viking missions to Mars.

Gore, R. "Sifting for Life in the Sands of Mars." *National Geographic*, January 1977 • This beautifully illustrated article summarized the finding of the Viking missions.

Hartmann, W. K. "What's New on Mars?" *Sky & Telescope*, May 1989 • This fine article by a noted planetary scientist discusses the current state of our knowledge about Mars.

McKenzie, D. P. "The Earth's Mantle." *Scientific American*, September 1983 • This article discusses the processes that occur in the mantle with particular attention to the convection currents in the upper 700 km of ductile rock.

Prinn, R. C. "The Volcanoes and Clouds of Venus." *Scientific American*, March 1985 • This excellent article discusses evidence for active volcanoes on Venus and describes how they affect the planet's atmosphere.

Saunders, R. S. "Venus: The Hellish Place Next Door." *Astronomy*, March 1990 • This beautifully illustrated article surveys our current understanding of the Venusian environment and compares it with Earth's.

Toon, O. B. "How Climate Evolved on the Terrestrial Planets." *Scientific American*, February 1988 • This superb article compares the evolution of the atmospheres of Venus, Earth, and Mars.

Van Andel, T. *New Views on an Old Planet*. Cambridge University Press, 1985 • This well-written summary of the Earth's evolution includes good discussions of continental drift, long-term climatic changes, and the evolution of life.

Saturn

Uranus

Earth

Neptune

Jupiter

The Jovian planets and Earth   *This montage shows Jupiter, Saturn, Uranus, Neptune, and Earth reproduced to the same scale. Jupiter and Saturn are composed primarily of hydrogen and helium, like the Sun. Note that the cloud features on Saturn are much less distinct than those on Jupiter. Uranus has a bland appearance, but Neptune looks somewhat like Jupiter. These photographs were taken by the Voyager spacecraft that flew past the Jovian planets in the late 1970s and 1980s. (Courtesy of S. P. Meszaros; NASA)*

8

The Jovian Planets

Jupiter is an active, vibrant, multicolored world, more massive than all the other planets combined. In this chapter, we see that some features in Jupiter's turbulent clouds are fleeting, whereas others, like the Great Red Spot, seem to endure year after year. We then examine Saturn with its spectacular system of thin, flat rings. These rings actually consist of thousands of ringlets composed of ice fragments and ice-covered rock. Jupiter and Saturn have similar interior structures and internal sources of heat, but for different reasons. We then turn to Uranus and Neptune. These outer two Jovian planets differ distinctly from the inner two in many ways. Uranus, which has a very bland appearance, is tipped on its side—its axis of rotation lies nearly in its orbital plane. This remarkable orientation suggests that Uranus may have been the victim of a staggering impact by a massive planetesimal. Neptune is a bluish world with markings that give it a Jupiterlike appearance. Neptune even has a huge storm reminiscent of Jupiter's Great Red Spot. Both Uranus and Neptune are orbited by a system of thin, dark rings.

More than any other single factor, temperatures throughout the young solar nebula dictated the final characteristics of the planets that orbit the Sun. In the warm, inner
regions of this ancient nebula, surviving dust grains consisted primarily of metals, silicates, and oxides. The temperature was simply too high for such volatile substances as
water, methane, and ammonia to condense substantially.
The four planets that formed close to the Sun were therefore composed almost entirely of rocky material. Their
surface gravities were too low and their surface temperatures too high to retain any of the abundant but lightweight hydrogen and helium gases that made up most of
the solar nebula.

The four Jovian planets are much farther from the Sun
than the terrestrial planets are (recall Figure 6-1). For example, Saturn is roughly ten times farther from the Sun than
the Earth is. In the past, as now, it was cold at these vast
distances. In the young solar nebula, the dust grains so far
from the protosun were covered with thick, frosty coatings
of frozen water, methane, and ammonia. These volatile substances thus became important constituents of the planets
that accreted in the outer reaches of the solar system.

Many astronomers think that the Jovian planets were
formed in a two-step process. First, accretion of ice-

coated dust grains led fairly quickly to the formation of
four large protoplanets, each several times the mass of the
Earth. Then the gravitational pull of these protoplanets
attracted and retained substantial quantities of hydrogen
and helium that existed in the cool outer reaches of the
solar nebula. Calculations suggest that this gathering of
lightweight gases became very efficient after the Jovian
protoplanets had grown beyond a certain mass. The final
result was the largest planets in the solar system.

8-1 Jupiter is a huge, colorful world composed largely of hydrogen and helium

Jupiter is huge: Its mass is 318 times greater than Earth's.
Indeed, the mass of Jupiter is $2\frac{1}{2}$ times the combined
masses of all the other planets, satellites, asteroids,
meteoroids, and comets in the solar system. Jupiter's
equatorial diameter is 11.2 times larger and its volume
about 1400 times larger than the Earth's.

Jupiter's average density can be computed from its
mass and size: It is only 1330 kg/m³. This low average
density is entirely consistent with the picture of a huge

Figure 8-1 Jupiter from Earth *Various belts and zones are
easily identified in this Earth-based view of the largest planet in
the solar system. The Great Red Spot was exceptionally prominent when this photograph was taken. (Courtesy of S. Larson)*

Figure 8-2 Jupiter from a spacecraft *This view was sent back
from Voyager 1 in 1979. Features as small as 600 km across can
be seen in the turbulent cloudtops of this giant planet. Complex
cloud motions surround the Great Red Spot. (NASA)*

Figure 8-3 Changes in the Great Red Spot These two views, taken $4\frac{1}{2}$ years apart, show major changes in Jupiter's clouds around the Great Red Spot. **(a)** This view was taken in 1974 at a distance of 545,000 km from the cloudtops. **(b)** This view was taken in 1979 from a distance of six million kilometers. Notice the dramatic increase in turbulent cloud activity that occurred between the two spaceflights. (NASA)

a   b

sphere of hydrogen and helium compressed by its own gravity. Observations from spacecraft indicate that by weight Jupiter is composed of approximately 82% hydrogen and 17% helium, with very small amounts of methane, ammonia, water vapor, and other gases.

Through an Earth-based telescope, Jupiter is a colorful, intricately banded sphere, as shown in Figure 8-1. Figure 8-2 is a close-up view from a spacecraft. The most prominent features are alternating dark and light bands shaded in subtle tones of red, orange, brown, yellow, and blue that are parallel to Jupiter's equator. The dark, reddish bands are called **belts,** the light-colored bands are called **zones.** These features are not the only conspicuous markings. A large, red-orange oval called the **Great Red Spot** is often visible in Jupiter's southern hemisphere. This remarkable feature, which has been observed since the mid-1600s, appears to be a long-lived storm in the planet's dynamic atmosphere. Many careful observers have reported smaller spots that last for only a few weeks or months in Jupiter's turbulent clouds.

Although it is the largest and most massive planet in the solar system, Jupiter has the fastest rate of rotation. At its equatorial latitudes, Jupiter completes a full rotation in only 9 hours 50 minutes 30 seconds. However, Jupiter is not a solid, rigid object. By observing features in the belts and zones, we see that the polar regions of the planet rotate a little more slowly than do the equatorial regions. Near the poles, the rotation period is about 9 hours 55 minutes 41 seconds. The first person to notice this **differential rotation** of Jupiter was the Italian astronomer Giovanni Domenico Cassini in 1690. You may recall that Cassini was the same gifted observer who first determined Mars's rotation rate.

Jupiter's colorful cloudtops are the turbulent, uppermost layer of its thick atmosphere. Are the observed rotation rates of these clouds representative of the rotation rates of deeper levels or of a solid central core? Intriguing clues come from radio waves emitted by Jupiter. In particular, radio emissions with wavelengths in the range of 3 to 75 cm vary slightly in intensity, with a period of 9 hours 55 minutes 30 seconds. This radio emission is believed to be directly associated with Jupiter's magnetic field, which is anchored deep inside the planet. In other words, radio observations reveal Jupiter's internal rotation period, which is slightly different from the atmospheric rotation we can observe through a telescope.

8-2 The Great Red Spot is a high-pressure anticyclone

During the 1970s, four American spacecraft flew past Jupiter and sent back spectacular close-up pictures of its dynamic atmosphere. Short-term changes in Jupiter's cloud cover are most apparent in the vicinity of the Great Red Spot. Over the past three centuries, Earth-based observers have also reported many long-term variations in the spot's size and color. At its largest, the Great Red Spot was so huge that three Earths could have fit side by side across it. At other times (as in 1976 and 1977), the spot faded from view. During two flybys in 1979, the Great Red Spot was only slightly larger than the Earth.

Figure 8-3 shows two contrasting views of the Great Red Spot. In 1974 the Great Red Spot was embedded in a broad white zone that dominated the planet's southern hemisphere. A few years later, however, the cloud structure had changed dramatically. A dark belt had broadened and encroached on the Great Red Spot from the north, and the entire region was apparently embroiled in much greater turbulence.

Careful examination of cloud motions in and around the Great Red Spot reveals that the spot rotates counterclockwise with a period of about six days. Furthermore, winds to the north of the spot blow to the west, whereas winds south of the spot move toward the east. The circulation around the Great Red Spot is thus like a wheel spinning between two oppositely moving surfaces (Figure 8-4). This surprisingly stable wind pattern has survived for at least three centuries.

The Great Red Spot is a high-pressure system that protrudes above the surrounding cloudtops. Anyone familiar with weather forecasting knows that the Earth's weather is dramatically affected by high-pressure and low-pressure systems. A **high-pressure system** (commonly called a "high") is simply a place where more than the usual amount of air happens to be located. This excess air weighs down on the Earth's surface, resulting in high atmospheric pressure on the ground. A **low-pressure system** (commonly called a "low") is a place having less than the normal amount of air, resulting in low atmospheric pressure on the ground. If we could see air, highs would look like bulges in the atmosphere (where extra amounts of air are piled up) and lows would look like depressions or troughs.

Gravity strongly influences the basic dynamics of the Earth's weather, causing the air to flow "downhill" from the high-pressure bulges into the low-pressure troughs. But because the Earth is rotating, the wind flow from the highs toward the lows is not along straight lines. Instead, the Earth's rotation deflects the winds, producing either a clockwise or counterclockwise flow about the high- and low-pressure regions.

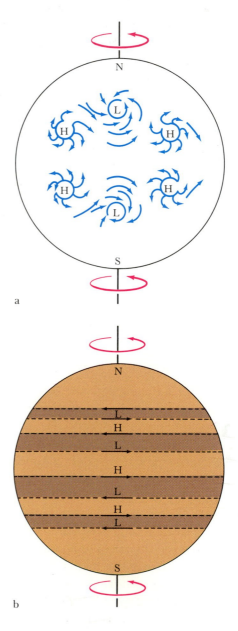

Figure 8-5 Wind-flow patterns of Earth and Jupiter (a) Because of the Earth's rotation, cyclonic wind flowing into a low-pressure region or anticyclonic winds flowing out of a high-pressure region rotate either clockwise or counterclockwise, depending on the hemisphere in which the weather system is located. (b) Jupiter's rapid rotation stretches low- and high-pressure systems all the way around the planet. Light-colored zones are high-pressure systems, and the dark-colored belts are regions of low pressure.

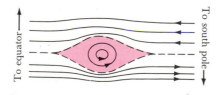

Figure 8-4 Circulation around the Great Red Spot The Great Red Spot spins counterclockwise, completing a full revolution in about six days. Meanwhile, winds to the north and south of the spot blow in opposite directions. Consequently, circulation associated with the spot resembles a wheel spinning between two oppositely moving surfaces. (Adapted from A. P. Ingersoll)

Figure 8-5*a* shows the resulting wind flow about highs and lows. In Earth's northern hemisphere, winds blowing toward a low-pressure region rotate counterclockwise about the low, forming a **cyclone.** Winds blowing away from a high-pressure region rotate clockwise about the high, forming an **anticyclone.** In the southern hemisphere, the directions of rotation are reversed: Cyclonic winds rotate clockwise and anticyclonic winds counterclockwise.

This basic pattern of wind flow applies to the atmosphere of any rotating planet. The six-day rotation of the Great Red Spot is counterclockwise. Thus the Great Red Spot is an anticyclone, a long-lasting high-pressure bulge in Jupiter's southern hemisphere.

Highs and lows also dominate the weather on Jupiter, but the planet's rapid rotation draws them into bands that completely encircle Jupiter, as shown in Figure 8-5*b*. The light-colored zones are the cloudtops of high-pressure systems stretched all the way around the planet. Similarly, dark-colored belts are low-pressure systems encircling the planet. High-speed winds mark the boundaries between the cyclonic flow of the belts and the anticyclonic flow in the zones.

Figure 8-6 A brown oval *Large brown ovals in Jupiter's northern hemisphere are caused by openings in the upper clouds that reveal warm, dark-colored gases below. The length of this oval is roughly equal to the Earth's diameter. The spacecraft* Voyager 1 *was four million kilometers from Jupiter when this picture was taken. (NASA)*

8-3 Computer studies explain some of the phenomena we see in Jupiter's clouds

The flyby pictures showed other anticyclones in Jupiter's southern hemisphere. These features appear as white ovals like those in Figure 8-3*b*. The wind flow in these ovals is clearly counterclockwise.

Most of the white ovals are observed in Jupiter's southern hemisphere. In contrast, brown ovals are more common in Jupiter's northern hemisphere (Figure 8-6). The white ovals are the high-altitude cloudtops of high-pressure systems, but the brown ovals result from holes in Jupiter's cloud cover that permit us to see down into warmer regions of its atmosphere. Like the Great Red Spot, a white oval is apparently long-lived; Earth-based observers have found them in the same locations since 1938. However, a brown oval lasts for only a year or two.

The distribution of ovals is best seen in computer-generated Figure 8-7, which shows how Jupiter would look if you were located directly over the planet's north or south pole. The regular spacing of cloud features such as ripples, plumes, and light-colored wisps is obvious.

Computer processing was also used to "unwrap" Jupiter, producing a map like the views of the planet in Figure 8-8. Note the changes that occurred during the four months between the flybys of *Voyager 1* and *Voyager 2*. The regular spacing of light-colored plumes is apparent in the equatorial regions of both pictures. The Great Red Spot moved westward and the white ovals moved eastward during the interval between the two flybys.

These beautiful Voyager photographs suggest that Jupiter's clouds are the result of incomprehensible turmoil, and you may wonder if there is anything constant in the Jovian atmosphere. Surprisingly, there is. Telescopic observations over the past 80 years and the Voyager data demonstrate that the wind speeds in the Jovian atmosphere are remarkably stable. Although Jupiter's colorful bands change quite rapidly, the underlying wind patterns do not.

Jupiter's persistent wind patterns consist of broad streams of counterflowing eastward and westward winds. Computer simulations involving whirlpools and eddies

a North pole

b South pole

Figure 8-7 The northern and southern hemispheres Computer processing was used to construct these views that look straight down onto Jupiter's north and south poles. (a) In the northern view, light-colored plumes are evenly spaced around the equatorial regions. Several brown ovals are visible. (b) In this southern view, the three biggest white ovals are separated from each other by almost exactly 90° of longitude. In both views, the banded structure of belts and zones is absent near the poles. The ragged black spots are areas not photographed by the spacecraft. (NASA)

a *Voyager 1* view

b *Voyager 2* view

Figure 8-8 A comparison of Voyager 1 and Voyager 2 views Computer processing produced these two "unwrapped" views of Jupiter from (a) Voyager 1 and (b) Voyager 2. Each view was aligned with respect to Jupiter's magnetic axis so that displacements to the right or left represent real cloud motions. Notice that the Great Red Spot moved westward while the white ovals moved eastward during the four months between the two flybys. (NASA)

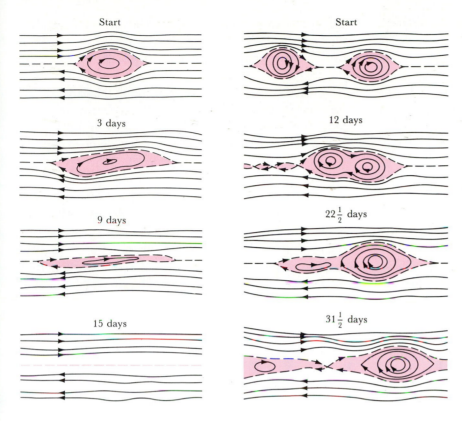

Start   Start

3 days   12 days

9 days $22\frac{1}{2}$ days

15 days $31\frac{1}{2}$ days

Figure 8-9 (left) The demise of an unstable vortex *In this computer simulation, a small vortex is rotating too slowly to remain intact. After slightly more than one week, the vortex is pulled apart between counterflowing zonal jets. (Adapted from A. P. Ingersoll)*

Figure 8-10 (right) The merging and maintenance of stable vortices *This computer simulation shows the collision and merger of two rapidly spinning, stable vortices. The result is a larger vortex and the ejection of material. Such mergers occur near the Great Red Spot. (Adapted from A. P. Ingersoll)*

caught between counterflowing streams help us understand both the long-term and short-term features in Jupiter's clouds. Figure 8-9, for example, shows the behavior of a small, unstable whirlpool, technically called a **vortex.** This whirlpool is spinning too slowly to remain intact and so is torn apart by the counterflowing winds. Larger, rapidly rotating vortices do survive, however, in these simulations. The white ovals and the Great Red Spot endure by simply rolling with the wind currents (recall Figure 8-4). Figure 8-10 shows a simulation in which two stable vortices merge to form a larger one. The long-lived white ovals apparently maintain themselves in this fashion.

8-4 Saturn's spectacular rings are composed of fragments of ice and ice-coated rock

The magnificent rings of Saturn make this planet one of the most spectacular objects in the nighttime sky for an amateur astronomer using a small telescope (Figure 8-11). Saturn is so far away, however, that our Earth-based telescopes can reveal only the coarsest and largest features.

Figure 8-11 Saturn from the Earth *This view is one of the best ever produced of Saturn by an Earth-based observatory. Sixteen original color images taken during the same night in 1974 with a 1.5-m telescope were combined to make this photograph. Note the prominent Cassini division in the rings, and the faint belts and zones in the Saturnian atmosphere. (NASA)*

In 1675 G. D. Cassini discovered a dark division in the rings that looks like a gap about 5000 km wide. This gap, called the **Cassini division**, separates the outer **A ring** from the brighter **B ring** closer to the planet. By the mid-1800s, astronomers using improved telescopes were able to detect a faint **C ring** (or crepe ring) just inside the B ring (Figure 8-12).

Earth-based views of the Saturnian ring system change dramatically as Saturn orbits slowly about the Sun (a Saturnian year is equal to 29.4 Earth years). This change can be observed because the rings, which lie in the plane of Saturn's equator, are tilted 27° from the plane of Saturn's orbit. Thus, over the course of a Saturnian year, the rings are viewed from various angles by an Earth-based observer (Figure 8-13). At one time, the observer looks "down" on the rings; half a Saturnian year later, the "underside" of the rings is exposed to view from Earth. At intermediate times, the rings are seen edge-on and they disappear entirely from our view, which tells us that the rings are very thin—less than 2 km in thickness according to recent estimates. The last edge-on presentation of Saturn's rings was in 1980, and the next will occur in 1995.

Astronomers have known for more than a century that Saturn's rings cannot possibly be solid, rigid, thin sheets of matter. The Scottish physicist James Clerk Maxwell proved mathematically in 1857 that such a broad, thin, rigid sheet would break apart, and he concluded that Saturn's rings are composed of "an indefinite number of unconnected particles." In fact, the rings consist of millions of tiny moonlets, each circling Saturn along its own individual orbit.

Because Saturn's rings are very bright, the particles that form the rings must be highly reflective. Astronomers had long suspected the rings to be made of ice and ice-coated rocks, but confirming evidence was not obtained until the early 1970s, when astronomers identified the spectral features of frozen water in the near-infrared spectrum of the rings. Additional measurements from Earth-based observatories and the Voyager spacecraft in the early 1980s tell us that the temperature of the rings ranges from −180°C (−290°F) in the sunshine to less than −200°C (−330°F) in Saturn's shadow. Water ice is in no danger of melting or evaporating at these temperatures.

The sizes of particles in Saturn's rings were determined from the way in which sunlight and radio waves are scattered by the rings. Specifically, Voyager scientists measured the brightness of Saturn's rings from many angles as the spacecraft flew past the planet. They also measured changes in radio signals received from Voyager spacecraft as they passed behind the rings. Analyses of the data reveal that the largest particles in Saturn's rings are roughly 10 m across. More abundant are snowball-sized particles about 10 cm in diameter. The smallest particles are just a few micrometers wide, smaller than snowflakes.

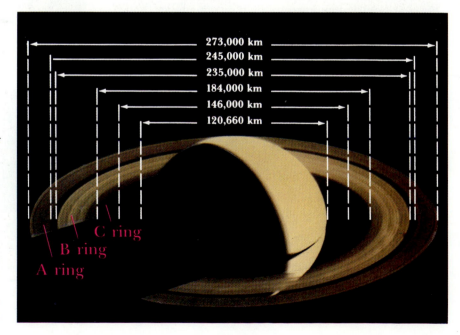

Figure 8-12 Saturn's classic rings Details of Saturn's rings are visible in this photograph sent back by Voyager 1. The equatorial diameter of Saturn as well as the diameters of the inner and outer edges of the rings are given in the overlay. Closest to the planet is the 19,000-km-wide C ring, so faint that it is almost invisible in this view. Next outward from Saturn is the broad, bright B ring, whose width is 25,500 km. The outermost ring that can be seen from Earth is the A ring, which is 14,000 km wide. The 5000-km-wide Cassini division lies between the B ring and the A ring. (NASA)

Figure 8-13 The changing appearance of Saturn's rings *Saturn's rings are tilted 27° from the plane of Saturn's orbit. Earth-based observers thus see the rings at various angles as* *Saturn moves around its orbit. Note that the rings, which are probably less than 2 km thick, seem to disappear entirely when they are viewed edge on. (Lowell Observatory)*

8-5 *Saturn's rings consist of thousands of closely spaced ringlets*

During the Voyager flybys of 1980 and 1981, the spacecraft sent back pictures showing unexpected details in the ring structures of Saturn. The broad rings were seen to consist of hundreds upon hundreds of closely spaced thin bands, or **ringlets,** of particles (Figure 8-14). Although intriguing suggestions have been proposed, scientists still do not understand just why Saturn's A, B, and C rings are divided into these thousands of ringlets.

In addition to the Cassini division, a second gap is visible in Figure 8-14, located in the outer portion of the A ring. This 270-km wide gap is named the **Encke division,** after the German astronomer Johann Franz Encke, who reported seeing it in 1838. Many astronomers have argued that Encke's report was erroneous, however, because his telescope did not have sufficient resolving power to produce an image of this narrow a gap in the rings. The first undoubted observation of this gap was made in the late 1880s with the newly constructed 36-inch refractor at the Lick Observatory in California.

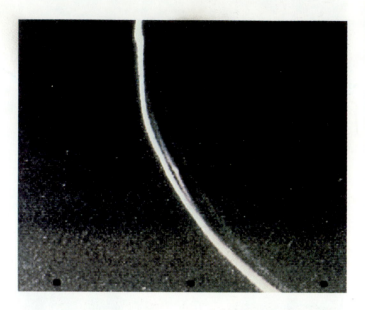

Figure 8-15 Details of the F ring *This* Voyager 1 *photograph of the F ring was taken from a distance of 750,000 km. The total width of the F ring is about 100 km. Within this span are several discontinuous strands, each roughly 10 km across. (NASA)*

The Voyager cameras also sent back the first high-quality pictures of the **F ring,** a thin ring visible just beyond the outer edge of the A ring in Figure 8-14. Close-up views revealed a startling and mysterious fact: The F ring is kinky and braided, actually consisting of several intertwined strands (Figure 8-15). One Voyager image shows a total of five strands, each about 10 km across. Voyager scientists were at a loss to explain this complex structure. The braids, kinks, knots, and twists in the F ring pose one of the most challenging puzzles in modern astronomy.

Through Earth-based telescopes, we see only the sunlit side of Saturn's rings. From this perspective, the B ring appears very bright, the A ring moderately bright, the C ring dim, and the Cassini division dark. The proportion of sunlight reflected back toward the Sun is directly related to the density of the fragments or particles in the ring. The B ring is bright because it has a high density of ice and rock fragments, whereas the darker Cassini division has a lower density of fragments.

Eight hours after it crossed from the northern to the southern side of the rings, *Voyager 1* took the photograph in Figure 8-16. The Sun was shining down on the northern side of the rings at that time, and so Figure 8-16 shows the sunlight that passes *through* the rings. As expected, the B ring looks darkest here because little sunlight gets through its high concentration of fragments, whereas the Cassini division looks bright because sunlight passes relatively freely through its low

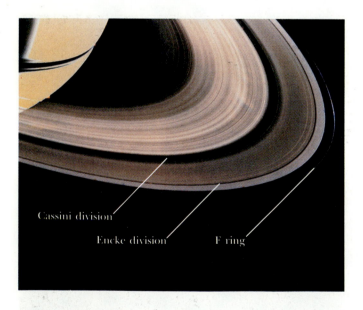

Figure 8-14 Saturn's rings from Voyager 1 Voyager 1 *took this view of Saturn's rings from a distance of about 1.5 million kilometers. Because the C ring scatters light differently from the A and B rings, it has a bluer color. The broad Cassini division is clearly visible, as is the narrow Encke division in the outer A ring. The very thin F ring is visible just beyond the outer edge of the A ring. (NASA)*

Figure 8-16 Details of the A ring *This view of the underside of Saturn's rings was taken by Voyager 1. Both the Encke division and the thin F ring are clearly visible toward the right side of the picture; the Cassini division and the outer edge of the B ring are on the left. The Cassini division appears bright in this view of the shaded side of the rings. (NASA)*

Figure 8-17 False-color view of ring details *Computer processing has severely exaggerated the subtle color variations in this view of the sunlit side of the rings from Voyager 2. Note that the C ring and the Cassini division appear bluish. Also note distinct color variations across both the A and B rings. (NASA)*

concentration of fragments. However, the fact that the Cassini division does appear bright is clear evidence that it does contain some fragments. If it contained no fragments at all, we would see the black of space through it.

Subtle color differences from one ring to the next give important clues about the chemical composition of the particles in the rings. These differences are clearly visible in Figure 8-17, in which the colors have been exaggerated by computer processing. The main chemical constituent is frozen water, but trace amounts of other chemicals (perhaps coating the surfaces of the ice particles) are probably responsible for the colors seen in this computer-enhanced view. These trace chemicals have not been identified. Nevertheless, the existence of color variations and their probable persistence over millions of years suggests that the icy particles do not migrate substantially from one ringlet to another.

8-6 Saturn's innermost satellites affect the appearance and structure of the rings

Astronomers have long known that one of Saturn's moons, Mimas, has an effect on the ring system. Mimas is a moderate-sized satellite that orbits Saturn every 22.6 hours (Figure 8-18). According to Kepler's third law,

Figure 8-18 Mimas *Mimas is the smallest and innermost of Saturn's six moderate-sized satellites. This view was taken by Voyager 1 at a range of nearly 500,000 km. The huge impact crater, named Herschel after the satellite's discoverer, is 130 km in diameter. Mimas itself is only 400 km in diameter. (NASA)*

particles in the Cassini division should orbit Saturn every 11.3 hours. Consequently, on every second orbit, particles in the Cassini division line up between Saturn and Mimas. During these repeated alignments, the combined gravitational forces of Saturn and Mimas cause small fragments to deviate from their original orbits. In this way, Mimas depletes the Cassini division of particles that would otherwise scatter sunlight back toward Earth. Earth-based astronomers therefore see the Cassini division as a dark band between neighboring rings.

The Voyager cameras also discovered three new ring systems: the D, E, and G rings. The **D ring** is Saturn's innermost ring system. It consists of a series of extremely faint ringlets located between the inner edge of the C ring and the Saturnian cloudtops. The **E ring** and the **G ring** both lie quite far from the planet, well beyond the outer edge of the A ring. Both of these outer ring systems are extremely faint, fuzzy, and tenuous. Each lacks the ringlet structure so prominent in the main ring systems. The E ring lies along the orbit of

Figure 8-20   The F ring and its two shepherds   Two tiny satellites (Prometheus and Pandora), each measuring about 50 km across, orbit Saturn on either side of the F ring. The gravitational effects of these two shepherd satellites focus and confine the particles in the F ring to a band about 100 km wide. This Voyager 2 picture was taken from a range of 10.5 million kilometers. (NASA)

Figure 8-19   Enceladus   This high-resolution image of Enceladus was obtained by Voyager 2 from a distance of 191,000 km. Ice flows and cracks strongly suggest that the surface has been subjected to recent geological activity. The youngest crater-free ice flows are estimated to be less than 100,000 years old. (NASA)

Enceladus, one of Saturn's icy satellites (Figure 8-19). Some scientists suspect that geysers on Enceladus are the source of ice particles in the E ring.

The Voyager cameras also discovered two tiny satellites that follow orbits on either side of the F ring (Figure 8-20). The gravitational forces of these two satellites keep the F ring particles in place. The outer satellite moves around Saturn at a slightly slower speed than that of the ice particles in the ring. As the ring particles pass by this satellite, they experience a tiny gravitational tug that tends to slow them down. These particles thus lose a little energy, which would cause them to fall into orbits a bit closer to Saturn. Meanwhile, the inner satellite is orbiting the planet at a somewhat faster rate than the F ring particles. As the satellite moves past the particles, its gravitational pull tends to speed them up, trying to nudge them into a slightly higher orbit. The combined effect of these two satellites focuses the icy particles into a well-defined narrow band about 100 km wide. Because of their confining influence, these two moons are called **shepherd satellites.**

A shepherd satellite that circles Saturn just beyond the outer edge of the A ring is responsible for the sharp outer edge of the A ring. As particles near the edge of the A ring pass by the slowly moving shepherd satellite, they feel a gravitational drag that slows them down slightly, preventing them from wandering into orbits farther from Saturn.

8-7 The atmospheres of Jupiter and Saturn both have three main cloud layers

It is warm deep inside the atmospheres of Jupiter and Saturn but cooler at their cloudtops. Infrared measurements of Jupiter confirm that the temperature rises with increasing depth below the Jovian cloud cover. This situation is analogous to the Earth's atmosphere, which is cool at the highest cloudtops but warmer near the ground.

Over the range of temperatures in the Jovian atmosphere, gases emit energy primarily as infrared radiation. Figure 8-21 shows nearly simultaneous photographs of Jupiter at infrared and visible wavelengths. In the infrared picture, the brighter parts of the image correspond to hotter temperatures. There is also a striking correlation between brightness in the infrared image and color in the visible-light image. In other words, the various colors in Jupiter's clouds correspond to differing temperatures and hence to differing depths in the atmosphere. (It is customary to discuss these features in terms of depths measured from the cloudtops rather than altitudes above the surface because the exact location of any solid or liquid surface on Jupiter is unknown.) Because brown clouds in visible-light images correspond to the brightest parts of the infrared picture, these clouds must be the warmest—hence the deepest—layers that we can see in the Jovian atmosphere. Whitish clouds form the next layer up, followed by red clouds in the highest layer.

The Jovian atmosphere has a minimum temperature of about −160°C (−260°F) at an altitude above the cloudtops where the atmospheric pressure is about 0.1 atm (Figure 8-22). A similar minimum occurs at the 0.1-atm level in Saturn's atmosphere, where the temperature is −180°C (−290°F). By analogy with the Earth's atmosphere (recall Figure 7-2), we can call this level the boundary between the stratosphere and the troposphere. As on the Earth, all the weather on Jupiter and Saturn takes place below the stratosphere.

Spectroscopic observations reveal that the atmospheres of Jupiter and Saturn contain methane (CH_4), ammonia (NH_3), and water vapor (H_2O). These compounds are the simplest combinations of carbon, nitrogen, and oxygen with hydrogen. From calculations of the behavior of these

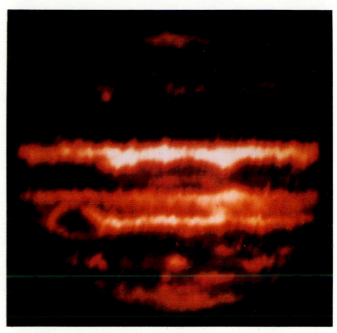

a

b

Figure 8-21 *Infrared and visible views of Jupiter* (a) *This infrared photograph was taken through the 200-inch Palomar telescope. The brightest parts of the image correspond to holes in the clouds where deeper and warmer regions of the Jovian atmosphere are visible. Dark parts of the image correspond to the cool cloudtops.* (b) *This image, in visible light, was taken by* Voyager 1 *at almost the same time as the infrared picture. Comparisons between the two photographs show that cloud color is correlated with depth in the Jovian atmosphere. The brown clouds are roughly 100°C warmer and 100 km lower than the red and whitish clouds.* (NASA)

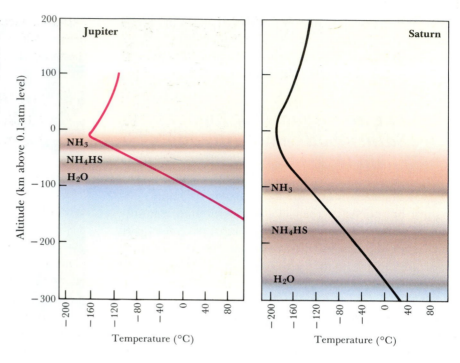

Figure 8-22   Temperature profiles of Jupiter and Saturn   *The structure of the upper atmospheres of Jupiter and Saturn is displayed in these graphs of temperature versus depth. Note that Saturn's atmosphere is more "spread out" than Jupiter's, which is a direct result of Saturn's weaker surface gravity. (Adapted from A. P. Ingersoll)*

Figure 8-23   Saturn's clouds from Voyager 1   *This view of Saturn's cloudtops was taken by* Voyager 1 *at a range of 1.8 million kilometers. Note that there is substantially less contrast between belts and zones on Saturn than there is on Jupiter. (NASA)*

chemicals under various conditions of temperature and pressure, scientists conclude that Jupiter and Saturn both have three main cloud layers. The uppermost layer, with red and whitish clouds, is composed of crystals of frozen ammonia. Deeper in the troposphere, ammonia (NH_3) and hydrogen sulfide (H_2S) combine to produce a middle cloud layer of ammonium hydrosulfide (NH_4SH) crystals. The third and deepest cloud layer has a bluish tint and is composed of water droplets and snowflakes of frozen water.

Although their atmospheres are similar in structure and composition, Saturn and Jupiter are not identical in appearance. Saturn's clouds lack the colorful contrast of Jupiter's. Nevertheless, some photographs do show faint stripes in Saturn's atmosphere similar to Jupiter's belts and zones (Figure 8-23). After computer processing to exaggerate Saturn's colors, details such as storm systems and ovals became visible (Figure 8-24).

The different appearances of Jupiter and Saturn are related to the different masses of the two planets. Jupiter's strong surface gravity compresses its atmosphere, thereby confining the three cloud layers to an altitude range of only 75 km. Saturn's somewhat weaker surface gravity subjects its atmosphere to less compression, so the same three cloud layers are spread out over a range of nearly 300 km (see Figure 8-22). The colors of Saturn's clouds are less dramatic than Jupiter's because the deeper layers are partly obscured by the thick atmosphere above them.

Figure 8-24 *Eddy currents in Saturn's atmosphere* *Computer processing exaggerates the colors in this* Voyager 2 *picture of Saturn's northern middle latitudes. The wavy line in the light blue ribbon is a pattern moving eastward at 150 m/s (340 mph). The dark oval and two puffy, blue-white spots below it are eddies drifting westward at roughly 20 m/s (45 mph). (NASA)*

By following features in the Jovian and Saturnian clouds, scientists have determined wind speeds in their upper atmospheres. Both planets exhibit counterflowing eastward and westward currents. However, Saturn's equatorial jet is much broader and faster than Jupiter's (Figure 8-25). In fact, wind speeds near Saturn's equator approach 500 m/s (1000 mph), approximately two-thirds the speed of sound.

8-8 Jupiter and Saturn emit more radiation than they receive from the Sun

Both Jupiter and Saturn have internal sources of energy. Each planet radiates more energy than it receives in the form of sunlight. The excess heat that escapes from Jupiter is residual energy left over from the formation of the planet. As gases from the solar nebula fell into the protoplanet, vast amounts of gravitational energy were converted into thermal energy that became trapped far below Jupiter's clouds. For the past 4.5 billion years, Jupiter has been slowly cooling off as this trapped energy escapes in the form of infrared radiation.

Saturn is both smaller and less massive than Jupiter. One would thus expect Saturn to cool more rapidly than Jupiter and hence to emit less energy today. But in fact Saturn radiates about $2\frac{1}{2}$ times as much energy as it receives from the Sun, whereas Jupiter emits only about $1\frac{1}{2}$ times as much. Why does Saturn emit so much more heat than Jupiter?

For some time before space probes were sent to the Jovian planets, astronomers had suspected that both Jupiter and Saturn have compositions similar to that of the

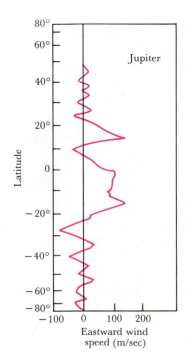

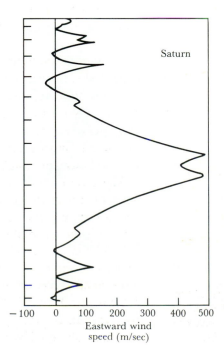

Figure 8-25 *Wind speeds on Jupiter and Saturn* *Average wind speeds on Jupiter and Saturn are here plotted for latitudes ranging from 80° north to 80° south. The positive numbers are eastward velocities; negative are westward velocities. Although both planets exhibit counterflowing currents, Saturn's equatorial zonal jet is much broader and faster than Jupiter's.*

original solar nebula and the Sun's atmosphere today. Each of these giant planets is massive enough and cool enough to have retained all of the gases that originally accreted from the solar nebula. The Voyager flybys did confirm that Jupiter has an abundance of elements (by weight, 82% hydrogen, 17% helium, and 1% all other elements) that is similar to the solar abundance. Surprisingly, however, the Voyager spacecraft reported that Saturn's atmosphere has less helium than expected. The chemical composition of Saturn's atmosphere, by weight, is 88% hydrogen, 11% helium, and only 1% all other elements.

A clever hypothesis links Saturn's apparent deficiency of helium to the excess heat that the planet radiates. According to this theory, Saturn did indeed cool more rapidly than Jupiter. This cooling triggered a process analogous to the development of a rainstorm here on Earth. When the air is cool enough, humidity in the Earth's atmosphere condenses into raindrops that fall to the ground. On Saturn, though, it is helium droplets that rain downward through the planet's atmosphere toward its core. Helium is deficient in Saturn's upper atmosphere simply because it has fallen farther down into the planet. Furthermore, as the helium droplets descend toward the planet's core, their gravitational energy is converted into thermal energy (heat) that eventually escapes from Saturn's surface.

The precipitation of helium from Saturn's clouds is calculated to have begun 2 billion years ago. The resulting release of energy adequately accounts for the extra heat radiated by Saturn since that time. Similar calculations for Jupiter indicate that only now it is reaching the stage where a significant amount of helium precipitation can begin in its outer layers. Saturn has therefore given us important clues about the probable course of Jupiter's future evolution.

8-9 Rapid rotation and liquid metallic hydrogen deep in their interiors endow Jupiter and Saturn with magnetic fields

Jupiter's rapid rotation profoundly affects the overall shape of the planet. Even a casual glance through a small telescope shows that Jupiter is slightly flattened, or oblate. The diameter across Jupiter's equator (143,800 km) is 6% percent larger than the diameter from pole to pole (135,200 km). Thus Jupiter is said to have an **oblateness** of 6%, or 0.06. Saturn is even more oblate than Jupiter

(examine Figure 8-26). Saturn's equatorial diameter is about 10% larger than its polar diameter, and so its oblateness is about 0.10.

If Jupiter and Saturn were not rotating, they would be perfect spheres. A massive, nonrotating object naturally settles into a spherical shape in which every atom on its surface experiences the same intensity of gravity aimed directly at the object's center. However, Jupiter and Saturn are rotating rather rapidly. Saturn's equatorial rotation period (10 hours 14 minutes) is only slightly longer than Jupiter's (9 hours 50 minutes). Because of rotation, every part of each planet also experiences an outward-directed "centrifugal force" that is proportional to each part's distance from the axis of rotation. Because equatorial regions are farther from the planet's axis of rotation than are polar regions, they are subjected to a stronger centrifugal force. Hence the planet's equatorial diameter is slightly larger than its polar diameter. This centrifugal stretching of the equatorial dimensions gives Jupiter and Saturn their characteristic oblate shapes.

At every point throughout each planet, the inward force of gravity is exactly balanced by the outward pressure of the compressed material plus the outward centrifugal effects of the planet's rotation. The shape and density of a planet adjust themselves to ensure that this balance is maintained.

The shape of a planet is an excellent indicator of its internal structure. Two planets with the same mass, aver-

Figure 8-26 Saturn from Voyager 2 *Saturn is the most oblate planet in the solar system. Saturn's equatorial diameter is 12,000 km larger than its diameter from pole to pole. Voyager 2 sent back this picture when the spacecraft was 34 million kilometers away from the planet. Faint belts and zones are clearly visible. (NASA)*

age density, and rotation will have slightly different oblateness if one planet has a compact core and the other does not. All other things being equal, a planet with a dense core will be more oblate than a planet without one.

Detailed calculations strongly suggest that 4% of Jupiter's mass is concentrated in a dense, rocky core. Jupiter's oblateness is in fact consistent with a rocky core nearly 13 times as massive as the entire Earth. Some of this core was probably the original "seed" around which proto-Jupiter accreted.

Jupiter's rocky core is probably not much bigger than the Earth, even though it is 13 times more massive than our planet. The tremendous crushing weight of the remaining bulk of Jupiter—equal to the mass of 305 Earths—compresses the core down to a sphere 20,000 km in diameter (Earth's diameter is 12,756 km). The pressure at Jupiter's center is about 80 million atmospheres, which squeezes the rocky material of Jupiter's core to a density of about 20,000 kg/m^3. The temperature at the planet's center is probably about 25,000 K. In contrast, the temperature at Jupiter's cloudtops is only 165 K.

Jupiter is composed largely of hydrogen. A hydrogen atom consists of a single proton orbited by a single electron. Deep inside Jupiter, however, pressures are so great that the electrons are stripped from their protons. The electrons, no longer bound to protons, are free to wander around and thereby create electric currents. In other words, the highly compressed hydrogen deep inside Jupiter behaves like a metal. Thus it is called **liquid metallic hydrogen.**

Detailed calculations strongly suggest that molecular hydrogen is transformed into liquid metallic hydrogen when the pressure exceeds three million atmospheres. This transition occurs at a depth of about 20,000 km below Jupiter's cloudtops. Thus, the internal structure of Jupiter consists of three distinct regions: a rocky core, surrounded by a 40,000-km-thick layer of liquid metallic hydrogen, surrounded in turn by a 20,000-km-thick layer of ordinary molecular hydrogen. The colorful cloud patterns we can see through telescopes are located in the outermost 100 km of the exterior layer of molecular hydrogen.

Information about the average density, oblateness, and the probable chemical composition of Saturn leads astronomers to infer that Saturn's interior structure resembles that of Jupiter: a solid, rocky core surrounded by a mantle of liquid metallic hydrogen, surrounded in turn by a layer of molecular hydrogen.

Saturn is less massive than Jupiter and thus the pressures inside it are not as high as those inside Jupiter.

Consequently, Saturn's rocky core is less compressed and larger than Jupiter's. Also, Saturn's weaker gravity is unable to convert as much hydrogen into a liquid metal as does Jupiter's. Thus the relative thicknesses of the layers differ in the models of the internal structures of these two planets (Figure 8-27). Saturn's rocky core is about

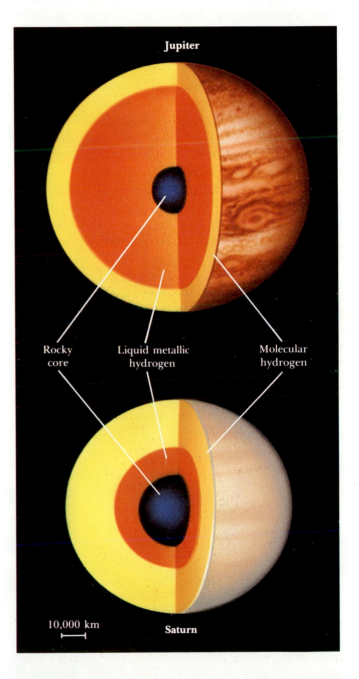

Figure 8-27 Internal structure of Saturn and Jupiter There are three distinct layers in the interiors of Jupiter and Saturn. Each planet's rocky core is surrounded by a layer of liquid metallic hydrogen, which in turn is enveloped in a thick layer of molecular hydrogen. Both diagrams are drawn to the same scale.

32,000 km in diameter, but its layer of liquid metallic hydrogen is only 12,000 km thick.

It is clear from Figure 8-27 that a large percentage of Jupiter's enormous bulk is electrically conductive liquid metal. Because of Jupiter's rapid rotation, electric currents in this thick layer of liquid metallic hydrogen generate a powerful magnetic field, in much the same way that liquid portions of the Earth's core produce the Earth's magnetic field. The intrinsic strength of Jupiter's magnetic field is 19,000 times greater than the Earth's, a direct result of Jupiter's massive liquid metallic region and its rapid rate of rotation.

Jupiter's powerful magnetic field surrounds the planet with an enormous magnetosphere, large enough to envelop the orbits of many of its moons. For Earth-based astronomers, the only evidence of this magnetosphere is a faint hiss of radio static. However, four spacecraft that journeyed to Jupiter in the 1970s revealed the awesome dimensions of the Jovian magnetosphere. The volume surrounded by the shock wave is nearly 30 million kilometers across. In other words, if you could see Jupiter's magnetosphere from the Earth, it would cover an area in the sky 16 times larger than the full Moon.

The inner regions of our magnetosphere are dominated by two huge Van Allen radiation belts (recall Figure 7-32) filled with charged particles. The same sort of belts would probably exist around Jupiter were it rotating slowly. But because Jupiter rotates rapidly, centrifugal forces spew out the particles into a huge electrically charged **current sheet** (Figure 8-28). This current sheet lies in the plane of Jupiter's magnetic equator. Jupiter's magnetic axis is inclined slightly from the planet's axis of rotation and the orientation of its magnetic field is the reverse of Earth's (a compass would point toward the south pole on Jupiter).

Saturn's mantle of liquid metallic hydrogen, like Jupiter's, produces a planetwide magnetic field. However, Saturn's magnetic field is weaker than Jupiter's. Saturn's slightly slower rotation and much smaller volume of liquid metallic hydrogen produce a magnetic field whose intrinsic strength is only about 0.03 times that of Jupiter's. Data from spacecraft indicate that Saturn's magnetosphere contains radiation belts similar to those of Earth instead of a huge Jupiterlike current sheet.

8-10 Earth-based observations suggest that Uranus and Neptune are similar to each other, but different from Jupiter and Saturn

Even through a large telescope, both Uranus and Neptune are dim, uninspiring sights. Each planet appears as a small, hazy, featureless disk with a faint greenish-blue tinge (Figures 8-29 and 8-30). Earth-based observations have provided only the most basic information about these two worlds simply because they are so far away.

At first glance, Uranus and Neptune appear to be twins. They have nearly the same size, mass, density, and chemical composition. In addition, the outer two Jovian planets are significantly different from the inner two. The large spheres of Jupiter and Saturn are composed primar-

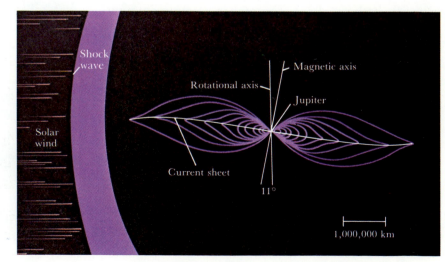

Figure 8-28 Jupiter's magnetosphere Jupiter's magnetosphere is enveloped in a shock wave, where the supersonic solar wind abruptly slows to subsonic speeds. Most of the particles of the solar wind are deflected around Jupiter in a turbulent region colored purple in this scale drawing. Particles that do become trapped inside Jupiter's magnetosphere are spewed out into a vast current sheet by the planet's rapid rotation. Jupiter's axis of rotation is inclined to its magnetic axis by about 11°.

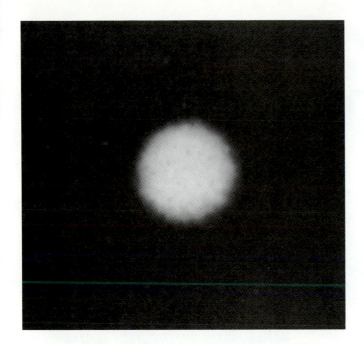

Figure 8-29 *Uranus Nearly three billion kilometers from the Sun, Uranus receives only $\frac{1}{400}$ the intensity of sunlight that we experience here on Earth. Uranus is therefore a dim and frigid world. At its brightest, Uranus appears as bright as the faintest stars we can see with our unaided eyes from Earth. Even though its diameter is about four times Earth's, Uranus always shows us a disk less than 4 arc sec across, which is roughly the size of a golf ball seen from a distance of 1 km. (New Mexico State University Observatory)*

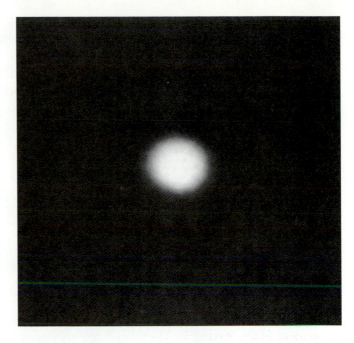

Figure 8-30 *Neptune At 4.5 billion kilometers from the Sun, Neptune's surface receives only $\frac{1}{900}$ the intensity of sunlight we receive on Earth. Neptune is therefore dimmer than Uranus. Although about the same size as Uranus, Neptune looks smaller through Earth-based telescopes because it is farther away. The maximum possible angular diameter of Neptune's disk is about 2 arc sec. That is roughly the same size as a penny seen from a distance of 1 km. (New Mexico State University Observatory)*

ily of hydrogen and helium, like the Sun. Uranus and Neptune, however, are distinctly smaller and less massive than either Jupiter or Saturn. If Uranus and Neptune also had solar abundances of the elements, their smaller masses would produce less compression and therefore lower average densities than those of Jupiter and Saturn. In fact, however, Uranus and Neptune have average densities comparable to or greater than those of Jupiter or Saturn. We therefore must conclude that Uranus and Neptune contain greater proportions of the heavier elements.

Astronomers who have calculated the internal structures of Uranus and Neptune agree that both planets must have substantial rocky cores like those at the centers of Jupiter and Saturn. Unlike Jupiter and Saturn, however, neither Uranus nor Neptune has the mass needed to achieve the pressure to produce liquid metallic hydrogen. Some scientists suspect Uranus and Neptune have superdense atmospheres that extend all the way down to their rocky cores. These atmospheres may contain extremely dense water clouds. Indirect evidence for such clouds comes from the apparent deficiency of ammonia

on both planets. This gas is easily dissolved in water, which would explain the low abundance of ammonia in the observable outer layers of both planets.

The outer layers of Uranus and Neptune are predominantly composed of hydrogen and helium in a gaseous state and at low density. The temperature in the upper atmospheres of both planets is so low (about 60 K, or −351°F) that methane and water there condense to form clouds of ice crystals. Methane freezes at the lowest temperature and thus forms the highest clouds. This gas efficiently absorbs red light, giving Uranus and Neptune their blue-green color.

8-11 *Voyager 2 revealed many details about Uranus, its rings, and its satellites*

After coasting for nearly $8\frac{1}{2}$ years through interplanetary space, *Voyager 2* arrived at Uranus in January 1986. During the few days that pictures and data poured in,

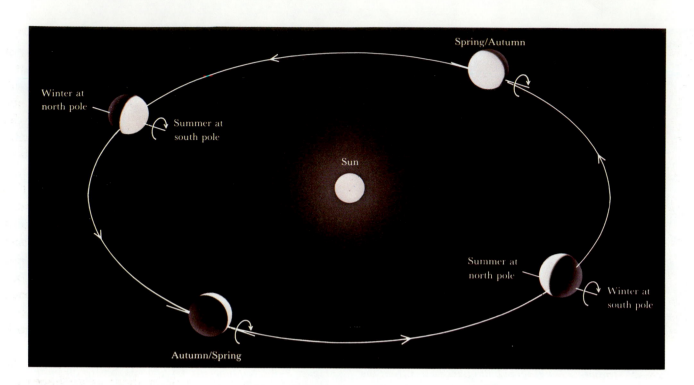

Figure 8-31 Exaggerated seasons on Uranus Uranus's axis of rotation is tilted so steeply that it lies nearly in the plane of the planet's orbit. Seasonal changes on Uranus are thus severely exaggerated. For example, during midsummer at Uranus's south pole, the Sun appears nearly overhead for many Earth years, while the planet's northern regions are subjected to a long, continuous winter night.

scientists learned more about this remote world than they had since Uranus was discovered two centuries ago.

It was already known from Earth-based observations that Uranus's axis of rotation lies very nearly in the plane of its orbit (Figure 8-31). Consequently, as Uranus moves along its 84-year orbit, the planet's north and south poles alternately point toward or away from the Sun, producing exaggerated seasons. During the summertime near Uranus's north pole, the Sun remains high above the horizon for many Earth-years while southern latitudes are subjected to a continuous, frigid winter night. Forty-two Earth-years later, the situation is reversed.

Uranus's north pole was aimed almost directly at the Sun as *Voyager 2* approached the planet. Thus Figure 8-32 shows its illuminated northern hemisphere. No clouds or any other atmospheric features are seen, but computer processing did reveal a smoglike haze over the pole.

Voyager 2's instruments discovered that Uranus has a magnetic field about 50 times stronger than Earth's. Furthermore, Uranus's magnetic axis (that is, the line joining the north and south magnetic poles) is tilted away from its axis of rotation by nearly 60°. This inclination is surprising because most planets have their magnetic and rotational axes nearly aligned. For example, Earth and Jupiter both have their magnetic axes inclined by about 11° from their rotational axes. Uranus is thus unique because of the unusual orientations of both its rotational and magnetic axes.

Uranus rotates differentially, as do Jupiter and Saturn. A day on Uranus is between 14.2 and 17.9 hours long, depending on the latitude. To determine the rotation period of the underlying body of the planet, scientists looked to Uranus's magnetic field, which is presumably anchored in the planet's core.

Voyager 2 detected regular changes in radio emission from Uranus's magnetosphere that repeated every 17.24 hours. These changes are caused by the motion of Uranus's oblique magnetic field as it is carried around by the planet's rotation. The magnetic field is presumably anchored deep inside the planet, and so 17.24 hours must be the rotation period of Uranus's core.

Revolving around Uranus in the plane of the planet's equator are numerous satellites and a system of thin, dark

Figure 8-32 Uranus from Voyager 2 *No distinctive cloud patterns are seen in any of the Voyager views of Uranus. The blue-green appearance of Uranus comes from methane in the planet's atmosphere, which absorbs red wavelengths from the incoming sunlight. Severe computer enhancement of the Voyager pictures does reveal faint cloud features, from which the rotation period of Uranus's atmosphere was determined to be about 16½ hours. (NASA)*

Prior to the collision, Uranus's satellites presumably orbited the upright planet in the plane of its equator, just like the regular satellites of the other Jovian planets. After the collision, however, tidal forces exerted by Uranus's equatorial bulge on the satellites disrupted their orbits. Only after many years and numerous collisions did the remaining chunks of rock and ice settle into the highly tilted plane of Uranus's equator, where we find them today.

Of all these moons, Miranda is the most fascinating because it is covered with unusual wrinkled and banded surface features (Figure 8-34). Miranda is the smallest of Uranus's five main satellites, and a shattering impact could have temporarily broken it into several pieces. Prior to that impact, Miranda's core was made up of dense rock and its outer layers of less dense ice. After the impact, blocks of debris drifted back together because of mutual gravitational attraction and formed a chaotic mix of rock and ice. The landscape we see today on Miranda is the result of huge, dense rocks trying to settle toward the satellite's center as blocks of less dense ice are forced upward toward the surface.

rings. Nine of these rings, ranging in width from 10 to 100 km, were discovered in 1977 when Uranus passed in front of a star. The star's light was momentarily blocked by each ring, which revealed their existence to astronomers. A picture taken while *Voyager 2* was in Uranus's shadow revealed numerous additional very thin rings (Figure 8-33).

Five of Uranus's satellites, ranging in diameter from 480 to nearly 1600 km, were known prior to the Voyager mission. *Voyager 2*'s cameras discovered ten additional satellites, each less than 50 km across. Several of these tiny moons are shepherd satellites whose gravity confines particles to the thin rings that circle Uranus.

Detailed examination of the Voyager photographs has led several planetary scientists to suggest that each of the Uranian moons may have had one or more shattering impacts. This suggestion agrees with the speculation that a catastrophic collision between Uranus and an Earth-sized object knocked Uranus on its side and set the planet's core rotating at an oblique angle, just as we see today.

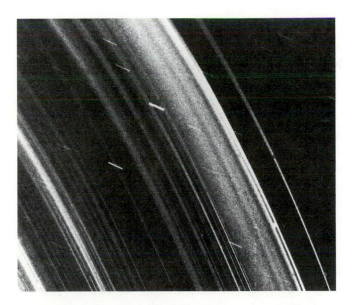

Figure 8-33 The rings of Uranus This view, taken when Voyager 2 *was in Uranus's shadow, looks back toward the Sun. Numerous fine dust particles between the main rings gleam in the sunlight. Uranus's rings are much darker than Saturn's, and this long exposure revealed many very thin rings and dust lanes not previously seen. The short streaks are star images blurred because of the spacecraft's motion during the exposure. (NASA)*

Figure 8-34 *Miranda* *The patchwork appearance of Miranda suggests that this satellite consists of huge chunks of rock and ice that came back together after an ancient, shattering impact by an asteroid or a neighboring Uranian moon. The curious banded features that cover much of Miranda are parallel valleys and ridges that may have formed as dense, rocky material sank toward the satellite's core. (NASA)*

8-12 Neptune is a cold, bluish world with Jupiterlike surface features and a Uranuslike interior

The arrival of *Voyager 2* at Neptune in August 1989 capped one of NASA's most ambitious and most successful missions. Through Earth-based telescopes, Neptune presents a tiny, featureless disk that is barely distinguishable from a star. Scientists were therefore overjoyed at the detailed, close-up pictures and wealth of data sent back to Earth by the spacecraft.

At first glance Neptune looks like a bluish Jupiter (Figure 8-35). Methane in Neptune's atmosphere absorbs longer visible wavelengths but not the shorter ones. Sunlight reflected from Neptune's clouds is thus depleted of reds and yellows, making the planet appear quite blue.

The most prominent feature in Neptune's atmosphere, called the **Great Dark Spot**, gives the planet its distinctly Jupiterlike appearance. The Great Dark Spot falls at about the same latitude on Neptune as the Great Red Spot falls on Jupiter and occupies a proportionate amount of

the planet's surface area. The wind flow in the Great Dark Spot is anticyclonic, just as it is in Jupiter's Great Red Spot. Both features are high-pressure systems out of which winds flow in a counterclockwise direction. This anticyclonic flow on Neptune is difficult to observe because high-elevation clouds hover over the Great Dark Spot. These clouds do not move along with the winds in the atmosphere and may be compared with the so-called lenticular clouds on Earth, which seem to hang over mountaintops. On Earth winds do blow through these clouds at high speeds, but the clouds do not seem to move. This illusion of permanence is actually a continuous process of creation and dissipation. Moisture carried by the winds condenses into visible ice crystals as it flows over the mountaintops, where it encounters cooler regions of the upper atmosphere. As the ice crystals descend the leeward side of the mountain, they meet warm air and turn back into water vapor. Neptune's whitish, cirruslike clouds are made of methane ice crystals. As prevailing winds on

Figure 8-35 *Neptune* *This view from* Voyager 2 *looks down on the southern hemisphere of Neptune. The Great Dark Spot, which is about the same size as the Earth, is near the center of this picture. Note the white, wispy methane clouds. Toward the lower left is a smaller dark spot, located 54° south of Neptune's equator. (NASA)*

Neptune carry the methane to elevated heights over the Great Dark Spot, the gas condenses into visible clouds.

The high elevation of Neptune's cirrus clouds was confirmed by photographs showing these clouds casting shadows on the main cloud deck (Figure 8-36). Knowing the angle of the Sun when the pictures were taken, scientists estimate that these clouds are 50 to 70 km above the cloud deck. Between the cirrus clouds and the cloud deck, Neptune's atmosphere is quite clear. For comparison, Earth's cirrus clouds hover about 7 to 9 km above sea level.

Neptune's belts and zones are much fainter than those on Jupiter. Most prominent is a broad, darkish band at high southern latitudes. Embedded in this band is a smaller dark spot, about a third the size of the Great Dark Spot. White, wispy clouds seem to hover over the smaller dark spot just as they do over the Great Dark Spot.

Temperatures in the Neptunian atmosphere range from about 60 K at the equator and the poles to about 50 K at mid-latitudes. This peculiar situation, wherein the poles and the equator are both warmer than the mid-latitudes, may be caused by the movement of gases in Neptune's upper atmosphere. At mid-latitudes, gases cool as they rise upward, but at the poles and equator the gases warm as they descend to lower elevations.

The dominant wind flow in the cloud layer that constitutes Neptune's visible surface is unlike that on any other Jovian planet. *Voyager 2* detected periodic variations in radio emission from Neptune every 16.1 hours. These emissions reveal the underlying rotation period of Neptune if we assume they are driven by the planet's magnetic field, which is embedded in its interior. Most of Neptune's cloud cover revolves about the planet more slowly than the planet rotates. Some cloud features take as long as 19.2 hours to go once around the planet. On Jupiter, Saturn, and Uranus, the clouds revolve around the planets more rapidly than the planets themselves rotate. Thus, Neptune's atmospheric circulation is retrograde.

Although its surface resembles that of Jupiter, Neptune's interior resembles that of Uranus. Since Uranus and Neptune have nearly the same size, mass, and density, it is reasonable to suppose that they have roughly the same chemical composition and internal structure. Both planets probably have a rocky core surrounded by a watery mantle and a thick atmosphere primarily composed of hydrogen and helium with an admixture of methane.

While probing Neptune's magnetosphere, *Voyager 2* made the surprising discovery that the planet's magnetic axis is inclined to its axis of rotation by 47°. This is also reminiscent of Uranus, whose magnetic axis is inclined by nearly 60° to its rotational axis. The magnetic fields of Uranus and Neptune have roughly the same strength. The magnetic fields of all four Jovian planets are oriented opposite to that of Earth; a compass on a Jovian planet points southward.

Like Uranus, Neptune is surrounded by a system of thin, dark rings (Figure 8-37). It is so cold at Uranus and Neptune that the ring particles can retain methane ice. Scientists speculate that eons of radiation damage have converted this methane ice into darkish carbon compounds, thus accounting for the low reflectivity of the rings.

Having completed its mission to the outer planets, *Voyager 2* is headed out of the solar system. For many years to come, scientists will continue to monitor the spacecraft's instruments as they probe the solar wind at extreme distances from the Sun. At present there are no plans to return to either Uranus or Neptune.

Figure 8-36 Cirrus clouds over Neptune *This picture shows vertical relief in Neptune's bright, methane cloud streaks. Voyager 2 photographed these clouds north of Neptune's equator near the terminator (the border between day and night on the planet). Note the shadows cast by the clouds onto the main cloud deck. (NASA)*

Figure 8-37 *Neptune's rings* *Two main rings are easily seen in this view alongside an over-exposed image of Neptune. Careful examination of this picture also reveals a faint inner ring. A faint sheet of particles, whose outer edge is located between the two main rings, extends inward toward the planet. (NASA)*

Summary

- Jupiter is by far the largest and most massive planet in the solar system.

- Jupiter and Saturn are primarily composed of hydrogen and helium. Both planets have an overall chemical composition very similar to that of the Sun.

- Because of their rapid rotation, Jupiter and Saturn are noticeably oblate, which provides important clues about their internal structure.

  Jupiter and Saturn both probably have rocky cores surrounded by a thick layer of liquid metallic hydrogen and an outer layer of ordinary hydrogen gas.

- The visible features of Jupiter (belts, zones, the Great Red Spot, ovals, and colored clouds) exist in the outermost 100 km of its atmosphere. Saturn has similar features, but they appear much fainter than on Jupiter.

  There are three cloud layers in the upper atmospheres of both Jupiter and Saturn. Saturn's are spread out over a greater altitude range than those of Jupiter, and so the colors of the Saturnian atmosphere are somewhat muted.

The colored ovals visible in the Jovian atmosphere represent gigantic storms, some of which (such as the Great Red Spot) are stable and persist for years. The ovals are cyclonic or anticyclonic storms at the boundaries between wind streams moving in opposite directions around the planet.

- Jupiter and Saturn both emit more heat than they receive from the Sun. Presumably Jupiter is still cooling. On Saturn, the precipitation of helium downward into the planet is probably the cause of its excess heat.

- Jupiter has a strong magnetic field created by currents in the metallic-hydrogen layer. Its huge magnetosphere contains a vast current sheet of electrically charged particles. Saturn's magnetosphere is similar to Jupiter's but has Earthlike radiation belts instead of a current sheet.

- Saturn is circled by a system of thin, broad rings lying in the plane of the planet's equator. Each of Saturn's major rings is composed of a great many narrow ringlets consisting of numerous fragments of ice and ice-coated rock.

Some of the ring boundaries are produced by shepherd satellites, whose gravitational pull restricts the orbits of the ring fragments.

- Uranus and Neptune are quite similar to each other in appearance, mass, size, and chemical composition.

  They each have a substantial rocky core possibly surrounded by an extremely dense, watery atmosphere.

- Uranus is unique in that its axis of rotation lies nearly in the plane of its orbit, producing greatly exaggerated seasons on the planet.

- Uranus has a system of thin, dark rings and five satellites similar to the moderate-sized moons of Saturn.

- Neptune's surface features resemble those of a bluish Jupiter, having a Great Dark Spot and faint belts and zones.

- Neptune's magnetic field is like Uranus's in that its magnetic axis is inclined steeply to the planet's axis of rotation.

- Like Uranus, Neptune is surrounded by a system of thin, dark rings. The low reflectivity of the ring particles may be due to radiation-damaged methane ice.

Review questions

1 Describe the appearance of Jupiter's atmosphere. Which features are long-lived and which are fleeting?

2 Describe the structure of Saturn's rings. What are they made of?

3 If Jupiter does not have any observable solid surface and its atmosphere rotates differentially, how are astronomers able to determine the planet's rotation rate?

4 Why do features in Saturn's atmosphere appear to be much fainter and "washed out" compared to features in Jupiter's atmosphere?

5 Why do astronomers believe that Jupiter does not have a large iron-rich core even though the planet possesses a strong magnetic field?

6 Compare and contrast Jupiter's magnetosphere with the magnetosphere of the Earth.

7 What is liquid metallic hydrogen? Which planets are presumed to contain this substance?

8 What is thought to be the source of Jupiter's internal heat? How does Jupiter's internal heat source differ from that inside Saturn?

9 Explain how shepherd satellites operate. Is *shepherd satellite* an appropriate term for these objects? Explain.

10 Could astronomers of antiquity see Uranus? If so, why do you suppose it was not recognized as a planet?

11 Describe the seasons on Uranus. Why are the Uranian seasons different from those on any other planet?

12 Briefly describe the evidence supporting the idea that Uranus was struck by a large planetlike object several billion years ago.

13 Why are Uranus and Neptune distinctly bluer than Jupiter and Saturn?

14 Compare the ring systems of Saturn and Uranus. Why were Uranus's rings unnoticed until the 1970s?

15 Compare and contrast the internal structures of Jupiter and Saturn with the internal structures of Uranus and Neptune. Can you propose an explanation to account for the differences between the inner and outer Jovian planets?

Advanced questions

16 Explain why Saturn is more oblate than Jupiter even though Saturn rotates more slowly.

17 What sort of experiment would you design in order to establish whether Jupiter has a rocky core?

18 When this textbook went to press, the *Galileo* spacecraft was still on its way to Jupiter. Consult such magazines as *Sky & Telescope* and *Science News* to determine the status of this mission. Have the problems with the spacecraft's antenna been solved?

19 As seen by Earth-based observers, the intervals between successive edge-on presentations of Saturn's rings alternate between 13 years 9 months and 15 years 9 months. Why do you think these two intervals are not equal?

20 NASA, the Jet Propulsion Laboratory, and the European Space Agency are currently planning a mission to Saturn that will place the *Cassini* spacecraft in orbit about the planet. Consult such magazines as *Sky & Telescope* and *Science News* to determine the status of this mission. Has the U.S. Congress approved funds for the mission? What is the estimated launch date? When will *Cassini* arrive at Saturn?

Discussion questions

21 Describe some of the semipermanent features in Jupiter's atmosphere. What factors influence their longevity? Compare and contrast these long-lived features with some of the transient phenomena seen in Jupiter's clouds.

22 Suppose that you were designing a mission to Jupiter that involved an airplanelike vehicle that would spend many days (months?) flying through the Jovian clouds. What observations, measurements, and analyses should this aircraft make? What dangers might it encounter and what design problems would you have to overcome?

23 NASA and the Jet Propulsion Laboratory have tentative plans to place spacecraft in orbit about Uranus and Neptune early in the twenty-first century. What kinds of data should be collected and what questions would you like to see answered by these missions?

For further reading

Beatty, J. "Getting to Know Neptune." *Sky & Telescope*, February 1990 • This beautifully illustrated article discussed many of the results from the *Voyager 2* flyby of Neptune.

———. "Report on the *Voyager* Encounters with Saturn." *Sky & Telescope*, January, October, and November 1981 • This series of articles boasts a magnificent collection of the best images from the Voyager flybys of Saturn.

Bennett, J. "The Discovery of Uranus." *Sky & Telescope*, March 1981 • This fascinating historical article describes Herschel's discovery of Uranus.

Berry, R. "Neptune Revealed." *Astronomy*, December 1989 • This summary of Voyager's results at Neptune includes an outstanding collection of photographs.

Burgess, E. *By Jupiter*. Columbia University Press, 1982 • This nontechnical book reviews our understanding of Jupiter in light of the Pioneer and Voyager flybys.

Burnham, R. "The Saturnian Satellites." *Astronomy*, November 1981 • This easily understood article discusses the various icy satellites of Saturn.

Carroll, M. "Project Galileo: The Phoenix Rises." *Sky & Telescope*, April 1987 • This article describes the long-delayed Galileo mission to Jupiter.

Elliot, J., and others. "Discovering the Rings of Uranus." *Sky & Telescope*, June 1977 • The team of astronomers who discovered Uranus's rings describe their airborne observations.

Gore, R. "Saturn: Riddle of the Rings." *National Geographic*, July 1981 • A discussion of some of the puzzling aspects of Saturn's rings is lavishly illustrated in the tradition of this famous periodical.

Ingersoll, A. "Jupiter and Saturn." *Scientific American*, December 1981 • Written shortly after the Voyager flybys, this enlightening article compares and contrasts the two largest planets.

———. "Uranus." *Scientific American*, January 1987 • This superb article summarizes the results of the *Voyager 2* flyby.

Morrison, D. *Voyages to Saturn*. NASA SP-451, 1982 • This excellent book written by a noted planetary scientist gives an insider's view of the Saturn flybys.

Morrison, D., and Samz, J. *Voyage to Jupiter*. NASA SP-439, 1980 • This superb book, which describes the Voyager missions to Jupiter, includes an excellent selection of color photographs.

Pollack, J., and Cuzzi, J. "Rings in the Solar System." *Scientific American*, November 1981 • This excellent article on the structure and evolution of planetary rings focuses primarily on Saturn but also discusses rings around Jupiter and Uranus.

Washburn, M. *Distant Encounters: The Exploration of Jupiter and Saturn*. Harcourt Brace Jovanovich, 1983 • This entertaining book describes the Pioneer and Voyager missions to the outer solar system.

Mercury

Moon

Io

Europa

Ganymede

Callisto

Titan

Triton

9

The Smaller Terrestrial Worlds

In this chapter we examine Mercury along with seven giant satellites large enough to qualify as terrestrial planets. We find that Mercury has a cratered, lunarlike surface but an Earthlike, iron-rich core with a magnetic field. We then follow the astronauts to the Moon and examine lunar rocks containing important clues about the Moon's history. Orbiting Jupiter are four more unique terrestrial worlds: Io with its sulfur volcanoes, Europa covered by a thin layer of shifting ice, and Ganymede and Callisto, each surrounded by a thick mantle of water and ice. At Saturn we find Titan, the only satellite with a substantial atmosphere, thicker even than Earth's. We then encounter Triton, the largest satellite of Neptune, and speculate that it is probably quite like Pluto, the smallest and most distant planet in the solar system. Surveying what we now know about the outer solar system, we realize that thousands of Plutolike worlds may orbit the Sun. The tilt of Uranus's axis may have resulted from a collision with such an object, while Triton may be a Plutolike world that long ago was captured into orbit about Neptune.

The smaller terrestrial planets Mercury, our Moon, and the largest satellites of Jupiter, Saturn, and Neptune are shown here to the same scale. Each of these worlds has its own unique characteristics and all are large enough to be classified as planets. Only Titan possesses a substantial atmosphere. (NASA)

In recent years spacecraft have revealed seven worlds that are roughly the same size as Mercury. Each of these seven giant satellites (recall Table 6-3) has its own unique geology. Along with Mercury, these seven worlds give us a new perspective on the variety of which nature is capable.

9-1 Mercury has a Moonlike surface and an Earthlike interior

Until 1974 we knew very little about Mercury, the small planet that formed in the warm inner regions of the solar nebula. Information about Mercury was difficult to obtain for two simple reasons: It is very small, and it is very near the Sun. Indeed, Mercury is so close to the Sun that most people (including many astronomers) have never seen it.

The best opportunities to see Mercury occur when the planet is as far from the Sun as it can be, at greatest eastern or western elongation. For a few days near the time of greatest eastern elongation, Mercury appears as an "evening star," hovering low over the western horizon for a short time after sunset. Near the time of greatest western elongation, it can be glimpsed as a "morning star," heralding the rising Sun in the brightening eastern sky.

Mercury travels around the Sun faster than any other object in the solar system, taking only 88 days to complete a full orbit. Thus Mercury passes through inferior conjunction at least three times a year, and you might expect to see Mercury occasionally silhouetted against the Sun in what is called a solar **transit.** Transits of Mercury across the Sun are not very common, though, because Mercury's orbit is tilted 7° to the plane of the Earth's orbit. As a

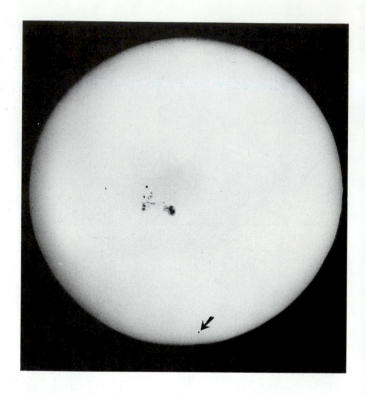

Figure 9-1 A transit of Mercury *Roughly a dozen transits of Mercury occur each century. This photograph shows the tiny planet silhouetted against the Sun during a transit on November 14, 1907. (Yerkes Observatory)*

result, Mercury usually lies well above or below the Sun at the moment of inferior conjunction. Only 14 solar transits of Mercury occurred during the entire twentieth century (Figure 9-1). The final one, scheduled for November 6, 1993, will be visible only from Australia and its immediate vicinity.

Figure 9-2 Earth-based views of Mercury *These two views are among the finest photographs of Mercury ever produced with an Earth-based telescope. Hazy markings are faintly visible on the tiny planet. (New Mexico State University Observatory)*

Figure 9-3 Mercury and our Moon *Mercury (left) and our Moon (right) are shown here to the same scale. Mercury's diameter is 4878 km and the Moon's is 3476 km. For comparison, the distance from New York to Los Angeles is 3944 km (2451 mi). Both worlds have a heavily cratered surface and virtually no atmosphere. Mercury has a substantial iron-rich core and a magnetic field; the Moon does not. Daytime temperatures at the equator on Mercury reach 430°C (800°F), hot enough to melt lead or tin. (NASA; Lick Observatory)*

Naked-eye observations of Mercury are best made at dusk or dawn, but the best telescopic views are obtained at midday when it is high above the degrading atmospheric effects near the horizon. The photographs in Figure 9-2, among the finest Earth-based views of Mercury ever recorded, were taken at midday. Because of its small size and nearness to the Sun, you cannot see much surface detail on Mercury through an Earth-based telescope. At best only a few faint, hazy markings can be identified.

We acquired our first detailed knowledge about Mercury's surface in 1974, when *Mariner 10* coasted to within 756 km (470 mi) of the planet's surface. As the spacecraft closed in on Mercury, scientists were surprised by the Moonlike pictures appearing on their television monitors. It became obvious that Mercury is a barren, desolate, heavily cratered world. Figure 9-3 shows a typical close-up view of Mercury and, for comparison, a picture of our Moon to the same scale.

Although our first impression is of a lunarlike landscape, closer scrutiny of Mercury's surface reveals some significant nonlunar characteristics. Lunar craters are densely packed, one overlapping the next. In sharp contrast, Mercury's surface has extensive plains between craters (compare Figures 9-4 and 9-5).

Astronomers believe that most craters on both Mercury and the Moon were produced during the solar system's first 700 million years. The strongest evidence comes from the direct analysis and dating of Moon rocks brought back by the Apollo astronauts. Debris remaining after the planets had formed rained down on these young worlds, gouging out most of the craters we see today.

Astronomers agree that the Moon and the terrestrial planets must have been completely molten spheres of liquid rock at first. After a few hundred million years, their surfaces solidified as the rock cooled. Nevertheless, large meteoroids could still easily puncture the thin, cooling crusts, allowing molten lava to well up from their interiors. Older craters were obliterated as seas of molten rock flooded portions of the planets' surfaces. Areas of extensive lava flooding are visible on the Moon today.

Figure 9-4 Mercurian craters and intercrater plains *This view of Mercury's northern hemisphere was taken by Mariner 10 at a range of 55,000 km (34,000 mi) from the planet's surface. Numerous craters and extensive intercrater plains appear in this photograph, which covers an area 480 km (300 mi) wide. (NASA)*

Figure 9-5 Lunar craters *This Earth-based photograph shows a portion of the Moon's southern hemisphere during the last quarter moon. Densely packed craters fill the view, which covers an area approximately 600 km (370 mi) wide. (The Observatories of the Carnegie Institution of Washington)*

The planets did not cool down at the same rate, however. A small planet can radiate its internal heat into space more easily and cool off more rapidly than a big planet. Because Mercury is larger than the Moon, it took longer for a thick, protective crust to form on Mercury than on the Moon. Throughout Mercury's early history, molten rock seeped up through cracks in its young, frail crust, and volcanism was probably pervasive. The resulting lava flows inundated many older craters, leaving behind the broad, smooth intercrater plains seen by *Mariner 10*.

Mariner 10 measured the surface temperature on Mercury and searched for traces of an atmosphere. Temperatures on Mercury vary from 700 K (800°F) at noon on the equator to 100 K (−280°F) at midnight. This 600-K temperature range is greater than that of any other planet or satellite in the solar system. Because of its high daytime surface temperature and low surface gravity, Mercury was not able to retain a substantial atmosphere.

The average density of Mercury (5430 kg/m³) is nearly the same as that of the Earth (5520 kg/m³). Yet typical rocks from the surfaces of the terrestrial planets have a density of only 3000 kg/m³ and are composed primarily of silicon and other lightweight elements. The higher average density of the Earth and Mercury is attributable to abundant quantities of iron that sank toward the planets' centers while the planets were still entirely molten. As we saw in Chapter 7, this process is called chemical differentiation. Because of gravity, dense elements sink toward a planet's center and force less dense material toward the surface. Chemical differentiation must have occurred during and immediately after the solar system formed, while the terrestrial planets were still entirely molten and internal mass motion could occur on a large scale.

Because Mercury's average density is slightly less than Earth's, you might suspect that Mercury's iron core is proportionally smaller than Earth's. This is not the case. Earth is 18 times more massive than Mercury. Because this larger mass is pushing down on the Earth's interior, the Earth's core is compressed much more than Mercury's. In fact, Mercury is the most iron-rich planet in the solar system, with iron accounting for 65 to 70% of its mass. The scale drawing in Figure 9-6 shows the interior structures of Mercury and Earth. Some astronomers speculate

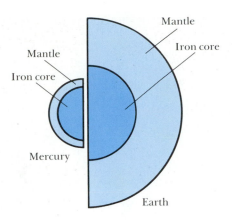

Figure 9-6   The internal structures of Mercury and Earth *Mercury is the most iron-rich planet in the solar system. Its iron core occupies an exceptionally large fraction of its interior.*

that Mercury's high abundance of iron may be the result of a devastating collision with a planetlike body during the final formative stages of the solar system 4.5 billion years ago. Such a collision could have stripped the planet of much of its rocky mantle while leaving the iron core largely intact.

Independent evidence of Mercury's large iron core came from *Mariner 10*'s magnetometers, which discovered that Mercury has a magnetic field. As we saw in Chapter 7, the Earth's magnetic field is produced by electric currents flowing in the liquid portions of our planet's iron core. These currents, carried around by the Earth's rotation, generate a planetwide magnetic field. Mercury rotates much more slowly than the Earth (59 days versus 24 hours), and most scientists had believed that such slow rotation would not produce a detectable magnetic field. They were surprised, therefore, to find that Mercury does have a weak magnetic field (the Earth's field is 100 times stronger).

Charged-particle detectors on *Mariner 10* mapped the structure of Mercury's magnetosphere. The results are shown in Figure 9-7. As we saw in Chapter 7 (recall Figure 7-32), when supersonic particles in the solar wind first encounter a magnetic field, they slow abruptly, producing a bow-shaped shock wave at the boundary where this sudden decrease in speed occurs. Just inside the shock wave is a turbulent region where most of the particles from the solar wind are swept around the planet. Because Mercury's magnetic field is not strong enough to capture particles permanently, it has nothing comparable to Earth's Van Allen belts.

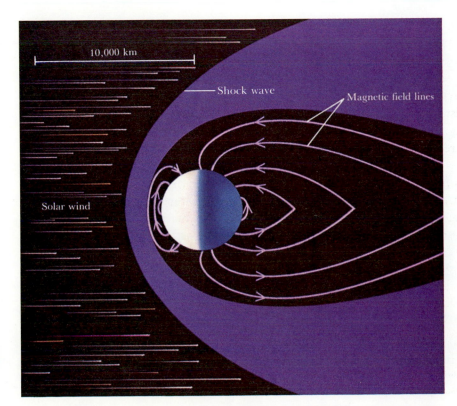

Figure 9-7   Mercury's magnetosphere *Mercury's weak magnetic field is just strong enough to carve out a cavity in the solar wind, preventing most of the impinging particles from striking the planet's surface directly. Most of the particles of the solar wind are deflected around the planet in a turbulent region colored purple in this scale drawing.*

9-2 *The Moon's early history can be deduced from the craters and plains visible on its surface*

Earth's own satellite consistently provides one of the most dramatic sights in the nighttime sky. The Moon is so large and so near that some of its surface features are readily visible to the naked eye. Even casual observation reveals that the same side of the Moon always faces the Earth. This is a stable situation resulting from the gravitational interaction between Earth and Moon.

With a small telescope you can see several different types of major lunar terrain (Figure 9-8). Most prominent are the large, dark, flat areas called **maria** (pronounced MAR-ee-uh). The singular form, **mare** (pronounced MAR-ay), meaning "sea" in Latin, was introduced in the seventeenth century when observers using early telescopes thought they saw large bodies of water on the Moon. In fact, bodies of liquid water cannot exist on our airless satellite. Because there is no atmospheric pressure, a lake

Figure 9-9 Details of Mare Tranquillitatis Close-up views of the lunar surface reveal numerous tiny craters and cracks on the maria. This photograph was taken from lunar orbit in 1969 by astronauts during a final photographic reconnaissance of potential landing sites. (NASA)

*Figure 9-8 **The Moon** Our Moon is one of seven large satellites in the solar system. The Moon's diameter (3476 km = 2160 mi) is slightly less than the distance from New York to San Francisco. This photograph is a composite of first-quarter and third-quarter views, and so elongated shadows enhance all surface features. (Lick Observatory)*

or ocean would boil furiously and evaporate rapidly as its atoms rushed to escape into the vacuum of space. The maria were actually formed by huge lava flows that inundated low-lying regions of the lunar surface 3.5 billion years ago. They have still retained their fanciful names, however, such as Mare Tranquillitatis (Sea of Tranquillity), Mare Nubium (Sea of Clouds), Mare Nectaris (Sea of Nectar), and Mare Serenitatis (Sea of Serenity).

The largest of the 14 maria is Mare Imbrium (Sea of Showers). Roughly circular, it measures 1100 km (700 mi) in diameter. Although the maria seem quite smooth in telescopic views from the Earth, close-up photographs by the Apollo astronauts reveal small craters and occasional cracks called **rilles** (Figure 9-9).

Perhaps the most familiar and characteristic features on the Moon are its craters. With an Earth-based telescope, some 30,000 of them are visible, ranging in size from 1 km to more than 100 km across. Following a tradition established in the seventeenth century, the most prominent craters are named after famous philosophers and scientists.

Figure 9-10 *Details of a lunar crater This photograph, taken from lunar orbit by Apollo 11 astronauts in 1969, shows a typical view of the Moon's heavily cratered far side. The large crater near the middle of the picture is approximately 80 km (50 mi) in diameter. Note the crater's central peak and the numerous tiny craters that pockmark the lunar surface. (NASA)*

Craters smaller than about 1 km in diameter cannot be seen from Earth, but photographs from lunar orbit reveal millions of craters that escape the scrutiny of Earth-based observers. Virtually all craters, both large and small, are the result of bombardment by meteoritic material.

Many of the youngest craters are surrounded by light-colored streaks called **rays** that were formed by material violently ejected during impact. In addition, many large craters have a pronounced central peak formed during a high-speed impact by a sizable meteoroid (Figure 9-10).

The flat, low-lying, dark maria cover only 15% of the lunar surface. The remaining 85% of the surface is light-colored, heavily cratered terrain at elevations generally higher than those of the maria. This second kind of terrain is called the **terrae**, or **highlands**. (*Terra* means "land" in Latin, and so in this fanciful terminology the entire lunar surface is covered by either "land" or "sea.")

One of the surprises arising from lunar exploration is that there are no maria on the Moon's far side, which consists entirely of heavily cratered highlands. Elevation measurements made by astronauts in lunar orbit revealed that the maria on the Moon's Earth-facing side are 2 to 5 km below the average lunar elevation. In contrast, the cratered terrae on the far side are typically at elevations up to 5 km above the average surface elevation. These elevation differences imply that the Moon's crust is thinner on the Earth-facing side than on the Moon's far side, as shown in Figure 9-11.

Large meteoroids easily punctured the thin, cooling crust on the Moon's Earth-facing side shortly after the solar system formed 4.5 billion years ago. Lava welled up, flooding the low-lying areas and producing the maria. Not much has happened since those ancient days; the entire lunar surface has remained almost unchanged for billions of years.

Because of the extraordinary thickness of the Moon's rigid lithosphere, no tectonic plate movement is possible. If the Moon has an iron-rich core, it does not endow our satellite with a magnetic field. Thus the Moon does not have a magnetosphere, and particles of the solar wind strike the lunar surface directly.

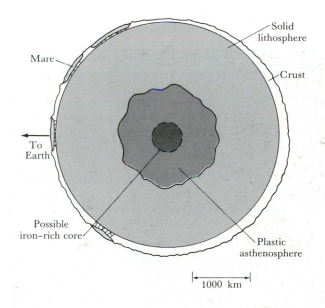

Figure 9-11 *The internal structure of the Moon Like the Earth, the Moon probably has a crust, a mantle, and a core. The lunar crust has an average thickness of about 60 km on the Earth-facing side but about 100 km on the far side. The crust and solid upper mantle form a lithosphere about 800 km thick. The plastic (nonrigid) asthenosphere probably extends all the way to the base of the mantle. If the Moon has an iron-rich core, it is solid and less than 700 km in diameter. Although the main features of the Moon's interior are analogous to those of the Earth's, the proportions are quite different. The information here is based on analyses of data from seismographs left on the Moon by astronauts.*

9-3 Lunar rocks were formed 3 to 4.5 billion years ago

There were six successful manned lunar landings. The first two, *Apollo 11* and *Apollo 12*, set down in maria. The remaining four (*Apollo 14* through *Apollo 17*) were made in progressively more challenging terrain, culminating in rugged mountains just east of Mare Serenitatis (Figure 9-12). The major factors in choosing the landing sites were concern for the astronauts' safety and the desire to explore a wide variety of geologically interesting features.

The Apollo astronauts brought back 382 kg (843 pounds) of lunar rocks that proved to be a very important source of information about the early history of the Moon and Earth. All the lunar rock samples appear to have formed through the cooling of molten lava. The samples are almost completely composed of the same minerals found in terrestrial volcanic rocks. In addition, the entire lunar surface is covered with a layer of fine powder and rock fragments produced by 4.5 billion years of relentless meteoritic bombardment. This layer, which ranges in thickness from 1 to 20 m, is called the **regolith** rather than

Figure 9-13 *Mare basalt* This 1.5-kg (3.5-pound) specimen of mare basalt was brought back by Apollo 15 astronauts. This particular sample is called a vesicular basalt because of the tiny holes, or vesicles, that cover 30% of the rock's surface. Gas must have been dissolved under pressure in the lava from which this rock solidified. When the lava reached the airless lunar surface, bubbles formed as the pressure dropped. Some of the bubbles were frozen in place as the rock cooled. (NASA)

Figure 9-12 *An Apollo astronaut on the Moon* The Moon is a desolate, barren, lifeless world. This typical view of the lunar surface shows an Apollo 17 astronaut near a large rock. Since the Moon has no atmosphere, lunar rocks have not been subjected to weathering and thus contain unaltered information about the early history of the solar system. From the six manned lunar landings between 1969 and 1972, astronauts brought back a total of 382 kg (843 pounds) of moon rocks. (NASA)

soil, because the term *soil* normally suggests the presence of decayed biological matter.

The astronauts who visited the maria discovered that these dark regions of the Moon are covered with basaltic rock similar to the dark-colored rocks formed by lava flows from volcanoes in Hawaii and Iceland. The rock of these low-lying lunar plains is called **mare basalt** (Figure 9-13).

In contrast to the dark maria, the lunar highlands are covered with a light-colored rock called **anorthosite** (Figure 9-14). On Earth, anorthositic rock is found only in such very old mountain ranges as the Adirondacks in the eastern United States. Anorthosite is richer in calcium and aluminum than the mare basalts are, which have more of the heavier elements such as iron, magnesium, and titanium. Anorthosite therefore has a lower density than basalt. The anorthositic magma apparently floated to the lunar surface when the Moon was molten, solidifying as it cooled to form the lunar crust. The denser mare basalts formed later from lava that oozed out of the interior to fill the mare basins.

By carefully measuring the abundances of trace amounts of radioactive elements in lunar samples, geologists confirmed that anorthosite is more ancient than the mare basalts. This result had been expected because the lunar highlands are densely cratered, whereas the basaltic surfaces of the maria show relatively few craters. Typical anorthositic specimens from the highlands are between 4.0 and 4.3 billion years old (one rock brought back by

Figure 9-14 Anorthosite The light-colored lunar terrae are covered with an ancient type of rock called anorthosite. Anorthositic rock is believed to be the material of the original lunar crust. This sample, called the "Genesis rock" by the Apollo 15 astronauts, who picked it up at the base of the Apennine Mountains, has an age of approximately 4.1 billion years. (NASA)

Apollo 17 is nearly 4.6 billion years old). All these ancient specimens represent material from the Moon's original crust. In contrast, all the mare basalts are between 3.1 and 3.8 billion years old. Apparently, the mare basalts solidified from lavas that gushed up from the Moon's mantle and flooded the mare basins between 3.1 and 3.8 billion years ago, just about the time the oldest rocks in the Earth's present surface layers were being formed.

The Apollo astronauts brought back many specimens of **impact breccias,** which are various rock fragments that have been cemented together by meteoritic impact (Figure 9-15). In addition to making breccias and churning up the regolith, meteoritic impacts also melt rocks to produce glass. Many lunar samples are coated with a thin layer of smooth, dark glass created when the surface of the rock suddenly melted and then rapidly solidified. Small black glass beads are also common in the lunar regolith. These glass spheres were presumably formed from droplets of molten rock hurled skyward by the impact of a meteoroid.

Many lunar samples bear the scars of meteoritic dust grains traveling at thousands of kilometers per hour, which produce tiny craters on moon rocks (Figure 9-16). Because of these tiny, glass-lined craters, Moon rocks often seem to sparkle when held in the sunshine.

Meteoritic bombardment is the only source of "weathering" for lunar rocks. The rate of this weathering is actually quite slow. Geologists estimate that it takes tens of millions

of years to wear away a layer of rock only 1 mm thick. Features formed 3 billion years ago are well preserved today, and the astronauts' footprints will remain sharply imprinted on the lunar surface for millions of years to come.

Although lunar rocks bear a strong resemblance to terrestrial rocks, there are some important differences. Every terrestrial rock contains some water, but lunar rocks are totally dry. There is absolutely no evidence that water ever existed on the Moon. Since the Moon lacks both an atmosphere and water, it is not surprising that the astronauts found no traces of life.

Volatile elements such as potassium and sodium melt and boil at relatively low temperatures, whereas **refractory elements** like titanium, calcium, and aluminum melt and boil at much higher temperatures. Compared to terrestrial rocks, lunar rocks have slightly greater proportions of refractory elements and slightly lower proportions of volatile elements. This fact implies that the Moon formed from Earthlike material baked at temperatures high enough to boil away some of the volatile elements, leaving the young Moon relatively enriched in refractory elements.

These characteristics of moon rocks are consistent with the collision-ejection theory of the Moon's formation discussed in Chapter 6. According to this theory, that a huge asteroid struck the Earth about 4.5 billion years ago. During that impact, a large quantity of vaporized rock was blasted out of the Earth (recall Figure 6-18). The temperature of the ejected rock was high enough to drive

Figure 9-15 A lunar breccia Meteoritic impacts can cement rock fragments together to form breccias. This impact breccia was brought back by Apollo 16 astronauts from the rim of a crater near their landing site. (NASA)

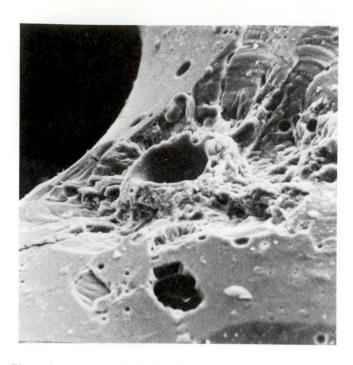

Figure 9-16 A microscopic crater *The upper surfaces of many moon rocks are covered with nearly microscopic craters produced by the impact of high-speed meteoritic dust grains. These glass-lined craters are typically less than 1 mm in diameter. (NASA)*

off water and other volatile chemicals. As rock fragments cooled and solidified in orbit about the Earth, they gradually accreted to form the Moon.

At first, the young Moon was largely molten because of heat from the impacts of rock fragments still falling onto the lunar surface and from the decay of radioactive isotopes. As the Moon cooled over the next few hundred million years, low-density lava floating on the lunar surface began to solidify into the anorthositic crust that exists today. The heavy barrage of large rock fragments ended about 4 billion years ago, with the final impacts producing the ancient craters that cover the lunar highlands.

At the end of this crater-making era, more than a dozen asteroid-sized objects, each possibly as large as 100 km across, rained down on the young Moon, blasting out the vast mare basins. Then, from 3.8 to 3.1 billion years ago, great floods of molten rock gushed up out of the lunar interior, filling the impact basins and creating the maria we see today.

Very little has happened on the Moon since those ancient times. A few fresh craters have been formed (Figure 9-17), but the world the astronauts visited has remained largely unchanged for over 3 billion years.

Figure 9-17 Eratosthenes *Eratosthenes is a young crater 61 km (38 mi) in diameter on the southern edge of Mare Imbrium. Another young crater, Copernicus, is near the horizon in this photograph, taken by the* Apollo 17 *astronauts in 1972. (NASA)*

9-4 The formation of the Galilean satellites probably mimicked the formation of the solar system

Galileo Galilei was the first person to see the four largest satellites of Jupiter. He called them the "Medicean stars" to attract the attention of the Medici family, who were wealthy Florentine patrons of the arts and sciences. To Galileo in 1610, they provided observational evidence supporting the heretical Copernican cosmology. For the modern astronomer, the Voyager flyby photos of 1979 revealed four extraordinary terrestrial worlds, now called the **Galilean satellites,** different from anything anyone had ever seen or imagined. They are named after the mythical lovers and companions of Zeus: Io, Europa, Ganymede, and Callisto.

When viewed through an Earth-based telescope, the Galilean satellites look like mere pinpoints of light. With patience, you can follow these four worlds as they orbit Jupiter. Because their orbital periods are fairly short—from 1.8 days for Io to 16.7 days for Callisto—major changes in the positions of the satellites from one night to the next are easy to observe.

The brightness of each of the moons varies slightly as it moves along its orbit. These variations can be attributed to dark and light surface areas that are alternately exposed to or hidden from view as the satellite rotates. Careful measurements show that the brightness of each satellite varies with a period equal to the satellite's orbital period. In other words, each Galilean satellite rotates exactly once on its axis during each orbit. We can thus conclude that each Galilean satellite keeps the same hemisphere perpetually facing Jupiter, just as our Moon keeps the same side facing the Earth.

Because of Voyager photographs, the diameters of the Galilean satellites could be measured accurately (Figure 9-18). The two inner Galilean satellites, Io and Europa,

Figure 9-18 *The Galilean satellites with our Moon and Mercury* Jupiter's four largest satellites are shown here along with our Moon and Mercury. All six worlds are reproduced to the same scale. Io has numerous active volcanoes and Europa has a smooth, icy surface; both are roughly the same size as our Moon. Ganymede and Callisto are each covered with a 1000-km-thick layer of ice; both are roughly the size of Mercury. (NASA)

Table 9-1   The Galilean satellites, Mercury, and the Moon

Name	Distance from Jupiter (km)	Orbital period (days)	Diameter (km)	Mass (Moon = 1)	Average density (kg/m³)
Io	412,600	1.77	3630	1.22	3570
Europa	670,900	3.55	3138	0.65	2970
Ganymede	1,070,000	7.16	5262	2.01	1940
Callisto	1,883,000	16.69	4800	1.47	1860
Mercury	——	——	4878	4.49	5430
Moon	——	——	3476	1.00	3340

are approximately the same size as our Moon. The two outer satellites, Ganymede and Callisto, are comparable in size to Mercury.

Deflections in the trajectories of the Voyager spacecraft as they passed the Galilean satellites provided important data from which their masses were computed. Europa, the smallest Galilean satellite, is also the least massive. Ganymede, the largest Galilean satellite, is the most massive. Data about the Galilean satellites are listed in Table 9-1.

As soon as reliable mass and diameter measurements were available, it became apparent that the average densities of the satellites are related to their distances from Jupiter. The innermost satellite, Io, has the highest average density (3570 kg/m³), slightly denser than our Moon. The next satellite, Europa, has an average density of 2970 kg/m³, which is not quite as dense as our Moon. Recalling that rocks in the Earth's crust typically have densities around 3000 kg/m³, it is reasonable to suppose that both Io and Europa are made primarily of rocky material.

The pattern of decreasing density with increasing distance from Jupiter continues with the outer two satellites. Ganymede and Callisto each have an average density of less than 2000 kg/m³, indicating that these two satellites are composed of roughly equal amounts of rock and ice.

Certain parallels exist between the arrangement of the Galilean satellites about Jupiter and the planets about the Sun. For example, moving outward from the Sun, average density declines from more than 5400 kg/m³ for Mercury to less than 700 kg/m³ for Saturn (recall Table 6-2). Scientists therefore began to suspect that the same general processes that had formed the solar system were at work during the formation of the Galilean satellites, though on a much smaller scale.

NASA scientists recently used computer simulations to determine the conditions necessary for the formation of the Galilean satellites. Included in the computations was the fact that Jupiter emits twice as much energy as it receives from the Sun. The scientists calculated that frozen water could be retained and incorporated into satellites at the distances of Ganymede and Callisto but that only rocky material would condense at the orbits of Io and Europa because of Jupiter's warmth. Jupiter's gravity and heat thus produced two distinct classes of Galilean satellites, just as warmth from the protosun caused a dichotomy between the small, rocky, dense inner planets and the huge, gaseous, low-density outer planets.

9-5   Io is covered with colorful deposits of sulfur compounds ejected from numerous active volcanoes

Within a few hours after Voyager 1 passed near Jupiter, Io loomed into view, and the probe began sending back a series of strange and unexpected pictures, such as the one shown in Figure 9-19. Baffled by what they were seeing, scientists jokingly compared Io to pizzas and rotten oranges.

A major clue to these puzzling vistas was uncovered several days after the Jupiter flyby, when a navigation engineer noticed a large umbrella-shaped cloud protruding from Io in one photograph—an erupting volcano! No one had expected to obtain photographs of erupting volcanoes on Io. After all, a probe making a single trip past the Earth would be very unlikely to catch a large volcano in the act of erupting.

Careful reexamination of the close-up photographs revealed eight giant ongoing eruptions. These volcanoes are named after gods and goddesses associated with fire

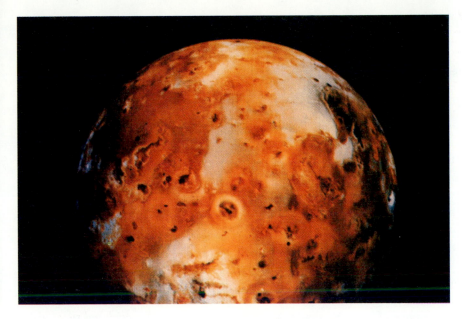

Figure 9-19 **Io** *This close-up view of Io was taken by Voyager 1. Notice the extraordinary range of colors from white, yellow, and orange to black. Scientists believe that these brilliant colors result from surface deposits of sulfur ejected from Io's numerous volcanoes. (NASA)*

in Greek, Norse, Hawaiian, and other mythologies. Figure 9-20 shows two views of the plume of Prometheus.

As it orbits Jupiter, Io is repeatedly caught in a gravitational tug-of-war between the huge planet on one side and the other Galilean satellites on the other. This gravitational battle distorts Io's orbit, varying its distance from Jupiter. As the distance varies, tidal stresses on Io alternately squeeze and flex the satellite. This constant tidal flexing in turn causes frictional heating of Io's interior. Calculations show that the heat pumped into Io this way

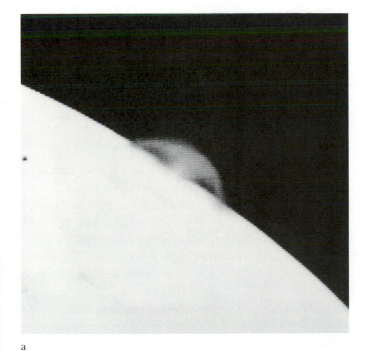

a

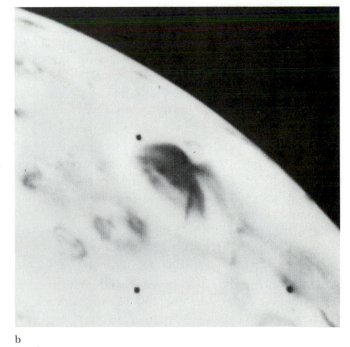

b

Figure 9-20 **Prometheus on Io** *These two views from Voyager 1, taken two hours apart, show details of the plume of the volcano called Prometheus rising 100 km above Io's surface. (a) The plume's characteristic umbrella shape is seen silhouetted against the blackness of space. (b) When viewed against the light background of Io's surface, jets of material give the plume a spiderlike appearance. (NASA)*

is equivalent to 2000 tons of TNT exploding every second. Eventually this energy makes its way to Io's surface, producing the numerous volcanoes.

The plumes and fountains of material spewing from Io's volcanoes rise to astonishing heights of 70 to 280 km above the surface. To reach these altitudes, the material must emerge from the volcanic vents with speeds between 300 and 1000 m/s, a speed much greater than that found in the most violent terrestrial volcanoes. For example, Vesuvius, Krakatoa, and Mount St. Helens had eruption velocities of only about 100 m/s. Scientists therefore began to suspect that Io's volcanoes operate in a fundamentally different way from volcanoes on Earth. The evidence of these differences came from Voyager's pictures and data.

No impact craters like those on our Moon were seen on Io. Material from the volcanoes apparently obliterates impact craters soon after they are created. The absence of craters indicates that Io's surface is extremely young—perhaps less than 100 million years old.

The Voyager cameras revealed numerous black dots on Io, which seem to be the volcanic vents from which the eruptions occur. These black spots are typically 10 to 50 km in diameter and form 5% of Io's surface. Lava flows radiate from many of these black dots (Figure 9-21), some of which have volcanic plumes.

Evidence supporting the volcanic nature of the black spots came from Voyager instruments that measured the intensity of infrared radiation across Io's surface. Some of the black spots have temperatures as high as 20°C, in sharp contrast to the surrounding surface temperature of only –146°C.

After they discovered widespread volcanic activity on Io, scientists soon concluded that sulfur ejected from the volcanoes is responsible for Io's brilliant colors. Sulfur is normally bright yellow, but when it is heated and then suddenly cooled, it can assume a range of colors from orange and red to black.

Voyager detected sulfur dioxide (SO_2) in the plumes from Io's volcanoes. Sulfur dioxide is an acrid gas commonly discharged from volcanic vents here on Earth. When this gas is released into the cold vacuum of space from eruptions on Io, it crystallizes into white snowflakes. It is likely that the whitish deposits on Io (see Figure 9-19) are sulfur dioxide frost or snow.

The abundant sulfur and sulfur dioxide on Io suggest an explanation for the mechanism of its volcanoes. Indeed, the term *volcano* may be the wrong word altogether. We have seen that material is ejected from volcanic vents on Io at much higher velocities than is observed in even the most explosive volcanic eruptions on Earth. In addition, Io's vents are not located at the tops of tall volcanic mountains. Many volcanoes on Earth and Mars have an easily recognizable conical shape with a caldera at the summit. Few of Io's calderas (the black spots) are associated with major topographical relief.

In all these respects, Io's volcanoes are more similar to terrestrial geysers than volcanoes. In a geyser, such as those in Yellowstone Park in Wyoming, water seeps down to volcanically heated rocks, changes suddenly to steam, and erupts explosively through a vent. If the Old Faithful geyser were to erupt under the low gravity and vacuum that surround Io, it would send a plume of water and ice to an altitude of 40 km.

Both sulfur and sulfur dioxide are molten at depths of only a few kilometers below Io's surface because of the heat generated by tidal flexing. Geologists have pointed out that sulfur dioxide could be the principal propulsive agent driving Io's eruptions. Just as the explosive conversion of water into steam produces a geyser on Earth, the sudden conversion of liquid sulfur dioxide into a high-pressure gas could produce an eruption on Io. Calculations indicate that this explosive expansion of sulfur dioxide could generate eruption velocities up to 1000 m/s.

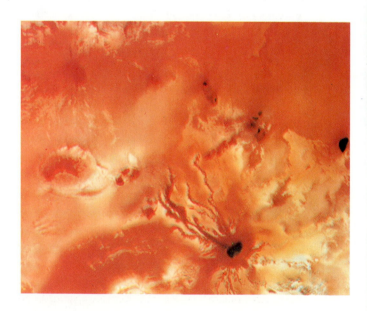

Figure 9-21 A volcanic center on Io *No impact craters are seen in close-up pictures such as this one taken by Voyager 1. Long, meandering lava flows radiate from many of the black dots that seem to be the sites of intense volcanic activity. This photograph covers an area 1000 by 800 km, approximately twice the size of California. (NASA)*

The material erupting from Io's geyserlike volcanoes is composed primarily of sulfur and sulfur dioxide. Voyager's instruments failed to detect any other abundant gases such as the water vapor and carbon dioxide emitted from terrestrial volcanoes. Io has apparently been completely outgassed by volcanic activity over hundreds of millions of years. Io has not been able to retain its volatile gases and has almost no atmosphere because its surface gravity is comparable to that of our Moon.

It is estimated that each volcano on Io ejects roughly 10,000 tons of material per second. Although this material is a hot mixture of molten sulfur and sulfur dioxide gas under high pressure as it gushes from a volcanic vent, the gas–liquid mixture rapidly cools and solidifies in the cold, nearly perfect vacuum around Io. It then takes about half an hour for the fine particles of sulfur dust and sulfur dioxide snow to fall back down onto the surface.

Altogether, Io's volcanoes and vents eject an estimated 100 billion tons of matter each year, producing a sulfur-rich layer 10 m thick over Io's entire surface annually. Thus the surface is constantly changing, and it is probably safe to say that there are no long-lived or even semipermanent features on Io.

Figure 9-22 Europa *Europa's ice surface is covered by numerous streaks and cracklike features that give the satellite a fractured appearance. The streaks are typically 20 to 40 km wide. This picture, taken by* Voyager 2, *reveals surface features as small as 5 km across. (NASA)*

9-6 Europa is covered with a smooth layer of ice, crisscrossed with many cracks

Voyager 1 did not pass near Europa, but *Voyager 2* captured the excellent view shown in Figure 9-22. Europa is a very smooth world with no mountains and very few craters, crisscrossed with a spectacular series of streaks and cracks. Most of the cracks appear to be filled with dark-colored material, but some cracks contain light-colored substances.

We have seen that Europa's average density is about 10% less than the average density of our Moon. Spectroscopic observations from Earth indicate that Europa has frozen water on its surface. These two facts suggest that Europa's surface may be covered with an ice layer 100 km thick. That would be consistent with its remarkable smoothness. An "ocean" of ice 100 km deep would certainly hide mountain ranges and other topographical features. But what causes the network of cracks, and why are impact craters so rare?

Tidal squeezing is responsible for the volcanism on Io, which gives Io many of its extraordinary characteristics. Europa is caught in a similar tidal tug-of-war, with Jupiter to one side and the two largest Galilean moons periodically passing on the opposite side. However, Europa is much farther from Jupiter than Io is, and the tidal effects on Europa are considerably weaker than those on Io.

The tidal flexing of Europa is responsible for the network of cracks that cover its surface. Some of the darkest streaks, in fact, follow paths along which the tidal stresses are calculated to be strongest. Although the tidal flexing of Europa is far too weak to produce volcanoes, it is thought to supply enough energy to jostle and churn the satellite's icy coating. This activity would explain why only a few small impact craters have survived to the present time, even though Europa's surface is much older than Io's.

The streaks on Europa are presumably caused by cracks in the icy coating through which water gushed up and then froze. These cracks are a few tens of kilometers wide. To accommodate all of them, Europa's surface area must have increased by 10% to 15% since the cracks first began to appear. It seems unreasonable to suppose that Europa is expanding like an inflating balloon, though. Apparently more is happening on Europa than meets the eye. Perhaps old surface material is somehow being pulled back down into the mushy layer below the ice coating and then recycled, just as the Earth's crust is pulled back down into the Earth's mantle in subduction zones (recall Figure 7-17). Europa's surface may represent a water-and-ice version of plate tectonics.

9-7 Ganymede and Callisto have heavily cratered, icy surfaces

Europa's density can be explained by 100 km of ice and water on top of an otherwise rocky world. A much thicker layer of water and ice must surround Ganymede and Callisto, however. To be consistent with average densities slightly less than 2000 kg/m^3, the rocky cores of these two outer satellites must be enveloped in mantles of water and ice nearly 1000 km thick.

Figures 9-23 and 9-24 show close-up views of Ganymede and Callisto. Both satellites have the kind of ancient, cratered surface normally associated with Moonlike landscapes. Of course, the craters on both worlds are of ice rather than rock.

Ganymede is the largest satellite in the solar system. It is 5262 km in diameter, slightly larger than Mercury. The largest single feature on Ganymede is a vast, dark, circular island of ancient ice called Galileo Regio (examine Figure

Figure 9-24 Callisto *This view from* Voyager 2 *shows the second largest Galilean satellite, Callisto. Its diameter is almost exactly the same as Mercury's. (NASA)*

Figure 9-23 Ganymede *This view from* Voyager 2 *of Ganymede shows the hemisphere that always faces away from Jupiter. The surface is dominated by a huge, dark, circular region called Galileo Regio, which is the largest remnant of Ganymede's ancient crust. (NASA)*

9-23). It measures 4000 km in diameter and covers nearly one third of the hemisphere of Ganymede that faces away from Jupiter. This surface feature is the only one on the Galilean satellites that can be detected with Earth-based telescopes.

Ganymede has two very different kinds of terrain, which are distinguished by both appearance and age (Figure 9-25). Dark, polygon-shaped regions are presumed to be the oldest surface features because they exhibit a high density of craters. Light-colored, heavily grooved terrain found between the dark angular islands is much less cratered and therefore younger.

It is easy to distinguish young craters on Ganymede. The youngest craters are surrounded by bright rays of freshly exposed ice. Older craters clearly have been covered with deposits of dark meteoritic dust. The most ancient craters are barely visible on Galileo Regio and on the other dark, angular island remnants of old crust. The degradation and near obliteration of the oldest craters probably also involved the slow plastic flow of Ganymede's icy surface.

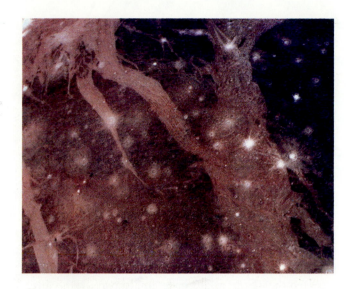

Figure 9-25 Young and old terrain on Ganymede This close-up view of Ganymede was taken by Voyager 2. *Features as small as 5 km across can be seen. Dark, angular islands of Ganymede's ancient crust are separated by younger, light-colored, grooved terrain. The southwest edge of Galileo Regio appears at the right side of the picture. (NASA)*

down 3 billion years ago as its cooling crust froze to unprecedented depths.

When Ganymede formed 4.5 billion years ago, it may have been completely covered with an ocean roughly 1000 km deep. During the next 200 million years, the water cooled and formed a thick coating of ice. Today this layer of solid ice is probably about 100 km thick. Beneath it lies a 900-km-thick slushy mantle of water and ice.

Callisto, Jupiter's outermost Galilean satellite, looks very much like Ganymede: numerous impact craters scattered over an ancient, dark, icy crust. There is one obvious difference, however—Callisto has no younger, grooved terrain. We thus infer that tectonic activity never began on Callisto. Perhaps because of its greater distance from Jupiter, the ocean that enveloped young Callisto 4.5 billion years ago froze more rapidly and to a greater depth than on Ganymede, forever preventing tectonic processes. Callisto's icy crust may in fact be several times thicker than Ganymede's, extending to depths of several hundred kilometers. It is bitterly cold on Callisto. Voyager's instruments measured a noontime temperature of −118°C (−180°F), which plunged to a nighttime low of −193°C (−315°F).

Craters also tell us about the history of the younger, light-colored terrain covered with numerous grooves. In some places the cratering is about as dense as that on the ancient crust, but in other places it is only one-tenth that amount. We thus suspect that this grooved terrain was formed over a long period of time. The process probably began quite early in Ganymede's history and continued through the period of intense meteoritic bombardment. The age of the grooved terrain therefore probably ranges from about 3.5 to 4.5 billion years.

High-resolution photographs such as Figure 9-26 show that this grooved terrain actually consists of parallel mountain ridges up to 1 km high and spaced 10 to 15 km apart. These features suggest that plate tectonics may have dominated Ganymede's early history. Water seeping upward through cracks in the original crust would freeze and force apart fragments of the original crust, producing jagged, dark islands of old crust separated by bands of younger, light-colored, heavily grooved ice. The cracks play a role in plate tectonics on Ganymede that is analogous to the role of the oceanic rifts on the Earth. But unlike thinner-crusted Europa, where tectoniclike activity may even occur today, tectonics on Ganymede bogged

Figure 9-26 Grooved terrain on Ganymede This picture, taken by Voyager 1, *shows an area roughly as large as the state of Pennsylvania. The smallest visible features are about 3 km across. Numerous parallel mountain ridges are spaced 10 to 15 km apart and have heights up to 1 km. (NASA)*

Figure 9-27 Callisto Numerous craters pockmark Callisto's icy surface, as seen in this mosaic of views from Voyager 1. *A huge impact basin called Valhalla dominates the Jupiter-facing hemisphere of this frozen, geologically inactive world. A series of concentric rings up to 3000 km in diameter surrounds the impact site. (NASA)*

Voyager 1 photographed a huge impact feature on Callisto's Jupiter-facing hemisphere (Figure 9-27). This feature, called the Valhalla Basin, consists of a large number of concentric rings 50 to 200 km apart with diameters ranging up to 3000 km. Valhalla was produced by an asteroid-sized object very early in the satellite's history.

The great age of Valhalla is inferred from both the presence of overlying impact craters and the absence of substantial vertical relief of the concentric rings. The Valhalla impact probably occurred around 4 billion years ago, when the satellite's young, relatively thin crust was still plastic enough to flow and reduce the height of the rings in the ice.

The Voyager pictures also showed traces of a Valhalla-like impact on Ganymede's Galileo Regio. Segments of a system of concentric rings cover a large portion of this dark island of ancient crust. However, no obvious impact feature is found at the center of this ring system. Apparently the development of grooved terrain completely obliterated the impact basin.

9-8 Titan has a thick, opaque atmosphere rich in methane, nitrogen, and hydrocarbons

Long before the Voyager flybys, astronomers knew Saturn's largest satellite to be an extraordinary world. It was discovered in 1655. By the early 1900s, several scientists suspected that Titan might have an atmosphere because it is cool enough and massive enough to retain heavy gases. Confirming evidence came in 1944, when astronomers discovered spectral lines of methane in the sunlight reflected from Titan. Titan is the only satellite in the solar system known to have an appreciable atmosphere.

Because of this atmosphere, Titan was a primary target for the Voyager missions. To everyone's chagrin, however, the Voyagers spent hour after precious hour sending back featureless images such as Figure 9-28. Titan's cloud cover is so thick that it blocks any view of the surface and allows very little sunlight to penetrate to the ground; the surface of Titan must be a dark, gloomy place.

In size, mass, and average density, Titan is quite similar to the largest Jovian satellites. We would thus expect its internal structure to resemble those of Ganymede and Callisto—a rocky core surrounded by a mantle of frozen water nearly 1000 km thick.

Titan's thick atmosphere distinguishes it from all other satellites. The atmospheric pressure at Titan's surface is 1.6 atm, or 60% greater than the atmospheric pressure at sea level on Earth, even though Titan's surface gravity is lower than the Earth's. Considerably more gas must be weighing down on Titan than on Earth. About ten times more gas lies above each square centimeter of Titan's surface than above Earth's surface.

What factors leading to the formation of Titan's atmosphere did not exist for Ganymede or Callisto? For one, Titan formed in a much cooler part of the solar nebula. In contrast to the materials available in the warmer conditions near Jupiter, the ices from which Titan accreted probably contained substantial amounts of frozen methane and ammonia. As Titan's interior became warm through the decay of naturally occurring radioactive isotopes, these ices vaporized, producing an atmosphere

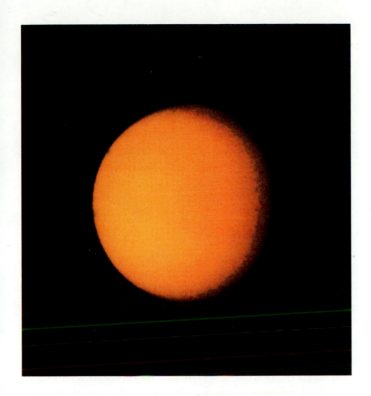

Figure 9-28 Titan This view of Titan was taken by Voyager 2. Very few features are visible in the thick, unbroken haze that surrounds this large satellite. The main haze layer is located nearly 300 km above Titan's surface. (NASA)

around the young satellite. Methane (CH_4) is stable in sunlight and thus remained in the atmosphere, but ammonia (NH_3) is easily broken down into nitrogen and hydrogen by the Sun's ultraviolet radiation. Because Titan's gravity is too weak to retain hydrogen, that gas escaped into space. Even today hydrogen is escaping from Titan at a substantial rate.

The breakup of ammonia and the resulting loss of hydrogen leaves Titan with an abundant supply of nitrogen. Voyager data suggest that roughly 90% of Titan's atmosphere is nitrogen. The two next most abundant gases are argon and methane.

The interaction of sunlight with methane induces chemical reactions that produce a variety of carbon–hydrogen compounds called **hydrocarbons.** Voyager's instruments detected small amounts of many hydrocarbons such as ethane (C_2H_6), acetylene (C_2H_2), ethylene (C_2H_4), and propane (C_3H_8) in Titan's atmosphere. Nitrogen combines with these hydrocarbons to produce other compounds such as hydrogen cyanide (HCN), some of which are the building blocks of the organic molecules on which life is based. However, there is little reason to suspect life

on Titan—its surface temperature is 95 K ($-288°F$)—but a more detailed study of its chemistry may shed light on the origins of life on Earth.

On Earth, the atmospheric pressure and temperature are near the **triple point** of water, meaning that water is found in all three phases: liquid, solid, and gas. The atmospheric pressure and temperature at Titan's surface are near the triple point of methane, however. Thus methane may play a role on Titan similar to that which water plays on Earth. Methane snowflakes may fall onto frozen methane polar caps, and methane raindrops may descend into methane rivers, lakes, and seas in warmer areas.

Some molecules are capable of joining together in long, repeating molecular chains to form substances called **polymers.** Many of the hydrocarbons and carbon–nitrogen compounds in Titan's atmosphere form such polymers. Droplets of some polymers remain suspended in the atmosphere to form the kind of mixture called an **aerosol** (Figure 9-29), but the heavier polymer particles settle down onto Titan's surface, probably covering it with a thick layer of sticky, tarlike goo. Some scientists

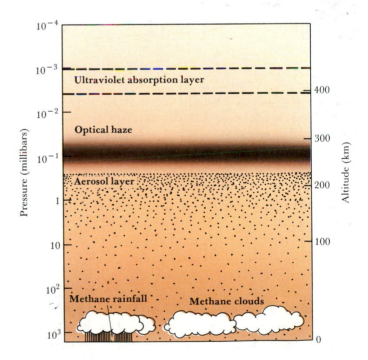

Figure 9-29 Titan's atmosphere The Voyager data suggest that Titan's atmosphere has three distinct layers. The uppermost layer absorbs ultraviolet radiation from the Sun. The middle layer is opaque to visible light. The lowest layer is an aerosol of suspended particles. Methane rain clouds may exist near Titan's surface, which is at a temperature of about 95 K. (Adapted from Tobias Owen)

estimate that the deposits of hydrocarbon sludge on Titan may be half a kilometer deep.

9-9 *Triton is a frigid, icy world with a young surface and a tenuous atmosphere*

Triton, the outermost giant satellite of the solar system, was discovered in 1846. Because of its great distance from Earth, however, astronomers were unable to learn much about this remote world until the *Voyager 2* flyby of Neptune in 1989. Even through a large telescope Triton is a faint, featureless, starlike pinpoint of light.

An important property of Triton, known since its discovery, is its retrograde orbit. Triton goes around Neptune opposite to the direction in which the planet rotates. No other giant satellite has such an orbit, and it is difficult to imagine how any satellite might form near a planet in an orbit opposing the direction of the planet's rotation. Only a few of the outer satellites of Jupiter and Saturn have retrograde orbits, and these bodies are probably captured asteroids. Some scientists have therefore hypothesized that Triton may have been captured long ago by Neptune's gravity.

The orbit of Triton is tilted by about 20° with respect to the plane of Neptune's equator, which in turn is tilted by about by about 29° to the plane of Neptune's orbit about the Sun. Because of its 16-hour rotation, Neptune is slightly oblate, possessing an equatorial bulge in the plane of its equator. The gravitational pull of this bulge causes the satellite's orbit to precess, resulting in significant changes in Triton's seasons over the years. There are years during which Triton experiences modest seasons like those on Earth, and others (as when *Voyager 2* passed by Triton in 1989) in which it has exaggerated, Uranuslike

seasons with the Sun shining down onto the satellite's south pole.

A high-resolution image of Triton's south polar region is shown in Figure 9-30. Note that very few craters are seen, indicating that some process must have obliterated the scars of numerous, ancient impacts left over from the first billion years of the solar system. A major modification of Triton's surface is consistent with the idea that the satellite once orbited the Sun on its own and was captured by Neptune, perhaps 3 or 4 billion years ago. Upon being captured, Triton most likely would have started off in a highly elliptical orbit, but today the satellite's orbit is remarkably circular. Triton's original elliptical orbit would have been "circularized" by tidal forces exerted on the satellite by Neptune's gravity. With each revolution about that original elliptical orbit, the changing distance to Neptune would have stretched and flexed Triton, causing its orbit to become more circular while depositing significant energy into the satellite itself. This energy would have melted much of the satellite's interior, and the resulting volcanism would have obliterated Triton's original surface features, including craters.

The pinkish ice that constitutes Triton's south polar cap is probably a layer of nitrogen frost. As summer arrives at Triton's south pole, this frost layer slowly evaporates, and many of the surface markings seen in Figure 9-30 may be the result of the northward flow of the resulting nitrogen gas. These winds are very tenuous, however, because Triton's atmosphere is so very thin. Atmospheric pressure on the satellite's surface is only 10^{-5} of that at sea level on Earth.

Triton exhibits some surface features seen on other icy worlds, such as long cracks resembling those on Europa and Ganymede. Other features unique to Triton are quite puzzling. For example, near the top of Figure 9-30, you

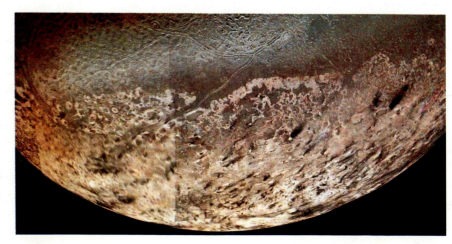

Figure 9-30 *Triton's south polar cap* Approximately a dozen high-resolution images were combined to produce this view of Triton's southern hemisphere. The pinkish polar cap is probably made of nitrogen frost. A notable scarcity of craters suggests that Triton's surface was either melted or flooded by icy lava after the era of bombardment that characterized the early history of the solar system. (NASA)

Figure 9-31 A frozen lake on Triton? Some scientists believe that this lakelike feature is the caldera of an ice volcano. The flooded basin is about 200 km wide and 400 km long, an area about the size of the state of West Virginia. (NASA)

can see a wrinkled terrain that resembles the skin of a cantaloupe. Triton also has a few frozen lakes like the one shown in Figure 9-31. Some scientists speculate that these lakelike features are the calderas of extinct ice volcanoes. A mixture of methane, ammonia, and water, which can have a melting point far below that of pure water, could have constituted a kind of "cold lava" on Triton. It is unlikely that any lava is flowing on Triton today, however, because the satellite is so very cold. Voyager instruments measured a surface temperature of 37 K (−395°F), making Triton the coldest world we have ever visited. Nevertheless, Voyager cameras did glimpse two plumes of gas extending up to 8 km above the satellite's surface. Perhaps these plumes are nitrogen gas escaping through vents or fissures warmed by the feeble summer Sun.

9-10 Pluto may be one of a thousand "ice dwarfs" that orbit the Sun at the outskirts of the solar system

Pluto was discovered in 1930 by Clyde W. Tombaugh, who used a wide-field camera to photograph sections of the sky. He recognized the ninth planet from the Sun as a faint starlike object that slowly shifts it position from night to night. Pluto's only moon, Charon, was discovered in 1978. The best pictures of Pluto and Charon have come from the Hubble Space Telescope (Figure 9-32).

The average distance between Charon and Pluto is a scant 19,700 km—less than $\frac{1}{20}$ the distance between the Earth and our Moon. Furthermore, Charon's orbital period of 6.3872 days is the same as the rotational period of Pluto. In other words, Pluto always keeps the same side facing Charon. As seen from the satellite-facing side of Pluto, Charon neither rises nor sets but instead seems to hover in the sky, as if perpetually suspended above the horizon.

From 1985 through 1990, the orbit of Charon was positioned so that Earth-based observers could view mutual eclipses of Pluto and its satellite. Astronomers used observations of these eclipses to determine that Pluto's diameter is 2300 km and Charon's 1190 km. For comparison, our Moon's diameter (3476 km) nearly equals the diameters of Pluto and Charon added together.

The average densities of Pluto and Charon, at about 2000 kg/m^3, are essentially the same as that of Triton (2070 kg/m^3). All three worlds are therefore probably composed of nearly the same proportion of rock and ice.

Pluto and Charon are remarkably alike in mass, size, density, and quite possibly chemical composition. Throughout the rest of the solar system, planets are always many times larger and more massive than any of their satellites. The exceptional similarities between Pluto and Charon suggest to some astronomers that this binary system may have formed when Pluto collided with a similar body. Perhaps chunks of matter were stripped from the second body, leaving behind a mass, now called Charon, that was vulnerable to capture by Pluto's gravity. Alternatively, perhaps Pluto's gravity captured Charon into orbit during a close encounter between the two worlds.

For either of these scenarios to be feasible, there must have been many Plutolike objects in the outer regions of the solar system. One astronomer estimates that there must have been at least a thousand Plutos in order for a collision or close encounter between two of them to have

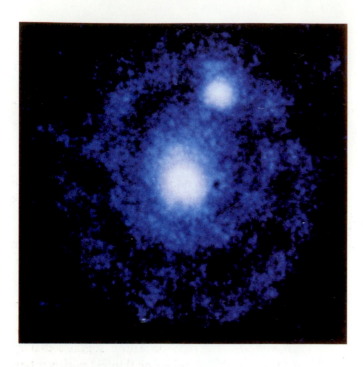

Figure 9-32 *Pluto and Charon   This picture of Pluto with its moon, Charon, was taken by the Hubble Space Telescope. Pluto and Charon are separated by only 19,700 km. The Pluto–Charon system deserves to be called a double planet, because these two objects resemble each other in mass and size more closely than do any other planet–satellite pair in the solar system. (NASA; European Space Agency)*

occurred at least once since the solar system formed 4.5 billion years ago.

Triton seems to offer evidence supporting the idea of many Plutos at the outskirts of the solar system. In addition to being quite like Pluto and Charon, Triton has a retrograde orbit that suggests it was gravitationally captured by Neptune. A gravitational capture of one body by another requires a very special set of orbital circumstances; the vast majority of close encounters between pas-

sing bodies only serves to deflect them from their original paths. For one successful capture of Triton by Neptune there must have been many unsuccessful near misses, which implies the existence of many Plutolike objects.

An overview of all the members of the solar system demonstrates that small objects are vastly more common than large objects. In other words, the population of solar system objects increases dramatically with decreasing mass. This trend suggests that for a thousand Plutolike worlds there must have been 10 to 50 icy, Earth-sized worlds in the vicinity of Uranus and Neptune. A direct hit by an object with roughly the mass of the Earth could have knocked Uranus on its side. None of these objects remain near the orbits of Neptune or Uranus today, because gravitational deflections by these two giant planets long ago catapulted all the remaining Plutolike worlds far from the Sun.

Some astronomers take issue with the idea of many Plutos. There really is no complete theory for the formation of Uranus and Neptune, they argue, and so speculation about Pluto and Charon seems pointless. Observers, however, may soon have the chance to test the theory by looking for these Plutolike worlds.

Any attempt to view distant Plutos with ordinary telescopes is futile because these tiny, remote bodies do not reflect enough sunlight to be seen. They would, however, be weak sources of infrared radiation. Even though these icy objects probably have surface temperatures of only 80 to 90 K, they are nevertheless warmer than the blank, background sky. Highly sensitive, wide-angle infrared telescopes may therefore be able to detect a population of Plutos at distances of about 100 AU from the Sun. An 8-meter infrared telescope currently planned for the summit of Mauna Kea in Hawaii will be an excellent tool for the search. If successful, astronomers will have to include a new class of objects in their theories about the formation of the solar system.

Summary

- Mercury's surface is pocked with craters like those of the Moon, but there are extensive, smooth intercrater plains. These features appear to have formed as the crust of the planet solidified.

- Surface temperatures on Mercury range from 100 to 700 K, the greatest range known on any of the planets.

- Mercury has an iron core much like that of the Earth.

- Mercury's magnetic field produces a magnetosphere that blocks the solar wind from the surface of the planet.

- The Earth-facing side of the Moon displays light-colored, heavily cratered highlands (terrae) and dark-colored, smooth-surfaced maria.

  The Moon's far side has no maria.

- Lunar rocks contain no water and also differ from terrestrial rocks by being relatively enriched in refractory elements and depleted in volatile elements.

- The Moon may have formed from rock torn from the Earth by a collision with a large asteroidlike object shortly after the solar system was formed.

  The Moon was molten in its early stages, and the anorthositic crust solidified from low-density magma that floated to the lunar surface. The mare basins were created later by the impact of planetesimals that released lava from the lunar interior.

  The Moon's surface has undergone very little change in the past three billion years.

- The inner two Galilean moons, Io and Europa, are roughly the size of our Moon and have densities similar to that of the Moon. The outer two Galilean moons, Ganymede and Callisto, are roughly the size of Mercury and are lower in density than the Moon or Mercury.

- Io is covered with a colorful layer of sulfur compounds deposited by frequent explosive eruptions from volcanic vents.

  Io's volcanic eruptions resemble terrestrial geysers. The energy heating Io's interior comes from tidal forces that flex the moon as it passes between the planet and the other large moons.

- Europa is covered with a smooth layer of frozen water that is crisscrossed by an intricate pattern of long cracks, probably produced by tidal flexing.

- The heavily cratered surface of Ganymede is composed of frozen water. Large polygons of dark, ancient surface are separated by regions of heavily grooved, lighter-colored, younger terrain. Plate tectonics apparently operated during the early history of Ganymede.

- Callisto also has a heavily cratered crust of frozen water, but plate tectonics apparently never operated on this moon, presumably because it quickly developed a thick, solid crust.

- The Galilean satellites probably formed through a process of accretion similar to the process that formed the solar system about the Sun, but on a smaller scale.

- The largest satellite of Saturn, Titan, has a dense nitrogen atmosphere in which methane may play a role similar to that of water on Earth.

  A variety of hydrocarbons are formed by the interaction of sunlight with methane, creating an aerosol layer in Titan's atmosphere and probably a thick sludge on its surface.

- The largest satellite of Neptune, Triton, moves in a retrograde orbit that suggests it may have been captured into orbit by Neptune's gravity.

  Triton is an icy world with a tenuous nitrogen atmosphere.

  The scarcity of craters on Triton suggests that its surface was either molten or flooded with icy lava sometime after the era of bombardment that left numerous craters on such worlds as Mercury and our Moon.

- Pluto, the smallest planet in the solar system, and its satellite Charon are icy worlds that resemble Triton.

  A population of Plutolike objects may exist far beyond the orbits of Uranus and Neptune.

Review questions

1 Why are naked-eye observations of Mercury best made at dusk or dawn, whereas telescopic observations are best made around noon?

2 Why do astronomers believe Mercury to be the most iron-rich planet in the solar system?

3 Compare the surfaces of Mercury and the Moon. How are they similar? How are they different?

4 How old are the lunar maria? How does this age compare with the age of the lunar highlands? What does this tell us about the early history of the lunar surface?

5 What is the lunar regolith?

6 What are the two most prevalent types of rock on the lunar surface? With what sort of lunar features or topology are these rocks associated?

7 What is an impact breccia?

8 What does a comparison of the chemical abundances of terrestrial rocks and lunar rocks tell us about the formation of the Moon?

9 Which Galilean satellites are nearly the same size as our Moon? Which are nearly the same size as Mercury?

10 How does the Galilean satellite system resemble the solar system? How is it different?

11 With all its volcanic activity, why doesn't Io possess an atmosphere?

12 Compare and contrast the surface features of the four Galilean satellites, discussing the relative geological activity and evolution of these satellites.

13 Why is it reasonable to conclude that the surface of Neptune's largest satellite is much younger than the surface of our Moon?

14 In what ways does Pluto resemble Charon and Triton? Why do these similarities inspire some astronomers to speculate that many Plutos orbit the Sun?

Advanced questions

15 Why is more lunar detail visible through a telescope when the Moon is near quarter phase than when it is at full phase?

16 How would you prove to someone that the Moon has no atmosphere?

17 Why do you suppose that no Apollo mission landed on the far side of the Moon?

18 Why is it reasonable to presume that none of the large satellites in the solar system, including our Moon, possess a substantial magnetic field?

19 Long before the Voyager flybys, Earth-based astronomers reported that Io appeared brighter than usual for a few hours after emerging from Jupiter's shadow. Based on what we know about the material ejected from Io's volcanoes, explain this brief anomalous brightening of Io.

20 Why are all the maria on the Earth-facing side of the Moon?

21 In what ways can Pluto and its moon be considered a "double planet"? Support your argument by calculating the ratio of the mass of each planet's largest satellite to the mass of the planet itself. What satellite–planet pair other than Pluto and Charon stands out in your list? (*Hint:* You

can simply look up the masses [in kilograms] of the various planets and their satellites in reference books. Alternatively, you could calculate these masses from the known average densities and diameters of these worlds. To do such calculations, you need to know that the volume of a sphere of radius r is given by the formula $\frac{4}{3}\pi r^3$. The mass of a planet equals its average density multiplied by its volume.)

22 When this textbook went to press, the *Galileo* spacecraft was still on its troubled way to Jupiter. Consult recent issues of such magazines as *Sky & Telescope* and *Science News* to learn about the current status of *Galileo*. Will it be able to fulfill its original mission of multiple flybys of the Galilean satellites?

Discussion questions

23 What evidence do we have that the surface features on Mercury were not formed during recent geological history?

24 Imagine that you are planning a lunar landing mission. What type of landing site would you select in order to obtain the most ancient rock specimens? Where might you land to search for evidence of recent volcanic activity?

25 Comment on the idea that without the presence of the Moon in our sky, astronomy would have developed far more slowly.

26 Compare the advantages and disadvantages of exploring the Moon with astronauts as opposed to mobile robots.

27 Speculate on the possibility that Europa, Ganymede, or Callisto might harbor some sort of marine life.

For further reading

Cadogan, P. *The Moon—Our Sister Planet*. Cambridge University Press, 1981 • This thoughtful and thorough introduction to lunar geology was written by a scientist who participated in the analyses of Moon rocks brought back by both Apollo astronauts and Soviet automated spacecraft.

Cooper, H. Apollo *on the Moon* and *Moon Rocks*. Dial, 1970 • These two books by a respected journalist give accounts of the Apollo Moon landings and the analysis of the Moon rocks that the astronauts brought back.

Cortright, E., ed. Apollo *Expeditions to the Moon*. NASA SP-350, 1975 • The story of the Apollo program is told by astronauts and other NASA personnel.

Cowen R. "Plutos Galore." *Science News*, September 21, 1991 • This brief article gives an excellent summary of Alan Stern's speculation that many Plutos orbit the Sun.

Dunne, J., and Burgess, E. *The Voyage of* Mariner 10: *Mission to Venus and Mercury*. NASA SP-424, 1978 • This NASA publication gives a fine overview of the entire *Mariner 10* mission.

French, B. "What's New on the Moon?" *Sky & Telescope*, March and April 1977 • These two articles give an excellent summary of what we have learned from the Moon rocks and the Apollo program.

Harrington, R., and Harrington, B. "The Discovery of Pluto's Moon." *Mercury*, January/February 1979 • This article gives authoritative insights into the discovery of Pluto's moon.

Hartmann, W. "The Moon's Early History." *Astronomy*, September 1976 • This article, written by a noted planetary scientist, describes the events that shaped the Moon.

———. "The Significance of the Planet Mercury." *Sky & Telescope*, May 1976 • This excellent article by a noted planetary scientist discusses the importance of what we have learned from the *Mariner 10* mission to Mercury.

Johnson, T. "The Galilean Satellites." In Beatty, J., and Chaikin, A., eds. *The New Solar System*. 3rd ed. Sky Publishing and Cambridge University Press, 1990 • A well-written, up-to-date summary of our current understanding of the Galilean satellites.

Johnson, T., and Soderblom, L. "Io." *Scientific American*, December 1983 • This article, which is devoted to volcanism on Io, includes an enlightening discussion of the properties of sulfur and sulfur dioxide.

Moore, P. *The Moon*. Rand McNally, 1981 • This comprehensive yet brief atlas covers the gamut with many attrac-

tive illustrations and an excellent collection of maps and photographs.

Morrison, D. "Four New Worlds: The Voyager Exploration of Jupiter's Satellites." *Mercury*, May/June 1980 • This article by a noted planetary scientist discusses what we have learned from the Voyager flybys.

Murray, B. "Mercury." *Scientific American*, May 1976 • This article is one of the best summaries of what we now know about the innermost planet.

Murray, B., and Burgess, E. *Flight to Mercury*. Columbia University Press, 1977 • This exciting book, which includes an excellent selection of photographs, intersperses the history of the *Mariner 10* mission with accounts of newsworthy events that often overshadowed the mission.

Owen, T. "Titan." *Scientific American*, February 1982 • This article, which discusses many details learned from the Voyager missions, speculates that the chemistry of Titan's atmosphere may resemble that of Earth before life arose.

Schmitt, H. "Exploring Taurus-Littrow: *Apollo 17*." *National Geographic*, September 1973 • This superb description of the final manned lunar landing in a hilly region called Taurus-Littrow was written by the astronaut-geologist who was there.

Soderblom, L. "The Galilean Moons of Jupiter." *Scientific American*, January 1980 • An excellent selection of photographs accompany this fine article, which summarizes the discoveries made during the Voyager flybys.

Strom, R. *Mercury: The Elusive Planet*. Smithsonian Institution Press, 1987 • This clear, well-written overview of the Mariner mission to Mercury is a superb nontechnical introduction to the innermost planet.

———. "Mercury: The Forgotten Planet." *Sky & Telescope*, September 1990 • This fascinating article explains some of the latest ideas about Mercury, including the theory that Mercury's large iron core resulted from a devastating impact 4.5 billion years ago.

10

Interplanetary Vagabonds

The planets are not the only objects that move in orbits about the Sun. In this chapter we discuss the asteroids, meteoroids, and comets, which are small but significant members of the solar system. Some meteoroids are fragments of asteroids while others contain important clues about conditions in the solar nebula 4.5 billion years ago. Likewise, comets are dusty chunks of ice that contain frozen samples of the solar nebula. When one of these "dirty snowballs" passes near the Sun, solar radiation vaporizes some of its ice, producing a long, flowing tail. After many such passages near the Sun, a comet's ice is depleted, leaving only a swarm of dust particles that occasionally rain down on the Earth in a meteor shower. Meteoroids and asteroids occasionally strike the Earth. Indeed, meteoroids and comets may have had a significant effect on our planet, including the extinction of more than one-half the species living on Earth some 63 million years ago. Even today life on Earth is threatened by the devastation that could be wrought by a wayward asteroid.

The head of Comet Halley *This photograph shows the bluish head of Comet Halley as it approached the Sun in 1985. Comet Halley orbits the Sun with an average period of 76 years. This image was constructed from three black-and-white photographs taken with red, blue, and green filters on the same telescope. Because the comet moved slightly with respect to the background stars, the images of the stars are elongated. (Anglo-Australian Observatory)*

Many rocks and chunks of ice that condensed out of the primordial solar nebula still orbit the Sun. Just as heat from the protosun produced two classes of planets—terrestrial and Jovian—two main types of interplanetary material were created. Near the Sun, interplanetary debris consists of rock fragments called asteroids or meteoroids. Far from the Sun we find dusty chunks of ice called comets.

10-1 Bode's law led to the discovery of asteroids between the orbits of Mars and Jupiter

In the late 1700s a young German astronomer, Johann Elert Bode, popularized a simple rule that describes the distances of the planets from the Sun. This rule is usually known today as **Bode's law**—an unfortunate name because it is not a physical law and was not developed by Bode. It had first been published in 1766 by Johann Titius, a German physicist and mathematician. Most astronomers now regard this "law" as merely a coincidence, but it did lead to the discovery of a large number of previously unknown objects that orbit the Sun.

Bode's rule for remembering the distances of the planets from the Sun goes like this:

1 Write down the sequence of numbers 0, 3, 6, 12, 24, 48, 96, ... (Note that each number after the second one is simply twice the preceding number.)

2 Add 4 to each number in the sequence.

3 Divide each of the resulting numbers by 10.

As shown in Table 10-1, the final result is a series of numbers that correspond remarkably well to the distances (in astronomical units) of the planets from the Sun.

Astronomers regarded Bode's rule as merely a useful trick for remembering the planetary distances—until 1781, when William Herschel discovered Uranus, whose average distance from the Sun is very near that predicted by Bode's scheme. Suddenly it seemed far more likely that Bode's rule might actually represent some physical property of the solar system.

Astronomers now looked with new interest at the location of the "missing planet" in the sequence of Bode's law—the gap between the orbits of Mars and Jupiter. Six German astronomers who jokingly called themselves the "Celestial Police" organized an international group to

Table 10-1 Bode's law

Bode–Titius progression	Planet	Actual distance from Sun (AU)
(0 + 4)/10 = 0.4	Mercury	0.39
(3 + 4)/10 = 0.7	Venus	0.72
(6 + 4)/10 = 1.0	Earth	1.00
(12 + 4)/10 = 1.6	Mars	1.52
(24 + 4)/10 = 2.8	?	
(48 + 4)/10 = 5.2	Jupiter	5.20
(96 + 4)/10 = 10.0	Saturn	9.54
(192 + 4)/10 = 19.6	Uranus	19.18
(384 + 4)/10 = 38.8	Neptune	30.06
(768 + 4)/10 = 77.2	Pluto	39.44

begin a careful search for the missing planet. Before their search got under way, however, the anticipated announcement came from Sicily.

The Sicilian astronomer Giuseppe Piazzi had been carefully preparing a map of faint stars in the constellation of Taurus. On January 1, 1801, he noticed a dim, previously uncharted star that shifted its position slightly over the next several nights. Later that year the orbit of this object was determined to lie between the orbits of Mars and Jupiter. At Piazzi's request, the object was named Ceres (pronounced SEE-reez) after the patron goddess of Sicily.

Ceres orbits the Sun once every 4.6 years at an average distance of 2.77 AU, which is in remarkable agreement with the distance given by Bode's law for the missing planet. Ceres is very small, however—its diameter is estimated to be a scant 914 km. Ceres thus does not qualify as a full-fledged planet, and astronomers continued the search.

In 1802 the German astronomer Heinrich Olbers discovered another faint, starlike object that moved against the background stars. He called it Pallas, after the Greek goddess of wisdom. Like Ceres, Pallas orbits the Sun every 4.6 years at an average distance of 2.77 AU. Pallas is even dimmer and smaller than Ceres, with an estimated diameter of only 522 km. Obviously, Pallas is not the missing planet either.

The discovery of these two small objects with similar orbits at the distance expected for the missing planet led astronomers to suspect that Bode's missing planet might have somehow broken apart or exploded. The search for other small objects therefore continued. Only two more were found—Juno and Vesta—until the mid-1800s, when telescopic equipment and techniques improved. Astronomers then began to stumble across many more such objects circling the Sun between the orbits of Mars and Jupiter. These objects are today called asteroids or **minor planets.**

The next major breakthrough came in 1891, when the German astronomer Max Wolf began using photographic techniques to search for asteroids. A total of 300 asteroids had been found up to that time, each painstakingly discovered by scrutinizing the skies for faint, uncharted stars whose positions shifted slowly from one night to the next. With the advent of astrophotography, however, the floodgates of data were opened. Astronomers could simply aim a camera-equipped telescope at the stars and take long exposures. If an asteroid happened to be in the field of view, it left a distinctive trail on the photographic plate because of its movement along its orbit during the time exposure (Figure 10-1). Using this technique, Wolf alone discovered 228 asteroids.

Although thousands of asteroids have been sighted, only 3000 have well-determined orbits. An additional 6000 asteroids have poorly known orbits, and the orbits of 20,000 more have never been determined. The orbits of all officially discovered asteroids are published annually in the famous Soviet catalogue *Ephemerides of Minor Planets.*

To become the official discoverer of an asteroid, you must do a lot more than just produce one photograph showing an asteroid trail whose path does not match any known orbit listed in the *Ephemerides.* You must track the asteroid long enough to compute an accurate and reliable orbit (important data may be available from colleagues who may have inadvertently photographed it on earlier occasions). Then you must prove the accuracy of the orbit by locating the asteroid again on at least one succeeding opposition. At that time, an official number will be assigned to your asteroid (Ceres is 1, Pallas is 2, and so forth). You will also be given the privilege of selecting a name for your asteroid.

Ceres is unquestionably the largest asteroid. With its diameter of 914 km, Ceres accounts for about 30% of the mass of all the asteroids combined. Only two others (Pallas and Vesta) have diameters greater than 500 km. About 30 other asteroids have diameters between 200 and 300 km, and 200 more are bigger than 100 km across.

Figure 10-1 Two asteroids *Asteroids are detected by their elongated trails on time-exposure photographs of the stars. The images of two asteroids are seen in this picture (arrows). Astronomers sometimes find asteroids accidentally while photographing various portions of the sky for other purposes. (Yerkes Observatory)*

Astronomers estimate that roughly 100,000 asteroids exist that are bright enough to appear on photographs taken from Earth. The vast majority are less than 1 km across. Like Ceres, Pallas, and Juno, most asteroids circle the Sun at distances between 2 and $3\frac{1}{2}$ AU. This region of the solar system between the orbits of Mars and Jupiter is called the **asteroid belt** (Figure 10-2). Asteroids whose orbits lie entirely within this region are called **belt asteroids.**

The combined matter of all the asteroids (including an estimate for those not yet officially known) would produce an object barely 1500 km in diameter, considerably smaller than our Moon. Because Bode's missing planet would not have been large enough to rank with the terrestrial planets, it seems more reasonable that the asteroids are debris left over after the solar system formed out of the solar nebula. Constant gravitational perturbations caused by the enormous mass of Jupiter probably kept planetesimals from accreting into larger objects in the region between Mars and Jupiter. The missing planet simply never had a chance to form.

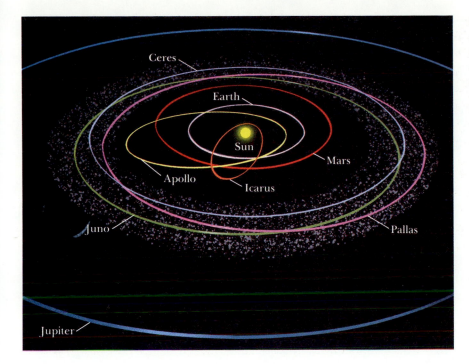

Figure 10-2 The asteroid belt Most asteroids orbit the Sun in a $1\frac{1}{2}$-AU-wide belt between the orbits of Mars and Jupiter. Of these asteroids, the orbits of Ceres, Pallas, and Juno are indicated. Also shown are the orbits of Apollo and Icarus, asteroids whose orbits cross Earth's orbit.

10-2 Jupiter's gravity affects the structure of the asteroid belt and captures asteroids along its orbit

The orbits of the asteroids are affected slightly by the gravitational pulls of the various planets. Most notable are the effects of Jupiter because of its large mass and proximity to the asteroid belt. Mars also perturbs asteroid orbits, but to a much lesser extent than Jupiter. The combined effect of such gravitational perturbations over the ages has caused the asteroids to prefer certain orbits while avoiding others. It would be a monumental task to compute the resulting distribution of asteroid orbits because the gravitational forces of many bodies would have to be included. Nevertheless, we can appreciate some basic characteristics of the asteroid belt by considering the effects of Jupiter alone.

Imagine a belt asteroid moving along its orbit between the orbits of Jupiter and Mars. Each time the faster-moving asteroid catches up with and passes massive Jupiter, it experiences a slight gravitational tug toward Jupiter. This tug alters the asteroid's orbit slightly. However, over the ages, these close passes occur at different points along the asteroid's orbit, and so their effects tend to cancel each other.

Now imagine an asteroid circling the Sun once every 5.93 years, which is exactly half of Jupiter's orbital period. On every second trip around the Sun, the asteroid finds itself lined up between Jupiter and the Sun again and again, always at the same location and with the same orientation. These gravitational effects add up to deflect the asteroid from its original 5.93-year orbit, leaving a gap in the asteroid belt. According to Kepler's third law, a period of 5.93 years corresponds to a semimajor axis of 3.28 AU. Because of Jupiter, there are no asteroids that orbit the Sun at this average distance.

Similarly, we would expect to find a gap corresponding to an orbital period of one-third Jupiter's period, or 3.95 years. Comparable gaps should exist for other simple relationships between the periods of asteroids and Jupiter. The data graphed in Figure 10-3 show that such gaps do exist. They are called **Kirkwood gaps** in honor of the American astronomer Daniel Kirkwood, who first drew attention to them.

Although Jupiter's gravitational pull depletes certain orbits in the asteroid belt, it captures asteroids at other locations much farther from the Sun. There are two points along Jupiter's orbit where the gravitational forces of the Sun and Jupiter work together to hold asteroids in orbit. These two locations are called the **Lagrange points** L_4 and L_5 in honor of the French mathematician whose

Figure 10-3 **The Kirkwood gaps** *This histogram displays the number of asteroids at various distances from the Sun. Notice that very few asteroids have orbits whose orbital periods correspond to simple fractions (such as $\frac{1}{2}$, $\frac{2}{7}$, $\frac{2}{5}$, and $\frac{1}{3}$) of Jupiter's orbital period. Gravitational perturbations caused by repeated alignments with Jupiter have deflected asteroids away from these orbits.*

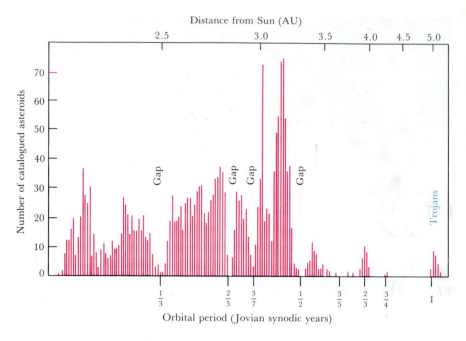

calculations revealed this gravitational effect. Point L_4 is located one-sixth of the way around Jupiter's orbit ahead of the planet, and point L_5 occupies a similar position behind the planet (Figure 10-4).

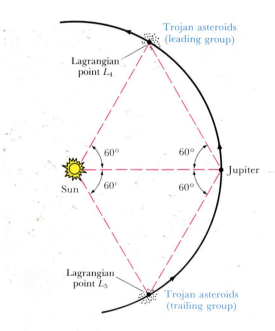

Figure 10-4 **The Trojan asteroids** *Asteroids are trapped at the two Lagrange points along Jupiter's orbit by the combined gravitational forces of Jupiter and the Sun. Asteroids at these locations are named after Homeric heroes of the Trojan War.*

The asteroids trapped at Jupiter's Lagrange points are called **Trojan asteroids,** each named after a hero of the Trojan War. Approximately four dozen Trojan asteroids have been catalogued so far, and some astronomers believe that there may be several hundred rock fragments orbiting near each Lagrange point.

10-3 Asteroids occasionally collide with each other and with the inner planets

In addition to the belt asteroids and the Trojan asteroids, there are other asteroids distinguished by highly elliptical orbits that bring them into the inner regions of the solar system. Occasionally one of these asteroids passes quite close to Earth. Figure 10-5 shows Eros as it passed within 23 million kilometers of our planet in 1931. In 1968 Icarus passed Earth at a distance of only 6 million kilometers. One of the closest near-misses in recent history occurred in 1937, when Hermes passed us at a distance of 900,000 km—only a little more than twice the distance to the Moon. A similar close call occurred on March 23, 1989, when an asteroid called 1989FC passed within 800,000 km of the Earth. If this asteroid had struck the Earth, the impact would have been equivalent to the explosion of a thousand 20-megaton hydrogen bombs.

During these close encounters, astronomers can examine the details of the asteroids. For example, an

Figure 10-5 Eros Eros is one of the asteroids that occasionally passes near the Earth. This photograph was taken in February 1931, when the distance between Eros and the Earth was only 23 million kilometers. The dimensions of Eros are roughly 10 × 20 × 30 km. This asteroid rotates with a period of 5.27 hours. (Yerkes Observatory)

There is ample evidence that interasteroid collisions successfully fragment asteroids into small pieces. In 1918, for instance, the Japanese astronomer Kiyotsugu Hirayama drew attention to groups of asteroids that share nearly identical orbits. These groupings presumably resulted from the fragmentation of parent asteroids.

The collision of kilometer-sized asteroids must be an awesome event. Typical collision velocities are estimated at 1 to 5 km/s (2000 to 11,000 mph), which is more than sufficient to shatter rock. Only in a high-velocity collision is there enough energy to shatter an asteroid permanently. In collisions at low velocities, the resulting fragments may not achieve escape velocity from each other and will reassemble because of their mutual gravitational attraction. Alternatively, several large fragments may end up orbiting each other. This is probably what happened to both Pallas and Victoria, which are **binary asteroids,** each consisting of a main asteroid and a large satellite.

Interasteroid collisions produce numerous chunks of rock, many of which eventually rain down on Mercury, Venus, Earth, and Mars. Fortunately for us, the vast majority of these asteroid fragments, usually called **meteoroids,** are quite small. On rare occasions, however, a large fragment does collide with our planet. The result is an **impact crater,** whose diameter depends on the mass and speed of the impinging object.

asteroid's brightness often varies periodically, presumably because different surfaces are turned toward us as the asteroid rotates. Periodic variations in brightness thus reveal the asteroid's rate of rotation. Typically asteroids rotate with periods in the range of 5 to 10 hours.

Careful scrutiny of an asteroid's variations in brightness can also reveal the asteroid's shape and dimensions. Only the largest asteroids like Ceres, Pallas, and Vesta have enough gravity to pull themselves into a spherical shape. Smaller asteroids permanently retain the odd shapes resulting from collisions with other asteroids. A small asteroid looks dim when seen end on but appears brighter when seen broadside. Measurements of an asteroid's brightness variations thus tell astronomers a lot about its shape.

In October 1991 the Jupiter-bound *Galileo* spacecraft passed near the asteroid Gaspra and sent back close-up views (Figure 10-6). Gaspra appears to be a survivor of numerous catastrophic collisions in which a parent body was broken down into smaller and smaller pieces. Indeed, a series of grooves on Gaspra's surface suggest that the asteroid was hit with nearly enough force to smash it to bits.

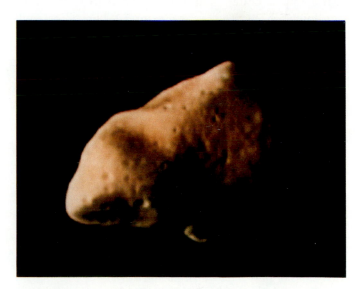

Figure 10-6 The asteroid Gaspra This photograph of Gaspra was taken in October 1991, when Galileo passed within 1600 km of the asteroid. Gaspra's cratered, irregular shape measures 16 × 12 km in this view. However, its overall dimensions are an estimated 12 × 20 × 11 km. Craters as small as 160 m across are visible on Gaspra's surface. (NASA)

Figure 10-7 The Barringer Crater *An iron meteoroid measuring 50 m across struck the ground in Arizona 25,000 years ago. The result was this beautifully symmetrical impact crater measuring 1.2 km in diameter and 200 m deep at its center. (Meteor Crater Enterprises)*

One of the most impressive and best-preserved terrestrial impact craters is the famous Barringer Crater near Winslow, Arizona. The crater measures 1.2 km across and 200 m deep (Figure 10-7). The crater was formed 25,000 years ago when an iron-rich object measuring about 50 m across struck the ground with a speed estimated at 11 km/s (25,000 mph). The resulting blast was equal to the detonation of a 20-megaton hydrogen bomb.

Iron is one of the more abundant elements in the universe (recall Table 6-4), as well as one of the most common rock-forming elements (recall Table 6-5), and so it is not surprising that iron is an important constituent of asteroids and their fragments called meteoroids. Another element, iridium, is common in iron-rich minerals but is rare in ordinary rocks. Measurements of iridium in the Earth's crust can therefore tell us about the rate at which meteoritic material has been deposited on the Earth over the ages.

Geologists Walter Alvarez and his physicist father, Luis Alvarez, from the University of California at Berkeley were involved in making such measurements in the late 1970s. Working at a site of exposed limestone in the Apennine Mountains in Italy, the Alvarez team discovered an exceptionally high abundance of iridium in a dark-colored layer of clay between the limestone strata (Figure 10-8). Following the announcement of this 1979 discovery, a comparable layer of iridium-rich material has been uncovered at a variety of sites around the world. In all cases, geological dating reveals that this apparently worldwide layer of iridium-rich clay is about 63 million years old.

Figure 10-8 The iridium-rich layer of clay *This photograph of strata in the Apennine Mountains of Italy shows a dark-colored layer of iridium-rich clay sandwiched between white limestone (below) from the late Mesozoic era and grayish limestone (above) from the early Cenozoic era. This iridium-rich layer may be the result of an asteroid impact that caused the extinction of the dinosaurs. The coin is the size of a U.S. quarter. (Courtesy of W. Alvarez)*

Paleontologists were quick to realize the profound significance of this particular date. Around 63 million years ago all the dinosaurs became extinct. In fact, at that time a staggering 65% of all the species on Earth disappeared within a relatively brief span of time.

The Alvarez discovery suggests a startling explanation for the dramatic extinction of more than half the life forms that inhabited our planet at the end of the Mesozoic era: Perhaps an asteroid hit the Earth. An asteroid 10 km in diameter slamming into the Earth would have thrown enough dust into the atmosphere to block out sunlight for several years. As plants died for lack of sunshine, the dinosaurs would have starved to death along with many other creatures in the vegetation food chain. The dust eventually settled, depositing an iridium-rich layer around the world. Tiny, rodentlike creatures that could ferret out seeds and nuts would have been prominent among the animals that managed to survive the holocaust, setting the stage for the rise of mammals in the Cenozoic era.

Some geologists and paleontologists are not yet convinced that such a meteoroid impact did produce the "great dying out" at the end of the Mesozoic era, but many scientists agree that this hypothesis fits the available evidence better than other explanations that have been offered.

10-4 Meteorites are classified as stones, stony irons, or irons, depending on their composition

A meteoroid, like an asteroid, is a chunk of rock in space. There is no official dividing line between meteoroids and asteroids, but the term *asteroid* is generally applied only to objects larger than a few hundred meters across.

A **meteor** is the brief flash of light (sometimes called a "shooting star") that is visible at night when a meteoroid strikes the Earth's atmosphere (Figure 10-9).

If a piece of rock survives its fiery descent through the atmosphere, the object that reaches the ground is called a **meteorite.** For thousands of years people have been finding specimens, and descriptions of meteorites appear in ancient Chinese, Greek, and Roman literature. Our ancestors placed special significance on these "rocks from heaven."

Meteorites are classified into three broad categories: stones, stony-irons, and irons. As their name suggests, **stony meteorites,** or **stones,** look like ordinary rocks at

Figure 10-9 A meteor *A meteor is produced when a piece of interplanetary rock or dust strikes the Earth's atmosphere at high speed. Exceptionally bright meteors, such as the one shown in this long exposure (notice the star trails), are usually called fireballs. (Courtesy of R. A. Oriti)*

first glance, but they are sometimes covered with a dark crust produced by the momentary melting of the meteorite's outer layers during its fiery descent through the atmosphere (Figure 10-10). When a stony meteorite is

Figure 10-10 A stony meteorite *Ninety-three percent of all meteorites that fall on the Earth are stones. Many freshly discovered specimens, like the one shown here, are coated with dark crusts. This particular stone fell in Texas. (From the collection of R. A. Oriti)*

Figure 10-11 A stone (cut and polished) When cut and polished, some stony meteorites are found to contain tiny specks of iron. This specimen was discovered in California. (From the collection of R. A. Oriti)

cut in two and polished, tiny flecks of iron are sometimes found in the rock (Figure 10-11).

Although stony meteorites account for an estimated 93% of all meteoritic material that falls on the Earth, stones are the most difficult specimens to find. If they go undiscovered and are exposed to the weather for a few years, they become almost indistinguishable from common terrestrial rocks. Meteorites with a high iron content are much easier to find because they can be located with a metal detector. Consequently, iron and stony-iron meteorites dominate most museum collections.

As their name suggests, **stony-iron meteorites** consist of roughly equal amounts of rock and iron. Olivine is the mineral commonly suspended in the matrix of iron, as in the sample shown in Figure 10-12.

Iron meteorites (Figure 10-13), or **irons**, account for nearly 6% of the material that falls on the Earth. Iron meteorites may contain from 10 to 20% nickel.

In 1808 Count Aloys von Widmanstätten discovered a conclusive test for the authenticity of the most common type of iron meteorite. About 75% of all iron meteorites are of a type that exhibit unique crystalline structure after being cut, polished, and briefly dipped into a dilute acid solution. These crystalline designs are appropriately called **Widmanstätten patterns** (Figure 10-14).

Widmanstätten patterns constitute conclusive proof of a meteorite's authenticity because iron–nickel crystals can grow to lengths of several centimeters only if the molten metal cools slowly over many millions of years. Widmanstätten patterns can thus never be found in counterfeit meteorites, or "meteorwrongs," as they are humorously called.

The existence of Widmanstätten patterns strongly suggests that some asteroids were partly molten for a substantial period after they formed. Furthermore, the original size of the parent asteroid can be estimated by calculating how much rock must have insulated the molten iron–nickel interior to produce its long-term cooling rate. The results of such calculations imply that typical iron meteorites are fragments of asteroids that were 200 to 400 km in diameter.

The three main types of meteorites may have come from different parts of a parent asteroid in the following manner. As soon as the asteroid had accreted from planetesimals 4.5 billion years ago, rapid decay of short-lived radioactive isotopes heated the asteroid's interior to temperatures above the melting point of rock. Over the next few million years, chemical differentiation occurred. As iron and other heavy elements sank toward the asteroid's center, they displaced the lighter elements (such as silicon) upward toward the asteroid's surface. After the asteroid

Figure 10-12 A stony-iron meteorite Stony-irons account for slightly less than 2% of all meteorites that fall on the Earth. This particular specimen is a variety of stony-iron called a pallasite. (From the collection of R. A. Oriti)

Figure 10-13 An iron meteorite *Irons are composed almost entirely of iron–nickel minerals. The surface of a typical iron is covered with thumbprintlike depressions caused by the melting away of the meteorite's outer layers during its high-speed descent through the atmosphere. This specimen was found in Australia. (From the collection of R. A. Oriti)*

Figure 10-14   Widmanstätten patterns   *When cut, polished, and etched with a weak acid solution, most iron meteorites exhibit interlocking crystals in designs called Widmanstätten patterns. This particular meteorite was found in Australia. (From the collection of R. A. Oriti)*

cooled and its core solidified, interasteroid collisions fragmented the parent body into meteoroids. Iron meteorites are specimens from the asteroid's core, stones being samples of its crust. Stony-irons presumably come from regions between an asteroid's core and crust.

Meteorites derived from the fragmentation of large asteroids have been subjected to substantial processing during the first billion years of the formation of the solar system. These meteoritic specimens are thus not representative of the primordial material from which the solar system was created. There is, however, a class of rare meteorites, called **carbonaceous chondrites,** that show no evidence of ever having been subjected to heating or melting.

About 6% of all the stones that fall on the Earth are carbonaceous chondrites. Their primordial nature is inferred from their high content of volatile compounds, sometimes including as much as 20% water. Furthermore, carbonaceous chondrites are rich in complex organic compounds. The water and volatile chemicals would have been driven out and the large organic molecules broken down if these meteorites had been subjected to any significant heating.

Shortly after midnight on February 8, 1969, the night sky around Chihuahua, Mexico, was illuminated by a brilliant blue-white light moving across the heavens. The dazzling display was witnessed by hundreds of people,

many of whom thought that the world was coming to an end. As the light moved across the sky, it exploded in a spectacular, noisy detonation that dropped thousands of rocks and pebbles over the terrified onlookers. Within hours, teams of scientists were on their way to collect specimens of a carbonaceous chondrite, collectively called *the Allende meteorite* after the locality (Figure 10-15).

Figure 10-15 A piece of the Allende meteorite *This carbonaceous chondrite fell near Chihuahua, Mexico, in February 1969. Note the meteorite's dark color, caused by a high abundance of carbon. Geologists believe that this meteorite is a specimen of primitive planetary material. The ruler is 6 inches long. (Courtesy of J. A. Woods)*

One of the most significant discoveries to come from the Allende meteorite was made by Gerald J. Wasserburg and his colleagues at the California Institute of Technology. They found unmistakable evidence of the former presence of a radioactive isotope of aluminum, ^{26}Al. This particular isotope has a half-life of only 720,000 years, which means that after 720,000 years, half of the ^{26}Al in a sample has changed into a stable isotope of magnesium, ^{26}Mg. On either an astronomical or geological time scale, ^{26}Al is quite short-lived.

Detectable amounts of ^{26}Al could have been included in the meteorite only if the radioactive aluminum had been created shortly before the solar system had formed. If only a few hundred million years had elapsed, virtually all the radioactive aluminum would have changed into magnesium before the meteorite formed. Astronomers were therefore faced with evidence of energetic nuclear processes that occurred in our vicinity roughly 4.5 billion years ago, about the time the Sun was born.

One of nature's most violent and spectacular explosions, called a supernova, occurs when a massive star dies. As we shall see in greater detail in Chapter 14, the doomed star blows itself apart in a cataclysm that hurls matter outward at tremendous speeds. During this detonation, violent collisions between nuclei produce a host of radioactive isotopes, including ^{26}Al. It thus seems inescapable that a supernova occurred very near the Sun's birthplace 4.5 billion years ago. In addition to contaminating the interstellar medium with radioactive ^{26}Al, the supernova's shock wave would have compressed the interstellar gas and dust, triggering the birth of the solar system.

Besides telling us about the creation of the solar system, the study of meteorites may shed light on the origin of life on Earth. **Amino acids,** the building blocks of proteins on which terrestrial life is based, are among the organic compounds found inside carbonaceous chondrites. Perhaps interstellar organic material falling on Earth played a role in the appearance of simple organisms on our planet nearly 4 billion years ago.

10-5 *A comet is a dusty chunk of ice that is partly vaporized as it passes near the Sun*

Heat from the protosun gave rise to two types of interplanetary material: rocky debris near the Sun, and loose conglomerations of ice and dust, called comets, far from

Figure 10-16 Comet West *A comet is always named after the person who first sees it. Astronomer Richard M. West first noticed this comet on a photograph taken with a telescope in 1975. After passing near the Sun, Comet West became one of the brightest comets in recent years. This photograph shows the comet in the predawn sky in March 1976. (Courtesy of H. Vehrenberg)*

the Sun. In contrast to asteroids, which travel around the Sun along roughly circular orbits that are largely confined to the asteroid belt and to the plane of the ecliptic, comets travel around the Sun along highly elliptical orbits inclined at random angles to the ecliptic. As a comet approaches the Sun, solar heat begins to vaporize the ices. The liberated gases surrounding the icy nucleus soon begin to glow, producing a fuzzy, luminous ball called the **coma** that can eventually expand to a million kilometers in diameter. Continued action by the solar wind and radiation pressure blows these luminous gases outward into a long, flowing **tail**. The result is one of the most awesome sights ever visible in the nighttime sky (Figure 10-16).

The solid part of a comet, called the **nucleus,** is a mixture of ice and dust typically measuring a few kilometers across. Frozen ammonia, methane, and water are the primary components of cometary ices. Harvard astronomer Fred L. Whipple, a pioneer in comet research, coined the term "dirty snowball" to describe a comet, reflecting the fact that in comets bits and pieces of dust and rocky material are mixed in with the ice. A close-up view of the nucleus of Halley's Comet is shown in Figure 10-17. The overall structure of a comet is diagrammed in Figure 10-18.

Not visible to the human eye is the **hydrogen envelope,** a tenuous sphere of gas surrounding the comet's nucleus and measuring as much as 20 million kilometers in diameter. Figure 10-19 shows two views of Comet Kohoutek: as it appeared to Earth-based observers in 1973 and as photographed by an ultraviolet camera from a rocket. From the ultraviolet view, astronomers first discovered the enormous extent of the hydrogen envelope.

Comets come in a wide range of shapes and sizes. For example, the comet shown in Figure 10-20 had a large,

2 km

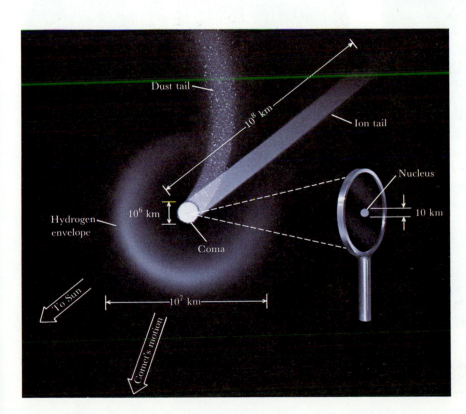

Figure 10-17 (above) **The nucleus of Comet Halley** *In March 1986 five spacecraft passed near Comet Halley. This close-up picture was taken by a camera on board the Giotto spacecraft and shows the potato-shaped nucleus of the comet. The nucleus is darker than coal and measures 15 km in the longest dimension and about 8 km in the shortest. The Sun illuminates the comet from the left. Two bright jets of dust extend 15 km from the nucleus toward the Sun, suggesting that major activity emanates from the sunlit side. (Max Planck Institut für Aeronomie)*

Figure 10-18 (left) **The structure of a comet** *The solid part of a comet (the nucleus) is roughly 10 kilometers in diameter. The coma can be as large as 100,000 to 1 million kilometers across, and the hydrogen envelope is typically 10 million kilometers in diameter. A comet's tail can be as long as 1 AU, long enough to reach all the way from the Earth to the Sun. (This drawing is not to scale.)*

Figure 10-19 Comet Kohoutek and its hydrogen envelope These two photographs of Comet Kohoutek are reproduced to the same scale. (Left) The comet in visible light. (Right) This view of the comet in ultraviolet wavelengths reveals a huge hydrogen cloud surrounding the comet's head. (Johns Hopkins University; Naval Research Laboratory)

Figure 10-20 The head of Comet Brooks This comet, named Comet Brooks after its discoverer, had an exceptionally large, bright coma. It dominated the night skies in October 1911. (Lick Observatory)

Figure 10-21 Comet Ikeya-Seki This comet, named after its two Japanese codiscoverers, dominated the predawn sky in late October 1965. Although its coma was tiny, its tail was 1 AU long. (Lick Observatory)

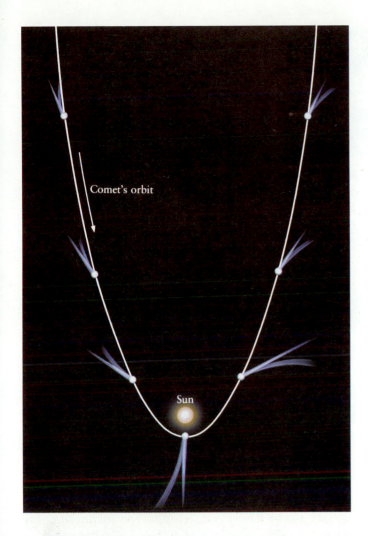

Figure 10-22 *The orbit and tail of a comet* *The solar wind and radiation pressure from sunlight blow a comet's dust particles and ionized atoms away from the Sun. Consequently, the comet's tail always points away from the Sun.*

by radiation pressure. The relatively straight ion tail can exhibit dramatic changes from night to night (Figure 10-23). The more amorphous dust tail is typically arched.

Astronomers typically discover at least a dozen comets every year. Some are short-period comets, which circle the Sun in less than 200 years. Like the famous Halley's Comet, they appear again and again at predictable intervals. The majority of comets discovered each year are long-period comets, however, which typically take 1 to 30 million years to complete one orbit of the Sun. These comets travel along extremely elongated orbits and consequently spend most of their time at distances of roughly 10^4 to 10^5 AU from the Sun.

Because astronomers discover long-period comets at a rate of roughly one per month, it is reasonable to suppose that there is an enormous population of comets out there around 50,000 AU from the Sun. This reservoir of cometary nuclei surrounding the Sun is called the **Oort cloud** after the Dutch astronomer Jan Oort, who first proposed its existence in the 1950s. Estimates of the number of "dirty snowballs" in the Oort cloud range as high as 12 billion. Only a large reserve of cometary nuclei can explain why we see so many long-period comets even though each one takes several million years to travel once around its orbit.

Comets cannot survive very many passages near the Sun. Eventually a comet's ices are completely vaporized

bright coma but a short, stubby tail. In contrast, the comet seen in Figure 10-21 had an inconspicuous coma, but its tail had an astonishing length of 1 AU, long enough to reach all the way from the Earth to the Sun.

It has long been known that comet tails always point away from the Sun (Figure 10-22), regardless of the direction of the comet's motion. In fact, the Sun usually produces two comet tails: an **ion tail** and a **dust tail**. Ionized atoms (that is, atoms missing one or more electrons) are swept directly away from the Sun by the solar wind. Micrometer-sized dust particles are blown away from the comet's coma

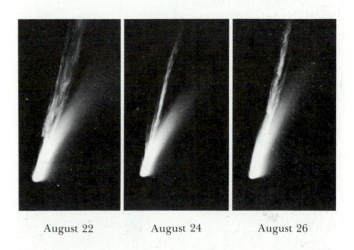

| August 22 | August 24 | August 26 |

Figure 10-23 *The two tails of Comet Mrkos* *Comet Mrkos dominated the evening sky in August 1957. These three views, taken at two-day intervals, show dramatic changes in the comet's ion tail. In contrast, the slightly curved dust tail remained fuzzy and featureless. (Palomar Observatory)*

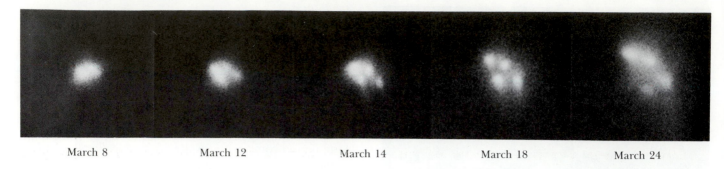

March 8   March 12   March 14   March 18   March 24

Figure 10-24 The fragmentation of Comet West Shortly after passing near the Sun in 1976, the nucleus of Comet West broke into four pieces. This series of five photographs clearly shows the disintegration of the comet's nucleus. Figure 10-16 shows a wide-angle view of this comet. (New Mexico State University Observatory)

and only a swarm of meteoritic dust and pebbles remains. A comet's nucleus will disintegrate more quickly if it happens to pass very close to the Sun. Figure 10-24 shows the breakup of the nucleus of a comet shortly after it passed the Sun at the distance of Mercury's orbit.

After a comet dies, its remaining dust and rock fragments spread out in a loose collection of debris that continues to circle the Sun along the comet's orbit. If the Earth's orbit happens to pass near or through the swarm, a **meteor shower** is seen on the Earth. Nearly a dozen meteor showers can be seen each year. Like comets, meteor showers are not confined to the plane of the ecliptic.

As incredible as it may seem, a total of 300 tons of extraterrestrial rock and dust is estimated to fall on the Earth each day. The fluffy, low-density material from comets burns up in the atmosphere, and only denser specimens related to asteroids typically reach the ground.

Nevertheless, there is evidence that a comet struck the Earth in the recent past.

On June 30, 1908, a spectacular explosion occurred over the Tunguska region of Siberia. Hundreds of square kilometers of forest were devastated (Figure 10-25), and the blast was audible 1000 km away. The explosion was equivalent to the detonation of a tactical nuclear warhead with the destructive power of several hundred kilotons of TNT.

The most likely explanation of this event is that a small comet (perhaps a 100-m fragment of the short-period Comet Encke) collided with the Earth. No impact crater was formed, and the trees at "ground zero" were left standing upright, but they were completely stripped of branches and leaves. This phenomenon is what would be expected from a loosely consolidated ball of cometary ices that vaporized with explosive force before striking the ground. The Tunguska event is a good example of the kind of devastation that can be wrought by interplanetary debris.

Figure 10-25 Aftermath of the Tunguska event In 1908 a piece of a comet's nucleus struck the Earth's atmosphere over the Tunguska region of Siberia. Trees were blown down for many kilometers in all directions from the impact site. (Courtesy of Sovfoto)

Summary

- Bode's law is a numerical sequence that gives the distances (in astronomical units) of the planets Mercury through Uranus from the Sun. This "law" inspired nineteenth-century astronomers to search for a planet in the gap between the orbits of Mars and Jupiter.

- Thousands of belt asteroids with diameters from a few kilometers to about 900 km circle the Sun between the orbits of Mars and Jupiter.

  Gravitational perturbations by Jupiter deplete certain orbits within the asteroid belt. The resulting gaps, called Kirkwood gaps, occur at simple fractions of Jupiter's orbital period.

  Jupiter's gravity also captures asteroids in two locations, called Lagrange points, along Jupiter's orbit.

- Some asteroids move in elliptical orbits that cross the orbits of Mars and Earth. Many of these asteroids will eventually strike one of the inner planets.

  An asteroid may have struck the Earth 63 million years ago, causing the extinction of the dinosaurs and many other species.

- Small rocks in space are called meteoroids. When a meteoroid enters the Earth's atmosphere, it produces a fiery trail called a meteor. If part of the object survives the fall, the fragment that reaches the Earth's surface is called a meteorite.

  Meteorites are grouped in three major classes according to their composition: iron, stony-iron, and stony meteorites.

  Rare stony meteorites called carbonaceous chondrites may be relatively unmodified material from the solar nebula. These meteorites often contain organic material and may have played a role in the origin of life on Earth.

- An analysis of isotopes in certain meteorites suggests that a nearby supernova explosion triggered the formation of the solar system 4.5 billion years ago.

- Comets are chunks of ice and rock fragments that generally move in a highly elliptical orbit about the Sun at a great inclination to the plane of the ecliptic.

As a comet approaches the Sun, its icy nucleus develops a luminous coma surrounded by a vast hydrogen envelope. An ion tail and a dust tail extend from the comet, pushed away from the Sun by the solar wind and radiation pressure.

Fragments of "burned-out" comets produce meteor swarms.

Millions of cometary nuclei probably exist in the Oort cloud some 50,000 AU from the Sun.

Review questions

1 What is Bode's law, and why is it not really a law?

2 Why are asteroids, meteoroids, and comets of special interest to astronomers who want to understand the early history of the solar system?

3 Describe the asteroid belt.

4 Why do you suppose that there are many small asteroids but only a few very large asteroids?

5 Can you think of another place in the solar system where there is a phenomenon similar to the Kirkwood gaps in the asteroid belt? Explain.

6 What are the Trojan asteroids, and where are they located?

7 Describe the three main classifications of meteorites. How might these different types of meteorites have been formed?

8 Suppose you found a rock that you suspect to be a meteorite. Describe some of the things you could do to see if it were a meteorite or a "meteorwrong."

9 Why do astronomers believe that meteoroids come from asteroids, whereas meteor showers are related to comets?

10 With the aid of a drawing, describe the structure of a comet.

11 What is the Oort cloud, and how might it be related to planetesimals left over from the formation of the solar system?

Advanced questions

12 Why do you suppose comets are generally brighter after passing perihelion?

13 Since there are different types of meteorites, would you expect different types of asteroids to exist also? What sorts of observations might an Earth-based astronomer make in order to discover chemical differences among asteroids?

14 Where on Earth might you find large numbers of stony meteorites that are not significantly weathered?

15 Some astronomers have recently argued that passage of the solar system through an interstellar cloud of gas could perturb the Oort cloud, causing many comets to deviate slightly from their original orbits. What might be the consequences for Earth?

Discussion questions

16 Suppose it were discovered that the asteroid Hermes had been perturbed in such a way as to put it on a collision course with Earth. Describe what you would do to counter such a catastrophe within the framework of present technology.

17 From the abundance of craters on the Moon and Mercury, we know that numerous asteroids and meteoroids struck the inner planets early in the history of the solar system. Is it reasonable to suppose that numerous comets also pelted the planets 3 to 4 billion years ago? Speculate about the effects of such a cometary bombardment, especially with regard to the evolution of the primordial atmospheres of the terrestrial planets.

For further reading

Beatty, J. K. "An Inside Look at Halley's Comet." *Sky & Telescope*, May 1986 • Published only a few months after the Comet Halley flybys, this article contains some fine photographs of the comet's nucleus as well as a first look at the data sent back by the spacecraft.

Dodd, R. *Thunderstones and Shooting Stars: The Meaning of Meteorites.* Harvard University Press, 1986 • This clear introduction to the recovery, classification, and study of meteorites was written by a noted geologist.

Hartmann, W. "The Smaller Bodies of the Solar System."

Scientific American, September 1975 • This article by a noted planetary scientist discusses asteroids and meteorites.

Knacke, R. "Sampling the Stuff of a Comet." *Sky & Telescope*, March 1987 • This article, the last of three in a special issue of *Sky & Telescope* describing results from the Comet Halley flybys, discusses the chemical composition of comets.

Kowal, C. *Asteroids.* Ellis Horwood/John Wiley, 1988 • This fascinating book covers the history and science of asteroids and speculates about future space missions to take advantage of asteroid resources.

Morrison, D. "Target Earth: It *Will* Happen." *Sky & Telescope*, March 1990 • This apocalyptic article by a noted planetary scientist presents strong arguments that the Earth is destined to experience devastating meteoritic impacts over the long term.

Olson, R. "Giotto's Portrait of Halley's Comet." *Scientific American*, July 1979 • This article describes historical sightings of Comet Halley, including that by Giotto di Bondone after whom a spacecraft was named.

Russell, D. "The Mass Extinctions of the Late Mesozoic." *Scientific American*, January 1982 • This article discusses evidence that an asteroid hit the Earth 63 million years ago, causing the extinction of the dinosaurs and many other species.

Sagan, C., and Druyan, A. *Comet.* Random House, 1985 • This beautiful volume on comet science and lore is stylishly written, well organized, and contains excellent illustrations.

Weissman, P. "Are Periodic Bombardments Real?" *Sky & Telescope*, March 1990 • This article looks at the idea of periodic meteoritic bombardment, which has been proposed to explain apparently periodic mass extinctions that punctuate the Earth's history.

———. "Realm of the Comet." *Sky & Telescope*, March 1987 • This article, the first of three in a special issue of *Sky & Telescope* describing results from the Comet Halley flybys, discusses the Oort cloud.

Whipple, F. "The Black Heart of Comet Halley." *Sky & Telescope*, March 1987 • This article, the second of three in a special issue of *Sky & Telescope* describing results from the Comet Halley flybys, discusses the structure and properties of the comet's nucleus.

———. *The Mystery of Comets.* Smithsonian Institution Press, 1985 • This appealing, popular-level book includes personal reminiscences by a distinguished astronomer who has spent most of his life studying comets.

———. "The Nature of Comets." *Scientific American*, February 1974 • The scientist who invented the "dirty snowball" model discusses the structure and behavior of comets.

11

Our Star

We conclude our exploration of the solar system by taking a close look at the Sun, which is a typical star. At the Sun's center, thermonuclear reactions convert hydrogen into helium to provide the energy by which the Sun shines. Ionized gases, photons, and magnetic fields interact on a colossal scale as this energy makes its way toward the solar surface, where it is radiated into space. We then explore the Sun's surface and find that it is the scene of a host of bewildering phenomena including sunspots, which are sites of concentrated magnetic fields. The number of sunspots varies with an 11-year period, revealing recurrent processes that are part of a more general 22-year solar cycle that affects the entire solar atmosphere. Vibrations of the solar surface furnish information about the Sun's interior. Because we can view the Sun from close range, studies of the Sun give us important insights into the nature of stars in general. For the clues that it can give us about the universe as well as for its importance to life on Earth, solar astronomy is an exciting and rewarding field of research.

The X-ray sun and the solar corona *This composite view of the Sun's outer atmosphere combines a white-light view taken during a solar eclipse visible in Hawaii with an X-ray image taken concurrently by a camera on board a rocket launched from New Mexico. The white-light photograph shows streamers extending quite far above the solar surface. The X-ray view shows hot gases near the solar surface in shades of yellow, orange, and red. By merging these two views, scientists can study coronal structures over a range of wavelengths and construct a three-dimensional picture of solar activity. (Courtesy of L. Golub and S. Koutchmy)*

The Sun is an average star. Its mass, size, surface temperature, and chemical composition lie roughly midway between the extremes exhibited by other stars. Unlike other stars, however, the Sun is available for close-up examination. Studying the Sun therefore offers excellent insights into the nature of stars in general.

Understanding the Sun is important to humanity because the Sun is our source of heat and light. Life would not be possible on Earth without the energy provided by the Sun. Even a small change in the Sun's size or surface temperature could profoundly alter conditions on the Earth, either melting the polar caps or producing another ice age.

Although the Sun is a commonplace star, it is a dramatic arena where we can observe the fascinating and complicated interaction of matter and energy on a colossal scale. Beautiful and bewildering phenomena occur as columns of hot gases gush up to the solar surface, interact with the Sun's magnetic field, and dissipate energy into the Sun's outer atmosphere. The source of all this energy lies buried at the Sun's center.

11-1 The Sun's energy is produced by thermonuclear reactions in the core of the Sun

During the nineteenth century, geologists and biologists found convincing evidence that the Earth must have existed in more or less its present form for hundreds of millions of years. This fact posed severe problems for physicists, because it seemed impossible to explain how the Sun could have been shining for so long, radiating immense amounts of energy into space. If the Sun were made of coal, for example, it could burn for only 5000 years.

An important key to the source of the Sun's energy was provided in 1905 by Albert Einstein's special theory of relativity. One of the implications of Einstein's theory is that matter and energy are interchangeable according to the simple equation

$$E = mc^2$$

In other words, a mass (m) can be converted into an amount of energy (E) equivalent to mc^2, where c is the speed of light. Because c is a large number and c^2 is huge, a small amount of matter can be converted into an awesome amount of energy.

Inspired by Einstein's work, astronomers began to wonder if the Sun's energy output might come from the conversion of matter into energy. But exactly what kind of mechanism would transform matter into energy?

In the 1920s the British astronomer Arthur Eddington speculated that temperatures at the center of the Sun must be much greater than had previously been thought. Under such high-temperature conditions hydrogen nuclei can fuse together to produce helium nuclei in a reaction that transforms a tiny amount of mass into a very large amount of energy. Because the nuclei fuse together under high temperatures, this process is called **thermonuclear fusion.**

Recall that the nucleus of a hydrogen (H) atom consists of a single proton. The nucleus of a helium atom (He) consists of two protons and two neutrons. In the nuclear process

$$4H \rightarrow He$$

two of the four protons from the hydrogen atoms are changed into neutrons to produce a single helium nucleus. This reaction also releases two positively charged electrons, called **positrons,** that carry off the electric charges relinquished by the transmuted protons. In addition, massless particles called **neutrinos** are also given off.

When hydrogen is converted into helium, matter is lost because the ingredients (four hydrogen nuclei) have a combined mass just slightly greater than the product (one helium nucleus). The mass lost during this reaction may be calculated as follows:

$$4 \text{ hydrogen atoms} = 6.693 \times 10^{-27} \text{ kg}$$
$$- 1 \text{ helium atom} = 6.645 \times 10^{-27} \text{ kg}$$
$$\overline{\text{mass lost} = 0.048 \times 10^{-27} \text{ kg}}$$

This lost mass is converted into energy in the amount predicted by the equation $E = mc^2$.

The Sun's mass, usually designated $M_\odot$ is 2×10^{30} kg (333,000 Earth masses), and the Sun's total power output, called its **luminosity** ($L_\odot$) is 3.9×10^{26} watts. To produce this luminosity, 600 million metric tons of hydrogen must be converted into helium within the Sun each second. This prodigious rate is possible because the Sun contains a vast supply of hydrogen—enough to continue the present rate of energy output for another 5 billion years.

This process of nuclear fusion, by which hydrogen is converted into helium at the Sun's center, is called

hydrogen burning, even though nothing is actually burned in the conventional sense of that word. The ordinary burning of wood, coal, or any flammable substance is a chemical process involving only the electrons that orbit the nuclei of atoms. Thermonuclear fusion is a far more energetic process that involves violent collisions between the nuclei of atoms.

Hydrogen burning occurs at the Sun's core, where it is very hot (15 million K). Normally the positive electric charge on protons is quite effective in keeping the protons far apart because like charges repel each other. But in the extreme heat of the Sun's center, the protons are moving so fast that they can penetrate each other's electric fields and stick together. In Chapter 13 we shall learn that most of the stars you can see in the sky have hydrogen burning occurring at their centers.

11-2 A theoretical model of the Sun can tell us how energy gets from the Sun's center to its surface

Although the Sun's interior is hidden from our view, we can use the laws of physics to calculate what is going on below the solar surface (Figure 11-1). The resulting **model** of the Sun tells us about the Sun's internal characteristics, such as its pressure, temperature, and density. A model of the Sun is usually displayed as a table or graph on which physical characteristics are given for various depths beneath the solar surface.

To develop a model of the Sun or of any other stable star, we first note that the Sun is not undergoing any dramatic changes. The Sun is not exploding or collapsing, nor is it significantly heating up or cooling off. The Sun is thus in balance both mechanically and thermally.

Mechanical balance, often called **hydrostatic equilibrium,** means simply that a star is supporting its own weight. Because of gravity, the tremendous weight of the Sun's outer layers pressing inward from all sides tries to make the star contract. However, as gravity compresses the star, gas pressure inside the star increases. The greater the compression, the higher the internal pressure. Hydrostatic equilibrium is achieved when the pressure at every depth within the star is exactly sufficient to support the weight of the overlying layers.

Thermal balance, often called **thermal equilibrium,** means simply that a star keeps shining. Vast amounts of energy escape from the Sun's surface each second. This

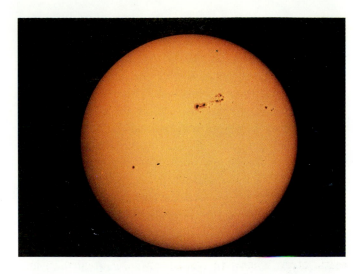

Figure 11-1 The Sun's surface *The Sun is the only star whose surface details can be examined through Earth-based telescopes. Astronomers always take great care to avoid severe damage to their eyes by viewing the Sun through extremely dark filters or by projecting the Sun's image onto a screen. The Sun is so bright that its rays, focused by the lens of an unprotected eyeball, can destroy the retina. Therefore, never look directly at the Sun. (Celestron International)*

energy is constantly resupplied by thermonuclear fusion in the Sun's core taking place at a steady, persistent rate that maintains thermal equilibrium. But exactly how is energy transported from the Sun's center to its surface?

Experience teaches us that energy always flows from hot regions to cooler regions. For example, if you heat one end of a metal bar with a blowtorch, the other end of the bar eventually becomes warm. This method of energy transport is called **conduction.** Conduction varies significantly from one substance to another (copper is a good heat conductor, wood is not), depending on the arrangement and interaction of the atoms. Because conditions inside stars like the Sun are not favorable for conduction, this process is not an efficient means of energy transport in ordinary stars. Nevertheless, conduction is important in very compact stars called white dwarfs, which are discussed in Chapter 14.

Two other means of energy transport—convection and radiative diffusion—do operate inside stars like the Sun, moving energy from a star's center to its surface. Convection involves the circulation of gases between hot and cold regions. Just as a hot-air balloon drifts skyward, so hot gases rise toward a star's surface while cool gases sink back down toward its center. The net effect of this physical movement of gases is to transfer heat energy from the center toward the surface.

In **radiative diffusion,** photons created in the ther-monuclear inferno at a star's center diffuse outward toward the star's surface. The paths of individual photons are quite random, because they are knocked about by atoms and electrons inside a star. The overall migration, however, is outward from the hot core, where the photons are constantly being created, toward the cooler surface, where they escape into space. In all, it takes roughly a million years for energy created at the Sun's center to reach the solar surface and finally escape as sunlight.

The concepts of hydrostatic equilibrium, thermal equilibrium, and energy transport can be expressed in the form of a set of mathematical equations that are collectively called the **equations of stellar structure.** These equations can be solved to give the ranges of the pressure, temperature, and density that must exist inside a star to maintain equilibrium.

Astrophysicists use high-speed computers to solve the equations of stellar structure, thus developing a detailed theoretical model of the structure of a star. The astrophysicist begins with astronomical data about the star's surface. For the Sun, such data would be entered as the Sun's surface temperature (5800 K), its luminosity (3.9 × 10^{26} watts), and its gas pressure and density (almost zero). The equations of stellar structure are used to calculate conditions layer by layer toward the star's center. In this way, the astrophysicist can discover how temperature, pressure, and density increase with increasing depth below the star's surface. This description constitutes a **stellar model.** We have learned through this technique that the temperature at the Sun's center is 15.5 million K, the pressure is 3.4×10^{11} atm, and the density is 150,000 kg/m^3.

Figure 11-2 presents a theoretical model of the Sun. The four graphs show how the Sun's luminosity, mass, temperature, and density vary from the Sun's center to its surface. For instance, the upper graph gives the percentage of the Sun's luminosity. Note that the luminosity rises to 100% at about one-quarter of the way from the Sun's center to its surface. This tells us that all the Sun's energy is produced within a volume extending out to $\frac{1}{4}$ solar radius (1 solar radius, which is the distance from the Sun's center to its surface, is nearly 700,000 km).

Also note that the mass rises to nearly 100% at about 0.6 solar radius from the Sun's center. Almost all of the Sun's mass is therefore confined to a volume extending only 60% of the distance from the Sun's center to its surface.

From the Sun's center out to 0.8 solar radius, energy is transported by radiative diffusion, as we have seen. This

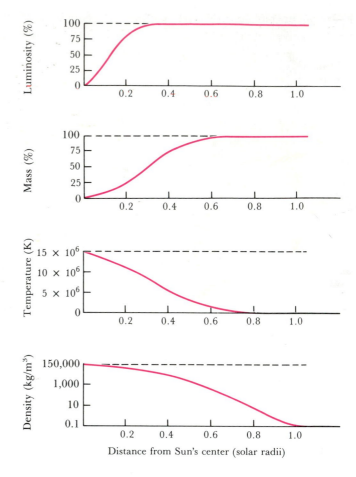

Figure 11-2 A theoretical model of the Sun *The Sun's internal structure is displayed here with graphs that show how the luminosity, mass, temperature, and density vary with the distance from the Sun's center. A solar radius (the distance from the Sun's center to its surface) equals 696,000 km.*

inner region is therefore called the **radiative zone.** From 0.8 solar radius to the Sun's surface, the density of matter in the Sun is so low that convection dominates the energy flow. We thus say that the Sun has a **convective zone,** or **convective envelope.** These aspects of the Sun's internal structure are sketched in Figure 11-3.

11-3 *The mystery of the missing neutrinos inspires speculation about the Sun's interior*

For every proton that changes into a neutron during hydrogen burning, a neutrino is released. Neutrinos have no electric charge and are generally believed to have no mass either, which means they resemble photons and

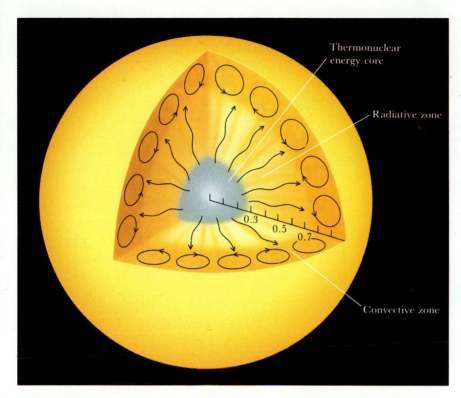

Thermonuclear
energy core

Radiative zone

0.3 0.5 0.7

Convective zone

Figure 11-3   The Sun's internal structure
Thermonuclear reactions occur in the Sun's core, which extends to a distance of 0.25 solar radius from the center. Energy from the core is transported outward via radiative diffusion to a distance of 0.8 solar radius. Convection is responsible for energy transport in the Sun's outer layers.

travel through space at the speed of light. (Some physicists theorize that neutrinos might have a tiny mass, probably less than $\frac{1}{10,000}$ the mass of an electron. If neutrinos do have mass, they travel at speeds less than that of light.)

Neutrinos are difficult to detect, because they do not interact much with ordinary matter. In fact, neutrinos interact so weakly and so infrequently with matter that they easily pass through the Earth as if it were not there. On very rare occasions, however, a neutrino strikes a neutron and converts it into a proton.

The Sun is also largely transparent to neutrinos, allowing these particles to stream outward, unimpeded, from its core. If astronomers could detect these particles, they would have an excellent means of probing the Sun's center. About 10^{38} neutrinos are produced at the Sun's center each second. This output is so huge that, here on Earth, roughly a hundred billion neutrinos pass through every square inch of your body every second! If astronomers could detect even a few of these particles, it might be possible to build a "neutrino telescope" that could be used to "see" the thermonuclear inferno that is hidden from ordinary view.

Inspired by such possibilities, Raymond Davis of the Brookhaven National Laboratory designed and built a large neutrino detector. This device uses 100,000 gallons of perchloroethylene cleaning fluid (C_2Cl_4) in a huge tank buried deep in a mine in South Dakota (Figure 11-4). Because matter is virtually transparent to neutrinos, most of the neutrinos from the Sun pass right through Davis's tank with no effect whatsoever. On rare occasions, though, a neutrino strikes the nucleus of one of the chlorine atoms in the cleaning fluid and converts one of its neutrons into a proton, creating a radioactive atom of argon. The rate at which this argon is produced tells us about the number of neutrinos from the Sun arriving at the Earth.

This experiment, which began in the mid-1960s, has been repeated with extreme care for more than 20 years. On the average, solar neutrinos create one radioactive argon atom every three days in Davis's tank. To the continuing consternation of astronomers, this rate corresponds to only one-third of the neutrinos predicted from the "standard model" of the Sun presented in Figure 11-2. Most astronomers now agree that Davis's experiment is probably not at fault, which compels them to reexamine their ideas about the Sun.

One possible solution to the "mystery of the missing solar neutrinos" is that there are different types of neutrinos, and perhaps Davis's experiment detects only one type. Some theorists predict that if neutrinos have

Figure 11-4 The solar neutrino experiment *This tank contains 100,000 gallons of perchloroethylene buried 1.5 km below ground at the Homestake Gold Mine in South Dakota. The Earth shields the tank from stray particles so that only solar neutrinos can convert chlorine atoms in the fluid into radioactive argon atoms. The rate at which argon atoms are created in the tank is a direct measure of the number of neutrinos coming from the Sun. (Courtesy of R. Davis, Brookhaven National Laboratory)*

mass, they may continually change from one type to another. This hypothetical phenomenon, called **neutrino oscillation,** might explain Davis's results.

Another possible explanation is that neutrinos are created in the Sun's core by nuclear reactions that are temperature-sensitive. The Sun's center would have to be only 10% lower in temperature to produce a drop in neutrino production that would bring the neutrino flux into agreement with Davis's experiment. Calculations using the equations of stellar structure, however, reveal that if you lower the Sun's central temperature by a million degrees, other obvious features of the Sun, such as its size and surface temperature, would differ from what we observe. The mystery of the missing solar neutrinos therefore remains a thorny problem that has astronomers questioning how well we really understand what is going on inside the Sun.

11-4 The photosphere is the lowest of three main layers in the Sun's atmosphere

Although astronomers often speak of the solar surface, the Sun really does not have a surface at all. As you move in toward the Sun, you encounter ever denser gases but no sharp boundary like the surface of the Earth or Moon. The Sun appears to have a surface (see Figure 11-1) because there is a specific layer in the Sun's atmosphere from which most of the visible light comes. This layer, which is about 300 to 400 km thick, is appropriately called the **photosphere** ("sphere of light"). The photosphere shines with a nearly perfect blackbody spectrum corresponding to an average temperature of 5800 K (recall Figure 5-4).

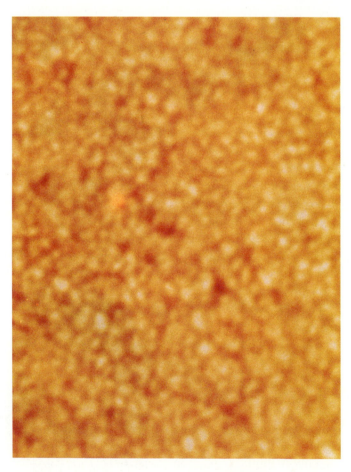

Figure 11-5 Solar granulation *High-resolution photographs of the Sun's surface reveal a blotchy pattern called granulation. Granules, each measuring about 1000 km across, are convection cells in the Sun's outer layers. (NOAO)*

The photosphere is the lowest of three layers that together constitute the Sun's atmosphere. Above the photosphere are two additional layers, the chromosphere and the corona, which are discussed later in this chapter. These upper layers are transparent to visible light; we can see through them down to the photosphere. We cannot, however, see through the shimmering gases of the photosphere, and so everything below the photosphere is called the Sun's interior.

Convection in the Sun's outer layers affects the appearance of the photosphere. Under good observing conditions with a telescope and using special dark filters to protect your eyes, you can often see a blotchy pattern called **granulation** (Figure 11-5). Each lightly-colored **granule** measures about 1000 km across and is surrounded by a darkish boundary.

In Chapter 5 we learned that relative motion between a light source and an observer affects the wavelengths of spectral lines (recall Figure 5-19), a phenomenon called the Doppler effect. The spectral lines of a source moving toward you are blueshifted, but those of a source moving away from you are redshifted. By carefully measuring the wavelengths of spectral lines in various parts of granules, astronomers can determine movements of the solar gases. Astronomers find that hot gases rise upward in the granules, cool off, spill over the edges of the granules, and plunge back down into the Sun along the intergranule boundaries. The difference in brightness between the center and the edge of a granule corresponds to a temperature drop of 300 K.

Granules are individual convection cells in the Sun's outer layers. Time-lapse photography shows that granules form, disappear, and then re-form in cycles lasting several minutes. At any one time, roughly four million granules cover the solar surface. Each granule occupies an area roughly equal to Texas and Oklahoma combined (about a million square kilometers).

The photosphere appears darker around the edge, or **limb,** of the Sun than it does toward the center of the solar disk (examine Figure 11-1). This phenomenon, called **limb darkening,** arises because we are looking obliquely at the photosphere near the edge of the disk. We therefore do not see as deeply into the Sun there as we do near the center of the disk. Near the limb, the gas we observe is cooler and thus appears dimmer than the deeper, warmer gas seen near the disk center.

11-5 The chromosphere is located between the photosphere and the Sun's outermost atmosphere

Immediately above the photosphere is a relatively cool, dim layer called the **chromosphere** ("sphere of color"), which is the second of the three major levels in the Sun's atmosphere. When the Moon blocks out the photosphere during a total eclipse, the chromosphere is visible as a pinkish strip around the edge of the dark Moon.

High-resolution images of the Sun's limb reveals that the chromosphere is composed of numerous spikes called **spicules** (Figure 11-6). Spicules are jets of gas that surge upward from the Sun. A typical spicule rises at the rate of 20 km/s, reaches a height of about 7000 km, then collapses and fades away after a few minutes. At any one time, roughly 300,000 spicules exist, covering a few percent of the Sun's surface.

Figure 11-6   Spicules and the chromosphere   Numerous spicules are seen in this photograph of the Sun's limb. The chromosphere consists of spicules, which are jets of cool gas that surge upward into warmer regions of the Sun's outer atmosphere. This photograph was taken though a filter that is transparent only to the wavelength of H$_\alpha$. (NOAO)

The spectrum of the chromosphere is dominated by emission lines. The characteristic pinkish color of the chromosphere is caused by the Balmer line H$_\alpha$ (recall Figure 5-17), which is the strongest emission line in the red region of the chromosphere's spectrum. The blue end of the chromosphere's spectrum is dominated by two bright emission lines, called the H and K lines of ionized calcium.

The spectrum of the photosphere is dominated by numerous absorption lines (recall Figure 5-5), including broad, dark lines at the wavelengths of H$_\alpha$ and the calcium H and K lines. The photosphere emits almost no light at these wavelengths. The chromosphere, however, is especially bright at these wavelengths. Astronomers can thus study details of the chromosphere by viewing the Sun through special filters that are transparent to light only at the wavelengths of H$_\alpha$ or the calcium lines.

A photograph of the solar surface taken through an H$_\alpha$ filter shows brushlike spicules protruding above the photosphere (Figure 11-7). Spicules are generally located on the boundaries between large, organized regions called **supergranules,** which are enormous convective cells in the photosphere. A typical supergranule is about 30,000 km in diameter and contains roughly a thousand ordinary granules. Gases slowly rise upward in the middle of a supergranule and move horizontally outward toward its edge, where they descend back into the Sun. This large-scale convection moves with speeds of roughly 0.5 km/s, which is about a tenth of the speed of gases churning in a granule.

11-6 *The corona is the outermost layer of the Sun's atmosphere*

The outermost region of the Sun's atmosphere is called the **corona.** It extends from the top of the chromosphere out to a distance of several million kilometers, where it merges into the solar wind. (As noted at the end of Chapter 6, the solar wind consists of high-speed protons and electrons constantly escaping from the Sun.) The three layers of the Sun's atmosphere are sketched schematically in Figure 11-8.

The total amount of visible light we receive from the solar corona is comparable to the brightness of the full

Figure 11-7 Supergranules *Numerous spicules are visible as dark, brushlike protuberances in this H$_\alpha$ photograph of the solar surface. Spicules are located along the irregularly shaped boundaries between supergranules. (NOAO)*

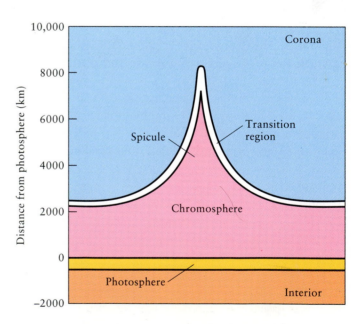

Figure 11-8 The solar atmosphere *The Sun's atmosphere has three distinct layers. The lowest, the photosphere, is roughly 300 to 400 km thick. The chromosphere extends to an altitude of about 2000 km above the photosphere, with spicules jutting up to 7000 km. The third, outer layer, the corona, extends many millions of kilometers out into space and merges with the solar wind. (Adapted from John A. Eddy)*

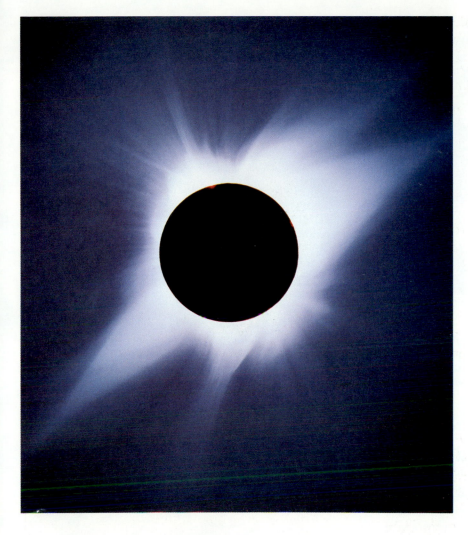

Figure 11-9 The solar corona This extraordinary photograph was taken during the total solar eclipse of July 11, 1991. Numerous streamers are visible, extending millions of kilometers above the solar surface. (Courtesy of R. Christen and M. Christen; Astro-Physics Inc.)

moon—only about one-millionth as bright as the photosphere. The corona can thus be seen only when the photosphere is blocked out during a total eclipse or in a specially designed telescope called a **coronagraph.** Figure 11-9 is an exceptionally detailed photograph of the corona taken during a total eclipse.

Around 1940 astronomers realized that the spectrum of the Sun's corona contains the emission lines of a number of highly ionized elements. For example, there is a prominent green line caused by the presence of Fe XIV (iron atoms each stripped of 13 electrons). Because extremely high temperatures are required to strip that many electrons from atoms, it is clear that the corona must be very hot. It is now known that coronal temperatures are in the range of 1 to 2 million K.

Astronomers do not understand why the corona is so hot. The churning gas motions in the Sun's convective envelope provide plenty of energy to heat the corona, but no one has been able to explain how that energy is transported to the corona. Some researchers suspect that sound waves or waves in the Sun's magnetic field carry energy from the photosphere to the corona, but astronomers do not understand how this happens.

Recent observations from space show that the Sun's corona exhibits complicated structure and activity. For example, in Figure 11-10 there is the huge, bubblelike disturbance called a **solar transient.** These short-lived protuberances erupt suddenly and expand rapidly outward through the corona. Solar transients probably occur as often as once a day, but they were unknown until the Sun was examined with a coronagraph carried aloft by Skylab.

X-ray photographs of the corona were also obtained during the Skylab missions. Recall from Chapter 5 that

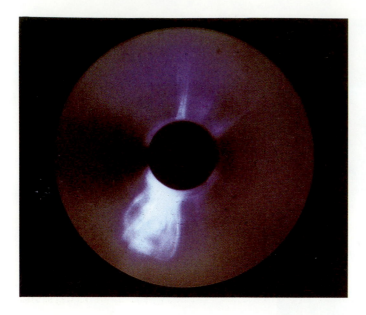

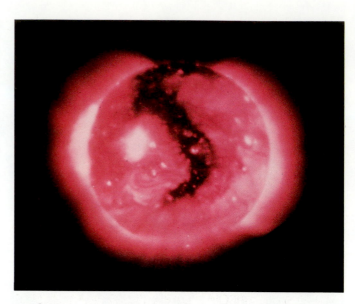

Figure 11-10 *A solar transient During the early 1970s, astronauts on Skylab discovered huge bubbles erupting outward through the corona. This solar transient was photographed in 1973. The leading edge of the bubble moved outward from the Sun at a speed of 500 km/s. (NASA and the High Altitude Observatory)*

Figure 11-11 *A coronal hole This X-ray picture of the Sun was taken by Skylab astronauts in 1973. A huge, dark, boot-shaped coronal hole dominates this view of the inner corona. Numerous bright points are also visible. (NASA; Harvard College Observatory)*

Wien's law tells us the temperature of a blackbody is inversely proportional to the dominant wavelength of the radiation it emits. The corona shines brightly at X-ray wavelengths because its temperature is in the millions of kelvin. The Skylab pictures reveal a very blotchy, irregular inner corona (Figure 11-11). Note the large dark area, called a **coronal hole** because it is nearly devoid of the usual hot, glowing coronal gases. Many astronomers suspect that coronal holes are the main corridors through which particles of the solar wind escape from the Sun.

In addition to dark coronal holes, X-ray photographs also reveal numerous bright spots that are hotter than the surrounding corona. Temperatures in these bright points occasionally reach 4 million K. Many of the bright coronal hot spots seen in X-ray pictures such as Figure 11-11 are located over sunspots.

11-7 Sunspots are only one of many phenomena associated with the 22-year solar cycle

Superimposed on the basic structure of the solar atmosphere are a host of phenomena that vary with a 22-year period. Irregularly shaped dark regions called **sunspots**

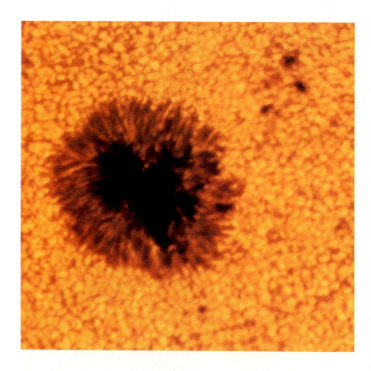

Figure 11-12 *A sunspot group This high-resolution photograph shows a mature sunspot group. The dark center of the spot is the umbra; the penumbra is less dark and has a feather-like appearance. Granulation is visible in the surrounding, undisturbed photosphere. (NOAO)*

(Figure 11-12) are the most easily recognized of these phenomena because they occur in the photosphere, the most visible part of the Sun's atmosphere. The dark, central core of a sunspot, called the umbra, is usually surrounded by a less dark border called the penumbra. Although sunspots vary greatly in size, typical sunspots measure a few tens of thousands of kilometers across.

Sunspots look dark against the solar surface because they are cooler than the surrounding photosphere (recall the Stefan–Boltzmann law). Temperatures in a typical sunspot are 4000 to 4500 K, or more than 1000 K cooler than in the undisturbed photosphere.

On rare occasions a sunspot group is so large that it can be seen with the naked eye. Chinese astronomers recorded such sightings 2000 years ago, and a huge sunspot group visible to the naked eye was seen in 1979. (ALWAYS USE SPECIAL DARK FILTERS OR OTHER MEANS TO PROTECT YOUR EYES WHEN VIEWING THE SUN! Looking directly at the Sun can cause blindness!) Of course, a telescope gives a much better view, and so it was not until Galileo that anyone had examined sunspots in detail. In fact, Galileo discovered that he could determine the Sun's rotation rate by following sunspots as they moved across the solar disk (Figure 11-13). He found that the Sun rotates once in about four weeks. Since a typical sunspot group lasts about two months, it can be followed for two solar rotations.

Careful observations demonstrate that the Sun does not rotate as a rigid body: The equatorial regions rotate more rapidly than the polar regions. A sunspot near the solar equator takes 25 days to go once around the Sun, but at 30° north or south of the equator a sunspot takes about 27 days to complete a rotation. The rotation period at 75° north or south of the equator is about 33 days, and

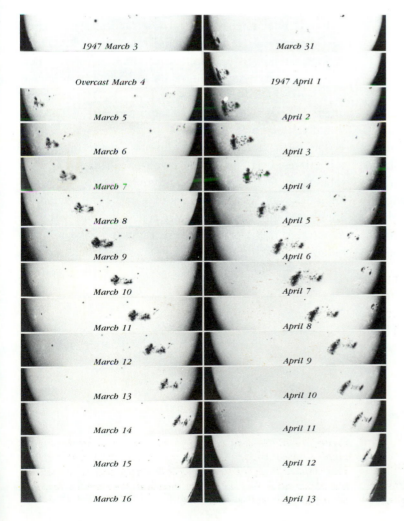

Figure 11-13 The Sun's rotation *By observing the same group of sunspots from one day to the next, Galileo found that the Sun rotates once in about four weeks. The equatorial regions of the Sun actually rotate somewhat faster than the polar regions. This series of photographs shows the same sunspot group over $1\frac{1}{2}$ solar rotations. (The Observatories of the Carnegie Institution of Washington)*

near the poles it may be as long as 35 days. This phenomenon is called differential rotation, because different parts of the Sun rotate at slightly different rates.

Observations over many years reveal that the number of sunspots changes periodically. In some years there are many sunspots, in others almost none. This pattern is called the **sunspot cycle.** As shown in Figure 11-14, the average number of sunspots varies with a period of about 11 years. A time of many sunspots is called a **sunspot maximum,** which occurred in 1970, 1980, and 1991. During a **sunspot minimum** the Sun is almost devoid of sunspots, as it was in 1965, 1976, and 1986.

Recent observations from the Solar Maximum Mission satellite demonstrated that the Sun's brightness is linked to the 11-year sunspot cycle. A device called a radiometer on board the spacecraft measured the intensity of light at all wavelengths arriving at the Earth. The data show that the Sun is brightest during sunspot maximum and dimmest during sunspot minimum. The overall change in the solar output is quite small, about 1 part in 2500.

At first this 11-year variation seems to pose a paradox. Why should the Sun be brightest in years when sunspots are most numerous? Apparently the light blocked by sunspots is more than made up for by an increase in bright patches, called **faculae,** that appear near sunspots and elsewhere on the solar surface.

In 1908 the American astronomer George Ellery Hale made the important discovery that sunspots are associated with intense magnetic fields on the Sun. When Hale focused a spectroscope on sunlight coming from a sunspot, he found that each spectral line in the normal solar spectrum is flanked by additional, closely spaced spectral lines not usually observed (Figure 11-15). This "splitting" of a single spectral line into two or more lines is called the **Zeeman effect** after the Dutch physicist Pieter Zeeman, who first observed it in 1896 in his laboratory. Zeeman

showed that the spectral lines split when the light source is inside an intense magnetic field. The more intense the magnetic field, the more the split lines are separated. The splitting of the spectral line in Figure 11-15 into three lines corresponds to a magnetic field roughly 5000 times more intense than the Earth's natural magnetic field.

Hale's discovery demonstrates that sunspots are areas where concentrated magnetic fields project through the hot gases of the photosphere. Because of the temperature, many atoms in the photosphere are ionized. As a result, the photosphere is a mixture of electrically charged ions and electrons, technically called a **plasma.** A plasma is an extremely good conductor of electricity that interacts vigorously with magnetic fields. Because a magnetic field restricts and constrains the motions of a plasma, a sunspot's intense magnetic field greatly inhibits the natural convective motions of the gases. Consequently, energy cannot flow freely upward from the Sun's convective zone, and the gases in this region of the photosphere cool off. This cool area emits less radiation than the unperturbed photosphere and thus appears dark in contrast to the surrounding solar surface.

Exotic phenomena occur around and above sunspots as a direct result of their intense magnetic fields. Huge, arching columns of gas called **prominences** often appear above sunspot regions (Figure 11-16). Some prominences hang suspended for days above the solar surface, while others blast material outward from the Sun at speeds of roughly 1000 km/s.

The most violent, eruptive events on the Sun, called solar flares, occur in complex sunspot groups. During a solar flare, temperatures in a compact region soar to 5 million K, and vast quantities of particles and radiation are blasted into space. The flare is usually over within 20 minutes. Ultraviolet and X-ray radiation take about 8 minutes to reach the Earth. High-energy particles arrive a

Figure 11-14 The sunspot cycle The number of sunspots on the Sun varies with a period of about 11 years. The most recent sunspot maximum occurred in 1991, and the most recent sunspot minimum occurred in 1986. The next sunspot minimum will probably occur in 1997.

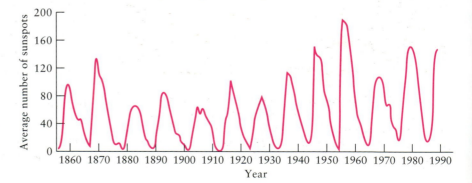

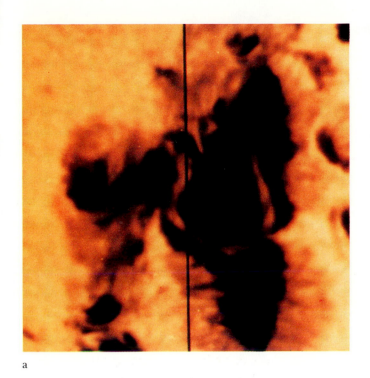

a

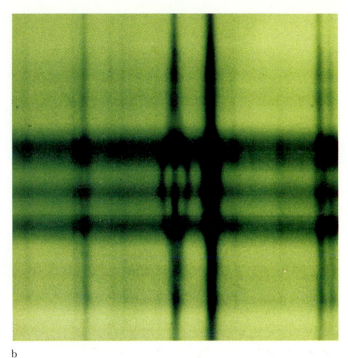

b

Figure 11-15 *Zeeman splitting by a sunspot's magnetic field* (a) *The black line drawn across the sunspot indicates the location toward which the slit of the spectroscope was aimed.* (b) *In the resulting spectrogram, one line in the middle of the normal* solar spectrum is split into three components. The separation between the three lines corresponds to a magnetic field roughly 5000 times stronger than the Earth's natural magnetic field. (NOAO)

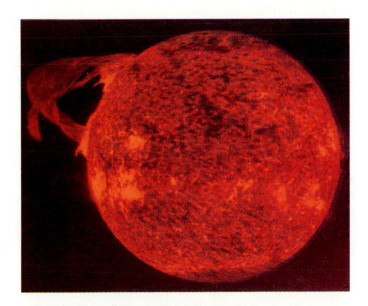

Figure 11-16 *A prominence   A huge prominence arches above the solar surface in this Skylab photograph taken in 1973. The radiation that exposed this picture is from singly ionized helium at a wavelength of 30.4 nm, corresponding to a temperature of about 50,000 K. (Naval Research Laboratory)*

day or two later and interfere with radio communications. Clouds of high-energy particles arrive a day or two later and often produce beautiful, shimmering lights in the night sky called **aurorae.** Most astronomers suspect that prominences and flares involve concentrated portions of the Sun's magnetic field.

Astronomers can construct artificial pictures called **magnetograms** that display the magnetic fields in the solar atmosphere by combining two photographs taken at wavelengths on either side of a magnetically split spectral line. The magnetogram in Figure 11-17 shows the entire solar surface along with an ordinary photograph of the Sun taken at the same time. Dark blue indicates the area of the photosphere covered by one (north) magnetic polarity, and yellow designates the area covered by the other (south) polarity.

Many sunspot groups are said to be **bipolar,** meaning that roughly comparable areas are covered by north and south magnetic polarities (Figure 11-18). The sunspots on the side of the group toward which the Sun is rotating are called the "preceding members" of the sunspot group; the

Figure 11-17 *Two views of the Sun* *The optical view of the Sun (right) was taken at the same time as the magnetogram (left). Notice how regions with strong magnetic fields are related to the locations of sunspots. Careful examination of the magnetogram also reveals that the magnetic polarity of the sunspots in the northern hemisphere is opposite that of those in the southern hemisphere. (NOAO)*

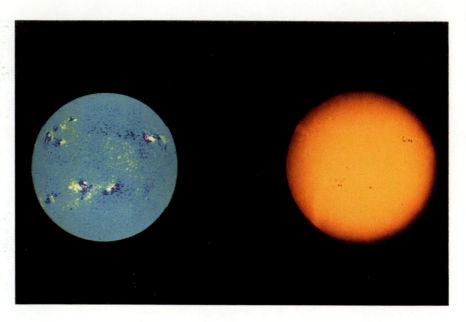

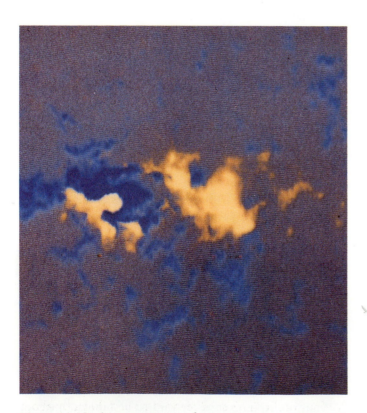

Figure 11-18 *A magnetogram of a sunspot group* *This artificially colored picture displays the intensity and polarity of the magnetic field associated with a large sunspot group. One side of a typical sunspot group has one magnetic polarity, and the other side has the opposite polarity. (NOAO)*

remaining spots, which follow behind, are called the "following members."

After years of studying the solar magnetic field, Hale was able to piece together a remarkable magnetic description of the solar cycle. First of all, Hale discovered that the preceding spots of all sunspot groups in one solar hemisphere have the same magnetic polarity. We now know that this polarity is the same as that of the hemisphere in which the group is located. In other words, in the hemisphere of the Sun's north magnetic pole, the preceding members of all sunspot groups have north magnetic polarity. In the south magnetic hemisphere, all the preceding members have south magnetic polarity.

In addition, Hale found that the polarity pattern completely reverses itself every 11 years. For instance, the hemisphere that has a north magnetic polarity at one solar maximum has a south magnetic polarity at the next solar maximum. For this reason astronomers prefer to speak of the **solar cycle,** which has a period of 22 years, rather than the 11-year sunspot cycle.

In 1960 the American astronomer Horace Babcock proposed a description that seems to account for many aspects of the 22-year solar cycle. Babcock's scenario, called a **magnetic dynamo** model, employs two basic properties of the Sun: its differential rotation and its convective envelope. Differential rotation causes the magnetic field in the photosphere to become wrapped around the

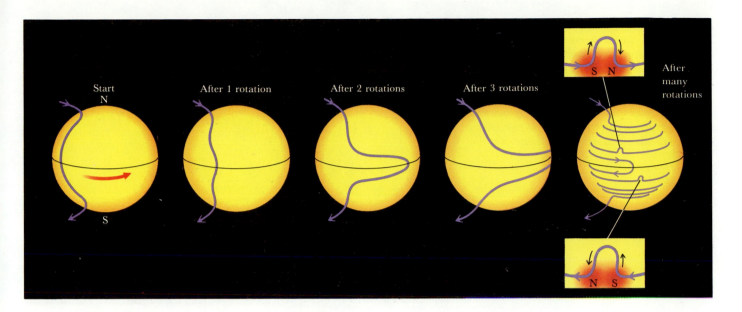

Figure 11-19 *Babcock's magnetic dynamo* *A possible partial explanation for the sunspot cycle involves the wrapping of a magnetic field around the Sun by differential rotation. Sunspots* *appear where the concentrated magnetic field breaks through the solar surface.*

Sun as shown in Figure 11-19. Convection in the photosphere causes the concentrated magnetic field to become tangled, and kinks erupt through the solar surface. Sunspots appear at locations where the magnetic field projects through the photosphere. Careful examination of how the magnetic field in Figure 11-19 is oriented reveals that the preceding members of a sunspot group should indeed have the same magnetic polarity as that of the hemisphere in which they are located.

There are many details that the magnetic dynamo model fails to explain. Most embarrassing is a general ignorance of the physics of sunspots. Astronomers still do not know what holds a sunspot together week after week. Calculations predict that a sunspot should break up and disperse as soon as it forms.

Our understanding of sunspots is further confounded by irregularities in the solar cycle. For example, the overall reversal of the Sun's magnetic field is often piecemeal and haphazard. One pole may reverse polarity long before the other so that for several weeks the Sun may have two north poles but no south pole at all. To make matters worse, there is strong historical evidence that all traces of sunspots and the sunspot cycle have vanished for many years. For example, virtually no sunspots were seen from 1645 through 1715. Similar sunspot-free periods apparently occurred at irregular intervals in earlier times.

11-8 *Astronomers can study the solar interior by measuring vibrations of the Sun's surface*

There are many reasons why solar astronomers would like to study the Sun's interior. For instance, the magnetic field that produces sunspots might be bits and pieces of ancient magnetism trapped deep inside the Sun as it formed 4.5 billion years ago. Other speculation includes the possibility that the Sun's core is rotating as much as three times faster than its surface. To search for the roots of the Sun's magnetic field and to study other deeply buried phenomena, astronomers need a way to probe the solar interior.

Geologists are able to study the Earth's interior structure by using a seismograph to record vibrations during earthquakes. Although there are no true sunquakes, the Sun does vibrate at a variety of frequencies, somewhat like a ringing bell. These vibrations, first noticed in 1960, are detected with sensitive Doppler shift measurements. Portions of the Sun's surface move up and down by about 10 km every 5 minutes (Figure 11-20). Slower vibrations with periods ranging from 20 minutes to nearly an hour were discovered in the 1970s. More recently a variety of long-period oscillations have been found.

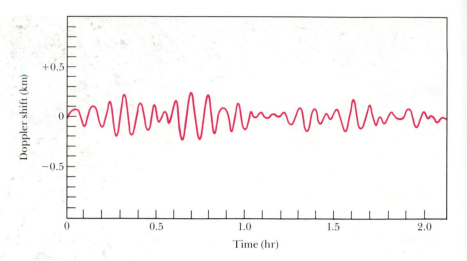

Figure 11-20 Vibrations of the Sun's surface
*Doppler shift measurements can be used to de-
termine the speed at which the Sun's photosphere bobs up and down. These data, which
extend over two hours, show numerous oscillations with a period of about five minutes. The
gradual variation in the size of the oscillations
is the result of interference between many
vibrations and wavelengths. (Adapted from
O. R. White)*

To conduct long-term studies of these slow pulsations,
astronomers have set up telescopes and launched balloon-
borne observatories at the South Pole, where the Sun can
be observed continuously for several months. This new
field of solar research is called **helioseismology.**

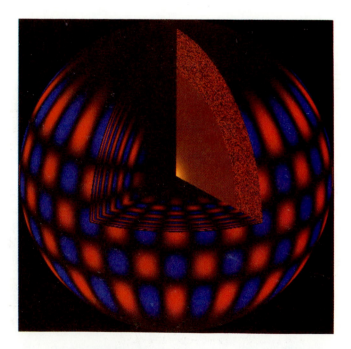

Figure 11-21 Sound waves resonating in the Sun *This computer-generated image shows one of the millions of ways in
which the Sun vibrates because of sound waves resonating in its
interior. The regions that are moving outward are colored blue;
those moving inward are red. The cutaway shows how deep
these oscillations are believed to extend. (National Solar
Observatory)*

The vibrations of the solar surface may be compared
with sound waves. If you were within the Sun's atmosphere, you would first notice a deafening roar, somewhat
like a jet engine, produced by the turbulence associated
with convection. Superimposed on this noise is a variety
of nearly pure tones. You would be unable to hear these
tones, however, for they are 16 octaves below the lowest
note on the piano keyboard.

Sound waves moving upward from the solar interior
are reflected by the solar surface, which acts somewhat
like a mirror to these waves. As a reflected sound wave
descends back into the Sun, the increasing density and
pressure bend the wave so severely that it is turned around
and aimed back out again toward the solar surface. In
other words, sound waves bounce back and forth between
the solar surface and layers deep within the Sun. These
sound waves can reinforce each other and resonate just
like sound waves inside an organ pipe.

There are millions of ways in which the Sun can oscil-
late as a result of sound waves resonating in its interior.
Figure 11-21 is a computer-generated illustration that
shows one such vibrational mode. Astronomers can
deduce detailed information about the solar interior from
measurements of these oscillations. For instance, it is possible to infer the rotation rate of the Sun's interior. By
comparing the speeds of sound waves that travel east to
west with those that travel west to east, astronomers
figured out the Sun's rotation rate at different depths
and latitudes. As shown in Figure 11-22, the Sun's pattern
of surface rotation persists throughout the outer 20% of
the Sun, where it is driven by large-scale convection. Further in convection ceases, and the Sun seems to rotate like
a rigid object with a period of about 27 days.

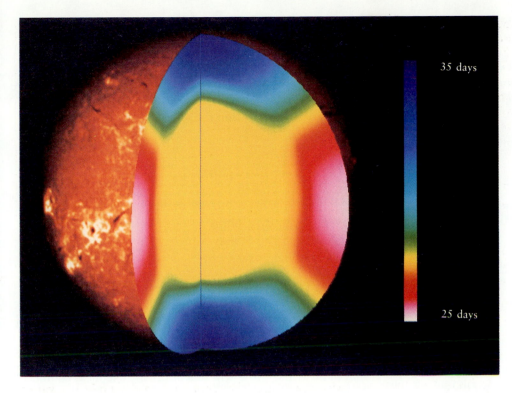

Figure 11-22 *Rotation of the solar interior* This cutaway picture of the Sun shows how the rate of solar rotation varies with depth and latitude. Colors represent rotation periods according to the scale given at the right. Note that the pattern of surface rotation, which varies from 25 days at the equator to 35 days near the poles, persists throughout the Sun's convective envelope. The Sun's radiative core seems to rotate like a rigid body. (Courtesy of K. Libbrecht; Big Bear Solar Observatory)

35 days

25 days

Summary

• The Sun's energy is produced by the thermonuclear process called hydrogen burning, in which four hydrogen nuclei fuse to produce a single helium nucleus and release energy.

   The energy released in a thermonuclear reaction comes from the conversion of matter into energy according to Einstein's equation $E = mc^2$.

   A thermonuclear reaction is a nuclear reaction that occurs only at very high temperatures; hydrogen burning reactions, for instance, occur only at temperatures of more than about 8 million K.

• A stellar model is a theoretical description of a star's interior derived from calculations based upon the laws of physics.

   The solar model suggests that hydrogen burning occurs in a core that extends from the Sun's center to about 0.25 solar radius.

Throughout most of the Sun's interior, energy moves outward from the core by radiative diffusion. In the Sun's outer layers, energy is transported to the Sun's surface by convection.

• The visible surface of the Sun is a layer at the bottom of the solar atmosphere called the photosphere. The gases in this layer shine with almost perfect blackbody radiation, and convection produces features called granules there.

   Above the photosphere is a layer of cooler and less dense gases called the chromosphere. Jets of gas called spicules rise up into the chromosphere along the boundaries of supergranules.

   The outermost layer of thin gases in the solar atmosphere is called the corona, which blends into the solar wind at great distances from the Sun. The gases of the corona are very hot but at low density. Solar transients, coronal streamers, and coronal holes are other features associated with the corona.

- Surface features on the Sun vary periodically in a 22-year solar cycle.

  Sunspots are relatively cool regions produced by local concentrations of the Sun's magnetic field. The average number of sunspots increases and decreases in a regular 11-year cycle.

  A solar flare is a brief eruption of hot, ionized gases from a sunspot group.

- The magnetic dynamo model suggests that many features of the solar cycle are caused by the effects of differential rotation and convection on the Sun's magnetic field.

- Astronomers probe the interior structure of the Sun by studying vibrations of the solar surface.

Review questions

1 Describe the dangers in attempting to observe the Sun. How have astronomers learned to circumvent these hazards?

2 Give an everyday example of hydrostatic equilibrium. Give an everyday example of thermal equilibrium.

3 Give some everyday examples of conduction, convection, and radiative diffusion.

4 What do astronomers mean by "a model of the Sun"?

5 Why do thermonuclear reactions occur only in the Sun's core?

6 What is hydrogen burning? Why is hydrogen burning fundamentally unlike the burning of a log in a fireplace?

7 Describe the Sun's interior, with attention to the main physical processes that occur at various depths within the Sun.

8 What is a neutrino, and why are astronomers so interested in detecting neutrinos from the Sun?

9 Describe the Sun's atmosphere. Be sure to mention some of the phenomena that occur at various altitudes above the solar surface.

10 Describe the three main layers in the solar atmosphere and how you would best observe them.

11 When do you think the next sunspot maximum and minimum will occur? Explain.

12 Why is the solar cycle said to have a period of 22 years, even though the sunspot cycle is only 11 years long?

13 How do astronomers detect the presence of a magnetic field in hot gases, such as those in the solar photosphere?

14 Explain why studying oscillations of the Sun's surface can give important, detailed information about physical conditions deep within the Sun.

Advanced questions

* 15 Using the mass and size of the Sun, calculate the Sun's average density. Compare your answer with the average densities of the Jovian planets. (*Hint:* The volume of a sphere of radius r is $\frac{4}{3}\pi r^3$.)

16 What would happen if the Sun were not in a state of hydrostatic or thermal equilibrium?

* 17 Assuming that the current rate of hydrogen burning in the Sun remains constant, what fraction of the Sun's mass will be converted into helium over the next five billion years? How will this affect the chemical composition of the Sun?

* 18 Calculate the wavelengths at which the photosphere, chromosphere, and corona emit the most radiation. Explain how the results of your calculations suggest the best way to observe these regions of the solar atmosphere. (*Hint:* Use Wien's law and assume that the average temperatures of the photosphere, chromosphere, and corona are 5800 K, 50,000 K, and 1.5×10^6 K, respectively.)

Discussion questions

19 Discuss the extent to which cultures around the world have worshiped the Sun as a deity throughout history. Why do you suppose there has been such widespread veneration?

20 Discuss some of the difficulties of correlating solar activity with changes in the terrestrial climate.

21 From 1645 to 1715 virtually no sunspots were seen, and northern Europe experienced a "Little Ice Age" during which record low temperatures were observed. Could these two phenomena be related? Explain.

22 Describe some advantages and disadvantages of observing the Sun (**a**) from space and (**b**) from the South Pole. What kinds of phenomena and issues do solar astronomers want to explore from both Earth-orbiting and Antarctic observatories?

For further reading

Bahcall, J. "Where Are the Solar Neutrinos?" *Astronomy,* March 1990 • This article examines the dilemma that the Sun seems to be emitting fewer neutrinos than theories predict.

Eddy, J. *A New Sun.* NASA SP-402, 1979 • This lavishly illustrated book gives a superb summary of our understanding of the Sun.

———. "The Case of the Missing Sunspots." *Scientific American,* May 1977 • This article by a noted solar astronomer discusses evidence for long periods when no sunspots were seen.

Foukal, P. "The Variable Sun." *Scientific American,* February 1990 • This up-to-date article discusses recent observations of the Sun's variability and its effects on Earth.

Frazier, K. *Our Turbulent Sun.* Prentice-Hall, 1983 • This well-researched report by a noted science writer describes the irregularity and variability of the Sun's energy output.

Friedman, H. *Sun and Earth.* Scientific American Library, 1986 • This excellent, beautifully illustrated book conveys the excitement and adventure of solar research.

Harvey, J., and others. "GONG: To See inside Our Sun." *Sky & Telescope,* November 1987 • This article describes the Global Oscillation Network Group (GONG), a worldwide program of scientists exploring the solar interior by means of the Sun's naturally occurring oscillations.

Leibacher, J. W., and others. "Helioseismology." *Scientific American,* September 1985 • This lucid article explains how astronomers can learn about the structure, composition, and dynamics of the Sun's interior from oscillations visible on its surface.

Levine, R. "The New Sun." In J. Cornell and P. Gorenstein, eds., *Astronomy from Space: Sputnik to Space Telescope.* MIT Press, 1983 • This article surveys the advantages of solar astronomy from space.

Mitton, S. *Daytime Star.* Scribner, 1981 • This easily readable book presents a completely nontechnical introduction to the Sun.

Nichols, R. "Solar Max: 1980–89." *Sky & Telescope,* December 1989 • This brief article describes the Solar Maximum Mission satellite, which made many important contributions to our understanding of the Sun during the 1980s.

Noyes, R. *The Sun, Our Star.* Harvard University Press, 1982 • This clear, well-presented introduction to the Sun includes interesting chapters on the climate and solar energy.

Robinson, L. "The Disquieting Sun: How Big, How Steady?" *Sky & Telescope,* April 1982 • This article describes controversial observations that the Sun is shrinking.

Taylor, M. "Observing from the South Pole." *Sky & Telescope,* October 1988 • This article discusses the work of astronomers who observe the Sun during the Antarctic summer.

Wallenhorst, S. "Sunspot Numbers and Solar Cycles." *Sky & Telescope,* September 1982 • This article discusses the counting of sunspots and the plotting of the solar cycle.

Wentzel, D. *The Restless Sun.* Smithsonian Institution Press, 1989 • This lucid, up-to-date, stimulating account of the Sun emphasizes the wealth of activity—from sunspots to flares—that our star so vividly displays.

Wolfson, R. "The Active Solar Corona." *Scientific American,* February 1983 • This article describes the dynamic processes that occur in the solar corona as rarified gases and magnetic fields interact on a grand scale.

12

The Nature of the Stars

Although they appear only as pinpoints of light, stars are huge spheres of glowing gas, much like our Sun. In this chapter, we see how astronomers have measured the distances to many of the nearer stars. Once a star's distance is known, its luminosity can easily be deduced. We also learn how astronomers determine the surface temperatures of stars from their spectra and colors. Stellar luminosities and surface temperatures come together in the all-important Hertzsprung–Russell diagram, which reveals the fundamental types of stars. Finally, we turn to the topic of binary stars, those surprisingly common systems in which two stars orbit each other. By observing the motions of the two stars in a binary, astronomers can determine stellar masses. All this information gives us important insights into the essential nature of stars and forms a foundation of our understanding of the heavens.

A cluster of stars By analyzing starlight, an astronomer can determine such details about a star as its surface temperature, chemical composition, and luminosity. This photograph clearly shows color differences in a star cluster called NGC 3293. Reddish stars are comparatively cool, with surface temperatures around 3000 K. They are also quite luminous and have very large diameters, typically 100 times as large as the Sun. Blue-white stars have much higher surface temperatures (15,000 to 30,000 K) and are roughly the same size as the Sun. (Anglo-Australian Observatory)

To the unaided eye the night sky is spangled with thousands of stars, each appearing as a pinpoint of light. A telescope reveals many thousands of other stars too faint to be seen with the naked eye, but every star appears only as a bright point of light. A star is a huge ball of hot gas like our Sun, held together by its own gravity. We now know that some stars are larger than our Sun, some smaller; some brighter, some dimmer. Some stars are hotter than the Sun, but others are cooler.

The quest for information about the masses, luminosities, surface temperatures, and chemical compositions of the stars has been a major preoccupation of twentieth-century astronomers. In recent years a remarkably complete picture has emerged. By understanding the stars, we gain insight into our relationship to the universe and our place in the cosmic scope of space and time.

12-1 Distances to nearby stars are determined by parallax

Looking up into the nighttime sky, you can see hundreds upon hundreds of stars. Some appear quite bright, but most are rather dim. As you gaze up at this starry panorama, one of the questions you might ask is, How far away are the stars? Are they nearby or extremely far away? The apparent brightnesses of the stars do not indicate their distances. A star that looks dim might actually be a brilliant star that is exceedingly remote. To determine the distances to the stars, astronomers use geometric techniques that involve painstaking measurements of stellar positions and motions.

The most straightforward way of measuring stellar distances is based on **parallax,** which is the apparent displacement of an observed object because of a change in the point of view. You experience parallax when nearby objects appear to shift their positions against a distant background as you move from one place to another (Figure 12-1). Stars exhibit the same phenomenon. As the Earth orbits the Sun, nearby stars appear to move back and forth against the background of the more distant stars.

The distance to a star can be determined by measuring the star's parallax. The parallax (p) of a star is half the angle through which the star's apparent position shifts as the Earth moves from one side of its orbit to the other (Figure 12-2). If the angle p is measured in seconds of arc, then the distance d to the star in parsecs is given by the equation

$$d = \frac{1}{p}$$

For example, a star whose parallax is $\frac{1}{2}$ arc sec is 2 pc from Earth. This simple relationship between parallax and distance in parsecs is one of the main reasons that astronomers usually measure cosmic distances in parsecs rather than light years. (Recall that 1 parsec equals 3.26 light years, and 1 ly is nearly 10^{13} km. See Figure 1-11.)

The first parallax measurement was made by the German astronomer-mathematician Friedrich Wilhelm Bessel. He found the parallax of 61 Cygni to be $\frac{1}{3}$ arc second, and so its distance is about 3 pc from Earth. The nearest star, Proxima Centauri, has a parallax of 0.77 arc sec, and thus its distance is 1.3 pc. The parallax method therefore involves the measurement of extremely tiny angles. For instance, the parallax of Proxima Centauri is comparable to the angular diameter of a dime seen from a distance of two miles.

Because parallaxes smaller than about $\frac{1}{20}$ arc sec are difficult to measure accurately from Earth-based observatories, the parallax method gives reliable distances only for stars nearer than about 20 pc. There are nearly 2000 stars within this range, half of which have had their parallaxes measured with high precision. Most of these nearby stars are far too dim to be seen with the naked eye. In

Figure 12-1 Parallax *Imagine looking at some nearby object (a tree) as seen against a distant background (mountains). If you move from one location to another, the nearby object will appear to shift its location with respect to the distant background scenery. This familiar phenomenon is called parallax.*

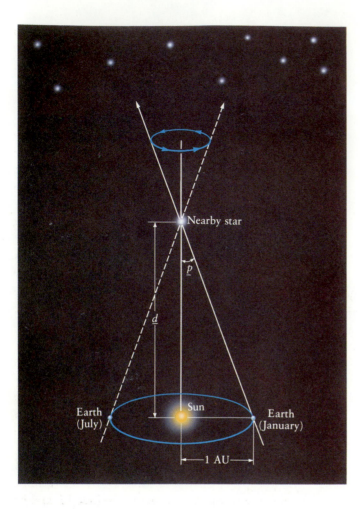

Figure 12-2 Stellar parallax *As the Earth orbits the Sun, a nearby star appears to shift its position against the background of distant stars. The angle p (the parallax of the star) is equal to the angular size of the radius of the Earth's orbit as seen from the star. The smaller the parallax (p), the larger the distance (d) to the star.*

of its mission to measure the distances to stars as far away as 500 pc. Since accurate knowledge of the distances to the stars is a crucial step in understanding the heavens, astronomers will increasingly turn to space-based observations during the 1990s.

12-2 A star's luminosity can be determined from its apparent magnitude and its distance from Earth

The system of magnitudes that astronomers use to denote the brightnesses of stars was invented in ancient Greece by the astronomer Hipparchus. The brightest stars Hipparchus saw in the sky he called first-magnitude stars. Those about one-half as bright he called second magnitude stars, and so forth, to sixth-magnitude stars, the dimmest ones he could see.

In the nineteenth century techniques were developed for measuring the amount of light arriving from a star, and since then astronomers have defined the magnitude scale more precisely. Measurements showed that a first-magnitude star is about 100 times as bright as a sixth-magnitude star. In other words, it would take 100 stars of magnitude +6 to provide as much total light as we receive from a single star of magnitude +1. As a result, the magnitude scale was redefined so that a magnitude difference of 5 corresponds exactly to a factor of 100 in the amount of light energy received. A magnitude difference of 1 therefore corresponds to a factor of 2.512 in light energy, because

$$2.512 \times 2.512 \times 2.512 \times 2.512 \times 2.512 = 100$$

Thus, for example, it takes about $2\frac{1}{2}$ third-magnitude stars to provide as much light as we receive from a single second-magnitude star.

Astronomers also extended the **magnitude scale** to describe the dimmer stars visible through their telescopes. For example, the dimmest stars visible through a pair of binoculars have a magnitude of +10. Through some of the largest telescopes in the world, it is possible to see stars as dim as magnitude +20. Photographs taken with long exposure times reveal even dimmer stars.

Astronomers use negative numbers to extend the magnitude scale so that it includes very bright objects. For example, Sirius—the brightest star in the sky—has a mag-

contrast, the majority of the familiar, bright stars in the nighttime sky are too far away to exhibit parallaxes measurable from the Earth's surface.

Parallax measurements made by an Earth-orbiting satellite would be unhampered by our atmosphere, permitting astronomers to determine the distances to stars well beyond the reach of ground-based observations. In 1989 the European Space Agency (ESA) launched a satellite called *Hipparcos* (an acronym for *HIgh Precision PARallax COllecting Satellite*), specifically designed to measure parallaxes as small as 0.002 arc sec. Although the satellite failed to achieve its proper orbit, astronomers are still hopeful that *Hipparcos* will be able to complete much

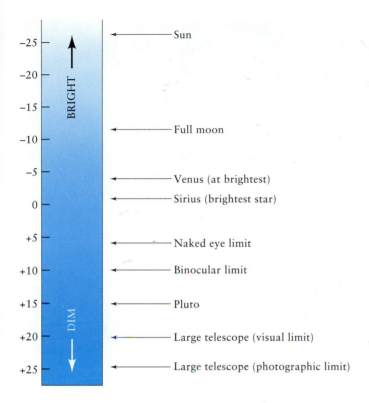

Figure 12-3   The apparent magnitude scale   *Astronomers denote the brightness of objects in the sky by the apparent magnitude of the objects. Most stars visible to the naked eye have magnitudes between +1 and +6. Photography through a large telescope can reveal stars as faint as magnitude +24.*

nitude of −1.5. At its brightest, Venus shines with a magnitude of −4.4. Of course, the Sun is the brightest object in the sky with a magnitude of −26.7.

Figure 12-3 illustrates the modern magnitude scale. These magnitudes are properly called **apparent magnitudes** because they describe how bright an object appears to an Earth-based observer. More precisely, apparent magnitude is a measure of the energy arriving at the Earth.

Apparent magnitudes do not tell us about the actual brightness of the stars. A star that looks dim in the sky might be a brilliant star that just happens to be extremely far away. To understand the stars, astronomers need to know how bright the stars really are. Absolute magnitude is a measure of a star's energy output.

The **absolute magnitude** of a star is defined as the apparent magnitude it would have if it were located at a distance of exactly 10 pc from the Earth. For example, if the Sun were moved to a distance of 10 pc from the Earth,

it would have an apparent magnitude of +4.8. Therefore the absolute magnitude of the Sun is +4.8. The absolute magnitudes of other stars range from roughly −10 for the brightest to +15 for the dimmest. The Sun's absolute magnitude is about in the middle of this range, suggesting that the Sun is an average star.

Absolute magnitude is a very informative quantity because it indicates the intrinsic brightness of a star. In contrast, apparent magnitude tells only how bright the star appears in the sky. To appreciate how astronomers determine the absolute magnitude of a star, you must first realize that the farther away a source of light is, the dimmer it appears.

Imagine a source of light like a light bulb or a star. As light moves outward from the source, it spreads out over increasingly larger regions of space, as shown in Figure 12-4. As the light spreads out, its brightness decreases. That is why the farther away a source of light is, the dimmer it appears. Specifically, the **inverse-square law** tells us that the apparent brightness of a light source is inversely proportional to the square of the distance between the source and the observer. For example, double the distance to a light source and its apparent brightness decreases by 2^2, or a factor of 4. Similarly, at triple the distance, the brightness decreases by 3^2, or a factor of 9.

Using the inverse-square law, astronomers have derived a mathematical equation relating three quantities: a star's apparent magnitude m, its absolute magnitude M, and its distance d from the Earth. If you know any two of these quantities (such as apparent magnitude and distance), you can calculate the third one (such as absolute magnitude). Thus astronomers measure the apparent magnitude of a nearby star, find its distance by measuring its parallax, then calculate its absolute magnitude.

Absolute magnitude is directly related to luminosity, the amount of energy escaping from a star's surface each second (usually expressed in watts). Many scientists prefer to speak of a star's luminosity rather than its absolute magnitude because luminosity is a direct measure of the star's energy output. A simple equation relates absolute magnitude to luminosity, and astronomers can convert from one to the other as they see fit.

For convenience, stellar luminosities are expressed in multiples of the Sun's luminosity ($L_\odot$), which is 3.90×10^{26} watts. The brightest stars (absolute magnitude of −10) have luminosities of $10^6\ L_\odot$. In other words, each of these stars has the energy output of a million Suns. The dimmest stars (absolute magnitude of +15) have luminosities of $10^{-4}\ L_\odot$.

Figure 12-4 The inverse-square law This drawing shows how the same amount of radiation from a light source must illuminate an ever-increasing area as distance from the light source increases. Because the light becomes spread out as it moves away from the source, the apparent brightness of the source decreases.

12-3 A star's color reveals its surface temperature

One of the first things you notice when comparing stars in the nighttime sky is their differences in apparent magnitude. More careful examination, even with the naked eye, reveals that stars also have different colors. For example, in the constellation of Orion, you can see the difference between reddish Betelgeuse and bluish Rigel (examine Figure 4-24).

A star's color is directly associated with its surface temperature through relationships such as Wien's law, which is discussed in Chapter 5 (review the blackbody curves in Figure 5-3). The intensity of light from a cool star peaks at long wavelengths, and so the star looks red (Figure 12-5a). A hot star's intensity curve is skewed toward short wavelengths, making the star look blue (Figure 12-5c). The maximum intensity of a star of intermediate temperature (such as the Sun) occurs near the middle of the visible spectrum, giving the star a yellow-white color (Figure 12-5b).

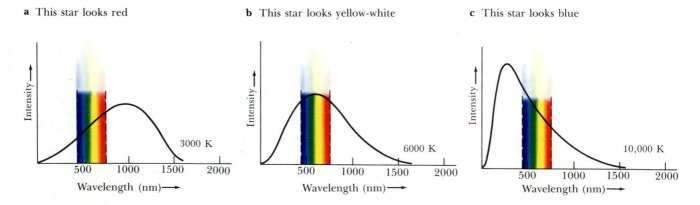

a This star looks red **b** This star looks yellow-white **c** This star looks blue

Figure 12-5 Temperature and color This diagram shows the relationship between the color of a star and its surface temperature. The intensity of light emitted by three hypothetical stars is plotted against wavelengths (compare Figure 5-3). The range of visible wavelengths is indicated. How a star's intensity curve is skewed determines the dominant color of its visible light.

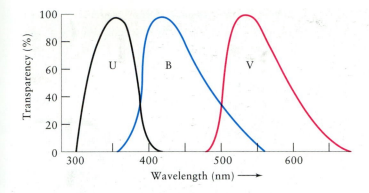

Figure 12-6 Light transmission through the UBV filters This graph shows the wavelength ranges over which standardized U, B, and V filters are transparent to light. The U filter is transparent to light from 300 to 400 nm, a range called the near-ultraviolet because it lies just beyond the violet end of the visible spectrum. The B filter is transparent from about 380 to 550 nm, and the V filter from about 500 to 650 nm.

In photometry, the astronomer aims a telescope at a star and measures the intensity of starlight three times, each time using a different filter. This procedure gives three apparent magnitudes for the star, usually designated by the capital letters U, B, and V. The astronomer then compares the intensity of starlight in neighboring wavelength bands by subtracting one magnitude from another to form the combinations (U – B) and (B – V), which are called the star's **color indices.** The UBV magnitudes and color indices for several representative stars are given in Table 12-1.

A color index tells you how much brighter or dimmer a star is in one wavelength band than in another. For example, the (B – V) color index tells you how much brighter or dimmer a star appears through the B filter than through the V filter.

Color index is important, because it tells you the star's surface temperature. If a star is very hot, its radiation is skewed toward the short-wavelength ultraviolet, which makes the star bright through the U filter, dimmer through the B filter, and dimmest through the V filter. The star Regulus (see Table 12-1) is such an example. Alternatively, if the star is cool, its radiation peaks at long wavelengths, making the star brightest through the V filter, dimmer through the B filter, and dimmest through the U filter. The stars Aldebaran and Betelgeuse are examples.

The graph in Figure 12-7 gives the relationship between the (B – V) color index and temperature. If you know a star's (B – V) color index, you can use this graph to find the star's surface temperature. For example, the Sun's (B – V) index is +0.62, which corresponds to a surface temperature of 5800 K.

To measure the colors of the stars accurately, astronomers have developed a technique called **photometry.** This technique uses a light-sensitive device (such as a CCD; recall Figure 4-17) at the focus of a telescope, with a standardized set of colored filters. The most commonly used filters are the **UBV filters,** each of which is transparent in one of three broad wavelength bands: the ultraviolet (U), the blue (B), and the central region (V) of the visible spectrum (Figure 12-6). The transparency of the V filter (V for "visual") mimics the sensitivity of the human eye.

Table 12-1 The UBV magnitudes and color indices of selected stars

Star name	V	B	U	(B – V)	(U – B)	Apparent color
Bellatrix (γ Ori)	1.64	1.42	0.55	–0.22	–0.87	Blue
Regulus (α Leo)	1.35	1.24	0.88	–0.11	–0.36	Blue-white
Sirius (α CMa)	–1.46	–1.46	–1.52	0.00	–0.06	Blue-white
Megrez (δ UMa)	3.31	3.39	3.46	+0.08	+0.07	White
Altair (α Aql)	0.77	0.99	1.07	+0.22	+0.08	Yellow-white
Sun	–26.78	–26.16	–26.06	+0.62	+0.10	Yellow-white
Aldebaran (α Tau)	0.85	2.39	4.29	+1.54	+1.90	Orange
Betelgeuse (α Ori)	0.50	2.35	4.41	+1.85	+2.06	Red

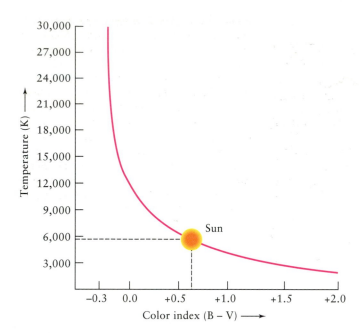

Figure 12-7 Blackbody temperature versus color index The (B − V) color index is the difference between the B and V magnitudes of a star. If the star is hotter than about 10,000 K, it is a very bluish star and its (B − V) index is less than zero. If a star is cooler than about 10,000 K, its (B − V) index is greater than zero. The Sun's (B − V) index is about 0.62, which corresponds to a temperature of 5800 K. After measuring a star's B and V magnitudes, an astronomer can estimate the star's surface temperature from a graph such as this one.

12-4 A star's spectrum is a clue to the star's surface temperature

The field of stellar spectroscopy was born in the 1860s when the Italian astronomer Angelo Secchi attached a spectroscope to his telescope and pointed it toward the stars. Secchi discovered that spectral lines are often observed in the spectra of stars. He was also the first astronomer to classify stellar spectra into **spectral types,** according to their appearance.

At first glance, stellar spectra seem to come in a bewildering variety. Some stellar spectra show prominent Balmer lines of hydrogen. Other spectra exhibit many absorption lines of calcium and iron. Still others are dominated by broad absorption features caused by molecules such as titanium oxide. To cope with this diversity, astronomers grouped similar stellar spectra in classes, or types. According to one classification scheme popular in the late 1800s, a star was assigned a letter from A through P, according to the strength of the Balmer hydrogen lines in the star's spectrum.

After Niels Bohr explained the structure of the hydrogen atom in the early 1900s (recall Figure 5-16), astronomers realized that the strength of the lines in a star's spectrum is directly related to the temperature of the gases in the star's outer layers.

To see why the appearance of a star's spectrum is profoundly affected by the star's surface temperature, consider hydrogen. Hydrogen is the most abundant element in the universe, accounting for about three-quarters of the mass of a typical star. However, hydrogen lines do not necessarily show up in a star's spectrum. Recall that hydrogen's Balmer lines are produced when an electron in the $n = 2$ orbit of hydrogen absorbs a photon having the energy to lift it to a higher orbit (recall Figure 5-17). If the star is much hotter than 10,000 K, high-energy photons pouring out of the star's interior knock electrons out of the hydrogen atoms in the star's outer layers. This process ionizes the hydrogen. When a hydrogen atom's only electron is torn away, no spectral lines can be produced.

Conversely, if the star is much cooler than 10,000 K, the majority of photons escaping from the star possess too little energy to boost many electrons up from the ground state to the $n = 2$ orbit of the hydrogen atoms. These unexcited atoms also fail to produce Balmer lines. In summary, in order to produce Balmer lines, a star must be hot enough to excite electrons out of the ground state but not hot enough to ionize the atoms. A stellar surface temperature of 10,000 K results in the strongest Balmer lines.

A prominent set of Balmer lines is thus a clear indication that a star's surface temperature is about 10,000 K. At other temperatures, the spectral lines of other elements dominate a star's spectrum. For example, at around 25,000 K the spectral lines of helium are strong because, at this temperature, photons have enough energy to excite helium atoms without tearing away the electrons.

When a hydrogen atom is ionized, its only electron is torn away and no absorption lines can be produced. An atom of a heavier element, however, has two or more electrons. When one electron is knocked away, the remaining electrons produce a new and distinctive set of spectral lines. For example, in stars hotter than about 30,000 K, one of the two electrons in a helium atom is torn away. The remaining electron produces a set of spectral lines that is different from the lines produced by un-ionized helium. When the spectral lines of singly ionized helium appear in a star's spectrum, we know that the star has a surface temperature greater than 30,000 K.

Astronomers designate an un-ionized atom with a Roman numeral I; thus H I is neutral hydrogen. A Roman numeral II is used to identify an atom with one electron missing, and so He II is singly ionized helium (He$^+$). Sim-

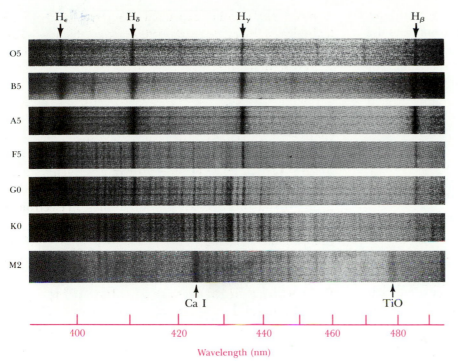

Figure 12-8 Principal types of stellar spectra *Each of these seven strips shows the spectrum of a star. Spectral lines of hydrogen, calcium, and titanium oxide are indicated. The hydrogen lines are strongest in A stars, which have surface temperatures of about 10,000 K. The spectra of G and K stars exhibit numerous lines caused by metals, indicating temperatures from 4000 to 6000 K. The broad, dark bands in the spectrum of an M star are caused by titanium oxide, which can exist only if the temperature is cooler than about 3500 K. (Courtesy of N. Houk, N. Irvine, and D. Rosenbush)*

ilarly, Si III is doubly ionized silicon (Si^{2+}), whose atoms are each missing two electrons.

In the early 1900s Annie Cannon and her colleagues at Harvard Observatory set up the spectral classification scheme we use today. Many of the A through P categories were dropped, and the remaining spectral types were reordered into the sequence **OBAFGKM**. This sequence has traditionally been memorized with the sexist mnemonic "Oh, Be A Fine Girl, Kiss Me!"

In the late 1920s Harvard astronomer Cecilia Payne and physicist Meghnad Saha succeeded in explaining precisely how a star's spectrum is affected by the star's surface temperature. In doing so, they demonstrated that the OBAFGKM sequence is actually a sequence in temperature. The hottest stars are O stars; their surface tempera-

tures are in excess of 35,000 K, and their spectra show He II and Si IV. M stars are the coolest stars; their surface temperatures of around 3000 K are so cool that atoms can stick together in molecules such as titanium oxide, whose spectral lines are prominent. Figure 12-8 shows representative examples of each spectral type.

Astronomers have found it useful to subdivide the original OBAFGKM sequence further. These finer steps are indicated by the addition of an integer from 0 through 9. Thus, for example, we have F8, F9, G0, G1, G2, . . . , G9, K0, K1, K2, ... The Sun, whose spectrum is dominated by singly ionized metals (especially Fe II and Ca II) is a G2 star.

The main characteristics of the modern spectral classification scheme are summarized in Figure 12-9, which

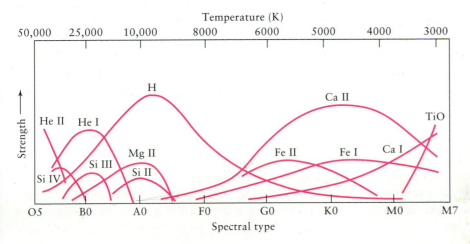

Figure 12-9 Spectral type and temperature *The strengths of the absorption lines of various elements are directly related to the temperature of the star's outer layers. For example, the Sun's spectrum has strong lines of singly ionized iron and calcium (Fe II and Ca II), corresponding to a spectral type of G2 and a surface temperature of about 5800 K. Note that hydrogen lines are strongest in A stars, whereas stars cooler than about 3500 K show absorption caused by titanium oxide.*

plots the strength of spectral lines against spectral type. This graph consolidates the information astronomers use to deduce the surface temperature and spectral type of a star from the intensity of the lines in its spectrum. For example, a star exhibiting strong Ca II and Fe I lines in its spectrum is a K5 star with a surface temperature around 4500 K.

12-5 The Hertzsprung–Russell diagram demonstrates that there are different kinds of stars

The first accurate measurements of stellar parallax were made in the mid-1800s, at about the same time that astronomers began observing stellar spectra. During the next half century, observing techniques improved, and the spectral types and absolute magnitudes of many stars became known.

Around 1905 the Danish astronomer Ejnar Hertzsprung pointed out that a regular pattern appears when the absolute magnitudes of stars are plotted against their color indices on a graph. Almost a decade later the American astronomer Henry Norris Russell independently discovered this regularity in a graph using spectral types instead of color indices. Plots of this kind are now known as **Hertzsprung–Russell diagrams,** or **H–R diagrams.**

Figure 12-10 is a typical Hertzsprung–Russell diagram. Each dot represents a star whose absolute magnitude and spectral type have been determined. Bright stars are near the top of the diagram, dim stars near the bottom. Hot stars (O and B stars) are toward the left side of the graph; cool stars (M stars) are toward the right side.

The most striking feature of an H–R diagram is that the data points are not scattered randomly all over the graph but are grouped in several distinct regions. The band stretching diagonally across the H–R diagram represents the majority of stars we see in the nighttime sky. This band, called the **main sequence,** extends from the hot, bright, bluish stars in the upper left corner of the diagram down to the cool, dim, reddish stars in the lower right corner. A star whose properties place it in this region of the H–R diagram is called a **main sequence star.** For example, the Sun (spectral type G2, absolute magnitude +4.8) is such a star.

Toward the upper right side of the H–R diagram is a second major grouping of data points. Stars represented by these points are both bright and cool. From Stefan's

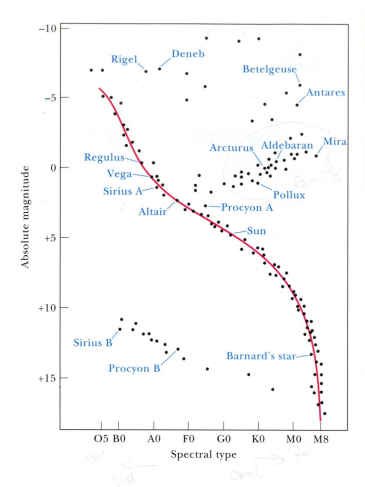

Figure 12-10 A Hertzsprung–Russell diagram *An H–R diagram is a graph on which the absolute magnitudes of stars are plotted against their spectral types. Each dot on this diagram represents a star in the sky whose absolute magnitude and spectral type have been determined. Some well-known stars are identified. Note that the data points are grouped in specific regions on the graph. This pattern reveals the existence of different types of stars in the sky: main sequence stars, giants, supergiants, and white dwarfs. The red curve indicates the location of the main sequence.*

law, we know that a cool object radiates much less light per unit of surface area than a hot object does. In order to be so bright, these stars must be huge, and so they are called **giants.** They are typically 10 to 100 times as large as the Sun and have surface temperatures around 3000 to 6000 K. Cooler members of this class of stars (those with surface temperatures from about 3000 to 4000 K) are often called **red giants** because they appear reddish in the nighttime sky. Aldebaran in the constellation Taurus and Arcturus in Boötes are examples of red giants that you can easily see with the naked eye.

A few rare stars are considerably bigger and brighter than typical red giants. These superluminous stars are appropriately called **supergiants.** Betelgeuse in Orion and Antares in Scorpius are examples of supergiants visible in the nighttime sky.

Finally, there is a third distinct grouping of data points toward the lower left corner of the Hertzsprung–Russell diagram. These stars are hot, dim, and small. They are appropriately called **white dwarfs.** These stars, which are roughly the same size as the Earth, can be seen only with the aid of a telescope.

There is another useful way to construct an H–R diagram. Instead of absolute magnitude, luminosity is plotted on the vertical axis of the graph. And instead of spectral type, the surface temperature is plotted on the horizontal axis. The resulting graph is still an H–R diagram, but the observational quantities (spectral type) have been replaced by calculated quantities (temperature).

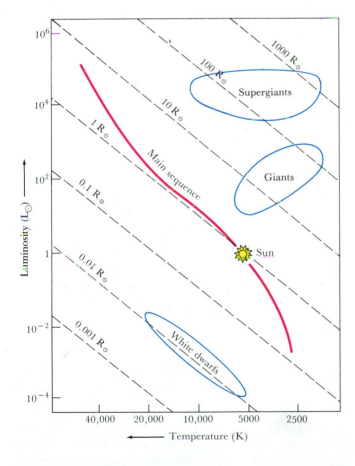

Figure 12-11   An H–R diagram and the sizes of stars   *On this H–R diagram, stellar luminosities are graphed against the surface temperatures of stars. Note that the various types of stars (main sequence, giants, supergiants, and white dwarfs) fall in the same regions of the graph as in Figure 12-10, where absolute magnitude is plotted against spectral type. The dashed diagonal lines indicate stellar radii. The Sun's size is midway between the largest and smallest stars we see in the sky.*

Figure 12-11 shows this type of H–R diagram. Note that the temperature scale on the horizontal axis of the graph increases toward the left, because Hertzsprung and Russell drew their original diagrams with O stars on the left and M stars on the right. (They made this choice because of the standard sequence OBAFGKM.) Having hot stars toward the left and cool ones toward the right is a convention that no one has seriously tried to change.

Also shown in Figure 12-11 are dashed lines that display the radii of stars. The symbol $R_\odot$ stands for 1 solar radius (that is, half the Sun's diameter) and is equal to 6.96×10^5 km, or roughly half a million miles. Notice that most red giants are 10 to 100 times as large as the Sun, whereas white dwarfs are only about $\frac{1}{100}$ the size of the Sun. Also notice that most main sequence stars are roughly the same size as the Sun.

The stars are classified into spectral types on the basis of the most prominent lines in their spectra. However, there are subtle differences even among the spectral lines of stars having the same spectral type. Based upon these minor differences, a system of **luminosity classes** was developed in the 1930s. Luminosity class I includes all the supergiants, and luminosity class V includes the main sequence stars. The intermediate classes distinguish giants of various luminosities, as indicated in Table 12-2. When the luminosity classes are plotted on the H–R diagram (Figure 12-12), they provide a useful subdivision of the star types in the upper right half of the diagram.

Astronomers commonly describe a star by combining its spectral type and its luminosity class into a sort of shorthand description, for example, calling the Sun a G2 V star. This notation supplies a great deal of information about the star, since its spectral type is correlated with its surface temperature and its luminosity class is correlated with its luminosity. Thus an astronomer knows immediately that a G2 V star is a main sequence star with a luminosity of about 1 $L_\odot$ and a surface temperature of

Table 12-2   Stellar luminosity classes

Luminosity class	Type of stars
I	Supergiant
II	Bright giant
III	Giant
IV	Subgiant
V	Main sequence

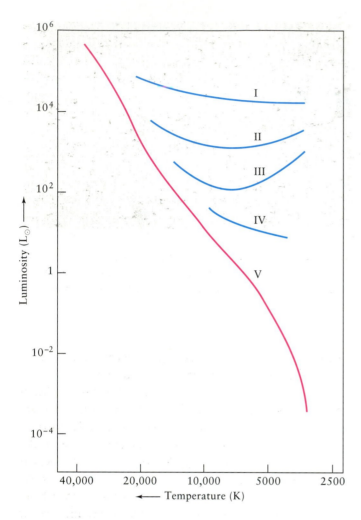

nearly 6000 K. Similarly, a description of Aldebaran as a K5 III star tells an astronomer that it is a red giant with a luminosity of around 500 L$_\odot$ and a surface temperature of about 4000 K.

An awareness of fundamentally different types of stars is the first important lesson to come from the H–R diagram. As we shall see in the following chapters, the different kinds of stars represent different stages of stellar evolution. We shall learn that a true appreciation of the H–R diagram requires an understanding of the life cycles of stars: how they are born, what happens as they mature, and what happens when they die.

12-6 Binary stars provide information about stellar masses

We now know something about the sizes, temperatures, and luminosities of stars. To complete our picture of the physical properties of stars, we need only to know their masses. There is, however, no practical and direct way to measure the mass of an isolated star observed in the sky.

Fortunately for astronomers, about half of the visible stars in the nighttime sky are not isolated individuals. Instead, they are members of multiple-star systems in which two or more stars orbit about each other. By observing how these stars orbit each other, astronomers can glean important information about the stellar masses.

A pair of stars located at nearly the same position in the night sky is called a double star. Thousands of double stars were observed and catalogued by William Herschel and his son John Herschel during the nineteenth century.

Figure 12-12 *Luminosity classes* *It is convenient to divide the H–R diagram into regions corresponding to luminosity classes. This subdivision permits finer distinctions between giants and supergiants. Luminosity class V encompasses the main sequence stars, including the dim red stars called red dwarfs toward the lower right side of the H–R diagram.*

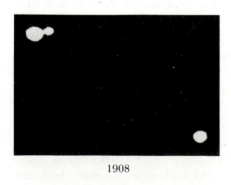

1908

1915

1920

Figure 12-13 *The binary star Kruger 60* *About one-half of the visible stars are double stars. This series of photographs shows the binary star Kruger 60 in the constellation Cepheus. The orbital motion of the two stars about each other is apparent.*

This binary system has a period of 44.52 years. The maximum angular separation of the stars is about 3.3 arc sec. Their apparent magnitudes are +9.8 and +11.4. (Yerkes Observatory)

Many of these double stars are in fact true **binary stars**, or **binaries**, pairs in which the two stars are actually orbiting each other.

In cases where astronomers can actually observe the two stars orbiting each other, a binary is called a **visual binary** (Figure 12-13). After many years of patient observation, astronomers can plot the orbit of one star about another in a binary pair (Figure 12-14).

A binary-star system is held together by gravity. Because the gravitational force between the two stars keeps them in orbit about each other, their orbital motions can be described by Newtonian mechanics. Specifically, their orbits obey Kepler's third law. For a binary system, Kepler's third law can be written as

$$M_1 + M_2 = \frac{a^3}{P^2}$$

where M_1 and M_2 are the masses of the two stars expressed in solar masses, P is the orbital period in years, and a is the semimajor axis (measured in astronomical units) of the elliptical orbit of one star about the other. With this equation an astronomer can calculate the sum

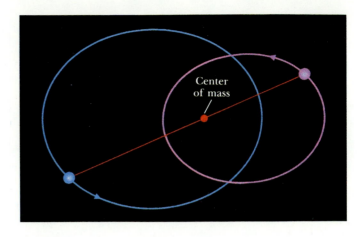

Figure 12-15 *Star orbits in a binary* Each star in a binary system follows an elliptical orbit about their common center of mass. The center of mass is always nearer the more massive of the two stars.

of the masses of the two stars in a binary if the binary's orbital period and semimajor axis are known.

Each of the stars in a binary system moves in an elliptical orbit about the **center of mass** of the system (Figure 12-15). This concept is analogous to two children on a seesaw. In order for the seesaw to balance properly, the center of mass of the two-child system must be located just above the support point, or fulcrum. As you no doubt know from experience, this center of mass is offset from the midpoint between the two children toward the heavier child. Similarly, the center of mass of a binary system could ostensibly be determined by placing the two stars at either end of a huge seesaw and determining where the fulcrum must be placed to balance the seesaw. The center of mass is always offset toward the more massive star.

In practice, the center of mass of a visual binary is determined by using the background stars as reference points. The separate orbits of the two stars can then be plotted as in Figure 12-15. The center of mass is located by finding the common focus of the two elliptical orbits. This information is needed to calculate the individual masses of the two stars.

Years of careful, patient observation of binaries have yielded the masses of many stars. As the data accumulated, an important trend began to emerge. For main sequence stars, there is a direct correlation between mass and luminosity: The more massive the star, the more luminous it is. This **mass–luminosity relation** can be conveniently displayed as a graph (Figure 12-16). Note that the range of stellar masses extends from $\frac{1}{10}$ of a solar mass to about 30 solar masses. The Sun's mass lies in the middle

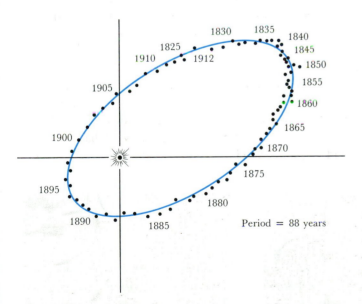

Figure 12-14 *The orbit of 70 Ophiuchi* After plotting the observations of a binary star over the years, astronomers can draw the orbit of one star with respect to the other. Once the orbit is known, Kepler's third law can be used to deduce information about the masses of the stars. This illustration shows the orbit of a faint double star in the constellation Ophiuchus. In plotting the orbit, either star may be regarded as the stationary one—the shape and size of the orbit will be the same in either case.

Figure 12-16 The mass–luminosity relation For main sequence stars, there is a direct correlation between mass and luminosity. The more massive a star, the more luminous it is.

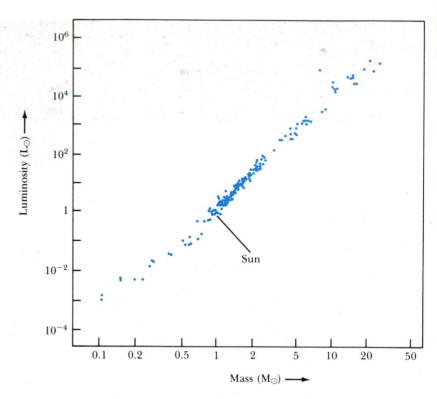

of this range, and so we see again that our star is ordinary and typical.

The mass–luminosity relation demonstrates that the main sequence on the H–R diagram is a progression in mass as well as in luminosity and surface temperature. The hot, bright, bluish stars in the upper left corner of the H–R diagram (see Figure 12-10) are the most massive main sequence stars in the sky. Likewise, the dim, cool, reddish stars in the lower right corner of the H–R diagram are the least massive. Main sequence stars of intermediate temperature and luminosity also have intermediate mass. This relationship of mass to the main sequence will play an important role in our later discussion of stellar evolution.

12-7 *The motion of a star affects the location of its spectral lines*

Many binary stars are scattered throughout our Galaxy, but only those that are nearby or that are widely separated can be distinguished as visual binaries. Star images in a remote binary are often blended together to produce a visual image that looks like a single star. Spectroscopy provides evidence that some stars that appear single are in fact binaries.

Occasionally spectral analysis yields incongruous spectral lines for some stars. For example, the spectrum of what appears at first to be a single star may include both strong hydrogen lines (characteristic of a type A star) and strong absorption bands of titanium oxide (indicating a type M star). Because a single star could not have the differing physical properties of these two spectral types, we conclude that this star is actually a binary system.

Spectroscopy can also be used to detect the movements of stars. As we saw in Chapter 5, the Doppler effect describes how the wavelength of light is affected by the relative motion between the source and the observer (recall Figure 5-19). If a source of light is coming toward you, you see a shorter wavelength than you would if the source were stationary. All the spectral lines in the spectrum of an approaching source are shifted toward the short-wavelength (blue) end of the spectrum. Conversely, all the spectral lines in the spectrum of a receding source are shifted toward the longer-wavelength (red) end of the spectrum. The size of the shift of a spectral line is proportional to the speed with which a source of light is moving toward or away from you. The greater the speed, the greater the shift.

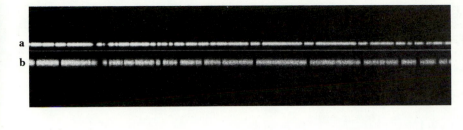

Figure 12-17 A spectroscopic binary A spectroscopic binary exhibits spectral lines that shift back and forth as the two stars revolve about each other. These two spectra show the behavior of the spectroscopic binary κ Arietis. **(a)** The stars are moving parallel to the line of sight (one star approaching Earth, the other star receding), producing two sets of shifted spectral lines. **(b)** Both stars are moving perpendicular to our line of sight. (Lick Observatory)

If the orbital speeds of the two stars in a binary are more than a few kilometers per second, the Doppler effect can be used to calculate important information about the binary, even though two separate stars cannot actually be observed. Such binaries yield a spectrum in which two complete sets of spectral lines shift back and forth. They are called **spectroscopic binary stars.** The regular, periodic shifting of the spectral lines is caused by the orbital motions of the stars as they revolve about their center of mass.

In many spectroscopic binaries, one of the stars is so dim that its spectral lines cannot be detected. The fact that the star is a binary is obvious, however, because its spectrum shows a single set of spectral lines that shift regularly back and forth. Such a **single-line spectroscopic binary** yields less information about its two stars than does a **double-line spectroscopic binary** like that shown in Figure 12-17.

Figure 12-17 shows two spectra of a spectroscopic binary taken a few days apart. In Figure 12-17a, two sets of spectral lines are visible, slightly offset in opposite directions from the normal positions of these lines. The spectral lines of the star moving toward the Earth are blueshifted; those of the other star (moving away from the Earth) are redshifted. A few days later the stars have progressed along their orbits so that one star is moving toward the left and the other toward the right. Because neither star is moving toward or away from the Earth, there is no Doppler shifting and both stars yield spectral lines at the same positions. That is why only one set of spectral lines appears in Figure 12-17b.

Significant information about the orbital velocities of the stars in a spectroscopic binary can be deduced from measuring shifts in spectral lines. This information is best displayed as a **radial velocity curve** graphing radial velocity versus time for the binary system (Figure 12-18). Radial velocity is the portion of a star's motion that is directed parallel to the line of sight between the Earth and the star.

In Figure 12-18, note that the wavy pattern repeats with a period of about 15 days, which is the orbital period of the binary. Also note that this pattern is displaced upward from the zero-velocity line by about 12 km/s, which is the overall motion of the binary system away from the Earth. Superimposed on this overall recessional motion are the periodic approaches and recessions of the two stars as they orbit about the center of mass.

The orbital speeds of the two stars in a binary are related to the masses of the stars by Kepler's laws and Newtonian mechanics. However, the individual masses of the two stars can be determined only if the tilt of their orbits is known. The angle of the orbits determines how much of the true orbital speeds of the stars appears as radial velocity measured from the Earth.

If the two stars are observed to eclipse each other, their orbits must be nearly edge-on, as viewed from the Earth. As we shall see next, individual stellar masses can be determined if a spectroscopic binary also happens to be an **eclipsing binary star.**

12-8 Light curves of eclipsing binaries provide detailed information about the stars

Some binary systems are oriented so that the two stars periodically eclipse each other as seen from Earth. Such eclipsing binaries can be detected even when the two stars cannot be resolved visually as two distinct images in a telescope. The apparent brightness of the image of the binary dims each time one star blocks out the other.

Using a light-sensitive detector at the focus of a telescope, an astronomer can measure light intensity from binaries very accurately. The data for an eclipsing binary

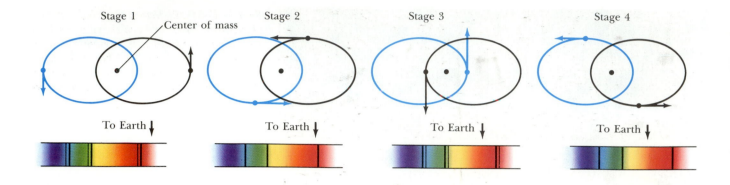

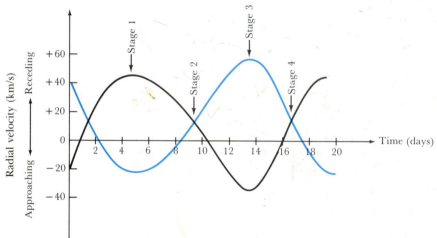

Figure 12-18 A radial velocity curve *The graph displays the radial velocity curve of the binary HD 171978. The drawings* *indicate the positions of the stars and their spectra at four selected moments during an orbital period.*

are most usefully displayed in the form of **light curves,** such as those shown in Figure 12-19. The overall shape of the light curve for an eclipsing binary reveals at a glance such facts as whether the eclipse is total or partial (compare Figures 12-19*a* and 12-19*b*).

The light curve of an eclipsing binary can yield a surprising amount of information. For example, the depths of the eclipse minima are related to the surface temperatures of the two stars. If an eclipsing binary is also a double-line spectroscopic binary, astronomers can calculate the mass and diameter of each star from analyses of both the light curves and the radial velocity curves. The maximum amount of information about stellar masses and sizes comes from such double-line spectroscopic/eclipsing binaries. These stars are rare, however, because the orbits of most spectroscopic binaries are tilted so that eclipses do not occur.

Additional details, such as tidal distortion, whereby the gravity of one star deforms the other, are revealed by the shape of the light curve (see Figure 12-19*c*). If one of the stars in an eclipsing binary is very hot, its radiation may create a "hot spot" on its cooler companion star. Every time this hot spot is exposed to our Earth-based view, we receive a little extra light energy, which produces a characteristic "bump" on the binary's light curve (see Figure 12-19*d*).

Information about stellar atmospheres can also be derived from light curves. Suppose that one star of a binary is a white dwarf and the other is a bloated red giant. By observing exactly how the light from the bright white dwarf is gradually cut off as it moves behind the edge of the red giant during the beginning of an eclipse, astronomers can infer the pressure and density in the upper atmosphere of the red giant.

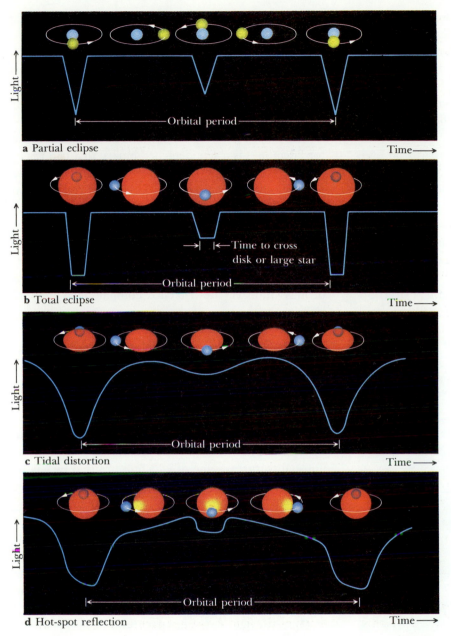

a Partial eclipse

b Total eclipse

c Tidal distortion

d Hot-spot reflection

Figure 12-19 Representative light curves of eclipsing binaries The shape of its light curve usually reveals many details about an eclipsing binary. Illustrated here are examples of (**a**) a partial eclipse, (**b**) a total eclipse, (**c**) tidal distortion, and (**d**) hot-spot reflection.

12-9 Mass transfer in close binary systems can produce unusual double stars

A single star leads a straightforward, birth-to-death existence, but exotic things can happen in a **close binary** system, where the stars are separated by only a few stellar diameters. In such binaries, the stars are so close together that the gravity of one can dramatically affect the appearance and evolution of the other.

As we shall see in the next chapter, a star expands dramatically and becomes a red giant as it grows old. When one member of a close binary becomes a giant, it can dump gas onto its companion. This process, called **mass transfer,** occurs only if the giant becomes sufficiently bloated and its companion star is near enough that the giant's outer layers can be gravitationally captured by the companion star.

In the mid-1800s the French mathematician Edward Roche pointed out that a figure-eight curve drawn around two stars in a binary can portray the gravitational domain of each star. This figure-eight-shaped boundary is often called the **critical surface,** and the two lobes of the critical surface are known as **Roche lobes.** The more massive star is always located inside the larger Roche lobe. Gases escaping a Roche lobe are no longer bound to that star and are free to leave the binary or to fall onto the companion star. When mass transfer occurs, gases flow across the point where the two Roche lobes touch.

In many binaries, the stars are so far apart that even during their red giant stage the stars' surfaces remain well inside their Roche lobes and little mass transfer occurs. Each star thus lives out its life independently as if it were single and isolated.

A binary system in which both stars are within their Roche lobes is referred to as a **detached binary** (Figure 12-20). If the two stars are relatively close together, when one star expands to become a giant it may fill or overflow its Roche lobe, in which case the system is called a **semidetached binary.** If both stars happen to fill their Roche

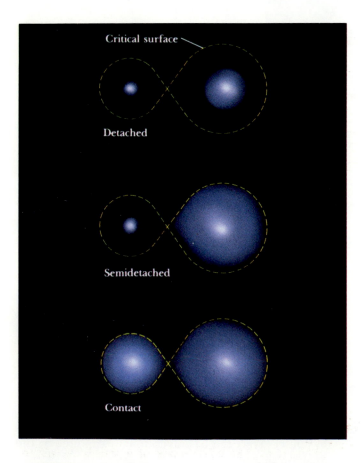

Figure 12-20 Detached, semidetached, and contact binaries *A double star is said to be detached, semidetached, or contact depending on whether either or both stars fill their Roche lobes. Mass transfer is often observed in semidetached binaries. The two stars in a contact binary share the same outer atmosphere.*

lobes, the system is called a **contact binary,** because the two stars actually touch and may share a common envelope of gas. Semidetached and contact binaries are most easily detected if they also happen to be eclipsing binaries, because their light curves then have a distinctly rounded appearance caused by these tidally distorted egg-shaped stars (recall Figure 12-19c).

The eclipsing binary called Algol (from an Arabic term for "demon") is a semidetached binary that can easily be seen with the naked eye in the constellation of Perseus. From Algol's light curve (see Figure 12-21a) astronomers have determined that the binary contains a giant that fills its Roche lobe. Sometime in the past, as this star expanded and became a giant, it dumped a significant amount of gas onto its companion. Astronomers theorize that the giant was originally the more massive star, but as a result of mass transfer, the detached companion is now more massive.

Mass transfer is still occurring in a semidetached eclipsing binary called β Lyrae in the constellation of Lyra, the Harp. Like Algol, β Lyrae contains a giant that fills its Roche lobe (Figure 12-21b). For many years astronomers were puzzled by the fact that the detached companion star in β Lyrae is severely underluminous, contributing virtually no light at all to the visible radiation coming from the system. Furthermore, the spectra of β Lyrae contain unusual features, some of which are caused by gas flowing between the stars and around the system as a whole.

The β Lyrae system was explained in 1963 when Su-Shu Huang of Northwestern University published his interpretation that the underluminous star in β Lyrae is enveloped in a huge rotating disk of gas captured from its bloated companion. The disk is so large and thick that it completely shrouds the secondary star, making it impossible to observe at visible wavelengths.

Observations made since the 1960s have confirmed the existence of a disk of gas, called an **accretion disk,** encircling the underluminous star in β Lyrae. The primary star is overflowing its Roche lobe, with gases streaming onto the disk at the rate of 10^{-5} $M_\odot$ per year. Ultraviolet spectra taken from spacecraft in the 1970s revealed that some gas is constantly escaping altogether from the system.

The fate of a semidetached system like Algol or β Lyrae depends primarily on how fast their stars evolve. If the detached star expands to fill its Roche lobe while the companion star is still filling its own Roche lobe, then the result is a contact binary. An example is W Ursae Majoris, in which two stars share the same photosphere (Figure 12-21c). In Chapters 14 and 15 we shall see that mass transfer onto dead stars produces some of the most extraordinary objects in the sky.

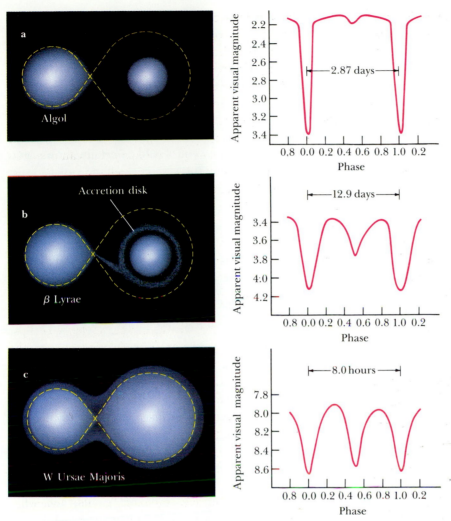

Figure 12-21 Three close binaries *Light curves for and sketches of three eclipsing binaries are shown. The "phase" denotes the fraction of the orbital period from one primary minimum to the next. (a) Algol is a semi-detached binary. (b) β Lyrae is a semidetached binary in which mass transfer has produced an accretion disk surrounding the detached star. (c) W Ursae Majoris is a contact binary.*

Summary

- Determining stellar distances is an important first step to understanding the nature of the stars.

    Distances to the nearer stars can be determined by parallax, the apparent shift of a star's location against the background stars while the Earth moves along its orbit.

- The apparent magnitude of a star is a measure of how bright the star appears to Earth-based observers. The absolute magnitude of a star is a measure of the star's true brightness and is directly related to the star's energy output, or luminosity.

    The absolute magnitude of a star is the apparent magnitude it would have if viewed from a distance of 10 parsecs. Absolute magnitudes are calculated from the star's apparent magnitude and distance.

    The luminosity of a star is the amount of energy escaping from the star each second.

- Astronomers use photometers to measure a star's brightness through a set of standard filters like the UBV filters. The color index of a star is the difference between magnitudes obtained with two different filters.

    The color indices of a star are a measure of its surface temperature.

- Stars are classified into spectral types (O, B, A, F, G, K, and M) based on major patterns of spectral lines in their spectra.

    The spectral type of a star is directly related to its surface temperature.

- The Hertzsprung–Russell (H–R) diagram is a graph plotting absolute magnitudes of stars against their spectral types (or, equivalently, luminosities against surface temperatures). The H–R diagram reveals the existence of such types of stars as main sequence stars, giants, supergiants, and white dwarfs.

- Binary stars are surprisingly common. Those that can be resolved as two distinct star images by an Earth-based telescope are called visual binaries.

    Each of the two stars in a binary system moves in an elliptical orbit about the center of mass of the system.

    The masses of the two stars in a binary system can be computed from measurements of the orbital period and orbital dimensions of the system.

- The mass–luminosity relation expresses a direct correlation between mass and luminosity for main sequence stars.

- Some binaries can be detected and analyzed, even though the system may be so distant or the two stars so close together that the two star images cannot be resolved with Earth-based telescopes.

    A spectroscopic binary is a system detected from the periodic shift of its spectral lines caused by the Doppler effect, as the orbits of the stars carry them alternately toward and away from the Earth.

    An eclipsing binary is a system whose orbits are viewed nearly edge-on from the Earth, so that one star periodically eclipses the other.

    Detailed information about the stars in an eclipsing binary can be obtained by studying its light curve.

- Mass transfer in a close binary system can affect the appearance and evolution of both stars.

    Mass transfer occurs when one star in a close binary overflows its Roche lobe.

Review questions

1 What is parallax? How do astronomers use parallax to measure the distances to stars? Why does this method work only with nearby stars?

2 What is the difference between apparent magnitude and absolute magnitude?

3 Briefly describe how you would determine the absolute magnitude of a nearby star.

4 Briefly describe how Wien's law and the Stefan–Boltzmann law can be used to deduce information about stars.

5 How and why is the spectrum of a star related to its surface temperature?

6 Describe the UBV filters and how an astronomer uses them to measure a star's surface temperature.

7 Explain why the color index of a star is related to its surface temperature.

8 What is the primary chemical component of most stars?

9 Which is the hottest star listed in Table 12-1? Which is the coolest?

10 Draw an H–R diagram and sketch the regions occupied by main sequence stars, red giants, and white dwarfs. Briefly discuss the different ways in which you could have labeled the axes of your graph.

11 How can observations of a visual binary lead to information about the masses of its stars?

12 What is a radial velocity curve? What kinds of stellar systems exhibit such curves?

13 What is the difference between a single-line and a double-line spectroscopic binary?

14 What is meant by the light curve of an eclipsing binary? What sorts of information can be determined from such a light curve?

15 What is the mass–luminosity relation? To what kind of stars does it apply?

16 What is a Roche lobe and what is its significance in close binary systems?

17 What is the difference between detached, semi-detached, and contact binaries?

Advanced questions

* 18 Van Maanen's star, named after the Dutch astronomer who discovered it, is a nearby white dwarf whose parallax is 0.232 arc sec. How far away is the star?

19 Sketch the radial velocity curve of a binary whose stars are moving in a circular orbit that is (a) perpendicular and (b) parallel to our line of sight.

20 Sketch the light curve of an eclipsing binary whose stars are moving along highly elongated orbits (a) with the major axes pointed toward the Earth and (b) with the major axes perpendicular to our line of sight.

* 21 Estimate the mass of a main sequence star that is 10,000 times as luminous as the Sun. What is the luminosity of a main sequence star whose mass is $\frac{1}{10}$ that of the Sun?

Discussion questions

22 Why do you suppose that stars of the same spectral type but different luminosity class exhibit slight differences in their spectra?

23 How might a star's rotation affect the appearance of its spectral lines?

24 Discuss the advantages and disadvantages of making stellar parallax measurements from a space telescope in a large solar orbit, say at the distance of Jupiter from the Sun.

For further reading

Ashbrook, J. "Visual Double Stars for the Amateur." *Sky & Telescope,* November 1980 • This informative article discusses techniques for observing double stars.

Evans, D., and others. "Measuring Diameters of Stars." *Sky & Telescope,* August 1979 • This article explains how the technique of interferometry is used to determine the diameters of stars.

Gingerich, O. "A Search for Russell's Original Diagram." *Sky & Telescope,* July 1982 • The article in the "Astronomical Scrapbook" section of *Sky & Telescope* gives entertaining insights about professional astronomy in the early 1900s.

Griffin, R. "The Radial-Velocity Revolution." *Sky & Telescope,* September 1989 • This article describes recent technological advances that have enabled astronomers to measure Doppler shifts with extreme precision.

Kaler, J. "Origins of the Spectral Sequence." *Sky & Telescope,* February 1986 • This superb article, which traces the history of OBAFGKM, includes many interesting historical insights into the work of the astronomers who struggled to develop reliable and meaningful schemes of spectral classification.

————. *Stars and Their Spectra.* Cambridge University Press, 1989 • This is an expanded version of Kaler's excellent series of articles on stellar spectroscopy that have appeared in *Sky & Telescope* over the past few years.

Mitton, J., and MacRobert, A. "Colored Stars." *Sky & Telescope,* February 1989 • This brief article in the "Celestial Calendar" section of *Sky & Telescope* gives a superb overview of star colors and includes a listing of extremely red stars and colorful double stars.

Nielsen, A. "E. Hertzsprung—Measurer of Stars." *Sky & Telescope,* January 1968 • This excellent historical sketch describes the work of one of the inventors of the H–R diagram.

Phillip, A., and Green, L. "Henry N. Russell and the H–R Diagram." *Sky & Telescope,* April 1978, May 1978 • These two articles give many fascinating insights into the life and times of Henry Norris Russell.

Tomkin, J., and Lambert, D. "The Strange Case of Beta Lyrae." *Sky & Telescope,* October 1987 • This article explores many fascinating details about the β Lyrae system.

Upgren, A. "New Parallaxes for Old: Coming Improvements in the Distance Scale of the Universe." *Mercury,* November/December 1980 • This well-written article describes astronomers' ongoing quest for accurate stellar distances.

13

Reflection and emission nebulae   *The two main bluish objects (toward the upper left side of this photograph) are reflection nebulae surrounding two young, hot, main sequence stars. Interstellar dust around these two stars efficiently reflects their bluish light. Several smaller reflection nebulae are scattered around the large reddish patch of ionized hydrogen gas. Dust mixed with the gas dilutes the intense red emission of the hydrogen atoms with a soft bluish haze. (Anglo-Australian Observatory)*

The Lives of Stars

Observations of stars, along with calculations of stellar models, have given astronomers an understanding of stellar evolution. In this chapter we learn that protostars form in cold clouds of interstellar dust and gas. A galaxy's spiral arms or an exploding supernova can trigger star formation in these huge clouds. Full-fledged main sequence stars are born when the temperature in the cores of contracting protostars becomes high enough to ignite hydrogen burning. We also examine the

remarkable transformation that stars undergo after consuming all the hydrogen in their cores. When hydrogen burning ceases, aging stars expand dramatically to become red giants. As a red giant expands, its core contracts and heats up. Eventually the star's central temperature becomes high enough to ignite the thermonuclear process of helium burning. Throughout these discussions, we find that it is enlightening to plot the evolution of individual stars and star clusters on H–R diagrams.

At a casual glance the heavens seem eternal and unchanging; the sky that we see at night is virtually indistinguishable from the sky our ancestors saw. But this permanence is an illusion. Stars emit huge amounts of radiation—expenditures that must produce changes and cause the stars to evolve. With careful observation and calculation, astronomers have assembled a theory of **stellar evolution.** This theory explains how stars are born in great clouds of interstellar gas and dust. They mature and grow old, and some eventually blow themselves apart in death throes that enrich interstellar space with new material for future stellar generations. The stars seem unchanging to human beings only because of the colossal time scales over which these changes occur. Major stages in the life of a star can last for millions or even billions of years.

13-1 *Protostars form in cold, dark nebulae*

Stars are created by the action of gravity on cold, dark clouds of interstellar gas and dust. As we learned in Chapter 5, the temperature of a gas is directly related to the average speed of its atoms and molecules. If an interstellar cloud is warm, its atoms are moving about so rapidly that there is no chance for a protostar to condense from the agitated gases. If the cloud's temperature is low, however, its atoms are moving slowly enough to allow denser portions of the cloud to contract gravitationally into clumps that collapse to form new stars.

Astronomers have discovered numerous cold, star-forming clouds scattered across the Milky Way. In some cases these clouds appear as dark regions silhouetted against a glowing background nebulosity, such as the famous Horsehead Nebula in Figure 13-1. In other cases they appear as dark blobs that obscure the background stars (Figure 13-2). These **dark nebulae** are sometimes called **Barnard objects,** after the American astronomer Edward Emerson Barnard, who discovered many of them around 1900.

A typical dark nebula contains a few thousand solar masses of gas spread over a volume roughly 30 light years across. The chemical composition of this material is the standard "cosmic abundance" of about 74% (by mass) hydrogen, 25% helium, and 1% heavier elements (recall Table 6-4). Infrared observations indicate that the cloud's internal temperature is about 10 K. At this temperature, the cloud is so cold and its atoms are moving so slowly that its gases do not produce enough internal pressure to support the cloud against its own weight. The cloud therefore contracts under the action of gravity and fragments into smaller lumps called **protostars.**

Figure 13-1 The Horsehead Nebula Dust grains block the light from the background nebulosity whose glowing gases are excited by ultraviolet radiation from young, massive stars. The nebula is located in Orion at a distance of roughly 1600 ly from Earth. The bright star to the left of center is Alnitak (ζ Orionis), the easternmost star in the "belt" of Orion. (Royal Observatory, Edinburgh)

Figure 13-2 A dark nebula *This dark nebula, called Barnard 86, is located in Sagittarius. It is visible in this photograph simply because it blocks out light from the stars beyond it. The cluster of bluish stars to the left of the dark nebula is NGC 6520. (Anglo-Australian Observatory)*

13-2 *Protostars evolve into young main sequence stars*

As early as the 1950s, astrophysicists performed calculations that describe the evolution of a protostar. At first, a protostar is merely a cool blob of gas several times larger than our solar system. Pressure inside the protostar cannot support all this cool gas, and so the protostar contracts. As it does so, gravitational energy is converted into thermal energy, which causes the gases to heat up and start glowing. After only a few thousand years of gravitational contraction, the surface temperature reaches 2000 to 3000 K. At this point the protostar is still quite large, so its glowing gases produce substantial luminosity. After only a thousand years of contraction, a protostar of 1 solar mass would be 20 times larger in diameter and 25 times brighter than the Sun.

Astrophysicists use high-speed computers and the equations of stellar structure (described in Chapter 11) to determine the conditions inside a contracting protostar. The results indicate how the protostar's luminosity and surface temperature change at various stages during its contraction; thereby we can plot the **evolutionary track** of the protostar on a Hertzsprung–Russell diagram.

Protostars are relatively cool when they begin to shine at visible wavelengths. Thus, the evolutionary tracks of protostars begin near the right side of the H–R diagram (Figure 13-3). However, continued gravitational contraction shifts protostars rapidly away from this region of the diagram. A protostar more massive than five Suns becomes hotter without much change in overall luminosity. The evolutionary tracks of massive protostars thus traverse the H–R diagram horizontally, from right to left. Less massive protostars become dimmer and their surface temperatures rise as they contract.

A protostar continues to shrink until the temperature at its center reaches a few million kelvin, when hydrogen burning begins. As we saw in Chapter 11, this thermonuclear process releases enormous amounts of energy. The outpouring of energy from hydrogen burning creates conditions inside the protostar that finally halt its gravitational contraction. Once hydrostatic and thermal equilibrium are established, a stable star is born. At this stage, the protostar's evolutionary track ends on the main sequence, as shown in Figure 13-3.

We now know that the main sequence represents stars in which hydrogen burning is occurring. This is a very stable state for most stars. For example, our Sun will remain on or near the main sequence, steadily burning hydrogen at its core, for a total of 10 billion years.

Note that the evolutionary tracks in Figure 13-3 end at locations along the main sequence that agree with the mass–luminosity relation (recall Figure 12-16). The most massive stars are the most luminous, while the least massive stars are the least luminous. Protostars less massive than about 0.08 solar masses never manage to develop the necessary pressures and temperatures to start hydrogen burning at their cores. These small protostars instead contract to become planetlike objects. Protostars with masses greater than about 80 solar masses (80 $M_\odot$) rapidly develop such extremely high temperatures that radiation pressure tends to disrupt them. Main sequence stars therefore have masses between about 0.08 and 80 $M_\odot$, with the high-mass stars being extremely rare.

The evolutionary tracks of protostars begin in the red giant region of the H–R diagram, but protostars are not red giants. An H–R diagram like that in Figure 12-10 shows where stars spend *most* of their lives. Protostars spend only a tiny fraction of their existence in the red giant region. The more massive a protostar is, the more rapidly it contracts and builds up the pressure and temperature required to ignite hydrogen burning at its core. For example, a 15-$M_\odot$ protostar takes only 10,000 years to become a main sequence star, whereas a 1-$M_\odot$

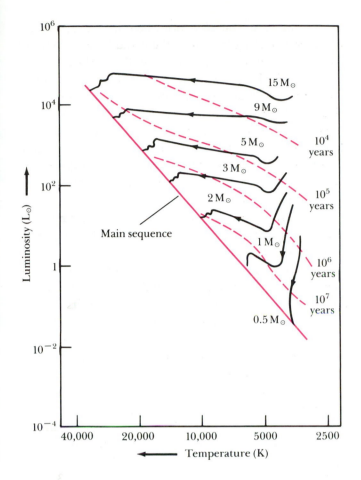

a

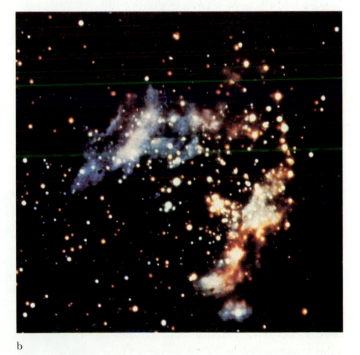

b

Figure 13-3   Pre-main-sequence evolutionary tracks   *The evolutionary tracks of seven stars having different masses are shown in this H–R diagram. The dashed lines indicate the stage reached after the indicated number of years of evolution. Note that all tracks terminate on the main sequence at points agreeing with the mass–luminosity relation. (Based on stellar model calculations by I. Iben)*

protostar takes a few tens of millions of years to do the same. By astronomical standards, these intervals are so brief that pre-main-sequence stars are quite transitory.

We are unlikely to observe the birth of a star at visible wavelengths, because its surrounding globule or interstellar cloud shields the protostar from our view. The vast amount of visible light emitted by the protostar is absorbed by interstellar dust in the surrounding **cocoon nebula,** which becomes heated to a few hundred kelvin. The warmed dust reradiates the energy at infrared wavelengths. Infrared observations can thus reveal what is going on inside a "stellar nursery."

Figure 13-4 shows views of a stellar nursery taken at visible and infrared wavelengths. The visible view

Figure 13-4 Newborn stars in the Swan Nebula (also called the Omega Nebula or M17) (a) *This image at visible wavelengths shows the Swan or Omega Nebula, so named because of its characteristic shape.* (b) *This infrared view, constructed from images taken at wavelengths of 1.2, 1.6, and 2.2 μm, reveals hundreds of stars that do not appear in the visible view. A comparison of the visible and infrared views demonstrates that many of the stars in this star-forming region are obscured by interstellar dust. (NOAO)*

(Figure 13-4*a*) is a familiar sight to many telescopic observers, but it tells only part of the story. The infrared picture (Figure 13-4*b*) shows hundreds of previously unseen stars, vividly demonstrating that most of the stellar nursery is hidden behind dust. Obscuring material is thickest toward the right (west) side of the nebula, where visible radiation is almost completely blocked, but numerous newborn stars shine brightly at infrared wavelengths.

13-3 Young star clusters are found in H II regions

From the evolutionary tracks of protostars in Figure 13-3, we can see that high-mass stars evolve more rapidly than low-mass stars. The massive main sequence stars, of spectral types O and B, are also the hottest, most luminous stars. Their surface temperatures are typically 15,000 to 35,000 K, and thus they emit vast quantities of ultraviolet radiation, as indicated by Wien's law.

The energetic ultraviolet photons from newborn massive stars easily ionize the surrounding hydrogen gas. For example, radiation from an O5 star can knock electrons from hydrogen atoms in a volume roughly 1000 ly across. This ionization has a dramatic effect on the nebula in which a cluster of stars is forming. While some hydrogen atoms are being knocked apart by ultraviolet photons, other hydrogen atoms are being reassembled as some of the free protons and electrons manage to get back together. During this recombination of hydrogen atoms, the captured electrons cascade downward through the atom's energy levels toward the ground state. These downward quantum jumps release numerous photons at many visible wavelengths, and the nebula begins to glow. Particularly prominent is the transition from $n = 3$ to $n = 2$, which produces H_α photons at 656 nm in the red portion of the visible spectrum (review Figure 5-17). Thus, the nebulosity around a newborn star cluster shines with a distinctive reddish hue.

Figure 13-5 shows one of these **emission nebulae**. Because these nebulae are predominantly ionized hydrogen, they are also called **H II regions**. The collection of a few hot, bright O and B stars near the core of the nebula that produces the ionizing ultraviolet radiation is called an **OB association**.

Observing individual stars in a young cluster can yield further information about stars in their infancy. Figure 13-6 shows a beautiful emission nebula surrounding the cluster called NGC 2264. By measuring each star's magnitude, color index, and distance, an astronomer can deduce its luminosity and surface temperature. The data for all the stars in the cluster can then be plotted on an H–R diagram, as shown in Figure 13-6. Note that the hottest stars, with surface temperatures around 20,000 K, are on the main sequence. These hot stars are the rapidly evolving, massive ones whose radiation is causing the surrounding gases to glow. The stars cooler than about

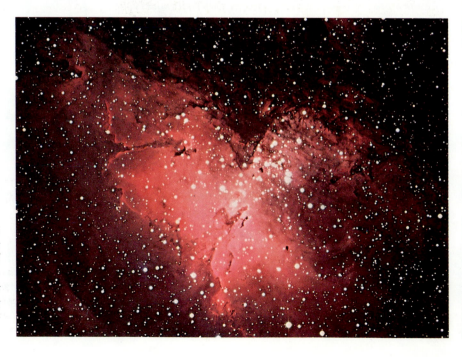

Figure 13-5 An H II region *Because of its shape, this emission nebula is called the Eagle Nebula. It surrounds the star cluster called M16 in the constellation of Serpens at a distance of 6500 ly from Earth. Several bright, hot O and B stars are responsible for the ionizing radiation that causes the gases to glow. (Anglo-Australian Observatory)*

10,000 K, however, have not yet quite arrived at the main sequence. These less massive stars, which are in the final stages of pre–main-sequence contraction, are just now beginning to ignite thermonuclear reactions at their centers. The locations of data points on Figure 13-6 suggest that the cluster is roughly two million years old.

Spectroscopic observations of the cooler stars in NGC 2264 show that many are vigorously ejecting gas, a very common phenomenon in most stars just before they reach the main sequence. Such gas-ejecting stars are called **T Tauri stars**, after the first example discovered in the constellation of Taurus. Some astronomers suggest that the onset of hydrogen burning is preceded by vigorous chromospheric activity marked by enormous spicules and flares that propel the star's outermost layers back into space. In fact, an infant star going through its T Tauri stage can lose as much as 0.4 $M_\odot$ before it settles down on the main sequence.

Figure 13-7 shows a young star cluster called the Pleiades in the constellation of Taurus that is easily visible to the unaided eye. In contrast to the H–R diagram for NGC 2264, nearly all the stars in the Pleiades are on the main sequence. The cluster's age is about 100 million years, which is how long it takes for the least massive stars to finally begin hydrogen burning in their cores.

Note the distinctly bluish color of the nebulosity around the Pleiades. This haze, called a **reflection nebula,** is caused by fine grains of interstellar dust that efficiently scatter and reflect blue light. Indeed, reflection nebulosity is blue for the same reason Earth's daytime sky is blue: Particles scatter short-wavelength light much more efficiently than longer-wavelength radiation. Blue light is therefore bounced around and reflected back toward us much more readily than is light of any other color.

A loose collection of stars such as the Pleiades or NGC 2264 is called an **open cluster** or a **galactic cluster.** Such

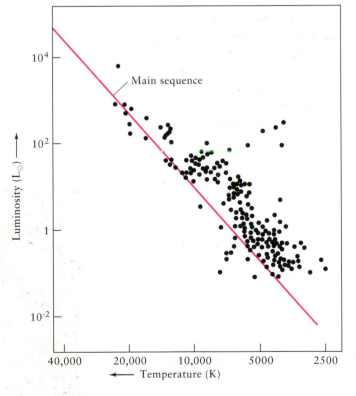

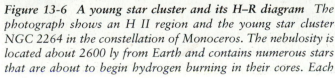

Figure 13-6 A young star cluster and its H–R diagram *The photograph shows an H II region and the young star cluster NGC 2264 in the constellation of Monoceros. The nebulosity is located about 2600 ly from Earth and contains numerous stars that are about to begin hydrogen burning in their cores. Each* dot plotted on the H–R diagram represents a star in this cluster *whose luminosity and surface temperature have been measured. Note that most of the cool, low-mass stars have not yet arrived at the main sequence. This star cluster probably started forming only 2 million years ago. (Anglo-Australian Observatory)*

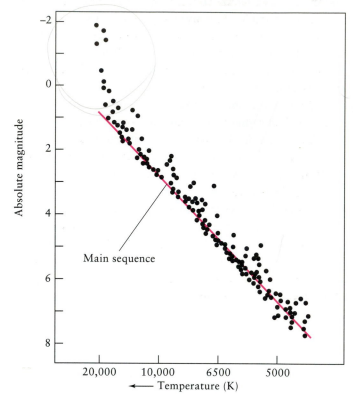

Figure 13-7 The Pleiades and its H–R diagram This open cluster called the Pleiades, which can easily be seen with the naked eye in the constellation of Taurus, is about 400 ly from Earth. Each dot plotted on the H–R diagram represents a star in the Pleiades whose absolute magnitude and surface temperature have been measured. Note that most of the cool, low-mass stars have arrived at the main sequence, indicating that hydrogen burning has begun in their cores. The cluster has a diameter of about 5 ly and is about 100 million years old. (Anglo-Australian Observatory)

clusters possess barely enough mass to hold themselves together by gravity. A star moving faster than the average speed will occasionally escape, or "evaporate," from a cluster. By the time the stars are a few billion years old, they may be so widely separated that a cluster no longer truly exists.

13-4 Star birth begins in giant molecular clouds

Astronomers agree that H II regions are stellar nurseries, but where do the H II regions come from? This question was finally answered in the 1970s when radio astronomers began discovering enormous clouds of gas scattered about our Galaxy.

As we saw in Chapter 6, hydrogen is by far the most abundant element in the universe. In the cold depths of interstellar space, hydrogen atoms combine to form hydrogen molecules (H_2). A molecule vibrates and rotates at specific frequencies dictated by the laws of quantum mechanics. As a molecule goes from one vibrational or rotational state to another, it emits or absorbs a photon. This process is analogous to what happens when an atom emits or absorbs a photon as an electron jumps from one energy level to another. Many interstellar molecules emit photons with wavelengths of a few millimeters. Consequently, in recent years radio telescopes tuned to wavelengths in this range have greatly increased our knowledge of the **interstellar medium,** or interstellar matter.

Although hydrogen molecules are scattered abundantly across space, they are difficult to detect. The hydrogen molecule consists of two atoms of equal mass joined together, and such molecules do not efficiently emit photons at radio frequencies. Radio astronomers can more easily detect asymmetric molecules such as carbon monoxide (CO), which consists of two atoms of unequal mass joined together. Carbon monoxide emits photons at a wavelength of 2.6 mm, corresponding to a transition between two rates of rotation of the molecule.

The presence of carbon monoxide is especially useful when probing the interstellar medium. From the known abundance of elements (recall Table 6-4) astronomers conclude that there are about 10,000 H_2 molecules for every CO molecule. Consequently, wherever astronomers detect strong emission from CO, they know that a considerable amount of hydrogen gas must also be present.

In mapping the locations of CO emission, astronomers soon realized that vast amounts of hydrogen are concentrated in **giant molecular clouds**. These clouds have masses in the range of 10^5 to 2×10^6 solar masses and diameters that range from 50 to 300 light years. The density inside one of these clouds is about 200 hydrogen molecules per cubic centimeter, which is several thousand times larger than the average density in interstellar space. Astronomers estimate that our Galaxy contains about 5000 of these enormous clouds.

The constellations of Orion and Monoceros include one of the most accessible regions of the sky for studying star formation and the interaction of young stars with the interstellar medium. Figure 13-8a shows a map of this region made with a radio telescope tuned to a wavelength of 2.6 mm. Note the extensive areas of the sky covered by giant molecular clouds. Comprehensive maps of CO emission, like that shown in Figure 13-8a, help astronomers understand how the large-scale structure of the interstellar medium is related to the formation of H II regions and OB associations.

Many galaxies, including our own and the one shown in Figure 13-9, have **spiral arms**, which are huge, arching lanes of glowing gas and stars. When we discuss details of our Galaxy in Chapter 16, we shall learn that spiral arms are caused by compression waves that squeeze the interstellar gases. When one of these giant molecular clouds passes through a wave, it is compressed, causing vigorous star formation to begin in the densest regions. As soon as massive stars form, they emit ultraviolet light that ionizes the surrounding hydrogen and an H II region is born. An

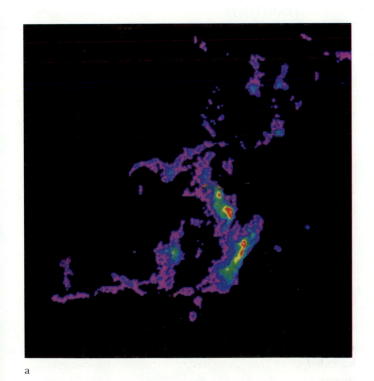

a

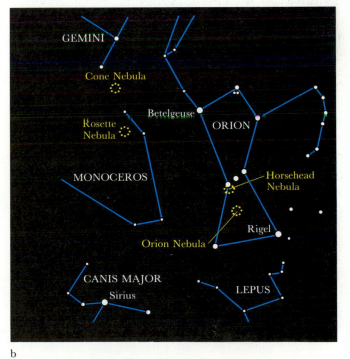

b

Figure 13-8 A map of carbon monoxide features in Orion (a) *This color-coded map of a large section of the sky shows the extent of giant molecular clouds in Orion and Monoceros. The intensity of carbon monoxide (CO) emission is displayed by colors in the order of the rainbow, from violet for the weakest to red for the strongest. Black indicates no detectable emission.*

(b) *This star chart covers the same area as the CO map in (a). The locations of four prominent star-forming nebulae are indicated. Note that the Orion and Horsehead nebulae are located at sites of intense CO emission. (Courtesy of R. Maddalena, M. Morris, J. Moscowitz, and P. Thaddeus)*

Figure 13-9   A spiral galaxy   This beautiful spiral galaxy, called NGC 2997, has two arching spiral arms outlined by numerous H II regions and clusters of young, hot stars. Each pinkish speck along the spiral arms of this galaxy is an H II region. The spiral arms of a galaxy are sites of vigorous star formation. (Anglo-Australian Observatory)

H II region is thus a small, bright "hot spot" in a giant molecular cloud. The famous Orion Nebula (Figure 13-10) is an example. Four hot, massive O and B stars at the heart of the Orion Nebula are responsible for the ionizing radiation that causes the surrounding gases to glow. The Orion Nebula is embedded in a giant molecular cloud whose mass is estimated at 500,000 solar masses.

The OB association at the core of the H II region affects the rest of the giant molecular cloud. Vigorous stellar winds, along with ionizing ultraviolet radiation from the O and B stars, carve out a cavity in the cloud. Because much of this outflow is supersonic, a shock wave forms where the outer edge of the expanding H II region impinges on the rest of the giant molecular cloud. This shock wave compresses the hydrogen gas through which it passes, stimulating a new round of star birth. The new O and B stars continue to power the expansion of the H II region still farther into the giant molecular cloud. Meanwhile, the older O and B stars left behind begin to disperse (Figure 13-11). In this way, an OB association "eats into" a giant molecular cloud, "spitting out" stars in its wake.

Infrared observations reveal many features that resemble protostars in the swept-up layer immediately behind the shock wave from an OB association. For instance, Figure 13-12 shows both optical and infrared views of the core of the Orion Nebula. Four O and B stars and glowing gas and dust dominate the center of the view at visible wavelengths. Infrared observations penetrate this obscuring material to reveal dozens of infrared objects that may be cocoons of warm dust enveloping newly formed stars.

13-5   Star birth is also triggered by supernova explosions that compress the interstellar medium

Presumably any mechanism that compresses interstellar clouds can trigger the birth of stars. As we shall see in detail in Chapter 14, a massive star can end its life with

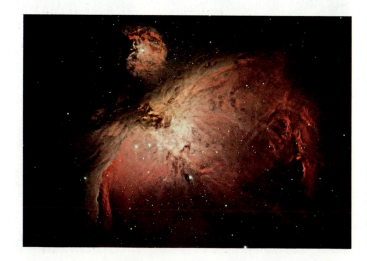

Figure 13-10   The Orion Nebula   This famous H II region can be seen with the naked eye. It is 1600 ly from Earth and has a diameter of roughly 16 ly. The mass of this nebula is about 300 solar masses. Four bright, massive stars at the center of the nebula produce the ultraviolet light that causes the gases to glow. These four stars, called the Trapezium, are separated from each other by only 0.13 ly. (Anglo-Australian Observatory)

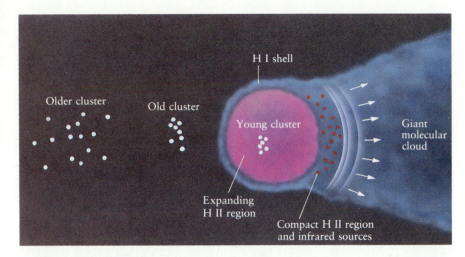

Figure 13-11 The evolution of an OB association *Ultraviolet radiation from young O and B stars produces a shock wave that compresses gas further into the molecular cloud, stimulating new star formation deeper into the cloud. Meanwhile, older stars are left behind.*

a violent detonation called a **supernova.** In a matter of seconds the core of the doomed star collapses, releasing vast quantities of particles and energy that blow the star apart. The star's outer layers are blasted outward into space at speeds of several thousand kilometers per second.

Astronomers find many nebulosities across the sky that are the shredded funeral shrouds of these dead stars. Such nebulae, like the Cygnus Loop shown in Figure 13-13, are called **supernova remnants.** Many supernova remnants have a distinctly arched appearance, as would be expected for an expanding shell of gas. This wall of gas typically moves away from the dead star at supersonic speeds. Its

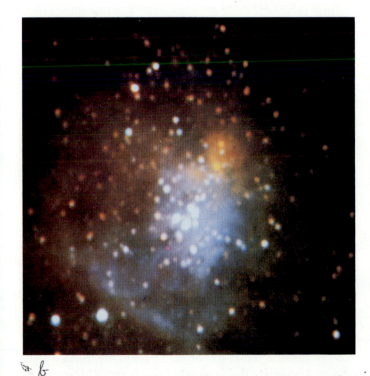

Figure 13-12 The core of the Orion Nebula (a) *This view at visible wavelengths shows the inner regions of the Orion Nebula (compare with Figure 13-10). At the center are four massive stars, called the Trapezium, which cause the nebula to glow.* (b) *This infrared view, also centered on the Trapezium, covers the* same area as (a). *Infrared radiation can penetrate interstellar dust more easily than can visible photons. Numerous infrared objects, many of which are probably new stars in the early stages of formation, are seen in the infrared view. (Anglo-Australian Observatory)*

Figure 13-13   A supernova remnant   This re-markable nebula, called the Cygnus Loop, is the remnant of a supernova explosion that oc-curred about 20,000 years ago. The expanding spherical shell of gas now has a diameter of about 120 ly. (Courtesy of H. Vehrenberg)

passage through the surrounding interstellar medium excites the atoms, causing the gases to glow.

Supersonic motion is always accompanied by a shock wave that abruptly compresses the gas through which it passes. If the expanding shell of a supernova remnant encounters an interstellar cloud, it can squeeze the cloud, stimulating star birth. As we learned in Chapter 10, there is evidence that the Sun was created in this fashion.

Our understanding of star birth has improved dramatically in recent years, primarily because of infrared and millimeter-wavelength observations. Nevertheless, many mysteries remain. For example, astronomers had generally assumed that there was a lot of interstellar dust shielding a stellar nursery from the disruptive effects of external sources of ultraviolet light, which is what we seem to find in our own Galaxy. However, in a neighboring galaxy called the Large Magellanic Cloud, there are young OB associations with virtually no dust. Does the process of star birth differ slightly from one galaxy to another?

Another problem is that different modes of star birth tend to produce different percentages of different kinds of stars. For example, the passage of a spiral arm through a giant molecular cloud tends to produce an abundance of massive O and B stars. In contrast, a shock wave from a supernova seems to produce fewer O and B stars, but many more of the less massive A, F, G, and K stars.

There may be additional mechanisms of star birth that have yet to be discovered. For example, a simple collision between two interstellar clouds should create new stars. When two such clouds collide, compression must occur

at the interface, with vigorous star formation to follow. Another possibility is that light from a star or group of stars may exert strong enough radiation pressure on the surrounding interstellar medium to cause compression, followed by star formation. The Rosette Nebula, shown in Figure 13-14, may be an example of this process.

In spite of unanswered questions, it is now clear that star birth involves mechanisms on a colossal scale that we have just begun to appreciate—from the deaths of massive stars to the rotation of the entire galaxy. The study of cold, dark stellar nurseries will certainly be an active and exciting area of astronomical research for many years to come.

13-6   When core hydrogen burning ceases, a main sequence star becomes a red giant

A main sequence star is a young star whose radiated energy comes from the thermonuclear process of hydrogen burning in its core. A main sequence star is in thermal equilibrium, with energy liberated in its core balanced by energy radiated from its surface. Eventually, however, all the hydrogen in the core of the star is used up. **Core hydrogen burning** then must cease, with dramatic effects upon the star's equilibrium, structure, and evolution.

We have seen that massive protostars quickly build up the necessary core temperature and pressure to ignite hydrogen burning and become main sequence O and B stars. Furthermore, these massive main sequence stars are the

Figure 13-14 *The core of the Rosette Nebula*
The Rosette Nebula is a large, circular emission nebula near one end of a sprawling giant molecular cloud in the constellation of Monoceros. Radiation from young, hot stars has blown gas away from the center of this nebula. Some of this gas has become clumped in dark globules that appear silhouetted against the glowing background gases. (Anglo-Australian Observatory)

most luminous stars; this rapid emission of energy must correspond to a rapid depletion of hydrogen in their cores. Thus, even though a massive O or B star contains much more hydrogen fuel than a less massive main sequence star, it consumes its hydrogen far more rapidly. The main sequence lifetime of a massive star is thus considerably shorter than that of a less massive star.

Hydrogen burning has continued in the Sun's core for the past 5 billion years. Initially the Sun's chemical composition by mass was roughly 75% hydrogen and 25% helium (plus a smattering of heavy elements). The ongoing fusion of hydrogen into helium in the Sun's core has dramatically altered the core composition, however. Indeed, there is now more helium than hydrogen at the Sun's center.

Enough hydrogen remains in the Sun's core for another 5 billion years of core hydrogen burning. Therefore, the Sun's total lifetime on the main sequence is 10 billion years. Table 13-1 shows how long other stars take to exhaust the supplies of hydrogen in their cores. Note that high-mass stars gobble up their hydrogen fuel in only a few million years, whereas low-mass stars take hundreds of billions of years to accomplish the same thing.

As the supply of hydrogen at a star's center dwindles, the star begins to have difficulty supporting the weight of its outer layers. This enormous weight pressing inward from all sides compresses the star's core slightly. The compressed gases become warmer, allowing hydrogen burning to move outward from the core. In other words, during its final years on the main sequence, a star makes a final

attempt to maintain hydrostatic and thermal equilibrium by enlarging its hydrogen-burning region. There is still plenty of fresh hydrogen surrounding the star's center. By tapping this supply, the star manages to eke out a few million more years on the main sequence.

Finally all the hydrogen in the core of an aging main sequence star is used up, so hydrogen burning ceases in the star's core. However, hydrogen burning continues in a thin spherical shell surrounding the core. This **shell hydrogen burning** initially occurs only in the hot region just outside the core, where the hydrogen fuel has not yet been exhausted.

No longer are thermonuclear reactions producing energy in the star's core. As a result, the heat flowing out of

Table 13-1   Main-sequence lifetimes

Mass ($M_\odot$)	Surface temperature (K)	Luminosity ($L_\odot$)	Time on main sequence (10^6 years)
25	35,000	80,000	3
15	30,000	10,000	15
3	11,000	60	500
1.5	7,000	5	3,000
1.0	6,000	1	10,000
0.75	5,000	0.5	15,000
0.50	4,000	0.03	200,000

Figure 13-15 The Sun today and as a red giant In about 5 billion years, when the Sun expands to become a red giant, its diameter will increase while its core becomes more compact. Today the Sun's energy is produced in a hydrogen-burning core whose diameter is about 300,000 km. When the Sun becomes a red giant, it will draw its energy from a hydrogen-burning shell surrounding a compact helium-rich core. The helium core will have a diameter of only 30,000 km.

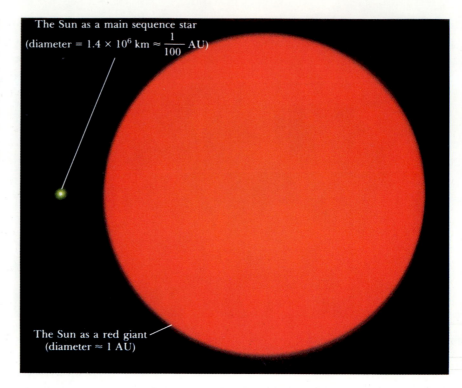

The Sun as a main sequence star
(diameter = 1.4×10^6 km $\approx \frac{1}{100}$ AU)

The Sun as a red giant
(diameter $\approx$ 1 AU)

the hot core is not being replaced. The core thus gradually contracts, converting gravitational energy into thermal energy to maintain thermal equilibrium. The hydrogen-burning shell slowly works its way outward from the original core, thereby dumping more helium onto the core, which continues to contract and heat up.

The high temperature of the star's core stimulates the hydrogen-burning shell, whose increased energy output increases the star's luminosity and causes the star's outer layers to expand, thereby increasing the overall size of the star. As the star's outer atmosphere expands farther and farther into space, its gases cool. Soon the temperature of the star's bloated surface has fallen to about 3500 K and its gases glow with a reddish hue. The star is then appropriately called a **red giant.**

The Sun will take about 5 billion years more to finish converting hydrogen into helium at its core. As the Sun's core contracts, its atmosphere will expand to envelop first Mercury and then Venus. As the red giant Sun swells to a diameter of about 1 AU, its surface temperature will decline to about 3500 K. As a full-fledged red giant (Figure 13-15), our star will shine with the brightness of 2000 Suns. Some of the inner planets will be vaporized, and the thick atmospheres of the outer planets will boil away to reveal their rocky cores. Thus, in its later years the aging Sun will destroy the planets that have accompanied it since its birth.

13-7 Helium burning begins at the center of a red giant

Helium is the "ash" of hydrogen burning. When a star first becomes a red giant, its hydrogen-burning shell surrounds a small, compact core of almost pure helium. In a moderately low-mass red giant, which the Sun will be 5 billion years from now, the dense helium core is about twice the size of the Earth and the star's bloated surface has a diameter roughly half that of the Earth's orbit.

At first no thermonuclear reactions occur in the helium-rich core of a red giant, because the temperature is too low to fuse helium nuclei. As the hydrogen-burning shell moves outward in the star, it adds mass to the helium core. The core slowly contracts, forcing the star's central temperature to climb.

When the central temperature reaches 100 million kelvin, **helium burning** is ignited at the star's center. This new thermonuclear reaction occurs in two steps. First, two helium nuclei combine. Then a third helium nucleus is added to this combination, resulting in a carbon nucleus:

$$3\,^4\text{He} \rightarrow\,^{12}\text{C} + \gamma$$

with the release of a gamma-ray photon (γ). Some of the carbon created in this process can fuse with an additional helium nucleus to produce oxygen:

$$^{12}C + {}^4He \rightarrow {}^{16}O + \gamma$$

Carbon and oxygen are the "ashes" of helium burning.

The creation of carbon and oxygen by helium burning releases energy, providing the aging star with a central energy source for the first time since leaving the main sequence. This energy source, properly called **core helium burning** because of its central location, establishes thermal equilibrium, thereby preventing any further gravitational contraction of the star's core. A mature red giant burns helium in its core for about 20% of the time that it spent burning hydrogen as a main sequence star. For example, in the distant future the Sun will consume helium in its core for about 2 billion years.

How helium burning begins at a red giant's center depends on the mass of the star. In high-mass stars (those with masses greater than about 3 $M_\odot$), helium burning begins gradually as temperatures in the star's core approach 100 million kelvin. In low-mass stars (those with less than 3 $M_\odot$), helium burning begins explosively and suddenly, in an event called the **helium flash.**

The helium flash occurs because of unusual conditions that develop in the core of a low-mass star on its way to becoming a red giant. To appreciate these conditions we must first understand how an ordinary gas behaves, then explore how the densely packed electrons at the star's center alter this behavior.

Under most circumstances, the gases inside a star act the way most gases do. If gases are compressed, they heat up, and if they expand, they cool down. This behavior serves as a "safety valve" to ensure that a star does not explode. For example, if energy production overheats the star's core, the core expands, cooling the gases and slowing the rate of thermonuclear reactions. Conversely, if too little energy is being created to support the star's overlying layers, the core contracts and the resulting increase in temperature speeds up the thermonuclear reactions and hence increases the energy output.

In a low-mass red giant, the core must undergo considerable gravitational compression to drive temperatures high enough to ignite helium burning. At the extreme pressures and temperatures deep inside the star, the atoms are completely torn apart into nuclei and electrons. In the star's highly compressed core, the free electrons are so closely crowded together that a law of physics called the **Pauli exclusion principle** comes into play. This principle, formulated in 1925 by the Austrian physicist Wolfgang Pauli, explains that two identical particles cannot simultaneously have the same set of attributes, like location and speed. In the submicroscopic world of atoms and particles, the Pauli exclusion principle is analogous to the idea that you can't have two things in the same place at the same time.

Just before the onset of helium burning, the electrons in the core of a low-mass star are so closely crowded together that any further compression would violate the Pauli exclusion principle. Because the electrons cannot be squeezed any closer together, they produce a powerful pressure that resists further contraction of the core.

This phenomenon, whereby closely packed particles resist compression in accord with the Pauli exclusion principle, is called **degeneracy**. Astronomers say that the helium-rich core of a low-mass red giant is degenerate and is supported by **degenerate-electron pressure**. This degenerate pressure, unlike the pressure of an ordinary gas, does not depend on temperature.

When the temperature in the core of a low-mass red giant reaches the level required for helium burning, energy begins to be released. The temperature therefore climbs, which causes helium burning to proceed more rapidly. However, because the pressure provided by the degenerate electrons is not affected by the temperature, the pressure does not change. Without the "safety valve" of increasing pressure, the star's core cannot expand and cool. The rising temperature causes the helium to burn at an ever-increasing rate, producing the helium flash.

As long as the star's core is degenerate, degenerate-electron pressure far exceeds the normal gas pressure produced by the nuclei and electrons. However, increasing temperature eventually makes the normal pressure become so large that it exceeds the degenerate-electron pressure. The electrons then start to behave like an ordinary gas. In summary, stars make a gas degenerate by compressing it; they convert it back into an ordinary gas by heating it. As soon as ordinary gas pressure predominates, the usual safety valve is once again in operation. The star's core therefore expands and cools, thereby terminating the helium flash. These events occur extremely rapidly; the helium flash is over in only a few seconds.

13-8 Evolutionary tracks on the H–R diagram reveal the ages of star clusters

It is very enlightening to follow the post-main-sequence evolution of mature stars by plotting their evolutionary tracks on the Hertzsprung–Russell diagram, as shown in

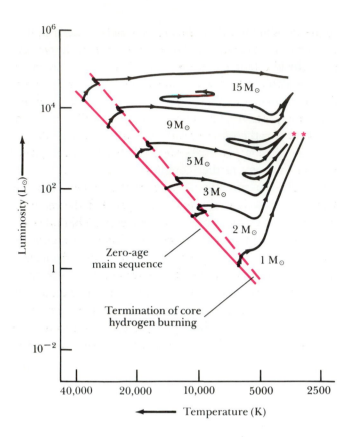

Figure 13-16 Post-main-sequence evolution *The evolutionary tracks of six stars are shown on this H–R diagram. In the high-mass stars, core helium burning ignites where the evolutionary tracks make a sharp turn in the red giant region of the diagram. The evolutionary tracks for low-mass stars (1 M$_\odot$ and 2 M$_\odot$) are shown only up to the points, indicated by the red stars, where the helium flash occurs at their centers. (Adapted from I. Iben)*

region of the H–R diagram. After the core helium burning begins, however, the evolutionary tracks back away from these temporary peak luminosities. The tracks then wander back and forth in the red giant region while the stars readjust to their new energy sources.

The evolutionary tracks of two low-mass stars also appear in Figure 13-16. However, these two tracks are shown only to the point where the helium flash occurs. The sudden flood of energy released by the helium flash does not affect the star's surface but rather is used up in altering conditions in the star's interior.

Immediately after the helium flash, the temperature of the low-mass star's superheated core is so high that its gas ceases to be degenerate. The star's core therefore expands, because it is again behaving like an ordinary gas. Temperatures around the expanding core fall, cooling the hydrogen-burning shell and reducing its energy output. As the star's energy output declines, its outer layers contract and heat up. Consequently, a post-helium-flash star should be both smaller and have a higher surface temperature than a red giant.

Examples of these post-helium-flash stars are found in old star clusters, which are called **globular clusters** because of their spherical shape. A typical globular cluster, like the one shown in Figure 13-17, contains up to a million stars in a volume 300 light years across. Astronomers know that such clusters are old because they con-

Figure 13-16. The **zero-age main sequence** (or **ZAMS**) is the location on the H–R diagram where stars first begin core hydrogen burning. In subsequent years, the evolutionary tracks slowly inch away from the ZAMS as the hydrogen-burning core grows in search of fresh fuel. The dashed line on Figure 13-16 shows the locations of the stellar models when all the core hydrogen has been consumed.

After core hydrogen burning ceases, the points representing high-mass stars move rapidly from left to right across the H–R diagram. During this transition, the star's core contracts and its outer layers expand in response to increased energy output from the star's hydrogen-burning shell. Although the star's surface temperature is decreasing, its surface area is increasing, so that its overall luminosity remains roughly constant.

Just before core helium burning begins, the evolutionary tracks of high-mass stars turn upward in the red giant

Figure 13-17 A globular cluster *A globular cluster is a spherical cluster that typically contains a few hundred thousand stars. This cluster, called M13, is located in the constellation of Hercules, roughly 25,000 ly from Earth. (U.S. Naval Observatory)*

tain no high-mass main sequence stars. If you measure the luminosity and surface temperature of many stars in a globular cluster and plot the data on an H–R diagram as shown in Figure 13-18, you find that the upper half of the main sequence is missing. All the high-mass main sequence stars have evolved long ago into red giants, leaving behind only low-mass, slowly evolving stars still undergoing core hydrogen burning.

The H–R diagram of a globular cluster typically shows a horizontal grouping of stars to the left of the center portion of the diagram (see Figure 13-18). These stars, called **horizontal branch stars,** are post-helium-flash, low-mass stars that have luminosities of about 50 $L_\odot$. In years to come, these stars will move back toward the red giant region as their fuel is devoured by core helium burning and shell hydrogen burning.

An H–R diagram of a cluster can be used to determine the age of the cluster. In the diagram for a very young cluster (review Figures 13-6 and 13-7), the entire main sequence is intact. As a cluster gets older, stars begin to leave the main sequence. The high-mass, high-luminosity stars are the first to become red giants as the main sequence starts to burn down like a candle. Over the years, the main sequence gets shorter and shorter. The top of the surviving portion of the main sequence is called the **turnoff point.** Stars at the turnoff point are just now exhausting the hydrogen in their cores, and their main sequence lifetime is equal to the age of the cluster (recall Table 13-1). For example, in the case of the cluster M55 (see Figure 13-18), 0.8-$M_\odot$ stars have just left the main sequence, and so the cluster's age is roughly 15 billion years.

Data for several star clusters are plotted on Figure 13-19, along with turnoff-point times from which the ages of the clusters can be estimated. The youngest clusters (those with their main sequences still intact) are open clusters in the disk of our Galaxy, where star formation is an ongoing process. Stars in these young clusters are said to be **metal-rich** because their spectra contain many prominent

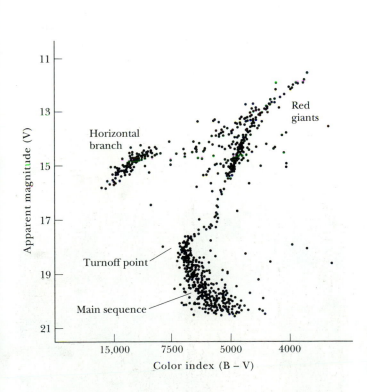

Figure 13-18 An H–R diagram of a globular cluster *Each dot on this graph represents a star in the globular cluster called M55 whose apparent magnitude and surface temperature have been measured. Note that the upper half of the main sequence is missing. The horizontal branch stars are low-mass stars that recently experienced the helium flash in their cores. (Adapted from D. Schade, D. VandenBerg, and F. Hartwick)*

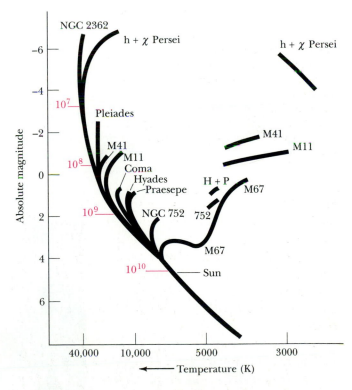

Figure 13-19 A composite H–R diagram *The black bands indicate where data from various star clusters fall on the H–R diagram. The ages of turnoff points (in years) are listed in red alongside the main sequence. The age of a cluster can be estimated from the location of the turnoff point, where the cluster's most massive stars are just now leaving the main sequence. (Adapted from A. Sandage)*

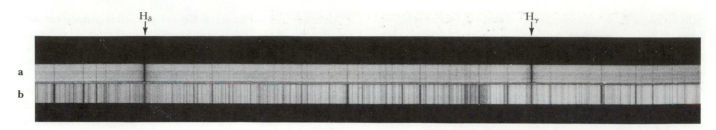

Figure 13-20 *Spectra of a metal-poor and a metal-rich star These spectra compare (a) a metal-poor star and (b) a metal-rich star (the Sun). Numerous spectral lines prominent in the solar spectrum are caused by elements heavier than hydrogen and* *helium. Note that corresponding lines in the metal-poor star's spectrum are weak or absent. Both spectra cover a wavelength range that includes H_γ and H_δ. (Lick Observatory)*

spectral lines of heavy elements. This material originally came from dead stars that exploded long ago, enriching the interstellar gases with the heavy elements formed in their cores. The Sun is a young, metal-rich star.

Most of the oldest clusters are globular clusters. Such clusters are generally located outside the disk of our Galaxy, and their spectra show only weak lines of heavy elements. These ancient stars are thus said to be **metal-poor.** They were created long ago from interstellar gases that had not yet been substantially enriched with heavy elements. Spectra of a metal-poor star and of the Sun are compared in Figure 13-20.

13-9 Red giants typically show mass loss

Both supergiants and red giant stars are so enormous that their bloated outer layers constantly leak gases into space. At times, this **mass loss** is quite significant (Figure 13-21).

Mass loss can be detected spectroscopically. Escaping gases coming toward us exhibit narrow absorption lines that are slightly blueshifted. According to the Doppler effect, this small shift toward shorter wavelengths corresponds to a speed of 10 km/s. This value is typical of the expansion velocities with which gases leave the tenuous outer layers of red giants. A typical rate of mass loss for a red giant is roughly 10^{-7} solar masses per year. For comparison, the Sun's mass loss rate is only 10^{-14} $M_\odot$ year.

Betelgeuse in the constellation of Orion is a good example of a red supergiant experiencing mass loss. Betelgeuse is 470 ly away and has a diameter roughly equal to the diameter of Mars's orbit. Recent spectroscopic observations show that this star is losing mass at the rate of 1.7×10^{-7} solar masses per year and is surrounded by a huge **circumstellar shell** that is expanding at 10 km/s. These escaping gases have been detected out to distances of 10,000 AU from the star. Consequently, the expanding circumstellar shell has an overall diameter of $\frac{1}{3}$ ly.

Figure 13-21 *A mass-loss star Old stars become red giants whose bloated outer atmospheres shed matter into space. This star is losing matter at a high rate and is surrounded by a reflection nebula caused by starlight reflected from dust grains. These dust grains may have condensed from material shed by the star. A typical red giant can lose 10^{-7} solar masses per year, and many are surrounded by circumstellar shells of matter they have shed. (Anglo-Australian Observatory)*

Figure 13-22 *A mass-loss supergiant star Beautiful nebulosity surrounds a supergiant star that is experiencing significant mass loss. The ejected material collides and interacts with the surrounding interstellar gas and dust, thereby producing the cosmic bubble seen here. This supergiant and its nebulosity are located in the constellation of Canis Major. (Anglo-Australian Observatory)*

Supergiant stars, which are brighter than 10^5 Suns, sustain mass loss throughout most of their existence, with mass-loss rates comparable to those of red giants. Figure 13-22 shows a supergiant star losing mass. In the next chapter, we shall see that dying stars also eject vast quantities of material into space. Nevertheless, mass loss from supergiants and red giants accounts for roughly one-fifth of all the matter returned by stars to the interstellar medium.

13-10 Many mature stars pulsate

After core helium burning begins, the evolutionary tracks of mature stars move across the middle of the H–R diagram. Figure 13-16 shows the evolutionary tracks of high-mass stars crisscrossing the H–R diagram. Post-helium-flash, low-mass stars on the horizontal branch also cross the middle of the H–R diagram as they return to the red giant region.

During these excursions across the H–R diagram, a star can become unstable and pulsate. In fact, there is a region on the H–R diagram between the main sequence and the red giant branch that is called the **instability strip** (Figure 13-23). When a star's evolutionary track carries it through this region, the star pulsates. As it does so, its brightness varies periodically.

Low-mass, post-helium-flash stars pass through the lower end of the instability strip as they move in the horizontal branch along their evolutionary tracks. These stars become **RR Lyrae variables,** named after the prototype in the constellation of Lyra. RR Lyrae variables all have periods shorter than one day, and all have roughly the same average brightness as stars on the horizontal branch. High-mass stars pass back and forth through the

upper end of the instability strip on the H–R diagram. These stars become **Cepheid variables.**

A Cepheid variable is recognized by the characteristic way in which its light output varies: rapid brightening

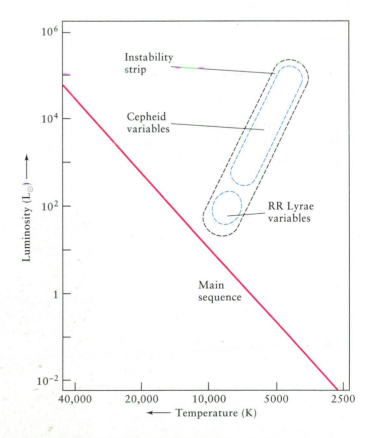

Figure 13-23 *The instability strip The instability strip occupies a region between the main sequence and the red giant branch on the H–R diagram. A star passing through this region along its evolutionary track becomes unstable and pulsates.*

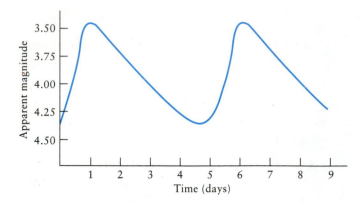

Figure 13-24 The light curve of a Cepheid variable This graph shows the light curve of δ Cephei, which is the prototype of an important class of variable stars. All Cepheid variables exhibit the same type of periodic changes in luminosity as shown here—rapid brightening followed by gradual dimming.

followed by gradual dimming. This behavior is most easily displayed in a light curve, a graph of a star's brightness plotted against time. The light curve of δ Cephei is shown in Figure 13-24. This star was the first pulsating variable to be discovered and serves as the prototype for this important class of stars, often simply called Cepheids.

A Cepheid variable brightens and fades because of a cyclic expansion and contraction of the star's outer layers. This behavior is deduced from spectroscopic observations. Spectral lines in the spectrum of δ Cephei shift back and forth with the same 5.4-day period as that of the magnitude variations. According to the Doppler effect, these shifts mean that the star's surface is alternately approaching and receding from us.

When a Cepheid variable pulsates, the star's surface oscillates up and down like a spring. During these cyclical expansions and contractions, the star's gases alternately heat up and cool down. Thus the characteristic light curve of a Cepheid variable results from both changing size and changing surface temperature.

Just as a bouncing ball eventually comes to rest, a pulsating star would soon stop pulsating without some sort of mechanism to keep its oscillations going. In 1941 the British astronomer Arthur Eddington explained that a Cepheid variable feeds energy into its pulsations by a valvelike action involving the periodic ionization and deionization of gas in the star's outer layers. Eddington's valve mechanism requires that the star be more opaque, or "light-tight," when compressed than when expanded. When the star is compressed, trapped heat pushes the star's surface outward. When the star is expanded, the heat escapes and so the star's surface, which is no longer supported, falls inward.

Cepheids are very important to astronomers because there is a direct relationship between a Cepheid's period and its average luminosity. Dim Cepheid variables pulsate rapidly, have periods of 1 to 2 days, and have an average brightness of a few hundred Suns. The most luminous Cepheids are the slowest variables, with periods of 100 days and average brightnesses equal to 10,000 Suns. This connection between period and brightness is called the **period–luminosity relation.** This relationship played an important role in determining the overall size and structure of the universe, as we shall see in Chapter 17.

The details of a Cepheid's pulsation depend on the abundance of heavy elements in its atmosphere. The average luminosity of metal-rich Cepheids is roughly four times greater than the average luminosity of metal-poor Cepheids having the same period. Thus there are two classes: **Type I Cepheids,** which are the brighter, metal-rich stars, and **Type II Cepheids,** which are the dimmer, metal-poor stars. The period–luminosity relation for both types of variables is shown in Figure 13-25.

Stellar pulsations can, in rare cases, be quite substantial. In some cases, the expansion velocity is so high that the star's outer layers are ejected completely. As we shall see in the next chapter, significant mass ejection accompanies the death of stars in a sometimes violent process that renews and enriches the interstellar medium for future generations of stars.

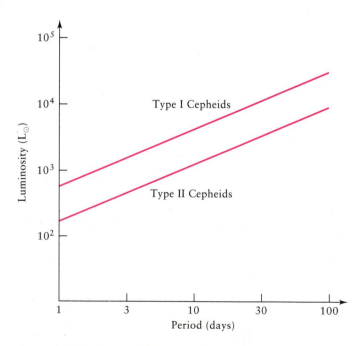

Figure 13-25 The period-luminosity relation The period of a Cepheid variable is directly related to its average luminosity. Metal-rich (Type I) Cepheids are brighter than the metal-poor (Type II) Cepheids.

Summary

- Enormous cold clouds of gas, called giant molecular clouds, are scattered about our Galaxy.

- Star formation begins when gravitational attraction causes protostars to coalesce within a giant molecular cloud. As a protostar contracts, its gases begin to glow. When its core temperature becomes high enough to begin hydrogen burning, the protostar becomes a main sequence star.

  The most massive protostars take the shortest time to become main sequence stars (O and B stars). They emit strong ultraviolet radiation that ionizes hydrogen in the surrounding cloud, creating reddish emission nebulae called H II regions.

- Ultraviolet radiation and stellar winds from the OB association at the core of an H II region create shock waves that compress the gas cloud, triggering formation of more protostars.

  Supernova explosions also compress gas clouds and trigger star formation.

- In the final stages of pre-main-sequence contraction, when hydrogen burning is about to begin in the core of a protostar, the star may undergo vigorous chromospheric activity that ejects large amounts of matter into space. Such gas-ejecting stars are called T Tauri stars.

- A collection of newborn stars is called an open cluster, or galactic cluster.

  Occasionally a rapidly moving star will escape, or "evaporate," from such a cluster.

- The more massive a star, the shorter its main sequence lifetime.

  The Sun has been a main sequence star for about 5 billion years and should remain so for about another 5 billion years. Less massive stars evolve more slowly and have longer lifetimes.

- Core hydrogen burning ceases when hydrogen is exhausted in the core of a main sequence star, leaving a core of nearly pure helium surrounded by a shell where hydrogen burning continues.

  Shell hydrogen burning adds more helium to the star's core, which contracts and becomes hotter. The outer atmosphere expands considerably and the star becomes a red giant.

- When the central temperature of a red giant reaches about 100 million kelvin, the thermonuclear process of helium burning begins. This process converts helium to carbon and oxygen.

  In a massive red giant, helium burning begins gradually. In a less massive red giant, it begins suddenly in a process called a helium flash.

- The age of a stellar cluster can be estimated by plotting its stars on an H–R diagram. The upper portion of the main sequence will be missing because more massive main sequence stars have become red giants.

  Relatively young stars are metal-rich; ancient stars are metal-poor.

- Supergiants and red giants undergo extensive mass loss, sometimes producing circumstellar shells of ejected material around the stars.

- When a star's evolutionary track carries it through a region called the instability strip in the H–R diagram, the star becomes unstable and begins to pulsate.

  Cepheid variables are high-mass pulsating variables exhibiting a regular relationship between period of pulsation and luminosity. RR Lyrae variables are low-mass pulsating variables with short periods.

Review questions

1 What is a giant molecular cloud, and what role do these clouds play in the birth of stars?

2 Why are low temperatures necessary for protostars to form inside dark nebulae?

3 What is an H II region?

4 Explain why thermonuclear reactions occur only at the center of a main sequence star and never on its surface.

5 What is an evolutionary track and how can such tracks help us interpret the H–R diagram?

6 Why do you suppose the vast majority of the stars we see in the sky are main sequence stars?

7 Draw the pre-main-sequence evolutionary track of the Sun on an H–R diagram. Briefly describe what was probably occurring throughout the solar system at various stages along this track.

8 On what grounds are astronomers able to say that the Sun has about 5 billion years remaining in its main sequence stage?

9 What will happen inside the Sun 5 billion years from now when it begins to turn into a red giant?

10 Draw the post-main-sequence evolutionary track of the Sun on an H–R diagram up to the point when the Sun becomes a red giant. Briefly describe what might occur throughout the solar system as the Sun undergoes this transition.

11 What does it mean when an astronomer says that a star "moves" from one place to another on an H–R diagram?

12 What are Cepheid variables and how are they related to the instability strip?

13 What is the helium flash?

14 How is a degenerate gas different from an ordinary gas?

15 Explain how and why the turnoff point on the H–R diagram of a cluster is related to the cluster's age.

16 Why do astronomers believe that globular clusters are made of old stars?

Advanced questions

17 If you took a spectrum of a reflection nebula, would you see absorption lines, emission lines, or no lines? Explain your answer.

18 What observations would you make of a star to determine whether its primary source of energy was hydrogen or helium burning?

* **19** How many 1.5-$M_\odot$ main sequence stars would it take to equal the luminosity of one 15-$M_\odot$ star?

* **20** How many times longer does a 1.5-$M_\odot$ star burn hydrogen at its core than does a 15-$M_\odot$ star?

21 Speculate on why a shock wave from a supernova seems to produce relatively few high-mass O and B stars compared to the lower-mass A, F, G, and K stars.

22 How would you distinguish a newly formed protostar from a red giant in view of their identical location on the H–R diagram?

23 What observational consequences would we find in H–R diagrams for star clusters if the universe had a finite age? Could we use these consequences to establish constraints on the possible age of the universe? Explain.

Discussion questions

24 What do you think would happen if the solar system passed through a giant molecular cloud? Do you think that the Earth has ever passed through such a cloud?

25 Speculate about the possibility of life forms and biological processes occurring in giant molecular clouds. In what ways might conditions in giant molecular clouds favor or hinder biological evolution?

For further reading

Blitz, L. "Giant Molecular Cloud Complexes in the Galaxy." *Scientific American*, April 1982 • A fine discussion of interstellar molecules highlights this article describing the structure and properties of giant molecular clouds.

Cohen, M. *In Darkness Born: The Story of Star Formation.* Cambridge University Press, 1988 • This well-written book presents an overview of our modern understanding of the early stages of stellar evolution.

Johnson, B. "Red Giant Stars." *Astronomy*, December 1976 • This well-written article surveys properties of red giant stars.

Kaler, J. B. "Journeys on the H–R diagram." *Sky & Telescope*, May 1988 • This brief article gives an excellent overview of evolutionary tracks that stars follow on the H–R diagram.

———. *Stars.* Scientific American Library, 1992 • This attractive book summarizes our current understanding of stars and stellar evolution.

Kippenhahn, R. *100 Billion Suns: The Birth, Life and Death of Stars.* Basic Books, 1983 • This classic book, written by a noted German astrophysicist, describes the life cycles of stars.

Lada, C. "Energetic Outflows from Young Stars." *Scientific American*, July 1982 • This article describes how radiation emitted by carbon monoxide molecules discloses bipolar outflow from newborn stars.

Robinson, L. "Orion's Stellar Nursery." *Sky & Telescope*, November 1982 • This article explores the Orion Nebula and its environs with the aid of high-resolution infrared observations.

Verschuur, G. *Interstellar Matters.* Springer-Verlag, 1988 • This superb book gives an entertaining and informative introduction to the interstellar medium, replete with interesting historical anecdotes.

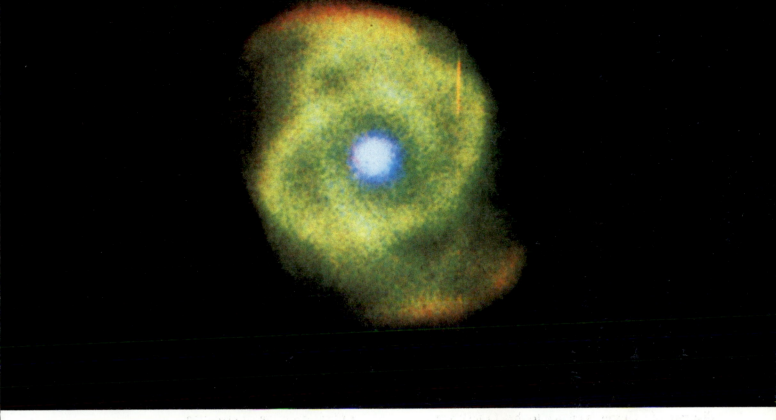

A planetary nebula Dying stars often eject their outer layers. A low-mass star can lose half its mass in a comparatively gentle process that produces a planetary nebula. The exposed stellar core typically has a surface temperature of about 100,000 K and is roughly one-tenth the size of the Sun. Ultraviolet radiation from the hot stellar core causes the surrounding gases to glow. The greenish color of this planetary nebula comes from oxygen ions. The central star in this nebula has exhausted all its nuclear fuel and is contracting to become a white dwarf. (Lick Observatory)

14

The Deaths of Stars

After experiencing old age as a red giant, a star approaches the end of its life. As stars die, they eject significant amounts of matter into space. In this chapter, we find that low-mass stars eject their outer layers relatively gently, producing planetary nebulae, whereas high-mass stars explode violently as supernovae. In 1987 a supernova visible to the naked eye gave astronomers an unprecedented opportunity to study the explosive death of a massive star. The core of a dead low-mass star contracts to become a white dwarf; this will be the fate of our Sun. The burned-out core of a high-mass star can collapse violently to become a neutron star. In examining white dwarfs and neutron stars, we discover that there are upper limits to the masses that these stellar corpses can have. Finally, we discuss the various fascinating forms in which neutron stars have been observed, such as pulsars, pulsating X-ray sources, and bursters.

From infancy through adulthood, a star leads a fairly placid life with hydrogen burning in its core. When the hydrogen is used up, the star's core contracts while its atmosphere expands, and so the star becomes a bloated red giant. In the years that follow, dramatic phenomena can occur as the star devours its remaining nuclear fuels and begins to die.

The ultimate fate of a dying star depends on its mass. As we shall see in the following pages, low-mass stars die gently, whereas high-mass stars die violently. Recent calculations on stellar evolution show that the dividing line between low-mass and high-mass stars is about 8 solar masses.

14-1 Low-mass stars die by gently ejecting their outer layers, creating planetary nebulae

Carbon and oxygen are the "ashes" of helium burning. After the helium flash in a low-mass red giant, substantial amounts of these two elements begin to accumulate at the star's center as a result of core helium burning. Eventually all the helium at the center of a low-mass star is used up, at which point core helium burning ceases. The star's core therefore contracts until stopped by the buildup of degenerate electron pressure in the core. This compression heats the helium-rich gases surrounding the degenerate carbon–oxygen core, and helium burning soon begins in a thin shell around the core. This process is called **shell helium burning.** The star's internal structure now consists of a helium-burning shell inside a hydrogen-burning shell, all within a volume roughly the size of the Earth. After a while, thermonuclear reactions in the hydrogen-burning shell temporarily cease, leaving the aging star with the structure shown in Figure 14-1.

When shell hydrogen burning first began, the outpouring of energy caused the star to expand and become a red giant. In describing the evolutionary track of the star on an H–R diagram, astronomers say that the star ascended the red giant branch for the first time. Then came the helium flash, at which time the star shifted over to the horizontal branch. But now the renewed outpouring of energy from shell helium burning causes the star to expand again. The star ascends the red giant branch for a second and final time to become a **red supergiant.** Such stars are also called **asymptotic giant branch stars,** or **AGB stars** for short, because of the region they occupy on an H–R diagram. These stars typically have diameters as big as the orbit of Mars and shine with the brightness of 10,000 Suns. A star experiencing this furious rate of energy loss cannot live much longer.

The star's impending death is signaled by instabilities that develop in its helium-burning shell. As with the helium flash discussed in the previous chapter, there is again a thermal runaway, but the details are quite different. The helium flash occurred because the star's core was degenerate. In contrast, the helium-burning shell is not compressed to a density high enough to be degenerate. Instead, a **helium-shell flash** occurs because the shell is thin. A slight increase in energy output from a thin shell increases the temperature of the helium there but does little to relieve the pressure from the star's overlying layers. This change further increases the energy output, which in turn drives up the temperature even more. This vicious circle produces a thermal runaway that ends only after the helium-burning shell becomes thick enough to relieve the pressure of the star's outer layers.

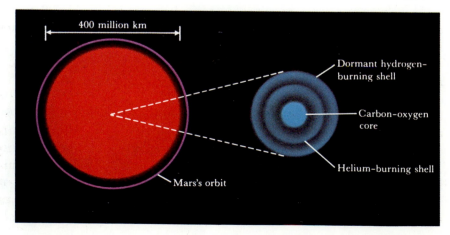

Figure 14-1 *The structure of an old low-mass star* Near the end of its life, a low-mass star becomes a red supergiant, with a diameter almost as large as the diameter of the orbit of Mars. The star's dormant hydrogen-burning shell and active helium-burning shell are contained within a volume roughly the size of the Earth.

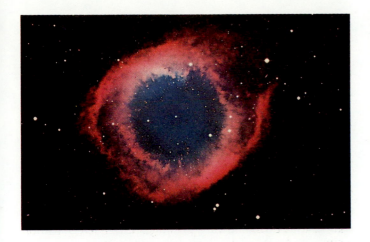

Figure 14-2 *The planetary nebula NGC 7293* This planetary nebula, often called the Helix Nebula, is located in the constellation of Aquarius (the Water Bearer). The star that ejected these gases is seen at the center of the glowing shell. The greenish color comes from oxygen ions, while the pink and red comes from nitrogen and hydrogen ions. This nebula, located about 700 ly from Earth, has an angular diameter equal to about half that of the full moon. (Anglo-Australian Observatory)

Figure 14-3 *The planetary nebula M27* This planetary nebula, called the Dumbbell Nebula because of its characteristic shape, is located in the constellation of Vulpecula (the Fox). Although most planetary nebulae are ring-shaped, many do have a dumbbell-like appearance caused by uneven expansion of the ejected gases. Perhaps a circumstellar disk of material inhibited the nebula's expansion in the equatorial plane of the dying star but not towards its poles. This nebula, located about 1250 ly from Earth, has an angular size equal to about one-quarter that of the full moon. (U.S. Naval Observatory)

During the flash, the helium shell's energy output jumps from 100 Suns to roughly 100,000 Suns in brief outbursts separated by relatively quiet intervals lasting about 300,000 years. During one of these outbursts, the dying star's outer layers can separate completely from the carbon–oxygen core. As the ejected material expands into space, dust grains condense out of the cooling gases. Radiation pressure from the star's hot, burned-out core acts on the specks of dust to continue propelling them outward, and the star sheds its outer layers altogether. A star can lose more than one-half of its mass in this fashion.

As a dying star ejects its outer layers, its hot core is exposed, emitting ultraviolet radiation intense enough to ionize the expanding shell of ejected gases. This ionization causes the gases to glow, producing a so-called **planetary nebula.** Planetary nebulae have nothing to do with planets. This unfortunate term was introduced in the eighteenth century when these glowing objects were thought to look like distant planets when viewed through small telescopes.

Many planetary nebulae, such as the one shown in Figure 14-2, have a distinctly spherical appearance arising from the symmetrical way in which the gases were ejected. In other cases, as in Figure 14-3, the rate of expansion is not the same in all directions, and the resulting nebula takes on an hourglass or dumbbell appearance.

Planetary nebulae are quite common. Astronomers estimate that there are 20,000 to 50,000 in our Galaxy alone. Spectroscopic observations of these nebulae show bright emission lines of hydrogen, oxygen, and nitrogen. From the Doppler shifts of these lines, astronomers conclude that the expanding shell of gas is moving outward from the dying star with speeds of 10 to 30 km/s. A typical planetary nebula has a diameter of roughly 1 ly, which means that it must have begun expanding about 10,000 years ago.

By astronomical standards, a planetary nebula is a very short-lived entity. After about 50,000 years, the nebula has spread over distances so far from the cooling central star that its nebulosity simply fades from view. The gases then mingle and mix with the surrounding interstellar medium. Astronomers estimate that all the planetary nebulae in the Galaxy return a total of 5 $M_\odot$ to the interstellar medium each year. This amounts to about 15% of all matter expelled by all sorts of stars each year. This contribution is so significant that planetary nebulae are thought to play an important role in the evolution of the Galaxy as a whole.

14-2 The burned-out core of a low-mass star cools and contracts to become a white dwarf

Stars less massive than about 8 M⊙ never develop the necessary central pressures or temperatures to ignite thermonuclear reactions that use carbon or oxygen as fuel. Instead, the process of mass ejection strips away the star's outer layers and exposes the carbon–oxygen core, which simply cools off.

The evolutionary tracks of three burned-out stellar cores are shown in Figure 14-4. The initial red supergiants had masses between 0.8 M⊙ and 3.0 M⊙. During their final spasms, these dying stars eject up to 60% of their matter. During the ejection phase, the outward appearance of these stars changes rapidly. The resulting changes in luminosity and surface temperature cause the points that represent such stars on the H–R diagram to race along their evolutionary tracks, sometimes executing loops corresponding to thermal pulses. Finally, as the ejected nebulae fade and the stellar cores cool, the evolutionary tracks of the dying stars take a sharp turn downward toward the white dwarf region of the diagram.

There is no possibility of igniting additional nuclear fuels inside one of these dead stars, and so the crushing weight of gases pressing inward from all sides severely compresses the stellar corpse. The density inside the dead star skyrockets until the electrons are so closely packed that they become degenerate. Because the degenerate-electron pressure is strong enough to support the star, the gravitational contraction halts and the star remains roughly the same size as the Earth. Such a star is called a **white dwarf.**

The density of matter in one of these Earth-sized stellar corpses is typically 10^9 kg/m^3. In other words, a teaspoonful of white dwarf matter brought to Earth would weigh 5 tons.

There is an upper limit to the mass that a white dwarf can have. This maximum mass is called the **Chandrasekhar limit** after Subrahmanyan Chandrasekhar at the University of Chicago, who pioneered theoretical studies of white dwarfs. The Chandrasekhar limit is equal to 1.4 M⊙; thus all white dwarfs have masses less than 1.4 M⊙. Above this limit, degenerate-electron pressure cannot support the weight of the star's matter pressing inward from all sides.

Many white dwarfs are found in the solar neighborhood, but all are too faint to be seen with the naked eye. One of the first white dwarfs to be discovered is a com-

panion to the bright star Sirius. The binary nature of Sirius was first deduced in 1844 by the German astronomer Friedrich Bessel, who noticed that the star was moving back and forth slightly, as if orbited by an unseen object. This companion, called Sirius B, was first glimpsed in 1862 (Figure 14-5). Recent satellite observations at ultraviolet wavelengths—where white dwarfs emit most of their light—demonstrate that the surface temperature of Sirius B is about 30,000 K.

As a white dwarf cools, both its luminosity and its surface temperature decline. As billions of years pass, white dwarfs get dimmer and dimmer as their surface temperatures drop toward absolute zero. This condition will be the final fate of our Sun: a cold, dark, dense sphere of degenerate gases rich in oxygen and carbon, about the size of the Earth.

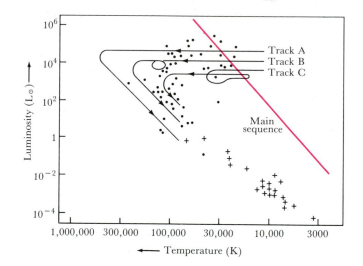

Evolutionary track	Supergiant mass (M⊙)	Mass of ejected nebula (M⊙)	White dwarf mass (M⊙)
Track A	3.0	1.8	1.2
Track B	1.5	0.7	0.8
Track C	0.8	0.2	0.6

Figure 14-4 Evolution from red supergiants to white dwarfs The evolutionary tracks of three low-mass red supergiants are shown as they eject planetary nebulae. The table gives the extent of mass loss in each case. The dots on this graph represent the central stars of planetary nebulae whose surface temperatures and luminosities have been determined. The crosses are white dwarfs for which similar data exist. (Adapted from B. Paczynski)

Figure 14-5 Sirius and its white dwarf companion Sirius, the brightest-appearing star in the sky, is actually a double star. The secondary star is a white dwarf, seen here at the "five o'clock" position, in the glare of Sirius. The spikes and rays around Sirius are created by optical effects within the telescope. (Courtesy of R. B. Minton)

14-3 High-mass stars die violently by blowing themselves apart in supernova explosions

High-mass stars end their lives very differently than do low-mass stars. A high-mass star is capable of igniting a host of additional thermonuclear reactions in its core. The more massive a star is, the greater the temperatures that can be achieved in its core by gravitational compression. High temperatures are required to initiate nuclear reactions involving heavy nuclei because the large electric charges of these nuclei exert strong forces tending to keep the nuclei apart. Only at the great speeds associated with high temperatures are the nuclei traveling fast enough to penetrate each other's repulsive electric fields and fuse together.

Recall that carbon and oxygen are the "ashes" of helium burning. When helium burning is finished at the core of a star, gravitational compression drives the star's central temperature up to 600 million kelvin, and **carbon burning** begins. This thermonuclear process produces elements such as neon and magnesium.

If a star is massive enough to drive its central temperature to 1.2 billion kelvin, **neon burning** begins. This process uses up the neon accumulated from carbon burning, further increasing the concentrations of oxygen and magnesium in the star's core.

If the central temperature of the star reaches about 1.5 billion kelvin, **oxygen burning** begins. The principal product of oxygen burning is sulfur. As the star consumes increasingly heavier nuclei, thermonuclear reactions produce many different elements. For instance, oxygen burning also produces isotopes of silicon, phosphorus, and more magnesium.

When a given nuclear fuel is exhausted in the core of a massive star, gravitational contraction to ever higher densities drives up the star's central temperature, thereby igniting the "ash" of the previous burning stage. Each successive thermonuclear reaction occurs with increasing rapidity. For example, detailed calculations for a 25-$M_\odot$ star demonstrate that carbon burning occurs for 600 years, neon burning for 1 year, but oxygen burning for only 6 months.

After half a year of core oxygen burning in a 25-$M_\odot$ star, gravitational compression forces the central temperature up to 2.7 billion kelvin, when **silicon burning** begins. This thermonuclear process proceeds so furiously that the entire core supply of silicon in a 25-$M_\odot$ star is used up in one day.

Silicon burning involves many hundreds of nuclear reactions. The major final product, or "ash," of this process is iron. Iron cannot fuel any further thermonuclear reactions, and so the sequence of burning stages ends. In order for an element to be a thermonuclear fuel, energy must be released when its nuclei collide and fuse. This energy comes from packing the neutrons and protons together more tightly in the ash nuclei than in the fuel nuclei. The protons and neutrons inside the iron nuclei are already so tightly bound together that no further energy can be extracted by fusing still more nuclei with iron.

The buildup of an inert, iron-rich core signals the impending violent death of a massive star. Surrounding this iron core, successive layers of shell burning consume the star's remaining reserves of fuel (Figure 14-6). The entire energy-producing region of the star is contained in a volume as big as the Earth, whereas the star's enormously bloated atmosphere is nearly as big as the orbit of Jupiter.

Because iron does not "burn," the electrons in the core must now support the star's outer layers by the brute

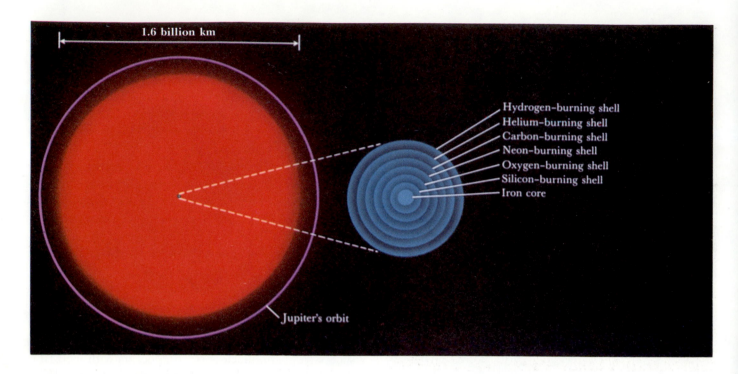

Figure 14-6 *The structure of an old high-mass star* *Near the end of its life, a high-mass star becomes a red supergiant almost as big as the orbit of Jupiter. The star's energy comes from six* *concentric burning shells, all contained within a volume roughly the same size as the Earth.*

strength of degeneracy pressure alone. Soon, however, the continued deposition of fresh iron from the silicon-burning shell causes the core's mass to exceed the Chandrasekhar limit. Electron degeneracy suddenly becomes unable to support the star's enormous weight, and the core begins to collapse.

Any star with a mass greater than 10 $M_\odot$ can develop an iron core that at some stage will exceed the Chandrasekhar limit, triggering a rapid series of cataclysms that will tear the star apart in a few seconds. Let us see how this happens in the death of a 25-$M_\odot$ star.

In a 25-$M_\odot$ star, degenerate-electron pressure fails when the density inside the iron core reaches 3 trillion kilograms per cubic meter. The core then immediately begins to collapse. Central temperatures promptly soar to almost inconceivable levels. In roughly a tenth of a second, the temperature exceeds 5 billion kelvin. Gamma-ray photons associated with this intense heat have so much energy that they begin to break up the iron nuclei in a process called **photodisintegration.**

Within another tenth of a second, as densities continue to climb, the electrons are forced to combine with protons to produce neutrons in a process called **neutronization.**

This nuclear transformation releases a flood of neutrinos, because a neutrino is created every time an electron combines with a proton to produce a neutron (Figure 14-7). As we saw in Chapter 11, neutrinos have no electric charge and are generally believed to have no mass either, which means they resemble photons and travel at the speed of light. The newly created neutrinos do not immediately escape from the star's core, however, because of the enormous density of matter that they encounter.

About one-quarter second after the collapse begins, the density in the core reaches 4×10^{17} kg/m^3, which is **nuclear density,** the density at which neutrons and protons are packed together inside nuclei. At nuclear density, matter is virtually incompressible. Thus, when the neutron-rich material of the core reaches this density, it suddenly becomes very stiff and the core collapse comes to an abrupt halt.

During this critical stage, the star's unsupported inner regions are plunging inward at speeds up to 15% of the speed of light. As this material crashes onto the now-rigid core, enormous temperatures and pressures develop, causing the falling material to bounce back. In just a fraction of a second, a wave of matter begins to move back out

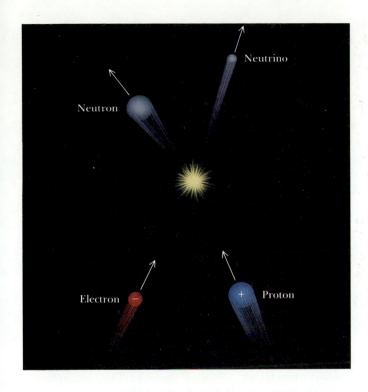

Figure 14-7 **The creation of a neutrino** *A collision between an electron and a proton can produce a neutron and a neutrino. The electron and the proton have equal and opposite electric charges; both the neutron and neutrino are electrically neutral. The neutrino may have a very tiny mass, much smaller than that of an electron, or it may be massless.*

Table 14-1 Evolutionary stages of a 25-$M_{\odot}$ star

Stage	Central temperature (K)	Central density (kg/m³)	Duration of stage
Hydrogen burning	4×10^7	5×10^3	7×10^6 years
Helium burning	2×10^8	7×10^5	5×10^5 years
Carbon burning	6×10^8	2×10^8	600 years
Neon burning	1.2×10^9	4×10^9	1 year
Oxygen burning	1.5×10^9	10^{10}	6 months
Silicon burning	2.7×10^9	3×10^{10}	1 day
Core collapse	5.4×10^9	3×10^{12}	0.2 second
Core bounce	2.3×10^{10}	4×10^{17}	Milliseconds
Explosive	About 10^9	Varies	10 seconds

toward the star's surface, propelled in part by the flood of neutrinos trying to escape from the star's core. This wave accelerates rapidly as it encounters less and less resistance, and soon it becomes a shock wave. After a few hours, this shock wave reaches the star's surface, by which time the star's outer layers have begun to lift away from the core. In this way a star becomes a **supernova.**

This particular description of the death of a 25-$M_{\odot}$ star is based on detailed computer calculations of an evolving stellar model. The key stages in such a star's evolution are summarized in Table 14-1.

The final stages in the evolution of other massive stars probably follow similar scenarios, although the details may be different. This particular 25-$M_{\odot}$ star ejects 24 $M_{\odot}$ of its mass, leaving behind a 1-$M_{\odot}$ corpse called a **neutron star.** Under slightly different conditions, a massive star might blow itself completely apart, leaving no corpse at all. For example, the bounce and subsequent shock wave might develop at the center of the star rather than at the surface of the neutronized core. Alternatively,

with a different set of initial conditions inside the star, a massive burned-out core might gravitationally collapse to form a **black hole.** This exotic object is discussed in the next chapter.

14-4 In 1987 a nearby supernova gave us a close-up look at the death of a massive star

As the outer layers of a massive dying star are blasted into space, the star's luminosity suddenly increases by a factor of 10^8, equivalent to a jump of 20 magnitudes in brightness. For a few days following the explosion, a supernova can shine as brightly as an entire galaxy.

On February 23, 1987, a supernova was discovered in the Large Magellanic Cloud, which is a companion galaxy to our own Milky Way. The supernova (designated SN 1987A because it was the first to be observed that year) occurred near an enormous H II region called 30 Doradus or the Tarantula Nebula because of its spiderlike appearance (Figure 14-8). The supernova was so bright that it could be seen with the naked eye.

A bright, nearby supernova is a rare event. In 1885 a supernova in the Andromeda Galaxy was just barely visible to the unaided eye. We have to go back to 1604 to find another supernova bright enough to have been seen without a telescope. SN 1987A gave astronomers the unique opportunity to study the death of a massive, nearby star using modern equipment.

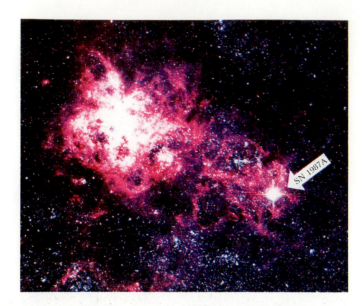

Figure 14-8 The supernova SN 1987A *In 1987 a supernova was discovered in a nearby galaxy called the Large Magellanic Cloud (LMC). This photograph shows a portion of the LMC that includes the supernova and a huge H II region called the Tarantula Nebula. At maximum brightness the supernova could be seen without a telescope by observers at southern latitudes. (European Southern Observatory)*

SN 1987A was unusual because it did not promptly rise to its expected maximum brightness. Instead, it stopped at only a tenth of the luminosity typical of an exploding massive star like the one described in Table 14-1. For the 85 days following the outburst, SN 1987A gradually brightened and then settled into a slow decline characteristic of an ordinary supernova.

Fortunately the doomed star had been observed prior to becoming a supernova, and these observations helped explain why SN 1987A was not altogether typical. The Large Magellanic Cloud is about 160,000 light-years from Earth—near enough to us that many of its stars have been individually observed and catalogued. The doomed star was identified as a B3 I supergiant (Figure 14-9). When this star was on the main sequence, its mass was about 20 $M_\odot$, although by the time it exploded it probably had shed a few solar masses.

SN 1987A was initially less luminous than expected because the doomed star was a blue supergiant when it exploded rather than a red supergiant. The evolutionary track for an aging 20-$M_\odot$ star wanders back and forth across the top of the H–R diagram, and so the star alternates between being a hot (blue) supergiant and cool (red) supergiant. The star's size changes significantly as it undergoes these changes in surface temperature. A blue supergiant is only 10 times larger in diameter than the Sun, but a red supergiant of the same luminosity would be 1000 times larger (recall Figure 12-11). Because the doomed star was relatively small when it exploded, it reached only a tenth of the brightness that a red supergiant would have.

Stellar model calculations suggest that the interior of the doomed star contained 6 $M_\odot$ of helium and 2 $M_\odot$ of heavier elements. At the time of detonation, the star's iron core had a mass of 1.5 $M_\odot$ and a temperature of 10 billion kelvin. About 0.15 $M_\odot$ of radioactive isotopes were also created during the explosion. Energy released by the decay of these isotopes contributed significantly to the supernova's brightness as it began to fade.

Figure 14-9 The doomed star and SN 1987A *The photograph on the left shows a small section of the Large Magellanic Cloud before the outburst, with the doomed star—a blue supergiant— identified by an arrow. The view on the right shows the supernova a few days after the explosion. (Anglo-Australian Observatory)*

Most of the energy of a supernova explosion is carried away by neutrinos, which are produced in great profusion during the collapse of the star's core. Neutrinos are very difficult to detect because they seldom interact with ordinary matter. In fact, neutrinos easily pass through the Earth as if it were not there.

Scientists have built neutrino detectors consisting of large tanks of water. Water (H_2O) contains many protons, the nuclei of hydrogen atoms. When a high-energy neutrino strikes a proton, it produces an electron that emits a flash of light called **Cerenkov radiation.** This type of radiation, first observed by the Russian physicist Pavel A. Cerenkov, occurs whenever a particle moves through water faster than light can. Such motion does not violate the tenet that the speed of light *in a vacuum* (3×10^8 m/s) is the ultimate speed limit in the universe. Light is slowed considerably as it passes through water, and high-energy particles can exceed this reduced speed without violating the laws of physics. Scientists try to detect neutrinos by observing Cerenkov flashes with light-sensitive photomultipliers mounted in the water (Figure 14-10).

Nearly a day before SN 1987A was observed in the sky, teams of scientists at neutrino detectors in Japan and the United States excitedly reported Cerenkov flashes from a burst of neutrinos. The Kamiokande II detector in Japan detected 12 neutrinos at about the same time that 8 were found by the IMB (Irvine-Michigan-Brookhaven) detector in a salt mine under Lake Erie. Neutrinos preceded the visible outburst because they escaped from the dying star before the shock wave from the collapsing core reached the star's surface. They were detected in the Northern Hemisphere, where the supernova is always below the horizon, after having passed through the Earth.

The ability to detect neutrinos from a supernova offers astronomers a new and exciting opportunity. Ordinary telescopic observations of a supernova explosion can show us only the expanding outer layers of the dying star. Even X-ray or radio observations fail to see through the hot gases being blasted into space. Thus, using ordinary techniques we cannot observe the extraordinary events occurring in and around the doomed star's core. However, neutrinos easily penetrate a star's outer layers. These particles carry information about the conditions under which they were created. By detecting these neutrinos and measuring their properties, astronomers can learn many details about the star's collapsing core.

About three and a half years after the detonation of SN 1987A, astronomers using the Hubble Space Telescope obtained a picture of the supernova that showed a ring of glowing gas around the exploded star (Figure 14-11). This gas is the relic of a hydrogen-rich stellar envelope that was

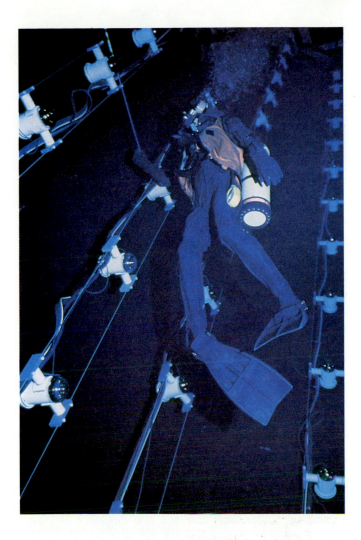

Figure 14-10 Inside a neutrino detector A diver is shown servicing one of the photomultiplier tubes in the Irvine-Michigan-Brookhaven (IMB) neutrino detector. Neutrinos from SN 1987A were detected by this apparatus when they struck protons in the water, producing brief flashes of light that were picked up by the photomultipliers. (Courtesy of K. S. Luttrell)

ejected by gentle stellar winds from the doomed star when it was a red supergiant, about 10,000 years ago. The diffuse gas was subsequently compressed into a narrow shell by high-speed stellar winds that flowed from the star when it became a blue supergiant. The gaseous ring seen in Figure 14-11 is glowing because it was ionized by the initial flash of ultraviolet radiation from the supernova. Astronomers predict that the material ejected from the supernova itself will strike the circumstellar shell sometime between 1997 and 2004. This collision will cause the shell of gas to brighten considerably and emit copious radiation at X-ray, ultraviolet, and visible wavelengths.

Astronomers have been rather lucky with SN 1987A. The doomed star had been studied and its distance from

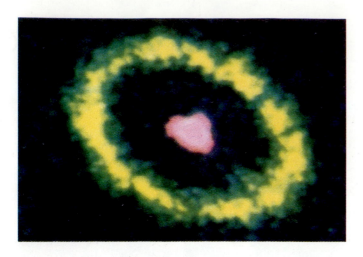

Figure 14-11 *A shell of gas around SN 1987A* *Intense radiation from the supernova explosion caused a circumstellar shell of gas around SN 1987A to glow. The luminescent ring seen in this view from the Hubble Space Telescope is about 1.3 ly in diameter. Between 1997 and 2004, high-speed gases ejected by the supernova will reach the ring and cause it to brighten significantly, but not enough to be visible to the naked eye. (NASA; ESA)*

Earth (160,000 ly) was known. The supernova was located in an unobscured part of the sky, and neutrino detectors happened to be operating at the time of the outburst. And finally, the supernova was unusual, which advanced our understanding of supernovae in general. Because SN 1987A is such an important object, astronomers will be monitoring its progress for years to come.

14-5 Accreting white dwarfs in close binary systems can also become supernovae

Astronomers often find supernovae in distant galaxies (Figure 14-12). Most of our understanding of supernovae comes from observing these remote outbursts. Such observations reveal that supernovae fall into two categories, designated Type I and Type II, which involve very different phenomena. A supernova's spectrum determines its type: hydrogen lines are prominent in Type II but absent in Type I.

Both types of supernovae begin with a sudden rise in brightness (Figure 14-13). A Type I supernova typically reaches an absolute magnitude of −19 at peak brightness, but a Type II supernova is usually about 2 magnitudes fainter. Type I supernovae then settle into a gradual decline that lasts for over a year, whereas the Type II light curve has a steplike appearance caused by alternating periods of steep and gradual declines in brightness.

A Type II supernova is caused by the death of a massive star. Gravitational energy powers a Type II supernova, because detonation is triggered by the collapse of the star's iron-rich core. Hydrogen lines dominate a Type II supernova's spectrum simply because that gas is abundant in the star's outer layers being blasted into space. SN 1987A was a Type II supernova.

A Type I supernova begins with a carbon–oxygen-rich white dwarf in a close, semidetached binary system. As we saw in Chapter 12, mass transfer can occur in a close

a

b

Figure 14-12 *A supernova* *Sometime during 1940, a supernova was seen in this galaxy in the constellation of Coma Berenices. (a) The galaxy before the outburst. (b) By the time*

this photograph was taken in 1941, the supernova (indicated by arrow) had faded from its maximum brightness. (The Observatories of the Carnegie Institution of Washington)

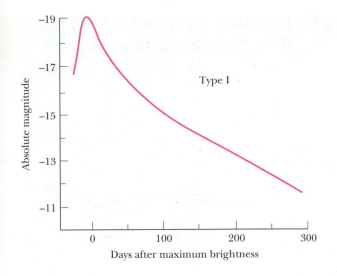

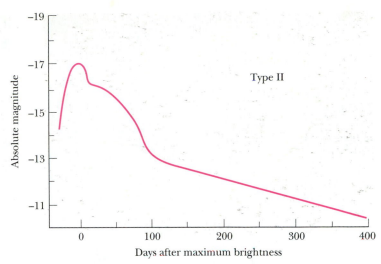

Figure 14-13 Supernova light curves A Type I supernova, which exhibits a gradual decline in brightness, is caused by an exploding white dwarf in a close binary system. A Type II super- *nova usually has alternating intervals of steep and gradual declines in brightness. Type II supernovae are caused by the explosive death of massive stars.*

binary if one star overflows its Roche lobe (recall Figures 12-20 and 12-21). To trigger a Type I supernova, a swollen red giant companion star dumps gas onto the white dwarf. When the white dwarf's mass gets close to the Chandrasekhar limit, carbon burning begins. Because the white dwarf is composed of degenerate matter, the usual "safety valve" involving pressure and temperature does not operate. In a catastrophic runaway process reminiscent of the helium flash, the rate of carbon burning skyrockets and the star blows up.

A Type I supernova is powered by nuclear energy, and the resulting spectacle in the sky is simply the fallout from a thermonuclear explosion, which produces a wide array of radioactive isotopes. Especially abundant is an unstable isotope of nickel that decays into a radioactive isotope of cobalt. The entire electromagnetic display of a Type I supernova, including the smooth decline of its light curve, directly results from the radioactive decay of nickel and cobalt. The fact that Type I and Type II supernovae reach nearly the same luminosity at maximum brightness is pure coincidence.

14-6 A supernova remnant is detectable at many wavelengths for many years after a supernova explosion

Astronomers find supernova remnants, the debris of supernova explosions, scattered across the sky. A beautiful example is the Veil Nebula, seen in Figure 14-14. The doomed star's outer layers were blasted into space with such violence that they are still traveling at supersonic speeds through the interstellar medium. As this expanding

Figure 14-14 The Veil Nebula This nebulosity is a portion of the Cygnus Loop (see Figure 13-13), which is the remnant of a supernova that exploded about 20,000 years ago. The distance to the nebula is about 1600 ly and the overall diameter of the loop is about 70 ly. (Palomar Observatory)

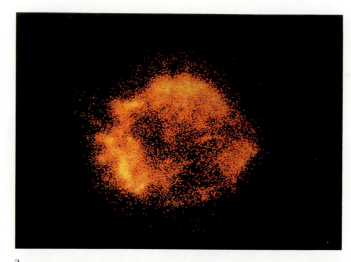

a

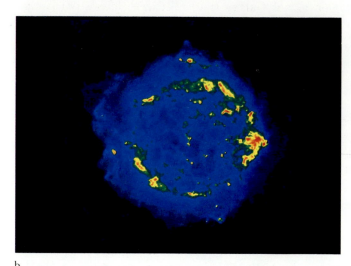

b

Figure 14-15 The Gum Nebula *The Gum Nebula is the largest known supernova remnant, spanning 60° of the sky, roughly centered on the southern constellation of Vela. The nearest portions of this expanding nebula are only 300 ly from the Earth. The supernova explosion occurred about 11,000 years ago, and its remnant now has a diameter of about 2300 ly. Only the central regions of the nebula are shown here. (Royal Observatory, Edinburgh)*

Figure 14-16 Cassiopeia A *Supernova remnants such as Cassiopeia A are typically strong sources of X rays and radio waves. (a) An X-ray picture of "Cas A" taken by the Einstein Observatory. (b) A corresponding radio image produced by the Very Large Array (VLA). The supernova explosion that produced this nebula occurred 300 years ago, about 10,000 ly from Earth. (Smithsonian Institution and the Very Large Array)*

shell of gas plows through the interstellar medium, it collides with interstellar atoms, which excites the gas and causes it to glow.

Many supernova remnants are quite large and cover sizable fractions of the sky. The largest is the Gum Nebula, with a diameter of 60° (Figure 14-15). This nebula looks so big because it is so close. Its near side is only about 300 ly from Earth. Studies of the nebula's expansion rate suggest that the supernova exploded around 9000 BC. It could have been witnessed by people then living in such places as Egypt and India. At maximum brilliancy, the exploding star probably reached a brightness equal to that of the Moon at first quarter.

Many supernova remnants are virtually invisible at optical wavelengths, but as the expanding gases collide with the interstellar medium, they radiate energy at a wide range of other wavelengths, from X rays through radio waves. For example, Figure 14-16 shows both X-ray and radio images of the supernova remnant Cassiopeia A. Optical photographs of this part of the sky reveal only a few small, faint wisps. Thus radio searches for supernova remnants are more fruitful than optical searches. Only two dozen supernova remnants have been found on photo-

graphic plates, but more than 100 remnants have been discovered by radio astronomers.

From the expansion rate of the nebulosity in Cassiopeia A, astronomers conclude that the supernova explosion occurred about 300 years ago. Although telescopes were in wide use by the late 1600s, no one saw the outburst. In fact, the last supernova seen in our Galaxy was observed by Johannes Kepler in 1604. In 1572 Tycho Brahe also recorded the sudden appearance of an exceptionally bright star in the sky. To find any other accounts of supernova explosions, we must delve into astronomical records that are almost a thousand years old.

Why have so few nearby supernovae been observed? Astronomers have seen more than 600 supernovae in distant galaxies. These observations suggest that in a typical galaxy like the Milky Way, Type I supernovae occur roughly once every 36 years, while Type II supernovae occur about once every 44 years. Thus it seems reasonable to suppose that about five supernovae occur in our Milky Way Galaxy each century. Where are they?

Vigorous stellar evolution in our Galaxy occurs primarily in the Galaxy's disk, where giant molecular clouds are located. Our Galaxy's disk is therefore the place where massive stars are born and supernovae explode. This region of our Galaxy is so filled with interstellar gas and dust, however, that we simply cannot see very far into space in directions occupied by the Milky Way. Supernovae probably do erupt every few years in remote parts of our Galaxy, but their detonations are hidden from our view by intervening interstellar debris.

14-7 Pulsars are rapidly rotating neutron stars with intense magnetic fields

The neutron was discovered during laboratory experiments in 1932. Within a year, two astronomers had predicted the existence of neutron stars. Inspired by the realization that white dwarfs are supported by degenerate-electron pressure, Fritz Zwicky at the California Institute of Technology and his colleague Walter Baade at Mount Wilson Observatory proposed that a highly compact ball of neutrons could similarly produce a powerful pressure. This **degenerate-neutron pressure**, like the degenerate-electron pressure, could also support a stellar corpse, perhaps even more massive than the Chandrasek-

har limit allows. "With all reserve," Zwicky and Baade theorized, "we advance the view that supernovae represent the transition from ordinary stars into neutron stars, which in their final stages consist of extremely closely packed neutrons." In other words, there could be at least two types of stellar corpses: white dwarfs and neutron stars.

This prophetic proposal was politely ignored by most scientists for years. After all, a neutron star must be a rather weird object. In order to transform protons and electrons into neutrons, the density in the star would have to be equal to nuclear density, about 10^{17} kg/m^3. Thus, a thimbleful of neutron star matter brought back to Earth would weigh 100 million tons. Furthermore, an object compacted to nuclear density would be very small. A 1-M$_\odot$ neutron star would have a diameter of only 30 km, about the same size as San Francisco or Manhattan. The surface gravity on one of these neutron stars would be so strong that the escape velocity would equal one-half the speed of light. All these conditions seemed so outrageous that few astronomers paid any serious attention to the subject of neutron stars—until 1968.

As a young graduate student at Cambridge University, Jocelyn Bell had spent many months assisting in the construction of an array of radio antennas covering $4\frac{1}{2}$ acres in the English countryside. By the fall of 1967 the instrument was completed, and Bell and her colleagues began detecting radio emissions from various celestial sources. In November, while scrutinizing data from the new telescope, Bell noticed that the antennas had detected regular "beeps" from one particular location in the sky. Careful repetition of the observations demonstrated that the radio pulses were arriving with a regular period of 1.3373011 seconds (Figure 14-17).

The regularity of this pulsating radio source was so striking that the Cambridge team suspected that they might be detecting signals from an advanced alien civilization. This possibility was soon discarded as several more of these pulsating radio sources, which soon came to be know as **pulsars,** were discovered across the sky. In all

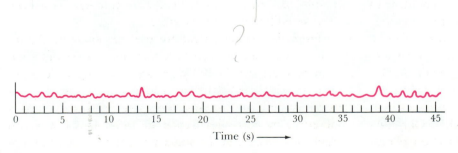

Figure 14-17 A recording of the first pulsar *This chart recording shows the intensity of radio emission from the first pulsar, discovered by British astronomers in 1967. Note that some pulses are weak and others are strong. Nevertheless, the spacing between the pulses is exactly 1.3373011 seconds. (Adapted from Antony Hewish)*

cases, the periods were extremely regular, from about 0.2 second for the fastest to about 1.5 seconds for the slowest.

When the discovery of pulsars was officially announced in early 1968, astronomers around the world began proposing all sorts of explanations for these regular radio pulsations. Many of these theories were bizarre, and arguments raged for months. However, by late 1968, all controversy was laid to rest with the discovery of a pulsar in the middle of the Crab Nebula.

In AD 1054 Chinese astronomers recorded the appearance of a supernova (they called it a "guest star") in the constellation of Taurus. When we turn a telescope toward this location, we find the Crab Nebula shown in Figure 14-18. This object looks like an exploded star and is a supernova remnant. The pulsar at the center of the Crab Nebula is called the Crab pulsar.

The fact that a pulsar is located in a supernova remnant tells us that pulsars are probably associated with dead stars. Before the discovery of pulsars, most astronomers believed all stellar corpses to be white dwarfs. There seemed to be a sufficient number of white dwarfs in the sky to account for all the stars that have died since our Galaxy was formed. It was thus generally assumed that all dying stars somehow manage to eject enough matter so that their corpses do not exceed the Chandrasekhar limit. The discovery of the Crab pulsar showed that these conservative ideas were wrong.

The Crab pulsar is one of the fastest pulsars ever discovered. Its period is 0.033 second, which means that it beeps 30 times each second. Calculations demonstrated that white dwarfs are too big and bulky to produce 30 signals per second. The existence of the Crab pulsar clearly indicates that the stellar corpse at the center of the Crab Nebula is much smaller and more compact than a white dwarf. Astronomers realized that they would have to face the prospect of neutron stars seriously.

What would a neutron star be like? As we have seen, a neutron star would be small and dense. It should also be rotating rapidly. All stars rotate, but most of them do so leisurely. For example, our Sun takes nearly a full month to rotate once about its axis. A collapsing star speeds up as its size shrinks, just as an ice skater doing a pirouette speeds up when she pulls in her arms. This phenomenon is a direct consequence of a law of physics called the **conservation of angular momentum,** which holds that the total amount of angular momentum in a system remains constant. An ordinary star rotating once a month would be spinning faster than once a second if compressed to the size of a neutron star.

In addition to having rapid rotation, we expect a neutron star to have an intense magnetic field. It is probably safe to say that every star has a magnetic field of some strength. In an average star, like our Sun, the strength is typically quite low because the magnetic field is spread

Figure 14-18 The Crab Nebula *This beautiful nebula, named for the armlike appearance of its filamentary structure, is the remnant of a supernova seen in AD 1054. The distance to the nebula is about 6000 ly, and its present angular size (4 by 6 arc min) corresponds to linear dimensions of about 7 by 10 ly. (Palomar Observatory)*

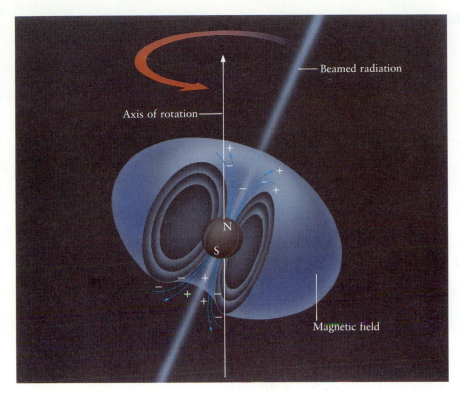

Axis of rotation

Beamed radiation

N
S

Magnetic field

Figure 14-19 A rotating, magnetized neutron star It is reasonable to suppose that a neutron star rotates rapidly and possesses a powerful magnetic field. Charged particles are accelerated near the star's magnetic poles and produce two oppositely directed beams of radiation. As the star rotates, the beams sweep around the sky. If the Earth happens to lie in the path of the beams, we see a pulsar.

out over millions upon millions of square kilometers of the star's surface. However, if a star of solar dimensions collapses down to a neutron star, its magnetic field becomes very concentrated, and its strength increases by a factor of a billion.

Finally, we would expect the axis of rotation of a typical neutron star to be inclined at some angle to the magnetic axis connecting the north and south magnetic poles (Figure 14-19), much in the same way that the Earth's magnetic and rotation axes are inclined to each other. The combination of a powerful magnetic field and rapid rotation operates like a giant electric generator to create intense electric fields near the star's surface. At its surface, there are plenty of protons and electrons, because the pressures there are too low to combine them into neutrons. The powerful electric fields acting on these charged particles cause them to flow out from the neutron star's polar regions along the curved magnetic field, as sketched in Figure 14-19. As the particles stream along the curved field, they are accelerated and emit energy. The result is two very thin beams of radiation pouring out of the neutron star's north and south magnetic polar regions.

A rotating, magnetized neutron star is somewhat like a lighthouse beacon. As the star rotates, its beams of radiation sweep around the sky. If the Earth happens to be located in the right direction, a brief flash can be observed each time the beam sweeps past our line of sight. This explanation for pulsars is often called the lighthouse model. Indeed, one of the stars at the center of the Crab Nebula is actually flashing on and off 30 times each second (Figure 14-20). Also visibly flashing is the Vela pulsar at the core of the Gum Nebula (see Figure 14-15). The Vela pulsar, with a period of 0.089 second, is the slowest pulsar ever detected at visible wavelengths.

The Crab pulsar is one of the youngest pulsars, its creation having been observed some 900 years ago. The Vela pulsar is also quite young: it was created about 11,000 years ago. Pulsars slow down as they get older. Only the very youngest pulsars are energetic enough to emit optical flashes along with their radio pulses. During the 1990s, astronomers will observe SN 1987A with the hope of seeing though the supernova's ejected gases to search for pulses that would indicate a rapidly spinning neutron star.

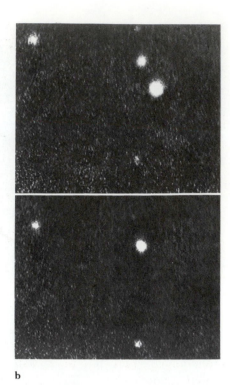

a

b

Figure 14-20 *The Crab pulsar* *A pulsar is located at the center of the Crab Nebula.* (**a**) *The Crab pulsar is identified by the arrow.* (**b**) *These two photographs show the Crab pulsar in its on and off states. Like the radio pulses, these visual flashes have a period of 0.033 second. (Lick Observatory)*

14-8 Pulsating X-ray sources are neutron stars in close binary systems

During the 1960s astronomers obtained tantalizing X-ray views of the sky during short rocket and balloon flights that briefly lifted X-ray detectors above the Earth's atmosphere. A number of strong X-ray sources were discovered, and each was then named after the constellation in which it is located. For example, Scorpius X-1 is the first X-ray source found in the constellation of Scorpius.

Astronomers were so intrigued by these preliminary discoveries that they built and launched *Explorer 42*, an X-ray-detecting satellite that could make observations 24 hours a day (Figure 14-21). The satellite was launched in 1970 from Kenya to place it in an equatorial orbit. In recognition of the hospitality of the Kenyan people, *Explorer 42* was renamed *Uhuru*, which means "freedom" in Swahili.

Uhuru gave us our first comprehensive look at the X-ray sky. As the satellite slowly rotated, its X-ray detectors swept across the heavens. Each time an X-ray source came into view, signals were transmitted to receiving stations on the ground. Before its battery and transmitter failed in early 1973, *Uhuru* had succeeded in locating 339 X-ray sources.

The discovery of pulsars was still fresh in everyone's mind when the *Uhuru* team discovered X-ray pulses coming from Centaurus X-3 in early 1971. Figure 14-22 shows data from one sweep of *Uhuru*'s detectors across Centaurus X-3. The pulses have a regular period of 4.84 seconds. A few months later similar pulses were discovered coming from a source called Hercules X-1, which has a period of 1.24 seconds. Because the periods of these two X-ray sources are so short, astronomers began to suspect that they had found rapidly rotating neutron stars.

It soon became clear, however, that systems such as Centaurus X-3 and Hercules X-1 are not ordinary pulsars like the Crab or Vela pulsars. Centaurus X-3 turns on and off periodically. Every 2.087 days, it turns off for almost 12 hours. This fact suggests that Centaurus X-3 is an eclipsing binary and that the X-ray source takes nearly 12 hours to pass behind its companion star.

Figure 14-21 **Uhuru (Explorer 42)** Uhuru *was a small satellite designed to detect astronomical sources of X rays. During three years of flawless operation, it observed more than 300 different X-ray-emitting objects across the sky. (NASA)*

The case for the binary nature of Hercules X-1 is even more compelling. It has an off state corresponding to a 6-hour eclipse every 1.7 days, and careful timing of the X-ray pulses shows a periodic Doppler shifting every 1.7 days. This information is direct evidence of orbital motion about a companion star: When the X-ray source is approaching us, its pulses are separated by slightly less than 1.24 seconds. When the source is receding from us, slightly more than 1.24 seconds elapse between the pulses.

Careful optical searches around the location of Hercules X-1 soon revealed a dim star named HZ Herculis. The apparent magnitude of this star varies between +13 and +15, with a period of 1.7 days. Because this period is exactly the same as the orbital period of the X-ray source, astronomers conclude that HZ Herculis is the companion star around which Hercules X-1 orbits.

Putting all the pieces together, astronomers now realize that systems such as Centaurus X-3 and Hercules X-1 are examples of double stars in which one is a neutron star. All these binaries have very short orbital periods. Conse-

quently, the distance between the ordinary star and its neutron star must be very small. This proximity enables the neutron star to capture gas escaping from the ordinary companion star.

To explain pulsating X-ray sources such as Centaurus X-3 or Hercules X-1, astronomers assume that the ordinary star either fills or nearly fills its Roche lobe. Either way, matter escapes from the star. This mass loss results from direct "Roche-lobe overflow" if the star fills its lobe, as in the case of Hercules X-1, or from a stellar wind if the star's surface lies just inside its lobe, as with Centaurus X-3 (Figure 14-23). A typical rate of mass loss from the ordinary star is roughly 10^{-9} $M_\odot$ per year.

The neutron star in a pulsating X-ray source, like an ordinary pulsar, rotates rapidly and has a powerful magnetic field inclined to the axis of rotation (recall Figure 14-19). Because of its strong gravity, a neutron star easily captures much of the gas escaping from its companion star. As the gas falls toward the neutron star, the magnetic field funnels the incoming matter down onto its

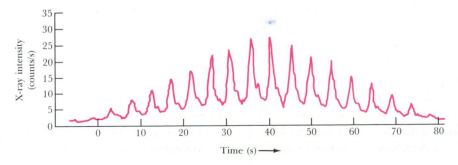

Figure 14-22 **X-ray pulses from Centaurus X-3** *This graph shows the intensity of X rays detected by* Uhuru *as Centaurus X-3 moved across the satellite's field of view. The successive pulses are separated by 4.84 seconds. The gradual variation in the height of the pulses from left to right was a result of the changing orientation of* Uhuru's *X-ray detectors toward the source as the satellite rotated. (Adapted from R. Giacconi and colleagues)*

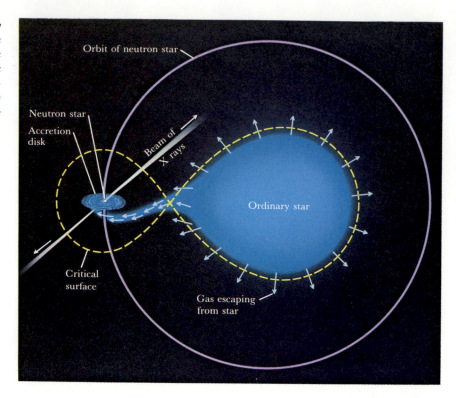

Figure 14-23 A model of a pulsating X-ray source Gas escaping from an ordinary star is captured by the neutron star. The in-falling gas is funneled down onto the neutron star's magnetic poles, where it strikes the star with enough energy to create two X-ray-emitting hot spots. As the neutron star spins, beams of X rays from the hot spots sweep around the sky.

north and south magnetic polar regions. The star's gravity is so strong that the gas is traveling at nearly half the speed of light by the time it crashes onto the star's surface. Because this violent impact creates hot spots at both poles with temperatures of about 10^8 K, these hot spots emit abundant X rays with a luminosity nearly 100,000 times brighter than the Sun. As the neutron star rotates, the beams of X rays from the polar caps sweep around the sky. If the Earth happens to be in the path of one of the two beams, we can observe a pulsating X-ray source. The pulse period is thus equal to the neutron star's rotation period. For example, the neutron star in Hercules X-1 is spinning at the rate of once every 1.24 seconds.

In a binary system, gas captured by a neutron star's gravity may go into orbit about the neutron star, as shown in Figure 14-23. The result is a rotating disk of material called an accretion disk. Accretion disks have been detected in many close binary systems where mass transfer is occurring (recall Figure 12-21*b*). Under certain circumstances, some bizarre things can happen.

With ordinary pulsating X-ray sources, like Hercules X-1, the rate at which gas falls onto the neutron star from the inner edge of the accretion disk is low enough to allow the resulting X rays to escape. If the companion star is dumping vast amounts of material onto the neutron star, however, the resulting energy cannot escape easily. Instead, tremendous pressures build up in the gases crowd-

ing down onto the neutron star. These pressures meet with strong resistance in the plane of the accretion disk, where newly arrived gases are constantly spiraling in toward the neutron star. The path of least resistance lies along the rotation axis of the accretion disk. Consequently, pressure around the neutron star is relieved by squirting gases outward in a perpendicular direction to the accretion disk. The result can be two powerful beams of high-velocity hot gases. This is apparently the explanation for the weird star SS433.

In the autumn of 1978, Bruce Margon and his colleagues at UCLA were observing SS433, which had been noted for strong emission lines in its spectrum. To everyone's surprise, the spectrum of SS433 contained several complete sets of spectral lines. One set was very redshifted from its usual wavelengths, while another set was comparably blueshifted. Somehow, SS433 is "coming and going at the same time." To make matters even more puzzling, the wavelengths of these redshifted and blueshifted lines change dramatically from one night to the next.

Astronomers had never seen anything like this, and soon many were observing SS433. By mid-1979, it was clear that the system's redshifted and blueshifted lines are actually moving back and forth across the spectrum of SS433 with a period of 164 days. Astrophysicists were quick to point out that the two sets of spectral lines could

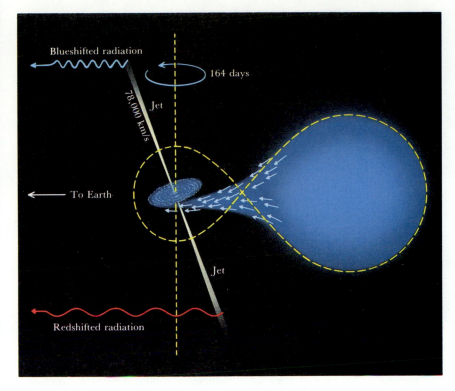

Figure 14-24 A model of SS433 Gas from a normal star is captured into an accretion disk about a neutron star. Two high-speed, oppositely directed jets of gas are ejected from the faces of the disk. Because the disk is tilted, the gravitational pull of the normal star causes the jets to precess with a period of 164 days.

be caused by two oppositely directed jets of gas, one tilted toward us and the other away from us. Furthermore, the 164-day variation could be explained by a precession, or "wobble," of the accretion disk and its two jets. As the two jets circle about the sky every 164 days, we see a periodic variation in the Doppler shift.

All these features come together in the model sketched in Figure 14-24. To explain the large redshifts and blueshifts discovered by Margon, gas in the two oppositely directed jets must have a speed of 78,000 km/s, roughly one-quarter the speed of light. In addition, the accretion disk must be tilted with respect to the orbital plane of the two stars of the binary system. Just as the tilt of the Earth's axis with respect to the plane of the ecliptic causes the Earth to precess, the tilt of the accretion disk results in the 164-day precession of the two jets.

Figure 14-25 shows four high-resolution radio views of SS433. Note the two oppositely directed appendages emerging from the central source. As we shall see in Chapter 18, many quasars and peculiar galaxies have a similar radio structure, though on a much larger scale. Because quasars are very far away, they are difficult to study. The real significance of SS433 may be that it gives us a miniature quasarlike object in our own celestial backyard.

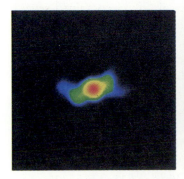
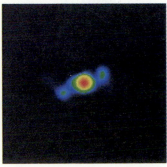
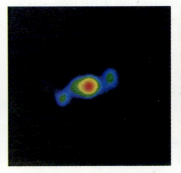
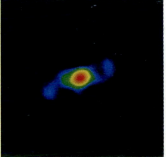

Figure 14-25 Four views of SS433 These four radio views, taken in early 1981, show jets of gas extending out to one-sixth of a light year on either side of SS433. Three-quarters of the radio emission comes from SS433 itself (the red central blob), which is located at the center of a supernova remnant 13,000 ly from Earth. (NRAO; Very Large Array)

a   b

Figure 14-26 Nova Herculis 1934 *These two pictures show a nova (**a**) shortly after peak brightness as a magnitude +3 star and (**b**) two months later, when it had faded to magnitude +12.*

Novae are named after the constellation and year in which they appeared. (Lick Observatory)

14-9 Explosive thermonuclear processes on white dwarfs and neutron stars produce novae and bursters

Low-mass stars are far more common than high-mass stars, and so white dwarfs are far more common than neutron stars. With all the bizarre and fascinating phenomena associated with neutron stars, you might be wondering if white dwarfs do anything more dramatic than simply cool off. The answer is definitely yes.

Occasionally a star in the sky suddenly brightens by a factor of between 10^4 and 10^6. This phenomenon is called a **nova** (not to be confused with a supernova, which involves a much greater increase in brightness). Novae are fairly common. Their abrupt rise in brightness is followed by a gradual decline that may stretch for several months or more (Figures 14-26 and 14-27).

Painstaking observations of many novae strongly suggest that all novae are members of close binary systems containing a white dwarf. Gradual mass transfer from the ordinary companion star (which presumably fills its Roche lobe) deposits fresh hydrogen onto the white dwarf. Because of the strong gravity, this hydrogen becomes compacted into a dense layer covering the hot sur-

face of the white dwarf. As more gas is deposited and compressed, the temperature in the hydrogen layer increases. Finally, when the temperature reaches about 10^7 K, hydrogen burning ignites throughout the layer, embroiling the white dwarf's surface in a thermonuclear holocaust that we see as a nova.

A similar phenomenon occurs with neutron stars. Beginning in late 1975, astronomers analyzing data from X-ray satellites realized that their instruments had detected sudden, powerful bursts of X rays from certain

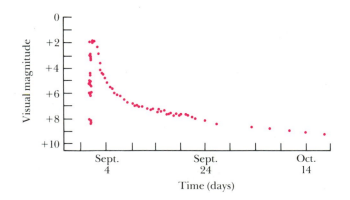

Figure 14-27 The light curve of Nova Cygni 1975 *This graph shows the history of a nova that blazed forth in the constellation of Cygnus in September 1975. The rapid rise in magnitude followed by a gradual decline is characteristic of all novae.*

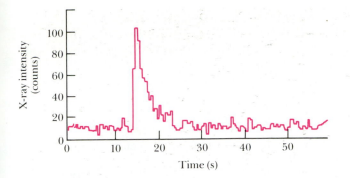

Figure 14-28 *X rays from a burster* *A burster emits X rays with a constant low intensity interspersed with occasional powerful bursts. This particular burst is typical. It was recorded on September 28, 1975, by an X-ray telescope on board the Astronomical Netherlands Satellite while the telescope was pointed toward the globular cluster NGC 6624. About one-third of all known bursters are located in globular clusters. (Adapted from Walter Lewin)*

objects in the sky. The record of a typical burst is shown in Figure 14-28. The source emits X rays at a constant low level, then suddenly, without warning, there is an abrupt increase, followed by a gradual decline. Typically an entire burst lasts for only 20 seconds. Sources that behave in this fashion are called **bursters**. Several dozen of them have been discovered, most located toward the center of our Galaxy.

Bursters, like novae, are believed to involve close binaries experiencing mass transfer. With a burster, however, the stellar corpse is a neutron star rather than a white dwarf. Gases escaping from the ordinary companion star fall onto the neutron star. The energy released as this gas crashes down onto the neutron star's surface produces the low-level X rays that are continuously emitted by the burster.

Most of the gas falling onto the neutron star is hydrogen, which becomes compressed against the hot surface of the star by the star's powerful surface gravity. In fact, temperatures and pressures in this accreting layer are so high that the arriving hydrogen is promptly converted into helium by the hydrogen-burning process. Constant hydrogen burning soon produces a layer of helium that covers the entire neutron star.

Finally, when the helium layer is about 1 m thick, helium burning ignites explosively, and we observe a sudden burst of X rays. In other words, whereas explosive hydrogen burning on a white dwarf produces a nova, explosive helium burning on a neutron star produces a burster. In both cases, the burning is explosive because the fuel is so strongly compressed against the star's surface that it is

degenerate, like the star itself. As we saw with the helium flash inside red giants, ignition of a degenerate thermonuclear fuel always involves a sudden thermal runaway because the usual "safety valve" between temperature and pressure is not operating.

In life as well as in death, the mass of a star determines its fate (Figure 14-29). Just as there is an upper limit to the mass of a white dwarf, there is also an upper limit to the mass of a neutron star. Above this limit, degenerate-neutron pressure cannot support the overpowering weight of the star's matter pressing inward from all sides. The Chandrasekhar limit for a white dwarf is 1.4 $M_\odot$, and the corresponding upper limit for a neutron star is about 3 $M_\odot$. What might happen if a dying massive star failed to eject enough matter to get below the upper limit for a neutron star? What, for example, might a 5-$M_\odot$ stellar corpse be like?

The gravity associated with a neutron star is so strong that the escape velocity is roughly one-half the speed of light. With a stellar corpse greater than 3 $M_\odot$, there is so much matter crushed into such a small volume that the escape velocity exceeds the speed of light. Because nothing can travel faster than light, nothing—not even light—can leave the dead star. The star therefore disappears from the universe, its powerful gravity leaving a hole in the fabric of space and time. The discovery of neutron stars thus inspired astrophysicists to examine seriously one of the most bizarre and fantastic objects ever predicted by modern science—the black hole.

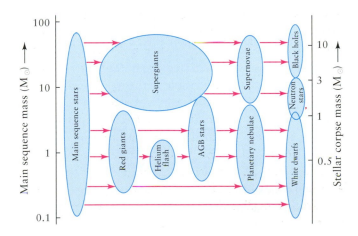

Figure 14-29 *A summary of stellar evolution* *The evolution of an isolated star depends on its mass. The scale at the left indicates the mass of a star when it is on the main sequence. The scale at the right gives the mass of the resulting stellar corpse. Stars less massive than about 8 $M_\odot$ can eject enough mass to become white dwarfs. High-mass stars can produce Type II supernovae and become neutron stars or black holes.*

Summary

- A low-mass star becomes a red giant when shell hydrogen burning begins. It becomes a horizontal branch star when core helium burning begins, and it becomes a red supergiant when the helium in its core is exhausted and shell helium burning begins.

    Thermal pulses in the helium-burning shell can eject the star's outer layers.

    Ultraviolet radiation from the hot carbon–oxygen core ionizes and excites the ejected gases, producing a planetary nebula.

- The burned-out core of a low-mass star becomes a dense sphere about the size of the Earth called a white dwarf.

    The maximum mass of a white dwarf (the Chandrasekhar limit) is 1.4 $M_\odot$.

- After exhausting its central supply of hydrogen and helium, a high-mass star undergoes a sequence of thermonuclear reactions in its core. These are carbon burning, neon burning, oxygen burning, and silicon burning. The star eventually develops an iron-rich core.

- A high-mass star can die in a supernova explosion that ejects most of the star's matter into space at very high speeds. This so-called Type II supernova is triggered by the gravitational collapse of the doomed star's core.

    If the core of the star survives the explosion, it may become a neutron star or even a black hole.

    Neutrinos were detected from the supernova SN 1987A, which was visible to the naked eye.

- An accreting white dwarf in a close binary system can also become a supernova when carbon burning ignites explosively throughout such a degenerate star. Such a detonation is called a Type I supernova.

- A neutron star is a very dense stellar corpse consisting of closely packed neutrons in a sphere roughly 30 km in diameter.

- A pulsar is a source of periodic radio pulses. It is caused by a rapidly rotating neutron star with a powerful magnetic field.

    Energy pours out of the north and south polar regions of the neutron star in intense beams that sweep around the sky.

- Some X-ray sources exhibit regular pulses. These objects are thought to be neutron stars in close binary systems with ordinary stars.

    Material from the ordinary star in a binary pair can fall onto the surface of its companion white dwarf or neutron star to produce a surface layer in which thermonuclear reactions can occur.

    Explosive hydrogen burning may occur in the surface layer of a companion white dwarf, producing the sudden increase in luminosity that we call a nova.

    Explosive helium burning may occur in the surface layer of a companion neutron star, producing the sudden increase in X-ray radiation called a burster.

Review questions

1 What is the difference between a red giant and a red supergiant?

2 Why is the temperature in a star's core so important in determining which nuclear reactions can occur there?

3 How is a planetary nebula formed?

4 What is a white dwarf?

5 What is the significance of the Chandrasekhar limit?

6 What is a neutron star?

7 Compare a white dwarf and a neutron star. Which of the two types of stellar corpses is most common?

8 On an H–R diagram, sketch the evolutionary track that the Sun will follow as it leaves the main sequence and becomes a white dwarf. Approximately how much mass will the Sun have when it becomes a white dwarf? Where will the rest of the mass have gone?

9 Why do you suppose that all the white dwarfs known to astronomers are relatively close to the Sun?

10 Why have radio searches for supernovae remnants been more fruitful than optical searches?

11 Why do astronomers believe that pulsars are rapidly rotating neutron stars?

12 What is the difference between Type I and Type II supernovae?

13 Compare a nova with a Type I supernova. What do they have in common? How are they different?

14 Compare a nova and a burster. What do they have in common? How are they different?

15 What is SS433?

16 Describe what radio pulsars, X-ray pulsars, and bursters have in common. How are they different manifestations of the same type of astronomical object?

Advanced questions

17 What prevents thermonuclear reactions from occurring at the center of a white dwarf? If no thermonuclear reactions are occurring in its core, why doesn't the star collapse?

18 Suppose you wanted to determine the age of a planetary nebula. What observations would you make, and how would you use the resulting data?

19 What reasons can you think of to explain why the rate of expansion of the gas shells in some planetary nebulae is not uniform?

20 What kinds of stars would you monitor if you wished to observe a supernova explosion from its very beginning? Look up tabulated lists of the brightest and nearest stars. Which, if any, of these stars are possible supernova candidates? Explain.

21 To determine accurately the period of a pulsar, astronomers must take the Earth's orbital motion about the Sun into account. Explain why.

*__22__ The distance to the Crab Nebula is about 2000 pc. When did it actually explode?

Discussion questions

23 Suppose that you discover a small glowing disk of light while searching the sky with a telescope. How would you observationally decide if this object is a planetary nebula? Could your object be something else? Explain.

24 During the weeks immediately following the discovery of the first pulsar, one explanation for them was that the pulses are signals from an extraterrestrial civilization. Why do you suppose astronomers discarded this idea?

For further reading

Bethe, H., and Brown, G. "How a Supernova Explodes." *Scientific American*, May 1985 • This enlightening article describes many fascinating details about the detonation of a Type II supernova.

Clark, D. *Superstars*. McGraw-Hill, 1984 • This layperson's introduction to supernovae blends history and modern science in nontechnical language.

Greenstein, G. *Frozen Star*. Freundlich Books, 1984 • This eloquent book skillfully blends a discussion of pulsars, neutron stars, and black holes with an accurate portrait of astronomical research as a human endeavor.

Kwok, S. "Not with a Bang But a Whimper." *Sky & Telescope*, May 1982 • This brief article discusses intriguing aspects of planetary nebulae and the role they play in the deaths of low-mass stars.

Lattimer, J., and Burrows, A. "Neutrinos from Supernova 1987A." *Sky & Telescope*, October 1988 • This article explains how a Type II supernova explosion produces a powerful burst of neutrinos and discusses the observations of such a burst from SN 1987A.

Margon, B. "The Bizarre Spectrum of SS433." *Scientific American*, October 1980 • The astronomer who discovered the unusual properties of SS433 discusses the star's spectrum and its interpretation.

Marschall, L. *The Supernova Story*. Plenum, 1988 • This superbly written book is an excellent choice for someone who wishes to learn more about supernovae, including SN 1987A.

Marschall, L., and Brecher, K. "Will Supernova 1987A Shine Again?" *Astronomy*, February 1992 • This beautifully illustrated article describes the exciting prediction that SN 1987A will brighten when ejecta from the supernova strike a shell of gas that surrounds the exploded star.

White, N. "New Wave Pulsars." *Sky & Telescope*, January 1987 • This article describes the discovery of "millisecond pulsars," which are neutron stars spinning 100 to 1000 times per second. These pulsars are believed to be members of close binary systems in which accretion onto a neutron star has sped up its rotation rate.

Woosley, S., and Weaver, T. "The Great Supernova of 1987." *Scientific American*, August 1989 • Two preeminent authorities on supernovae teamed up to write this superb article, which describes many fascinating details of SN 1987A.

15

A black hole in a double star system *This artist's rendition shows the close binary system that contains Cygnus X-1. Cygnus X-1 is a strong source of X rays and is widely believed by astronomers to be a black hole. Gas from the companion star is captured into orbit about the black hole, forming an accretion disk. As gases spiral in toward the black hole, they are heated to high temperatures. At the inner edge of the accretion disk, the gases are so hot that they emit X rays. (Courtesy of D. Norton, Science Graphics)*

Black Holes

A dying high-mass star can give rise to a stellar corpse so massive that it is doomed to collapse to a single point of infinite density. Such an object, called a black hole, is predicted by Einstein's general theory of relativity. A black hole is a strange but simple object. It is a place of inconceivably intense gravity from which nothing—not even light—can escape. Matter that falls into a black hole literally disappears forever from the universe. Nevertheless, a black hole is an uncomplicated object because its structure is completely specified by only three quantities: its mass, electric charge, and angular momentum. In recent years astronomers have found evidence that certain binary systems may contain black holes. Each of these binaries is a powerful source of X rays presumably produced by hot gases in an accretion disk surrounding the black hole.

Suppose that the mass of a dying star's burned-out core exceeds three solar masses (3 $M_\odot$). This mass is well above the Chandrasekhar limit, so degenerate-electron pressure cannot support the stellar corpse. Similarly, because 3 $M_\odot$ is also above the mass limit for neutron stars, degenerate-neutron pressure also cannot support the crushing weight of the burned-out matter pressing inexorably inward toward the dead star's center. If it can become neither a white dwarf nor a neutron star, what might this massive stellar corpse become?

As we saw in the previous chapter, a typical neutron star consists of roughly 1 $M_\odot$ of matter compressed by its own gravity to nuclear density in a sphere roughly 30 km in diameter. The star's gravity is so strong that the escape speed from its surface is one-half the speed of light.

A massive stellar corpse, whose weight overpowers degenerate-neutron pressure, easily compresses its matter to densities greater than nuclear density. It does not take much further compression to cause the escape velocity to exceed the speed of light. For example, if 3 $M_\odot$ of matter is squeezed inside a sphere 18 km in diameter, the escape velocity from the object becomes greater than the speed of light. Nothing can travel faster than the speed of light, and thus nothing—not even light—can manage to escape from the dead star. The star has disappeared from the observable universe, although some of its effects can be detected.

15-1 The general theory of relativity describes gravity in terms of the geometry of space and time

To appreciate fully the nature of massive stellar corpses, we must use the best theory of gravity at our disposal. The gravitational field around one of these massive dead stars is so strong that Isaac Newton's theories fail to describe it accurately. We must turn instead to Albert Einstein's general theory of relativity.

According to the classical physics of Newton, space is perfectly uniform and fills the universe like a rigid framework. Time passes at a monotonous, unchanging rate. It is always possible to measure how fast you are moving through this rigid fabric of space and time, and to calculate how your observations depend on your state of motion. Significant differences between observations made by a stationary person and by a moving person arise at speeds near the speed of light.

In 1905 Albert Einstein began a revolution in physics with his **special theory of relativity.** His goal was to reformulate electromagnetic theory so that it did not depend on the motion of an observer. In other words, using the special theory of relativity, both you on Earth and a friend in a rocketship traveling near the speed of light would have the same logical, complete description of electricity and magnetism, devoid of any pitfalls or paradoxes caused by your relative motions.

Einstein was guided in his thinking by one lofty idea: that neither our location in space and time nor our motion through space and time should prejudice our description of physical reality. This meant that Einstein had to abandon traditional, rigid notions of space and time. For example, imagine your friend whizzing across the solar system in a rocketship while you remain here at rest on the Earth. Einstein proved that in order for both of you to agree on the same coherent description of reality, you must say that your friend's clocks have slowed down and her rulers have shrunk. Specifically, an observer always finds that moving clocks seem to be slowed and moving rulers seem to be shortened in the direction of motion.

After developing the special theory of relativity, Einstein turned his attention to gravity. He began by demonstrating that it is not necessary to think of gravity as a force. According to Newton's theory, an apple falls to the ground because the force of gravity pulls the apple down. Einstein pointed out that the apple would behave exactly the same in free space far from any gravity if the floor were accelerating upward to meet the apple, as sketched in Figure 15-1.

This example of Einstein's **principle of equivalence** explains that, in a small volume of space, the downward pull of gravity can be accurately and completely duplicated by an upward acceleration of the observer. The two people in their closed compartments in Figure 15-1 have no way of telling who is at rest on the Earth and who is in the elevator moving upward at a constantly increasing speed.

This approach allowed Einstein to focus entirely on motion, rather than force, in discussing gravity. From his special theory of relativity, he knew exactly how rulers and clocks are affected by motion, and he could thus describe gravity entirely in terms of its effects on space and time. Far from a source of gravity, the acceleration is small and the effect on clocks and rulers is therefore small; nearer a source of gravity, the acceleration is greater and the distortion of clocks and rulers is greater. In this way, Einstein "generalized" his special theory to arrive at his general theory of relativity.

Figure 15-1 The equivalence principle The equivalence principle asserts that you cannot distinguish between being at rest in a gravitational field and being accelerated upward in a gravity-free environment. This idea was an important step in Einstein's development of the general theory of relativity.

The general theory of relativity describes gravity entirely in terms of the geometry of space and time. Far from a source of gravity, space is flat and clocks tick at their normal rate. Near a source of gravity, space is curved and clocks slow down. The stronger the gravity, the greater are these distortions of the shape of space and the flow of time.

In a weak gravitational field, Einstein's general relativity theory gives the same results as Newton's classical theory. But in stronger gravity, such as that near the Sun's surface, the two theories predict different results. For examples, recall the precession of Mercury's perihelion and the deflection of a light ray grazing the Sun, discussed in Chapter 3 (recall Figures 3-16 and 3-18). In these and other situations, general relativity has withstood many tests. However, since all these tests have been performed in gravitational fields much weaker than that near a black hole, general relativity may not be an accurate theory for describing very strong gravity. Nevertheless, since general relativity is by far the most elegant and accurate description of gravity ever devised, scientists have used it to predict the many extraordinary properties of black holes.

15-2 A black hole is a very simple object that has only a "center" and a "surface"

Imagine a dying star too massive to become either a white dwarf or a neutron star. The overpowering weight of the star's burned-out matter pressing inward from all sides causes the star to contract rapidly. The strength of gravity around the collapsing star increases dramatically as the star's matter is compressed to enormous densities inside the rapidly shrinking sphere. According to the general theory of relativity, distortions of space and time become increasingly pronounced around the dying star. Finally, the escape velocity from the star's surface equals the speed of light, and thus the star disappears from the universe. At this stage, space has become so severely curved that a hole is punched in the fabric of the universe. The dying star disappears into this hole in space, leaving behind only a black hole.

The geometry of space around a black hole is sketched in Figure 15-2. Note that space is flat far from the hole

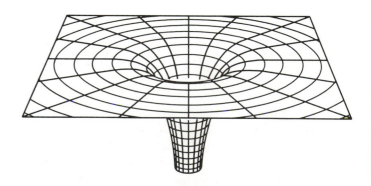

Figure 15-2   The geometry of a black hole This diagram shows how the shape of space is distorted by the gravitational field of a black hole. Far from the hole, gravity is weak and space is flat. Near the hole, gravity is strong and space is highly curved.

because gravity is weak there. Near the hole, however, gravity is strong, and space is severely curved.

The location in space where the escape velocity from the black hole equals the speed of light is called the **event horizon.** This sphere is sometimes thought of as the "surface" of the black hole. Once a massive dying star collapses inside its event horizon, it permanently disappears from the universe. The term *event horizon* is in fact quite appropriate, because this surface is literally a horizon in the geometry of space beyond which we cannot see any events.

In addition to making space curve, gravity causes time to slow down. If you stood at a safe distance and watched your friend fall toward a black hole, you would note that the friend's clocks tick more and more slowly. In fact, you would conclude that at the event horizon the clocks stop entirely.

Once a dying star has contracted inside its event horizon, no forces in the universe can prevent the complete collapse of the star down to a single point at the center of the black hole. The star's entire mass is crushed to infinite density at this point, called the **singularity.** The singularity corresponds to the "center" of a black hole.

The structure of a black hole is therefore very simple. As sketched in Figure 15-3, it has only two parts: a singularity (the center) surrounded by an event horizon (the surface). The distance between the singularity and the event horizon is the **Schwarzschild radius,** named after the German astronomer Karl Schwarzschild, who solved Einstein's equations of general relativity in 1916 in a way that revealed the properties of a black hole.

To understand why the complete collapse of such a doomed star is inevitable, think about your own life here on Earth, far from any black holes. You have the freedom to move as you wish through the three dimensions of space: up and down, left and right, or forward and back. But you do not have the freedom to move at will through the dimension of time. Whether we like it or not, we are all carried inexorably from the cradle to the grave.

Inside a black hole, powerful gravity distorts the shape of space and time so severely that the directions of space and time become interchanged. In a limited sense, inside a black hole you can have the freedom to move through time. This seeming gain does you no good, however, because you lose a corresponding amount of freedom to move through space. Whether you like it or not, you will be dragged inexorably from the event horizon toward the singularity. Just as no force in the universe can prevent the forward march of time from past to future outside a black hole, no force in the universe can prevent the inward

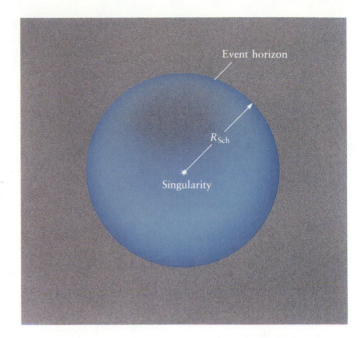

Figure 15-3   The structure of a black hole *A black hole has only two parts: a singularity surrounded by an event horizon. The distance between the singularity and the event horizon is called the Schwarzschild radius, denoted R_{Sch}. Inside the event horizon, the escape velocity exceeds the speed of light, so the event horizon is a one-way surface. Things can fall in, but nothing can get out.*

march of space from event horizon to singularity inside a black hole.

At the singularity, the strength of gravity is infinite, and so the curvature of space and time is infinite. In other words, space and time at the singularity are all jumbled up. Space and time do not exist there as separate, identifiable entities.

This confusion of space and time has profound implications for what goes on inside a black hole. All the laws of physics require a clear, distinct background of space and time. Without this identifiable background, we could not speak rationally about the arrangement of objects in space or the ordering of events in time. Because space and time are mixed up at the center of a black hole, the singularity does not obey the laws of physics. The singularity behaves in a random and capricious fashion, totally devoid of rhyme or reason.

Fortunately, we are shielded from the singularity by the event horizon. Although irrational things happen at the singularity, none of the effects manage to escape to the outside universe. Consequently, the outside universe remains understandable and predictable.

The irrational, random behavior of the singularity is so disturbing to physicists that, in 1969, the British mathematician Roger Penrose and his colleagues proposed the so-called **law of cosmic censorship:** "Thou shalt not have naked singularities." In other words, every singularity must be completely surrounded by an event horizon. If a naked singularity could exist, it would affect the universe in unpredictable and random ways.

15-3 *The structure of a black hole can be completely described with only three numbers*

In addition to shielding us from singularities, the event horizon prevents us from ever knowing much about anything that has fallen into a black hole. For example, there is no way we could ever discover the chemical composition of the massive star whose collapse produced a black hole. Even if someone went into a black hole to make measurements or chemical tests, there is no way the observer could get any of this information back to the outside world. Indeed, a black hole is an "information sink," because in-falling matter carries many properties about it (chemical composition, texture, color, shape, size) that are forever removed from the universe.

Because a black hole removes information from the universe, there is no way that this information can affect the structure or properties of the hole. For example, consider two black holes—one made from the gravitational collapse of 10 $M_\odot$ of iron, the other made from the gravitational collapse of 10 $M_\odot$ of peanut butter. Obviously, very different substances went into the creation of these two black holes. Once their event horizons formed, however, both the iron and the peanut butter permanently disappeared from the universe. As seen from the outside, the two holes look absolutely identical, making it impossible for us to tell which ate the peanut butter and which ate the iron. In this way, a black hole is unaffected by the information it destroys.

Is there anything we can know or measure about a black hole? In other words, what properties characterize a black hole?

First, we can measure the mass of a black hole. One way to do this would be to place a satellite in orbit about the hole. Kepler's third law relates a satellite's orbital period and semi-major axis to the mass around which the satellite moves. Thus, after measuring the size and period of the satellite's orbit, we can use this law to determine the mass of the hole. This mass is equal to the total mass of all the material that has gone into the hole.

Incidentally, science fiction abounds with nasty rumors that black holes are evil things that go around gobbling everything in the universe. Not so! The bizarre effects created by highly warped space and time are limited to a volume extending only a few million kilometers from the hole—less than the distance from the Sun to Mercury. Farther from the hole, gravity is weak enough that Newtonian physics can adequately describe everything. For example, at a distance of only a few astronomical units from a 10-$M_\odot$ black hole, the behavior of gravity is identical to that around any ordinary 10-$M_\odot$ star.

In addition to the total mass, we can also measure the total electric charge possessed by a black hole. Like the gravitational force, the electric force is a long-range interaction, making its effects felt in the space around the hole. Appropriate equipment on a space probe passing near the hole could measure the intensity of the electric field around the hole, and the hole's electric charge could be determined from this information.

In reality, we would not expect a black hole to possess any appreciable electric charge. For instance, if a hole did happen to start off with a sizable positive charge, it would vigorously attract vast numbers of negatively charged electrons from the interstellar medium, which would soon neutralize the hole's charge. For this reason, astronomers generally do not consider electric charge when discussing real black holes.

Although a black hole might have a tiny electric charge, it cannot have any magnetic field whatsoever. It can be mathematically proven from Einstein's equations that a black hole cannot possess the geometry that a magnetic field would require. During the creation of a black hole, the collapsing star would probably possess an appreciable magnetic field, but it would be radiated away in the form of electromagnetic and **gravitational waves** before the dead star could settle down inside its event horizon. Since gravitational waves are ripples in the overall geometry of space, some physicists are exploring the possibility of observing the creation of black holes by detecting the bursts of **gravitational radiation** emitted by massive, dying stars as they collapse.

In addition to its mass and electric charge, we can also measure a black hole's total angular momentum. Because of the conservation of angular momentum, we expect a

Figure 15-4 The ergosphere *A rotating black hole is surrounded by a region called the ergosphere, where the dragging of space and time around the hole is so severe that it is impossible for anything to remain at a fixed location. Because the ergosphere is outside the event horizon, this bizarre region is accessible to us and can be traversed by astronauts or asteroids without having them disappear into the black hole.*

black hole to be spinning very rapidly. Einstein's theory makes the startling prediction that this rotation causes space and time to be dragged around the hole. A spinning black hole is therefore surrounded by space that rotates with the hole. In fact, around the event horizon of every rotating black hole, a region exists where this dragging of space and time is so severe that it is impossible to stay in the same place. No matter what you do, you get pulled around the hole along with the rotating geometry of space and time. This region, where it is impossible to be at rest, is called the **ergosphere** (Figure 15-4).

To measure a black hole's angular momentum, we could place two satellites in orbit about the hole. Suppose that one satellite circles the hole in the same direction the hole rotates, the other in the opposite direction. One satellite is thus carried along with the rotating geometry of space and time, but the other is constantly fighting its way "upstream." The two satellites will therefore have dif-

ferent orbital periods. By comparing these two periods, the total angular momentum of the hole can be deduced.

And that is all. A black hole possesses no qualities other than mass, charge, and angular momentum. This simplicity is the essence of the famous **no-hair theorem,** first formulated in the early 1970s: "Black holes have no hair." Any and all additional properties carried by the matter that fell into the hole have disappeared from the universe and thus have no effect on the structure of the hole.

15-4 A black hole distorts the images of background stars and galaxies

Finding black holes in the sky is a difficult business. Obviously, since light cannot escape from a black hole, you cannot observe a black hole in the sense that you can see a star or a planet. The best you can hope for is to detect the effects of a black hole's powerful gravity.

One option to pursue is the distortion that a black hole creates in the appearance of objects behind it in space. For example, suppose that the Earth, a black hole, and a background star are in nearly perfect alignment, as sketched in Figure 15-5. Because of the warped space around the black hole, there are two paths along which light rays can travel from the background star to us here on Earth. Thus, we should see two images of the star.

The distortion of background images by a powerful source of gravity is called a **gravitational lens.** It is virtually the only way we can hope to find an isolated black hole in our Galaxy. Unfortunately, in order for this effect to be noticeable, the alignment between the Earth, the black hole, and a remote star must be almost perfect. Without nearly perfect alignment, the secondary image of the background star is too faint to be noticed.

No one has ever found a gravitational lens within our Galaxy. However, several gravitational lenses have been discovered far from our Galaxy that involve the remote, luminous objects called quasars. As we shall see in Chapter 18, a quasar is a very bright starlike object. Typical quasars shine as brightly as a hundred galaxies and are located a few billion light years from Earth. Nearly 4000 quasars have been discovered across the sky, so that, on the average, one quasar is found in roughly every 10 square degrees.

Figure 15-5 A gravitational lens A black hole can deflect light rays from a distant star so that an observer sees two images of the star. Several so-called gravitational lenses have been discovered in which light from a remote quasar is deflected by an intervening galaxy. No gravitational lenses caused by black holes have yet been discovered, however.

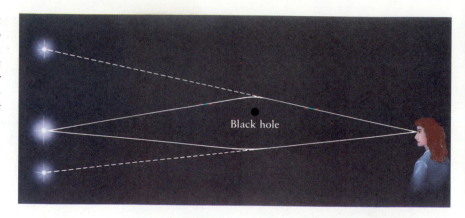

In 1979 astronomers were surprised to find two quasars separated by only 6 arc sec. They took spectra of both quasars and discovered that they were nearly identical. They thus concluded that they were looking at two images of the same quasar. Further observations revealed that there was a galaxy between the quasar images (Figure 15-6). The gravitational field of this galaxy bends the light from the remote quasar and thus acts like a gravitational lens.

By 1990 nearly two dozen candidates for gravitational lenses had been reported, of which six cases were quite convincing. All involve a remote quasar whose starlike image is split by a galaxy located between us and the quasar. When such a galaxy deflects the light from a remote quasar, three or four images can be produced,

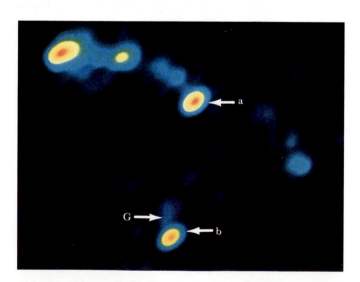

Figure 15-6 A "double" quasar Two images, (a) and (b), of the same quasar are seen in this radio view. Radio waves from the distant quasar are deflected to either side of a massive galaxy located between us and the quasar. A faint image of the deflecting galaxy (G) is seen directly above image (b). The jetlike feature protruding from the upper image (a) does not appear alongside the lower image because the jet is too far away from the required quasar-galaxy-Earth alignment. (VLA; NRAO)

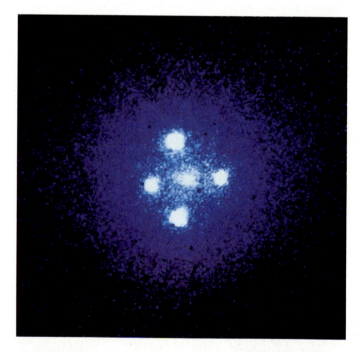

Figure 15-7 The Einstein cross This photograph from the Hubble Space Telescope shows the gravitational lensing of a quasar in the constellation of Pegasus. The quasar, about 8 billion light-years from Earth, is seen as four separate images surrounding a galaxy that is only 400 million light-years away. The diffuse image at the center of this Einstein cross is the core of the intervening galaxy. (NASA and ESO)

a

b

c

Figure 15-8 The gravitational lensing of a galaxy A com-puter was used to calculate these views of a galaxy seen through a gravitational lens consisting of a massive black hole (8×10^{12} $M_\odot$) located halfway between us and the galaxy. (a) An undis-torted view of the galaxy. (b) A view of the galaxy seen with a nearly perfect alignment between the galaxy, the black hole, and the Earth. (c) A view of the galaxy seen with the black hole displaced slightly from perfect alignment. (Courtesy of E. Falco, M. Kurtz, and M. Schneps; Smithsonian Astrophysical Observatory)

because the galaxy's mass is spread over a volume rather than being concentrated at a point, as in the case of a black hole. A fine example of four images is the "Einstein cross" in Figure 15-7. Searches for more gravitational len-ses are under way.

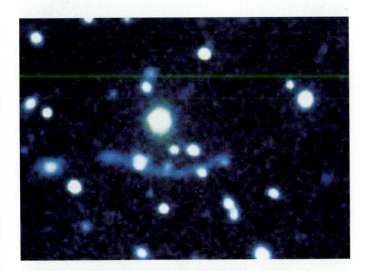

Figure 15-9 A giant luminous arc This luminous arc is about 300,000 light-years long and is located in a cluster of galaxies about 5 billion light-years from Earth. Almost every fuzzy spot on this picture is a galaxy. Spectroscopic observations indicate that this arc is the image of an extremely remote galaxy or quasar that has been stretched out into an arc by the gravita-tional field of an intervening galaxy. (NOAO)

Computers can be used to simulate the gravitational lensing of an extended object like a galaxy (Figure 15-8). The resulting images are elongated arcs that partly en-circle the location of the deflecting black hole. These arcs connect to form a ringlike image when the gal-axy, the black hole, and the observer are in nearly perfect alignment.

Several fine examples of arched gravitational images were discovered in the late 1980s. In 1987 Roger Lynds at Kitt Peak National Observatory and Vahe Petrosian of Stanford University found a huge luminous arc more than 300,000 ly long in a remote cluster of galaxies (Figure 15-9). This arc is now known to be the light from an extremely remote galaxy or quasar that has been stretched out into a curved image by the gravitational field of an intervening galaxy. In the early 1990s, Anthony Tyson at Bell Telephone Laboratories showed that detailed obser-vations of these distorted, arc-shaped background images can be used to map the mass of foreground galaxies

In 1988, Jacqueline Hewitt of MIT reported her obser-vations of a radio source that she believes is a ringlike image of a remote radio galaxy. Albert Einstein was first to point out that a ring-shaped image would be seen if a massive body were located directly between us and a remote source of light, and so the object shown in Figure 15-10 is called an "Einstein ring." Several additional ex-amples of huge arcs and Einstein rings have recently been identified.

Figure 15-10 The Einstein ring *A gravitational lens should produce a ringlike image if the background light source, the observer, and the deflecting galaxy or black hole are perfectly aligned. This radio object in the constellation of Leo is the first known example of the so-called Einstein ring. (VLA; NRAO)*

15-5 Black holes have been detected in close binary star systems

Binary stars offer the best chance of finding black holes in our Galaxy. For instance, if a black hole were to capture gas from its companion star, the fate of this material might reveal the existence of the hole.

Shortly after the *Uhuru* satellite was launched in the early 1970s, astronomers became intrigued with an X-ray source called Cygnus X-1. The source is highly variable and irregular. Its X-ray emission flickers on time scales as short as a hundredth of a second. One of the fundamental concepts in physics is that nothing can travel faster than the speed of light. Because of this limitation, an object cannot vary its brightness or flicker faster than the time required for light to travel across the object. Because light travels 3000 km in a hundredth of a second, Cygnus X-1 must be smaller than the Earth.

Cygnus X-1 occasionally emits radio radiation, and in 1971 radio astronomers succeeded in identifying it with the star HDE 226868 (Figure 15-11). Spectroscopic observations promptly revealed that HDE 226868 is a B0 supergiant with a surface temperature of about 31,000 K. Such stars do not emit significant amounts of X rays, and thus HDE 226868 alone cannot be Cygnus X-1. Because double stars are very common, astronomers began to suspect that the visible star and the X-ray source are in orbit about each other.

Figure 15-11 HDE 226868 *This star is the optical companion of the X-ray source Cygnus X-1. The star is a B0 supergiant located about 8000 ly from Earth. Many astronomers agree that Cygnus X-1 is probably a black hole. This photograph was taken with the 200-in. telescope at Palomar. (Courtesy of J. Kristian, The Observatories of the Carnegie Institution of Washington)*

Further spectroscopic observations soon showed that the spectral lines in the spectrum of HDE 226868 shift back and forth with a period of 5.6 days. This behavior is characteristic of a single-line spectroscopic binary; the companion of HDE 226868 is too dim to produce its own set of spectral lines. The clear implication is that HDE 226868 and Cygnus X-1 are the two components of a double-star system.

The B0 supergiant HDE 226868 is estimated to have a mass of about 30 $M_\odot$. As a result, Cygnus X-1 would have a mass of about 7 $M_\odot$; otherwise, it would not exert enough gravitational pull to make the B0 star wobble by the amount deduced from the periodic Doppler shifting of its spectral lines. Seven solar masses is too large for Cygnus X-1 to be either a white dwarf or a neutron star, so the only remaining possibility is a black hole.

This is not a firm conclusion, however, because HDE 226868 might be undermassive for its spectral type, which would imply a somewhat lower mass for Cygnus X-1. In addition, uncertainties in the distance to the binary system could further reduce estimates of the mass of Cygnus X-1. If all these uncertainties combined in just the right way, the estimated mass of Cygnus X-1 could be pushed down to about 3 solar masses. Thus there is a very slim chance that Cygnus X-1 might contain the most massive possible neutron star rather than a black hole.

If Cygnus X-1 does contain a black hole, the X rays do not come directly from the hole itself. Gas captured from HDE 226868 goes into orbit about the hole, forming an accretion disk about 4 million kilometers in diameter (Figure 15-12). As material in the disk spirals in toward the hole, friction heats the gas to temperatures approaching 2 million K. In the final 200 km above the hole, these extremely hot gases emit the X rays that we detect with our satellites. Presumably, the X-ray flickering is caused by small hot spots on the rapidly rotating inner edge of the accretion disk. In this way, the black hole's existence is announced by doomed gases just before they plunge to oblivion.

In the early 1980s a binary system similar to Cygnus X-1 was identified in the nearby galaxy called the Large Magellanic Cloud. The X-ray source, called LMC X-3, exhibits rapid fluctuations just like those of Cygnus X-1. LMC X-3 circles a B3 main-sequence star every 1.7 days. From its orbital data, astronomers conclude that the mass of the compact X-ray source is probably about 6 $M_\odot$, which would therefore seem to be a black hole. Once again, however, this conclusion is not firm. Some astronomers argue that dilution of the visible star's light by the accretion disk surrounding the compact object may have led to an overestimate of the object's mass.

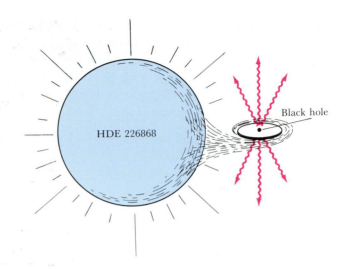

Figure 15-12 The Cygnus X-1 system *A stellar wind from HDE 226868 pours matter onto an accretion disk surrounding a black hole. The infalling gases are heated to high temperatures as they spiral in toward the hole. At the inner edge of the disk, just above the black hole, the gases become so hot that they emit vast quantities of X rays. An artist's rendition of this system is seen on the opening page of this chapter.*

Another black-hole candidate is a spectroscopic binary in the constellation of Monoceros that contains the flickering X-ray source A0620-00. The visible companion of A0620-00 is an orange dwarf star of spectra type K, which orbits the X-ray source every 7.75 hours. From orbital data, astronomers conclude that the mass of A0620-00 must be greater than 3.2 solar masses, and more probably is about 9 solar masses.

Perhaps the most convincing black-hole candidate is the spectroscopic binary called V404 Cygni, which consists of an X-ray source orbited by a low-mass G or K star. Doppler-shift measurements reveal that the line-of-sight velocity of the visible star varies by more than 400 km/s as it orbits its unseen companion every 6.47 days. These data make the mass of the companion at least 6.26 $M_\odot$. This firm lower limit is much higher than the lower limits for the black-hole candidates in Cygnus X-1, LMC X-3, or A0620-00, all of which just might contain 3-$M_\odot$ neutron stars rather than black holes. In comparison, the argument for a black hole in V404 Cygni seems to be quite persuasive.

There are several other rapidly flickering X-ray sources in binary systems; all are excellent black hole candidates. With these objects, a great deal of observational effort is necessary to rule out all non–black-hole explanations of the data. Only then can we feel confident that additional black holes have been discovered.

A black hole is one of the most bizarre and fantastic objects ever to emerge in modern physical science. Although the idea of black holes initially met with skepticism, it is seems quite clear that some of the stars we see in the sky are doomed to disappear from the universe someday, leaving only black holes behind. Even more astounding is the idea that enormous black holes, containing millions or even billions of solar masses, are located at the centers of many galaxies and quasars. As we shall see in the next chapter, one of these monstrosities may even be lurking at the center of our Milky Way, only 25,000 ly from the Earth.

Summary

- The general theory of relativity asserts that gravity causes space to become curved and time to slow down. These effects are significant only in the vicinity of large masses or very compact objects.

- If a stellar corpse has a mass greater than about 3 $M_\odot$, gravitational compression will make the object so dense that the escape velocity exceeds the speed of light. The corpse then contracts rapidly to a single point called a singularity.

   The singularity is surrounded by a surface called the event horizon, where the escape velocity equals the speed of light. Nothing—not even light—can escape from the region inside the event horizon.

- A black hole—a singularity surrounded by an event horizon—has only three physical properties: mass, electric charge, and angular momentum.

   In the case of a rotating black hole, a region called the ergosphere surrounds the event horizon. In the ergosphere, space and time themselves are dragged along with the rotation of the black hole.

- The general theory of relativity predicts the existence of gravitational radiation. Gravitational waves are ripples in the overall geometry of space and time that are produced by moving masses.

- The gravitational field of a black hole or a galaxy distorts the images of background objects. Several examples of such gravitational lenses have been discovered.

- Some binary star systems are thought to contain a black hole. In such a system, gases captured from the companion star by the black hole emit detectable X rays.

Review questions

1 Under what circumstances are degenerate-electron pressure and degenerate-neutron pressure incapable of preventing the complete gravitational collapse of a dead star?

2 In what way is a black hole blacker than black ink or a black piece of paper?

3 If the Sun suddenly became a black hole, how would the Earth's orbit be affected?

4 According to general relativity, why can't some sort of yet undiscovered degenerate pressure prevent the matter inside a black hole from collapsing all the way down to the singularity?

5 What is the law of cosmic censorship?

6 What is a gravitational lens?

7 What is the no-hair theorem?

8 Why do you suppose that all the black hole candidates mentioned in the text are members of very-short-period binary systems?

9 If light cannot escape from a black hole, how can we detect X rays from such objects?

Advanced questions

10 As a binary system loses energy by emitting gravitational waves, why do its members speed up, and why does the period of the system become shorter?

11 If more massive stars evolve and die before less massive stars, why do some black hole candidates have lower masses than their normal stellar companions?

12 Under what circumstances might a white dwarf or neutron star in a double-star system become a black hole?

Discussion questions

13 Describe the kinds of observations you might make in order to locate and identify black holes.

14 Speculate on the effects you might encounter on a trip to the center of a black hole.

For further reading

Chaffee, F. "The Discovery of a Gravitational Lens." *Scientific American*, November 1980 • This article, written soon after the first gravitational lens was discovered, describes the lens phenomenon with the aid of excellent diagrams.

Jeffries, A., and others. "Gravitational Wave Observatories." *Scientific American*, June 1987 • This article, which includes an excellent description of gravitational radiation, explains how lasers can be used to detect gravitational waves.

Kaufmann, W. *Black Holes and Warped Spacetime*. W. H. Freeman and Company, 1979 • This slim book gives a nontechnical overview of black holes and general relativity.

———. *Cosmic Frontiers of General Relativity*. Little, Brown, 1977 • This book describes many fascinating aspects of black holes, including views that might greet an astronaut who plunges through a wormhole connecting our universe with another.

McClintock, J. "Do Black Holes Exist?" *Sky & Telescope*, January 1988 • This article presents evidence for black holes in certain X-ray binaries, especially A0620-00.

Parker, B. "In and around Black Holes." *Astronomy*, October 1986 • This well-written article describes some of the fascinating properties of black holes.

Schild, R. "Gravity Is My Telescope." *Sky & Telescope*, April 1991 • This fascinating article describes recent discoveries of gravitational lenses, including the phenomenon of "microlensing"—the momentary brightening of a quasar's image caused by the perfect alignment between the Earth, the quasar, and a star in the lensing galaxy.

Thorne, K. "The Search for Black Holes." *Scientific American*, December 1974 • This superb article describes various aspects of black holes that may be useful in searches for these elusive objects.

Turner, E. "Gravitational Lenses." *Scientific American*, July 1988 • Beautiful illustrations make this article on gravitational lensing especially clear.

Our Galaxy   *This wide-angle photograph, taken from Australia, spans 180° of the Milky Way, from the Southern Cross at the left to Cygnus at the right. The center of the Galaxy is in the constellation of Sagittarius, in the middle of this photograph. Figure 16-1 is a comparable photograph showing the northern Milky Way. (Courtesy of D. di Cicco)*

16

The Milky Way Galaxy

The Milky Way, that hazy band of myriad faint stars stretching across the night sky, is our edge-on view of the disk of our own Galaxy. Throughout the twentieth century, astronomers have struggled to determine the size, shape, and rotation of our Galaxy. We now realize that we live in a vast disk-shaped assemblage of stars, gas, and dust roughly 80,000 ly in diameter. Huge spiral arms gracefully arching outward from the Galaxy's nucleus pose many puzzles for astronomers. Apparently, gravitational interactions with a

nearby galaxy or random bursts of star formation within our own Galaxy can compress the interstellar medium and give rise to spiral structure. Most mysterious is the remarkable source of energy at the very center of the Galaxy. Vast amounts of radiation pour from this compact source, which may be a supermassive black hole. By exploring our Galaxy, we gain important insights into the properties of galaxies in general, thereby preparing ourselves to widen our perspective on the universe and ask fundamental questions on a cosmic scale.

On a clear, moonless night, far from city lights, you can often see a hazy, luminous band stretching across the sky. Ancient peoples devised fanciful myths to account for this "milky way" among the constellations. Today we realize that this hazy band is actually our view from inside a vast disk-shaped assemblage of several hundred billion stars that includes the Sun.

In studying the Milky Way, we explore the universe on a grand scale. Instead of examining individual stars, we look at an entire system of stars. And instead of focusing on the location and life of an isolated star, we look at the overall arrangement and history of a huge stellar community of which the Sun is a member.

16-1 The Sun is located in the disk of the Galaxy, about 25,000 ly from the galactic center

Galileo was the first person to look at the Milky Way through a telescope. He immediately discovered that it is composed of countless dim stars. The Milky Way stretches all the way around the sky in a continuous band that is almost perpendicular to the plane of the ecliptic. Figure 16-1 is a wide-angle photograph showing roughly half of the Milky Way.

Because the Milky Way completely encircles us, astronomers in the eighteenth century began to suspect that the Sun and all the stars in the sky are part of an enormous disk-shaped assemblage called the **Milky Way Galaxy**. In the 1780s William Herschel attempted to deduce the Sun's location in the Galaxy by counting the number of stars in 683 regions of the sky. He reasoned that the greatest density of stars should be seen toward the Galaxy's center and a lesser density seen toward the edge of the Galaxy. Because Herschel found roughly the same density of stars all along the Milky Way, he concluded that we are at the center of the Galaxy. We now know that he was wrong.

The reason for Herschel's mistake was discovered in the 1930s by R. J. Trumpler. While studying star clusters, Trumpler discovered that more remote clusters appear unusually dim—more so than would be expected from their distance alone. Trumpler therefore concluded that interstellar space must not be a perfect vacuum: It contains dust that absorbs light from distant stars. Like the stars themselves, this obscuring material is concentrated in the plane of the Galaxy.

Great patches of this interstellar dust are clearly visible in wide-angle photographs, such as the one shown in Figure 16-1. At optical wavelengths, the center of the Galaxy is totally obscured from our view. The absorption of starlight by interstellar dust misled Herschel. Because he was actually seeing only the nearest stars in the Galaxy, he had no true idea of either the enormous size of the Galaxy or the vast number of stars concentrated around the galactic center.

Because interstellar dust is concentrated in the plane of the Galaxy, the absorption of starlight is strongest in those parts of the sky covered by the Milky Way. However, our view is relatively unobscured to either side of the Milky Way. Knowledge of our position in the Galaxy came from observations of globular clusters in these unobscured portions of the sky. Before we turn to those observations, we must review the nature of variable stars.

Figure 16-1 The Milky Way This wide-angle photograph shows the Milky Way extending from Sagittarius on the left to Cassiopeia on the right. Note the dark lanes and blotches. This mottling is caused by interstellar gas and dust that obscure the light from background stars. (Steward Observatory)

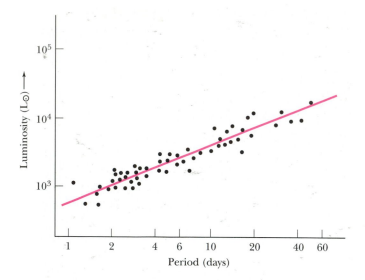

Figure 16-2 The period–luminosity relation *This graph shows the relationship between the periods and luminosities of classical (Type I) Cepheid variables. Each dot represents a Cepheid whose brightness and period have been measured. The line is the "best fit" to the data. (Adapted from H. C. Arp)*

Figure 16-3 The globular cluster M55 *The arrows indicate three RR Lyrae variables in this globular cluster located in the constellation of Sagittarius. From the average apparent brightness (as seen in this photograph) and the average true brightness (known to be roughly 100 Suns), astronomers have deduced that the distance to this cluster is 20,000 ly. (Harvard Observatory)*

In 1912 the American astronomer Henrietta Leavitt published her important discovery of the period–luminosity relation for classical (Type I) Cepheid variables. As we saw in Chapter 13 (recall Figure 13-25), Cepheid variables are pulsating stars that periodically vary in brightness. Leavitt studied numerous Cepheids in the Small Magellanic Cloud (a small galaxy very near the Milky Way) and found that their periods of light variation are directly related to their average luminosities (Figure 16-2). Today astronomers realize that there are two kinds of Cepheid variables: the metal-rich Type I Cepheids, which Leavitt studied, and the metal-poor Type II Cepheids. As shown in Figure 13-25, the Type II Cepheids are slightly dimmer than the Type I.

The period–luminosity law is a very important tool in astronomy because it can be used to determine distances. For instance, suppose you find a Cepheid variable. By measuring its period and using a graph such as Figure 16-2, you promptly discover the star's average luminosity. This measure of the star's true brightness can easily be expressed as an absolute magnitude. Meanwhile, you observe the star's apparent magnitude. Since you now know both the apparent and the absolute magnitudes, you can calculate the star's distance.

Shortly after Leavitt's discovery, Harlow Shapley, a young astronomer at the Mount Wilson Observatory in southern California, became very interested in the family of pulsating stars known as RR Lyrae variables, which are quite similar to Cepheid variables. RR Lyrae variables are commonly found in globular clusters (Figure 16-3). Shapley guessed that these variables are simply short-period classical Cepheids like those Leavitt had studied.

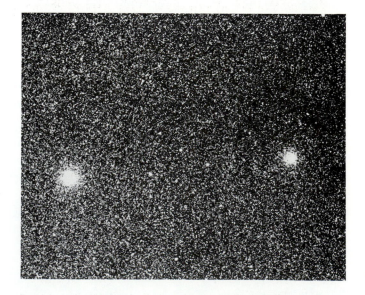

Figure 16-4 A view toward the galactic center *More than a million stars fill this view, which covers a relatively clear "window" just 4° south of the galactic nucleus in Sagittarius. Two prominent globular clusters are also seen. There is surprisingly little obscuring matter in this tiny section of the sky. (NOAO)*

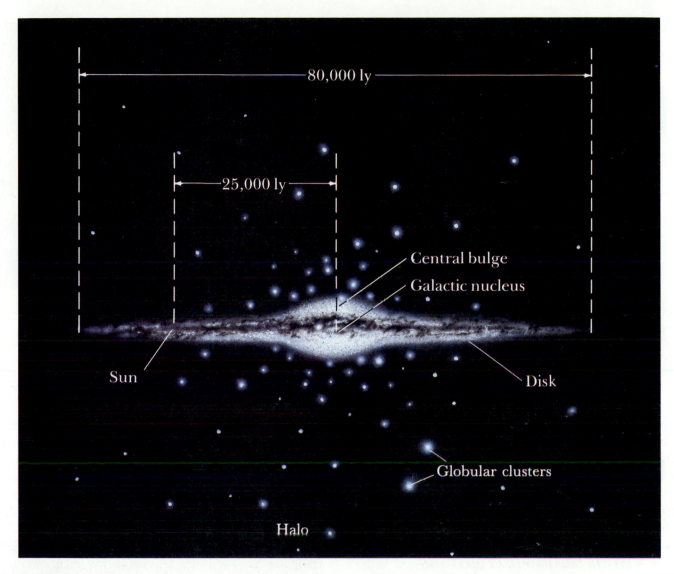

Figure 16-5 Our Galaxy (edge-on view) *There are three major components to our Galaxy: a thin disk, a central bulge, and a halo. The disk contains gas and dust along with young,* *metal-rich stars. The halo is composed almost exclusively of old, metal-poor stars. The central bulge around the galactic nucleus is a mixture of metal-rich and metal-poor stars.*

By 1915 Shapley had noticed a peculiar property of globular clusters. Ordinary stars and open star clusters are rather uniformly spread along the Milky Way. The majority of the globular clusters that Shapley studied, however, were located in one-half of the sky, widely scattered around the portion of the Milky Way in Sagittarius. Figure 16-4 shows two globular clusters in this part of the sky.

Shapley used the period–luminosity relation to determine the distances to the then-known 93 globular clusters in the sky. From their directions and distances, he mapped out the three-dimensional distribution of these clusters in space. By 1917 Shapley had discovered that the globular clusters form a huge spherical system that is not centered on the Earth. The clusters are instead centered about a

point in the Milky Way toward the constellation of Sagittarius. Shapley then made the bold conjecture, which was subsequently confirmed, that the globular clusters outline the true size and extent of the Galaxy.

Since Shapley's pioneering observations, many astronomers have ventured to measure the distance from the Sun to the **galactic nucleus** at the center of our Galaxy. A variety of techniques, including recent radio observations of gas clouds orbiting the galactic center, suggest a distance of 25,000 ly from the Sun to the galactic nucleus.

The distance to the center of the Galaxy establishes a scale from which the dimensions of other features can be determined. The **disk** of our Galaxy is about 80,000 ly in diameter and about 2000 ly thick (Figure 16-5). The

Figure 16-6 *Edge-on view of a spiral galaxy*
If we could view our Galaxy edge-on from a
great distance, it would probably look like this
galaxy in the constellation of Coma Berenices. A
thin layer of dust and gas is clearly visible in the
plane of the galaxy. Also note the reddish color
of the bulge that surrounds the galaxy's nucleus.
(U.S. Naval Observatory)

galactic nucleus is surrounded by a spherical distribution of stars, called the **central bulge,** that is about 15,000 ly in diameter. The spherical distribution of globular clusters defines the **halo** of the Galaxy. If we could view our Galaxy edge-on from a great distance, it would probably look somewhat like NGC 4565, shown in Figure 16-6.

16-2 The spiral structure of the Galaxy has been determined from radio observations

Because interstellar dust effectively obscures our optical views in the plane of the Galaxy, a detailed understanding of the structure of the galactic disk had to await the development of radio astronomy. Because of their long wavelengths, radio waves easily penetrate the interstellar medium without being scattered or absorbed. As we shall see in this section, radio observations reveal that the Galaxy has spiral-shaped concentrations of gas and dust called **spiral arms,** unwinding from the galactic center in a shape reminiscent of a pinwheel.

We have seen that hydrogen is by far the most abundant element in the universe. By looking for concentrations of hydrogen gas, we should therefore detect important clues about the structure of the disk of the Galaxy. Unfortunately, the major electron transitions in the hydrogen atom (recall Figure 5-18) produce photons at ultraviolet and visible wavelengths that do not penetrate the interstellar medium. What hope do we have of detecting all this hydrogen?

In addition to mass and charge, particles such as protons and electrons possess a tiny amount of angular momentum commonly called **spin.** An electron or a proton can be crudely visualized as a tiny spinning sphere. According to the laws of quantum mechanics, the electron and proton in a hydrogen atom can be spinning either in the same or opposite directions (Figure 16-7), but they can have no other spin orientations. If the electron flips over from one configuration to the other, the hydrogen atom must gain or lose a tiny amount of energy. For example, in going from parallel to antiparallel spins, the atom emits a low-energy photon whose wavelength is 21 cm. In 1951 a team of astronomers succeeded in detecting the faint hiss of 21-cm radio static from interstellar hydrogen.

The detection of 21-cm radio radiation was a major breakthrough that permitted astronomers to map the structure of the galactic disk. To see how this is done, suppose that you aim your radio telescope along a particular line of sight across the Galaxy as sketched in Figure 16-8. Your radio receiver picks up 21-cm emission from hydrogen clouds at points 1, 2, 3, and 4 (point S is the location of the Sun). However, the radio waves from these various clouds are Doppler shifted by slightly different amounts because they are moving at different speeds as they travel around the Galaxy.

The various Doppler shifts that occur cause the 21-cm radiation to be smeared out over a range of wavelengths.

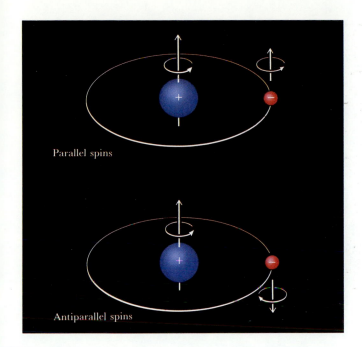

Figure 16-7 Electron spin and the hydrogen atom In the lowest orbit of the hydrogen atom, the electron and the proton can be spinning either in the same or opposite directions. When the electron flips over, the atom either gains or loses a tiny amount of energy. This energy is either absorbed or emitted as photons with wavelengths of 21 cm.

Because radio waves from gas clouds in different parts of the Galaxy arrive at our radio telescopes with slightly different wavelengths, it is possible to sort out the various gas clouds and thus produce a map of the Galaxy, such as that shown in Figure 16-9.

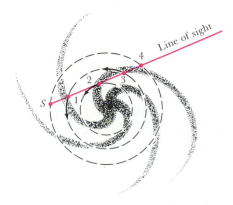

Figure 16-8 A technique for mapping the Galaxy Hydrogen clouds at different locations along our line of sight are moving at different speeds. Radio waves from the various gas clouds are therefore subjected to slightly different Doppler shifts, permitting radio astronomers to sort out the gas clouds and map the Galaxy.

Figure 16-9 A map of the Galaxy This map, based on radio telescope surveys of 21-cm radiation, shows the distribution of hydrogen gas in a face-on view of the Galaxy. Many hints of spiral structure are seen. The Sun's location is indicated by a white arrow near the top of the map. Details in the large, blank, wedge-shaped region toward the bottom of the map are unknown because gas in this part of the sky is moving perpendicular to our line of sight and thus does not exhibit a detectable Doppler shift. (Courtesy of G. Westerhout)

A 21-cm map of our Galaxy reveals numerous arched lanes of neutral hydrogen gas but gives only a vague hint of spiral structure. Photographs of other galaxies (Figure 16-10) show spiral arms outlined by bright stars and emission nebulae. As we saw in Chapter 13, these features are indicative of active star formation. Thus, the best way to chart the spiral structure of our Galaxy is to map the locations of star-forming complexes marked by H II regions, molecular clouds, and massive, hot, young stars in OB associations.

Interstellar absorption limits the range of visual observations in the plane of the Galaxy to less than 10,000 ly from the Earth. Nevertheless, there are enough OB associations and H II regions visible in the sky to plot the spiral arms in the vicinity of the Sun. As we saw in Chapter 13 (recall Figure 13-8), carbon monoxide is a good

a

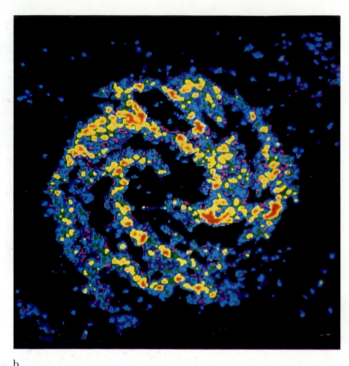

b

Figure 16-10 A spiral galaxy This galaxy, called M83, is in the southern constellation of Centaurus about 12 million light-years from Earth. (a) This photograph at visible wavelengths clearly shows the spiral arms illuminated by young stars and glowing H II regions. (b) This radio view at a wavelength of 21 cm shows the emission from neutral hydrogen gas. Note that the spiral arms are more clearly demarcated by hot stars and H II regions than by 21-cm radio emission. (Anglo-Australian Observatory; VLA, NRAO)

"tracer" of molecular clouds. Radio observations of this molecule have recently been used to chart remote regions of the Galaxy. Taken together, these observations indicate that our Galaxy has four major spiral arms and several short arm segments (Figure 16-11).

The Sun is located on a relatively short arm segment called the Orion arm, which includes the Orion Nebula (see Figure 13-10) and neighboring sites of vigorous star formation in that constellation (see the photograph on page 1). Two major spiral arms border either side of the Sun's position. The Sagittarius arm is on the side toward the galactic center. This is the arm you see during the summer months when you look at the portion of the Milky Way stretching across Scorpius and Sagittarius (see the photograph on the first page of this chapter). During winter, when our nighttime view is directed away from the galactic center, we see the Perseus arm. The remaining two major spiral arms are usually referred to as the Centaurus arm and the Cygnus arm, neither of which can be seen at visible wavelengths.

16-3 Moving at half a million miles per hour, the Sun takes 200 million years to complete one orbit of our Galaxy

The presence of spiral arms suggests that galaxies rotate. However, detecting and measuring the rotation of our Galaxy has been a difficult business.

Radio observations of 21-cm radiation from hydrogen gas give important clues about our Galaxy's rotation. By measuring Doppler shifts, astronomers can determine the speed of objects parallel to our line of sight across the Galaxy. These observations clearly indicate that our Galaxy does not rotate like a rigid body but rather exhibits differential rotation: Stars at different distances from the galactic center travel at different orbital speeds about the Galaxy.

Further clues to this rotation come from examining the motions of stars in the sky. Because of differential rota-

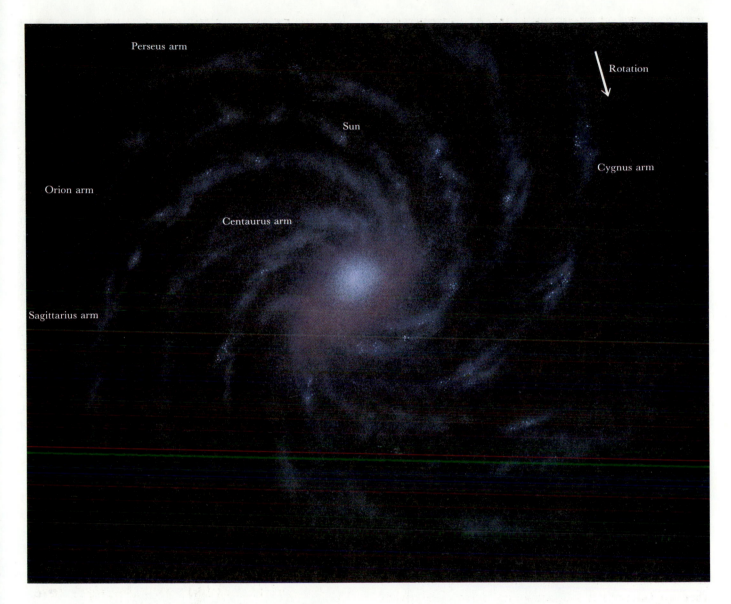

Perseus arm

Rotation

Sun

Cygnus arm

Orion arm

Centaurus arm

Sagittarius arm

Figure 16-11   Our Galaxy (face-on view)   Our Galaxy has four major spiral arms and several shorter segments of arms. The Sun is located on the Orion arm, between two major spiral arms. The Galaxy's diameter is about 80,000 ly, and the Sun is about 25,000 ly from the galactic center.

tion, the Sun is like a car on a circular freeway with the fast lane on one side and the slow lane on the other. As sketched in Figure 16-12, stars in the fast lane are passing the Sun and thus appear to be moving in one direction, while stars in the slow lane are being overtaken by the Sun and therefore appear to be moving in the opposite direction.

Unfortunately, like the 21-cm observations, this study reveals only how fast things are moving relative to the Sun. Of course, the Sun itself is moving. To get a complete picture of the Galaxy's rotation, we must find out how fast the Sun is traveling.

A method of computing this speed was proposed by the Swedish astronomer Bertil Lindblad. Not all the stars in the sky move in the orderly pattern in Figure 16-12. Globular clusters and stars in the halo of our Galaxy do not participate in the general rotation of the Galaxy, but instead have more or less random motions. Using the

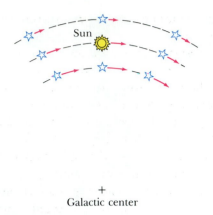

+
Galactic center

Figure 16-12 Differential rotation of the Galaxy Stars at different distances from the galactic center move at different speeds. As a result, stars on one side of the Sun's orbit about the Galaxy seem to be overtaking us, while stars on the other side seem to be lagging behind.

average of these random motions as a background, astronomers estimate that the Sun moves along its orbit about the galactic center at a speed of 230 km/s, or about one-half million miles per hour.

Given the Sun's speed and its distance from the galactic center, astronomers can calculate the Sun's orbital period. Traveling at one-half million miles per hour, our Sun takes 200 million years to complete one trip around the Galaxy. This result demonstrates how vast our Galaxy is.

Since we know the Sun's orbit around the Galaxy, we can use Kepler's third law to estimate the mass of the

Galaxy. Putting in the numbers, we obtain a mass of about 9.4×10^{10} M$_\odot$.

However, this estimate is too low, because Kepler's law gives us only the mass inside the Sun's orbit. The matter exterior to the Sun's orbit does not affect the Sun's motion and thus cannot be calculated from Kepler's law. Obviously, though, there is matter out there. In recent years, astronomers have been astonished to discover just how much matter apparently lies beyond the orbit of the Sun.

Because we know the true speed of the Sun, we can convert the Doppler shifts measured by radio astronomers into actual speeds of the spiral arms. This computation gives us a **rotation curve,** a graph of the speed of galactic rotation plotted outward from the galactic center (Figure 16-13).

According to Kepler's third law, the orbital speeds of stars or gas clouds beyond the confines of most of the Galaxy's mass should decrease with their distance from the Galaxy's center, just as the orbital speeds of the planets decrease with increasing distance from the Sun. Instead, galactic orbital speeds continue to climb well beyond the visible edge of the galactic disk, which means that we still have not detected the actual edge of our Galaxy. A surprising amount of matter must therefore be scattered around the edges of the Galaxy. In fact, our Galaxy's mass could easily be at least 6×10^{11} M$_\odot$. To make matters even more mysterious, this outlying matter is dark and does not show up on photographs. Astronomers suspect that this dark matter is spherically distributed all around the Galaxy along with the globular clusters. Thus the

Figure 16-13 The Galaxy's rotation curve This graph plots the orbital speeds of stars and gas in the Galaxy out to a distance of 60,000 ly from the galactic center. The dashed line labeled "Keplerian orbits" indicates how the rotation curve should decline beyond the edge of the Galaxy. The fact that the rotation curve continues to rise proves that vast quantities of subluminous matter surrounds the Galaxy.

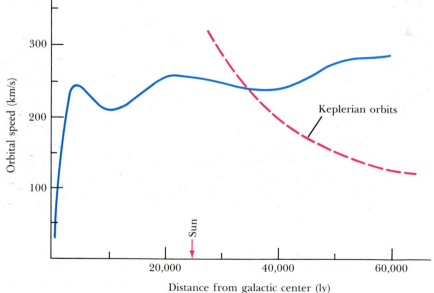

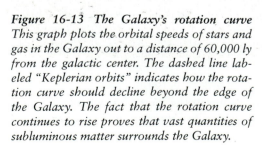

Galaxy's halo is more massive than previously expected. The nature of this dark mass—whether black holes, gas, or dim stars—is, so far, a complete mystery.

16-4 *Self-sustaining star formation can produce spiral arms*

That spiral arms should exist at all is another mystery that confounded astronomers for many years. Many galaxies exhibit beautiful arching arms outlined by brilliant H II regions and OB associations (recall Figure 16-10). As we think through the effects of a galaxy's rotation, a dilemma arises. All spiral galaxies have rotation curves similar to our own. As Figure 16-13 demonstrated, the velocity of stars and gases is fairly constant over a large portion of a galaxy's disk. However, the farther away that stars are from a galaxy's center, the farther they must travel to complete one orbit of the galaxy. Thus, stars and gases in the outskirts of a galaxy take much longer to complete an orbit than does material near the galaxy's center. Consequently, the spiral arms should eventually "wind up," wrapping themselves tightly around the nucleus. After a few galactic rotations, the spiral structure should disappear altogether.

The appearance of spiral arms varies widely. In some galaxies, called **flocculent spirals** from the word meaning "fleecy," the spiral arms are broad, fuzzy, chaotic, and poorly defined (Figure 16-14a). In other galaxies, called **grand-design spirals,** the spiral arms are thin, delicate, graceful, and well defined (Figure 16-14b). This range of shapes suggests that more than one mechanism can give rise to the spiral structure of a galaxy.

The theory of **self-propagating star formation** provides a straightforward explanation of spiral arms that takes into account the galaxy's differential rotation. Imagine that star formation begins in a dense interstellar cloud somewhere in the disk of a galaxy that does not yet have spiral arms. As soon as hot, massive stars form, their radiation compresses nearby interstellar gas, triggering the formation of additional stars in that gas. The massive stars also become supernovae that produce shock waves, which further compress the surrounding interstellar medium, thus encouraging still more star formation. As the star-forming region grows, the galaxy's differential rotation drags the inner edges ahead of the outer edges.

a

b

Figure 16-14 **Variety in spiral arms** *The differences from one spiral galaxy to another suggest that more than one mechanism can create spiral arms.* (a) *This galaxy has fuzzy, poorly defined spiral arms.* (b) *This galaxy has thin, well-defined spiral arms. (Courtesy of P. Seiden, D. Elmegreen, B. Elmegreen, and A. Mobarak; IBM)*

This conglomeration of newly formed stars, which includes bright O and B stars and glowing nebulae that make these areas quite conspicuous, soon becomes stretched out in the form of a spiral arm.

Spiral arms produced by bursts of star formation come and go more or less at random across a galaxy. Bits and pieces of spiral arms appear where star formation has recently begun but fade and disappear at other locations where all the massive stars have died off. Such galaxies thus have a chaotic appearance with poorly defined spiral arms, as was the case in Figure 16-14a. We must turn to another explanation to account for the orderly spiral structure of other galaxies.

16-5 Spiral arms can also be produced by density waves that sweep across a galaxy

In the 1920s Bertil Lindblad proposed that the spiral arms of a galaxy are a persistent pattern that moves among the stars. An analogous phenomenon is the movement of waves on the ocean. As the waves move across the surface of the water, the individual water molecules simply bob up and down in little circles. The waves are simply a pattern that moves across the water; no water actually travels with the wave pattern. Lindblad, in fact, used the term **density waves** in describing a possible cause of spiral structure.

This density-wave theory was greatly elaborated upon and mathematically embellished by the American astronomers C. C. Lin and Frank Shu in the mid-1960s. Lin and Shu argued that density waves passing through the disk of a galaxy cause material to "pile up" temporarily. A spiral arm is therefore simply a temporary enhancement, or compression, of the material in a galaxy.

This situation is analogous to a traffic jam. Imagine workers painting a line down a busy freeway. The cars normally cruise at 55 mph, but the crew of painters causes a temporary bottleneck. The cars must slow down temporarily to avoid hitting anyone. As seen from the air, there is a noticeable congestion of cars around the painters. An individual car spends only a few moments in the traffic jam before resuming its usual speed. The traffic jam in the vicinity of the painters lasts all day long, however, inching its way along the road. The traffic jam, seen from an airplane, is simply a temporary enhancement of the number of cars in a particular location.

To better understand how a density wave operates in a galaxy, think once again about the ocean. If the water molecules were left completely undisturbed, the surface of the ocean would be perfectly smooth. In reality, however, these molecules are constantly buffeted by small disturbances, called perturbations, such as the wind. A perturbation pushes one molecule, which pushes the next one, which pushes the next, and so on. The result is a water wave. Individual molecules on the surface of the ocean move in tiny elliptical paths as the wave pattern passes across the water (Figure 16-15a).

In a galaxy, the stars are separated by such vast distances that they virtually never collide. Nevertheless, stars do interact because they are affected by each other's gravity. In water waves or sound waves, molecular forces are responsible for orchestrating the motions of molecules. In a galaxy, the force of gravity controls the interactions among stars.

Seen from above, the undisturbed orbit of a star about the center of a galaxy would be a nearly perfect circle.

a Water wave

b Star orbit

Figure 16-15 *Water waves and stellar orbits* (a) *In a water wave, each molecule revolves about a point on the undisturbed water level in a tiny ellipse.* (b) *Similarly, a small disturbance in the orbit of a star can cause the star to revolve in tiny ellipses about its original orbit. The actual path of the star is a precessing ellipse. (Adapted from A. Toomre)*

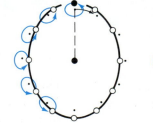
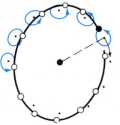
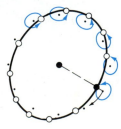

However, the motions of other matter in the galaxy produce small gravitational perturbations that cause the star to deviate from its undisturbed orbit. Just as a water molecule bobs up and down on the surface of the ocean, a star oscillates back and forth about its undisturbed orbit. Lindblad demonstrated that these oscillations can be described by thinking of the star as attached to a tiny epicycle. As sketched in Figure 16-15b, the star revolves counterclockwise around the epicycle while the epicycle itself is moving clockwise along the undisturbed path. The final path of the star lies along an ellipselike curve that slowly rotates. Of course, the gravity of this star affects the motions of its neighbors, creating a wave disturbance that propagates from one stellar orbit to the next.

The precessing elliptical orbits of stars in a galaxy are not randomly oriented, as sketched in Figure 16-16a. Instead, there is a correlation between orbits because of their gravitational interaction: Each precessing elliptical orbit is tilted with respect to its neighbor through a specific angle. The result, shown in Figure 16-16b, is a beautiful spiral pattern.

This spiral pattern arises in locations where the ellipses are bunched closest together. Of course, stars are randomly distributed along their orbits. With correlated orbits, however, some stars happen to get close together along huge, arching spiral arms. Spiral arms are seen where the stars' orbits are closest together for longest stretches.

A temporary enhancement in the number of stars has a profound effect on the interstellar gas and dust. The presence of the extra stars increases the gravitational attraction along the spiral. This force has almost no effect on the massive stars, which simply continue to lumber along their orbits. However, the atoms and molecules in the interstellar medium are readily sucked into the gravi-

tational well along the spiral forming the crest of the density wave.

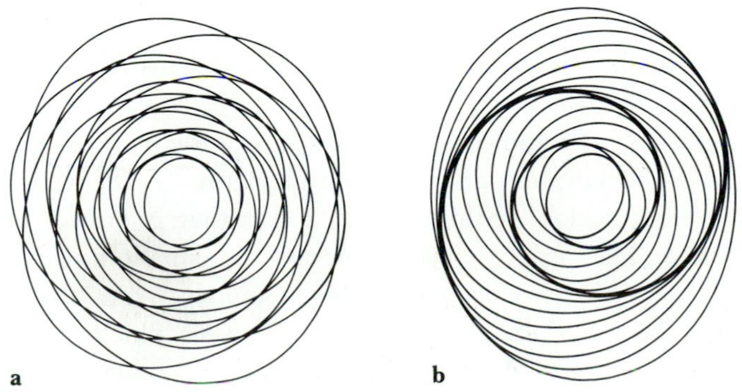

As the spiral patterns precess, the density waves move through the material of a galaxy at a speed roughly 30 km/s slower than the speed of the stars and interstellar medium. On its own, however, the interstellar gas can transport a disturbance, such as a slight compression, at a speed of only 10 km/s, which is the speed of sound in the interstellar medium. Thus, the density wave is supersonic, because its speed through the interstellar gas is greater than the speed of sound in that gas. As happens with a supersonic airplane traveling through the air, a shock wave builds up along the leading edge of the density wave. Shock waves are characterized by a sudden, abrupt compression of the medium through which they move, which is why you hear a "sonic boom" from a supersonic airplane.

The density-wave theory explains many of the properties of spiral structure. As spiral density waves sweep through the plane of a galaxy, they recycle the interstellar medium. Old gas and dust left behind from dead stars are compressed into new nebulae to form new stars. The sprawling dust lanes alongside the string of emission nebulae outlining a spiral arm attest to the recent passage of a compressional shock wave (examine Figure 16-14b). Because the material left over from the deaths of ancient stars is enriched in heavy elements, new generations of stars are more metal-rich than were their ancestors.

Density waves expend an enormous amount of energy to compress the interstellar gas and dust. There must be a driving mechanism that keeps density waves going. Grand-design spirals usually have a nearby companion galaxy. The asymmetric gravitational field of such a companion pulls on the gas, stars, and dust of a galaxy in a way that generates density waves. In the next chapter we shall further explore how gravitational interactions between galaxies induce spiral structure.

Figure 16-16 *The origin of spiral density waves* Both drawings have exactly the same number of ellipses, each representing the orbit of a star. (a) Randomly oriented ellipses. (b) Adjacent ellipses are inclined to each other at a constant angle.

16-6 *Infrared and radio observations are used to probe the galactic nucleus, whose nature is poorly understood*

The nucleus of our Galaxy is an active, crowded place. The number of stars in Figure 16-4 gives a hint of the stellar congestion there. If you lived on a planet near the

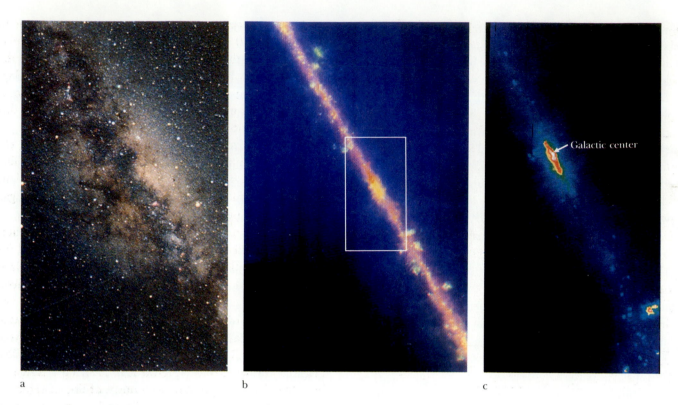

a   b   c

Figure 16-17 *The galactic center* (a) *This wide-angle view at visible wavelengths shows a 50° segment of the Milky Way centered on the nucleus of the Galaxy.* (b) *This wide-angle infrared view from IRAS covers the same area as* **a.** *Black represents the dimmest regions of infrared emission, with blue the next dimmest, followed by yellow and red; white represents the* strongest emission. The prominent band across this photograph is a layer of dust in the plane of the Galaxy. Numerous knots and blobs along the plane of the Galaxy are interstellar clouds of gas and dust heated by nearby stars. (c) *This close-up infrared view of the galactic center covers the area outlined by the white rectangle in* **b.** *(Courtesy of D. di Cicco; NASA)*

galactic center, you could see a million stars as bright as Sirius, the brightest single star in our own night sky. The total intensity of starlight from all those nearby stars would be equivalent to 200 of our full moons. In effect, night would never really fall on a planet near the center of the Galaxy.

Because of the severe interstellar absorption of light at visual wavelengths, some of our most important information about the galactic center comes from infrared and radio observations. Figure 16-17 shows three views looking toward the center of the Galaxy. Figure 16-17a is a wide-angle photograph at visible wavelengths covering a 50° segment of the Milky Way through Sagittarius and Scorpius. The same field of view at infrared wavelengths, from the IRAS satellite, is seen in Figure 16-17b. The prominent band across this infrared image is a thin layer of dust in the plane of the Galaxy. The numerous knots and blobs along the dust layer are interstellar clouds

heated by young O and B stars. Figure 16-17c is an IRAS view of the galactic center. Numerous streamers of dust (in blue) surround the galactic nucleus. The strongest infrared emission (in white) comes from Sagittarius A, which is a grouping of several powerful sources of radio waves. One of these sources, called Sagittarius A*, is believed to be the galactic nucleus.

Details of Sagittarius A can be seen in the high-resolution infrared photograph shown in Figure 16-18, which covers a region at the galactic center that is only 1 ly across. The various colors represent the emission at different infrared wavelengths. The prominent white object near the top of the view is a red supergiant, which shines as brightly as 100,000 Suns. Most of the other bright objects are red giants. The bluish Y-shaped feature is located at the galactic nucleus, centered on Sagittarius A*.

Radio observations give a different picture of the center of our Galaxy. In 1960 Doppler shift measurements of

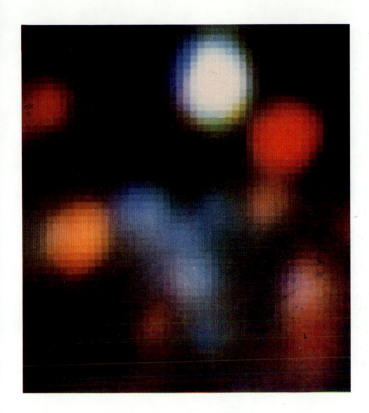

Figure 16-18 The galactic center This composite infrared photograph shows details of the galactic center. The galactic nucleus is located at the center of the bluish Y-shaped feature just below the middle of the photograph. The area covered is only about 1 ly across. (Anglo-Australian Observatory)

21-cm radiation revealed two enormous arms of hydrogen. One arm, which is located between us and the galactic center, is approaching us at a speed of 53 km/s. The other arm is on the other side of the galactic nucleus and is receding from us at a rate of 135 km/s. The total amount of hydrogen in these expanding arms is at least several million solar masses. Something quite extraordinary must have happened about 10 million years ago to expel such an enormous amount of gas from the center of the Galaxy.

In addition to analyzing 21-cm radiation from neutral hydrogen gas, astronomers also detect radio noise coming from the galactic center. This radio emission, which is produced by high-speed electrons spiraling around a magnetic field, is called **synchrotron radiation.** In spite of its small size, Sagittarius A is one of the brightest sources of synchrotron radiation in the entire sky.

Some of the most detailed radio images of the galactic center come from the Very Large Array (VLA). Figure

16-19*a* is a wide-angle view of Sagittarius A covering an area about 250 ly across. Huge filaments perpendicular to the plane of the Galaxy stretch 200 ly northward of the galactic center, then abruptly arch southward toward Sagittarius A. The orderly arrangement of these filaments suggests that a magnetic field may be controlling the distribution and flow of ionized gas.

The inner core of Sagittarius A, shown in Figure 16-19*b,* covers an area about 30 ly across. A pinwheel-like feature surrounds Sagittarius A* at the center of this view. One of the arms of this pinwheel is part of a ring of gas and dust orbiting the galactic center.

The motion of the gas clouds around the galactic center can be deduced from Doppler shift measurements of infrared spectral lines, such as emission from singly ionized neon. In the late 1970s, astronomers discovered that neon emission lines are severely broadened, perhaps from the orbital speed of the gas around the galactic nucleus. The radiation from gas coming toward us is blueshifted, whereas the radiation from receding gas is redshifted. The result is a smeared-out spectral line covering a range of wavelengths corresponding to a range of line-of-sight velocities. On one side of the galactic nucleus, gas is coming toward us at speeds up to 200 km/s, but on the other side it is rushing away from us at the same speeds.

Something must be holding this high-speed gas in orbit about the galactic center. Using Kepler's third law, astronomers estimate that 10^6 $M_\odot$ is needed to prevent this gas from flying off into interstellar space. The observed broadening of spectral lines suggest that an object with the mass of a million Suns is concentrated at Sagittarius A*. This object must be extremely compact—much smaller than a few light-years across. Many astronomers argue that an object this massive and this compact could only be a black hole. Because of its enormous mass, it is called a **supermassive black hole.** As we shall see in Chapter 18, astronomers find extraordinary activity also occurring at the nuclei of many other galaxies, which indicates the possibility of supermassive black holes at their centers.

Many astronomers disagree with the idea of a supermassive black hole at the center of our Galaxy. For instance, the high-resolution infrared view of the galactic center in Figure 16-18 shows nothing to suggest the presence of a supermassive black hole.

Astronomers are still groping for a better understanding of the galactic center. During the coming years, observations from Earth-orbiting satellites as well as from radio telescopes on the ground will certainly add to our knowledge and perhaps elucidate the nature of the core of the Milky Way.

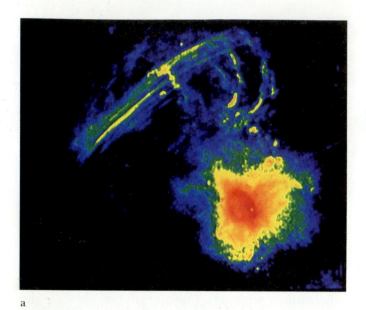

a

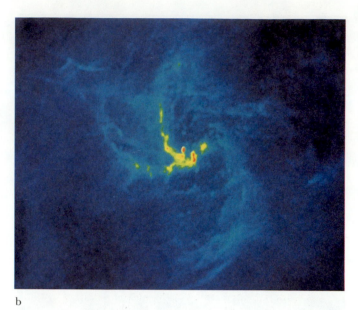

b

Figure 16-19 Two radio views of the galactic nucleus These two pictures show the appearance of the center of our Galaxy at radio wavelengths. The strongest radio emission is shown in red, weaker emission being colored green through blue. (a) This view covers an area of the sky about the same size as the full moon, corresponding to a distance of 250 ly across. The parallel fila-ments may be associated with a magnetic field. The galactic nucleus is toward the lower right, at the center of the strongest emission. (b) This high-resolution view shows details of the galactic center covering an area 30 ly across. The pinwheel-like structure is centered on Sagittarius A*. (VLA, NRAO; Courtesy of K. Y. Lo and N. Killeen, VLA, NCSA)

Summary

- Our Galaxy has a disk about 80,000 ly in diameter and about 2000 ly thick, with a high concentration of interstellar dust and gas in the disk.

  The galactic nucleus is surrounded by a spherical distribution of stars called the central bulge; the entire Galaxy is surrounded by a spherical distribution of globular clusters called the halo of the Galaxy.

  OB associations, H II regions, and molecular clouds in the galactic disk outline huge spiral arms.

  From studies of the rotation of the Galaxy, astronomers estimate that the total mass of the Galaxy is about 6×10^{11} $M_\odot$, with much of this mass being in some nonvisible form.

- The Sun is located about 25,000 ly from the galactic nucleus, between two major spiral arms.

  The Sun moves in its orbit at a speed of about one-half million miles per hour and takes about 200 million years to complete one orbit about the center of the Galaxy.

- Interstellar dust obscures our view at visual wavelengths along lines of sight that lie in the plane of the galactic disk.

  Hydrogen clouds can be detected despite the intervening interstellar dust by the 21-cm radio waves emitted in the spin-flip transition.

  The galactic nucleus has been studied at infrared and radio wavelengths, which also pass readily through intervening interstellar dust.

- According to the theory of self-propagating star formation, spiral arms are caused by the birth of stars over an extended region of the Galaxy. Differential rotation of the Galaxy stretches the star-forming region into an elongated arch of stars and nebulae.

- According to the density-wave theory, spiral arms are caused by density waves that sweep around the Galaxy.

   Each star moves in a precessing ellipse about the galactic nucleus; because the orbits are correlated, a spiral pattern is created.

   The gravitational field of this spiral pattern compresses the interstellar clouds through which it passes, thereby triggering the formation of the OB associations and H II regions that illuminate the spiral arms.

- Infrared and radio observations have revealed many details of the galactic nucleus, but astronomers are still puzzled by the processes occurring there.

   A supermassive black hole with a mass of about 10^6 $M_\odot$ possibly exists at the galactic center.

Review questions

1 Why do you suppose that the Milky Way is far more prominent in July than in December?

2 How would the Milky Way appear to us if our solar system were located at the edge of the Galaxy?

3 What observations led Harlow Shapley to conclude that we are not at the center of the Galaxy?

4 Explain why globular clusters spend most of their time in the galactic halo, even though their eccentric orbits take them close to the galactic center.

5 How is 21-cm radiation produced by hydrogen atoms? What do astronomers learn about our Galaxy from observations of that radiation?

6 Why do astronomers believe that vast quantities of dark matter surround our Galaxy?

7 What is the difference between a flocculent spiral galaxy and a grand-design spiral galaxy?

8 Briefly describe how the theory of self-propagating star formation accounts for the existence of spiral arms in galaxies.

9 Briefly describe how the density-wave theory accounts for the existence of spiral arms in galaxies.

10 What is synchrotron radiation?

11 Why are there no massive O and B stars in globular clusters?

12 How would you estimate the total number of stars in the Galaxy?

13 What evidence suggests that a supermassive black hole might be located at the center of our Galaxy?

14 If there is a supermassive black hole at the center of our Galaxy, why hasn't it swallowed all the stars in the Galaxy?

Advanced questions

15 What can you surmise about galactic evolution from the fact that the galactic halo is dominated by metal-poor stars while metal-rich stars are predominantly found in the galactic disk?

16 Why don't astronomers detect 21-cm radiation from the hydrogen in giant molecular clouds?

17 Describe the rotation curve you would get if the Galaxy rotated like a rigid body.

*** 18** Approximately how many times has the solar system orbited the center of the Galaxy since the Sun and planets were formed?

19 Compare the apparent distribution of open clusters, which contain young stars, with the distribution of globular clusters relative to the Milky Way. Can you think why open clusters were originally referred to as galactic clusters?

*** 20** The Galaxy is about 25,000 pc in diameter and 600 pc thick. If supernovae occur randomly in the Galaxy at the rate of about five each century, how often on the average would we expect to see a supernova within 300 pc (1000 ly) of the Sun?

21 Speculate on the reasons for the rapid rise in the Galaxy's rotation curve (see Figure 16-13) at distances close to the galactic center.

Discussion questions

22 From what you know about stellar evolution, the interstellar medium, and the density-wave theory, explain the appearance and structure of the spiral arms of grand-design spiral galaxies.

23 What observations would you propose in order to determine the nature of the hidden mass in our Galaxy's halo?

For further reading

Bok, B. "A Bigger and Better Milky Way." *Astronomy*, January 1984 • This entertaining and informative article was written by a renowned Dutch-American astronomer who made significant contributions to our modern understanding of the Milky Way.

———. "The Milky Way Galaxy." *Scientific American*, March 1981 • This article paints a clear picture of how astronomers have struggled to understand the structure and dynamics of our Galaxy.

Bok, B., and Bok, P. *The Milky Way*. 5th ed. Harvard University Press, 1981 • This classic book presents an up-to-date, in-depth survey of the Milky Way.

Chaisson, E. "Journey to the Center of the Galaxy." *Astronomy*, August 1980 • This well-written article describes many of the mysteries associated with the galactic center.

Geballe, T. "The Central Parsec of the Galaxy." *Scientific American*, July 1979 • This article explains how astronomers explore the swirling mass of stars, gas, and dust that envelop the galactic center.

Kraus, J. "The Center of Our Galaxy." *Sky & Telescope*, January 1983 • This brief article explains how radio astronomers investigate the galactic center.

Palmer, E. "Unveiling the Hidden Milky Way." *Astronomy*, November 1989 • Beautiful illustrations grace this article, which explains how radio astronomers peer through interstellar gas and dust to trace the spiral structure of the Galaxy.

Weaver, H. "Steps toward Understanding the Large-Scale Structure of the Milky Way." *Mercury*, September/October 1975, November/December 1975, January/February 1976 • This wonderful series of articles by a noted astronomer traces the development of our understanding of the Galaxy.

A spiral galaxy This galaxy, called NGC 1365, is the largest and most impressive member of a cluster of galaxies about 60 million light-years from Earth. NGC 1365 is classified as a barred spiral because of the "bar" crossing through its nucleus. From the ends of the bar two distinct spiral arms branch off. The yellow color of the nucleus and the bar shows that these parts of the galaxy are dominated by old, relatively cool stars. The blue color of the arms is caused by light from young, hot stars. (European Southern Observatory)

17

Galaxies

The universe is populated with galaxies, which come in many shapes and sizes. Some are disk-shaped, like our own Milky Way, with arching spiral arms that are active sites of star formation. Others are featureless, ellipse-shaped agglomerations of stars, virtually devoid of interstellar gas and dust. Most galaxies are grouped in clusters, which stretch across the universe in huge, lacy patterns. Galaxies occasionally collide, producing such spectacular phenomena as a violent burst of star formation and enormous trails of stars and gas cast outward into intergalactic space. These collisions may be responsible for the shapes and sizes of many galaxies. Most importantly, remote galaxies are receding from us at speeds that are proportional to their distances from our Galaxy. This relationship between distance and recessional velocity, called the Hubble law, is a powerful tool for the astronomer, as well as an indication that we live in an expanding universe.

William Parsons was the third earl of Rosse in Ireland. He was rich, he liked machines, and he was fascinated with astronomy. Accordingly, he set about building gigantic telescopes. In February 1845, his pièce de résistance was finished. This telescope's massive mirror measured 6 ft in diameter and was mounted at one end of a 60-ft tube controlled by cables, straps, pulleys, and cranes. For many years, this triumph of nineteenth-century engineering enjoyed the reputation of being the largest telescope in the world.

With this new telescope, Lord Rosse examined many of the nebulae discovered and catalogued by William Herschel. Lord Rosse observed that some of these nebulae have a distinct spiral structure. Perhaps the best example is M51 (also called NGC 5194, the 5194th object in the *New General Catalogue,* which is a list of all the nebulae and star clusters observed by William Herschel, his son John Herschel, and others).

Lacking photographic equipment, Lord Rosse had to make drawings of what he saw. Figure 17-1 shows a drawing he made of M51, and Figure 17-2 shows a modern photograph. Views such as this inspired Lord Rosse to echo the famous German philosopher Immanuel Kant, who in 1755 had suggested that these objects might be "island universes"—vast collections of stars far beyond the confines of the Milky Way.

Many astronomers did not subscribe to this notion of island universes. A considerable number of the objects listed in the *New General Catalogue* were in fact nebulae and star clusters scattered throughout the Milky Way, and it seemed just as likely that these intriguing spiral nebulae could also be members of our Galaxy.

The astronomical community became increasingly divided over the nature of spiral nebulae. Finally, in April 1920, a debate was held at the National Academy of Sciences in Washington, D.C. On one side was Harlow Shapley, a young, brilliant astronomer renowned for his recent determination of the size of the Milky Way Galaxy. Shapley believed the spiral nebulae to be relatively small, nearby objects scattered around our Galaxy like the globular clusters he had studied. Opposing Shapley was Heber D. Curtis of the Lick Observatory near San Jose, California. Curtis championed the island universe theory, arguing that each of these spiral nebulae is a rotating system of stars much like our own Galaxy.

The Shapley–Curtis debate generated much heat but little light. Nothing could be decided, because no one had any firm evidence to demonstrate exactly how far away the spiral nebulae were. Astronomy desperately needed a definitive determination of the distances to a spiral nebula. Such a measurement was the first great achievement of a young lawyer who abandoned a Kentucky law practice and moved to Chicago to study astronomy. His name was Edwin Hubble.

Figure 17-1 Lord Rosse's sketch of M51 *Using a large telescope of his own design, Lord Rosse was able to distinguish the spiral structure of this "spiral nebula." (Courtesy of Lund Humphries)*

Figure 17-2 The spiral galaxy M51 *This spiral galaxy in the constellation of Canes Venatici is often called the Whirlpool Galaxy because of its distinctive appearance. Its distance from Earth is about 20 million light-years. The "blob" at the end of one of the spiral arms is a companion galaxy. (NOAO)*

Figure 17-3 The Andromeda Galaxy (called M31 or NGC 224) This nearby galaxy covers an area of the sky roughly five times as large as the full moon. Under good observing conditions, the galaxy's bright central bulge can be glimpsed with the naked eye in the constellation of Andromeda. The distance to the galaxy is 2.2 million light-years. The white rectangle outlines the area shown in Figure 17-4. (Palomar Observatory)

Figure 17-4 Cepheid variables in the Andromeda Galaxy Two Cepheid variables are identified in this view of the out-skirts of the Andromeda Galaxy. Because these stars appear so faint (although they are in fact quite luminous), Hubble success-fully demonstrated that the "Andromeda Nebula" is extremely far away. (Palomar Observatory)

17-1 Edwin Hubble discovered the distances to galaxies and devised a system for classifying them

Edwin Hubble joined the staff of the Mount Wilson Observatory in Pasadena, California, and in 1923 he took a historic photograph of the "Andromeda Nebula," one of the spiral nebulae around which controversy raged. A modern photograph appears in Figure 17-3. Hubble carefully examined his photographic plate and discovered what he first thought to be a nova. Referring to previous plates of that region, he soon realized that the object was actually a Cepheid variable star. Further scrutiny over the next several months revealed many other Cepheids, two of which are identified in Figure 17-4.

As we saw in the previous chapter, Cepheid variables help astronomers determine distances. From the period–luminosity relation (recall Figure 13-25), astronomers can determine the average absolute magnitude of a Cepheid variable. From both the absolute magnitude and the apparent magnitude seen in the sky, a star's distance can be deduced.

Cepheid variables are intrinsically very bright. They typically have luminosities of a few thousand Suns. Hubble realized that for these luminous stars to appear as dim as they do on his photographs of the "Andromeda Nebula," they must be extremely far away. Straightforward calculations using modern data on Cepheids

Sa   Sb   Sc

Figure 17-5 Various spiral galaxies (face-on views) Edwin Hubble classified spiral galaxies according to the winding of the spiral arms and the size of the central bulge. Three examples are *shown here. (Sa courtesy of R. Schild; Sb and Sc courtesy of P. Seiden)*

demonstrate that M31 is 2.2 million light-years away, thus proving it to be not a traditional nebula but an enormous stellar system located far beyond the Milky Way. Today the "Andromeda Nebula" is properly called the Andromeda Galaxy.

Hubble's results, which were presented at a meeting of the American Astronomical Society on December 30, 1924, settled the Shapley–Curtis debate once and for all. The universe was recognized to be far larger and populated with far bigger objects than anyone had thus far seriously imagined. Hubble had discovered the realm of the galaxies.

There are millions of galaxies all across the sky, and they can be seen in every unobscured direction. A typical spiral galaxy contains 100 billion stars and measures 100,000 light-years in diameter.

Hubble found that galaxies can be classified into four broad categories: spirals, barred spirals, ellipticals, and irregulars. These categories form the basis for the **Hubble classification.** Both M51 and M31 (Figures 17-2 and 17-3) are examples of spiral galaxies, characterized by arched lanes of stars and glowing nebulae. While studying spiral galaxies, Hubble noted that this class could be further subdivided according to the size of the central bulge and the winding of the spiral arms (Figure 17-5). Spirals with tightly wound spiral arms and a prominent, fat central bulge are called **Sa galaxies.** Those with moderately wound spiral arms and a moderate-sized central bulge,

such as M51 and M31, are **Sb galaxies.** Finally, loosely wound spirals with a tiny central bulge are **Sc galaxies.**

Fortunately, the size of the central bulge is correlated with the degree of winding of the spiral arms, and so we can classify even those spiral galaxies that are viewed nearly edge on. For instance, M104 (Figure 17-6a) must be an Sa galaxy because of its huge central bulge. An Sb galaxy (Figure 17-6b) has a smaller central bulge. The tiny central bulge of an Sc (Figure 17-6c) is hardly noticeable at all in an edge-on view.

In **barred spiral galaxies** the spiral arms originate at the ends of a bar running through the galaxy's nucleus rather than from the nucleus itself (Figure 17-7). As with ordinary spirals, Hubble subdivided barred spirals according to the size of their central bulge and the winding of the spiral arms. An **SBa galaxy** has a large central bulge and tightly wound spiral arms. Likewise, a barred spiral with a moderate central bulge and moderately wound spiral arms is an **SBb galaxy,** and an **SBc galaxy** has loosely wound spiral arms and a tiny central bulge.

The development of a bar across a galaxy's nucleus is apparently related to the motions of the stars in the galaxy's disk. Computer simulations suggest that gravitational interactions between the stars in a huge rotating disk can cause the stars to "pile up" along a barlike structure. A bar is thus a fleeting occurrence, which may explain why ordinary spiral galaxies outnumber barred spirals.

a   b   c

Figure 17-6 Various spiral galaxies (edge-on views) (a) Be-cause of the large size of its central bulge, this galaxy is classified as an Sa. If we could see it face on, we would find that the spiral arms are tightly wound around a voluminous bulge. (b) Note the smaller central bulge in this Sb galaxy. (c) This galaxy is an Sc because of its insignificant central bulge. (Sa from NOAO; Sb and Sc courtesy of R. Schild)

Elliptical galaxies, so named for their distinctly ellipti-cal shapes, have no spiral arms. Hubble subdivided ellip-tical galaxies according to how round or flattened they look. The roundest elliptical galaxies are called **E0 gal-axies,** the flattest are **E7 galaxies.** Elliptical galaxies with intermediate amounts of flattening receive intermediate designations (Figure 17-8).

The Hubble scheme classifies galaxies entirely by their appearance from our Earth-bound view. An E1 or E2 galaxy might actually be a very flattened disk of stars that just happens to be viewed face-on, and a cigar-shaped E7 galaxy might look spherical when viewed end-on. Statisti-cal studies have shown, however, that elliptical galaxies of various actual degrees of ellipticity do exist.

Elliptical galaxies look far less dramatic than their spiral and barred spiral cousins because they are virtu-ally devoid of interstellar gas and dust. In addition, there is no evidence of young stars in most elliptical galaxies,

SBa   SBb   SBc

Figure 17-7 Various barred spiral galaxies As with spiral galaxies, Edwin Hubble classified barred spirals according to the winding of their spiral arms and the size of the central bulge. Three examples are shown here. (SBa and SBc courtesy of P. Seiden; SBb courtesy of R. Schild)

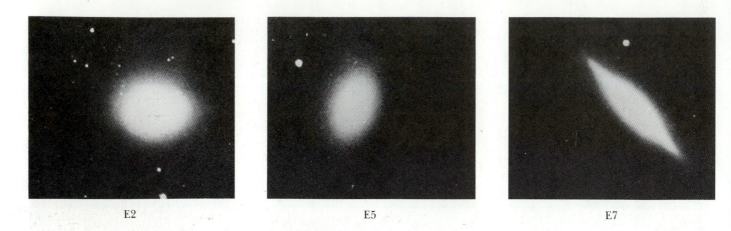

E2   E5   E7

Figure 17-8 Various elliptical galaxies Hubble classified elliptical galaxies according to how round or flat they looked. An E0 galaxy is round; a very flat elliptical galaxy is an E7. Three examples are shown here. (Yerkes Observatory)

and there is no material from which stars could have recently formed. Star formation in elliptical galaxies ended a long time ago.

Elliptical galaxies exist in an enormous range of sizes and masses. Both the biggest and the smallest galaxies in the universe are ellipticals. Figure 17-9 shows an example of a **giant elliptical galaxy.** This huge galaxy is about 20 times larger than the Milky Way and is located near the middle of a large cluster of galaxies in the constellation of Coma Berenices (Berenice's Hair).

Giant ellipticals are rather rare, but **dwarf elliptical galaxies** are quite common. Dwarf ellipticals are only a fraction the size of their normal counterparts and contain so few stars—only a few million—that these galaxies are

Figure 17-9 A giant elliptical galaxy This huge elliptical galaxy sits near the center of a rich cluster of galaxies in the constellation of Coma Berenices. Many normal-sized galaxies surround this giant elliptical, which is about 2 million light-years in diameter. (Courtesy of R. Schild)

Figure 17-10 A dwarf elliptical galaxy This nearby elliptical, called Leo I, is about 1 million light-years from Earth. It is a satellite of the Milky Way and is thus a member of the cluster to which we belong. (Anglo-Australian Observatory)

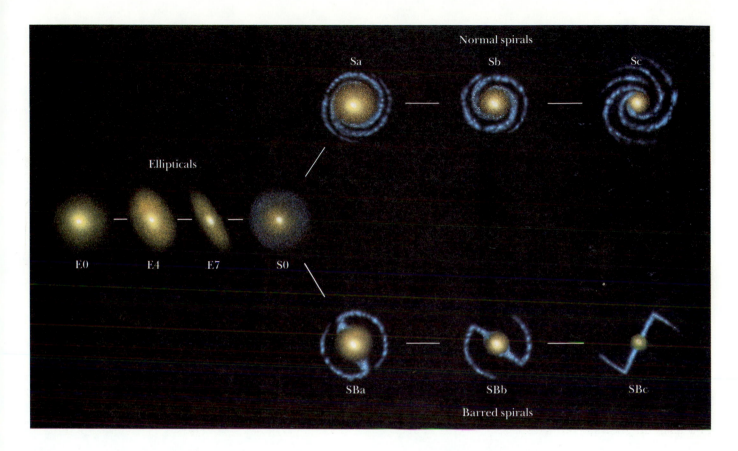

Figure 17-11 *Hubble's tuning fork diagram* Hubble summarized his classification scheme for regular galaxies with this tuning fork diagram. An S0 galaxy is a transitional type between ellipticals and spirals.

transparent. You can actually see straight through the center of a dwarf galaxy and out the other side, as shown in Figure 17-10.

Edwin Hubble connected the three types of galaxies—spirals, barred spirals, and ellipticals—in his now-famous tuning fork diagram (Figure 17-11). According to this scheme, S0 galaxies are a transition type between ellipticals and the two kinds of spirals. Although they look somewhat like ellipticals, S0 galaxies have both a central bulge and a disk like spiral galaxies.

Hubble found some galaxies that cannot be classified as spirals, barred spirals, or ellipticals. He called these **irregular galaxies.** Examples include the Large Magellanic Cloud (LMC) and the Small Magellanic Cloud (SMC), both of which can be seen with the naked eye from southern latitudes.

Telescopic views of the Magellanic clouds easily distinguish individual stars (Figures 17-12 and 17-13). Note that the SMC does not exhibit any geometric symmetry characteristic of spirals or ellipticals and is therefore a

Figure 17-12 *The Large Magellanic Cloud (LMC)* At a distance of only 160,000 light-years, this irregular galaxy is the nearest companion of our Milky Way Galaxy. Note the huge H II region (called the Tarantula Nebula or 30 Doradus) toward the left side of the photograph. Its diameter of 800 light-years and mass of 5 million Suns makes it the largest known H II region. (ROE/AAT Board)

Figure 17-13 *The Small Magellanic Cloud (SMC)* The SMC *is only slightly farther away from us than the LMC is. Because of its sprawling, asymmetrical shape, the SMC is classified as an irregular galaxy. Note that the SMC is rich in young blue stars. (ROE/AAT Board)*

true irregular. However, the LMC does apparently have a barlike structure. The Magellanic clouds are the nearest members of a cluster of galaxies to which the Milky Way and Andromeda galaxies belong.

17-2 *Galaxies are grouped in clusters that are members of superclusters*

Galaxies are not scattered randomly across the universe but are grouped in **clusters**. A typical cluster, called the Fornax cluster because it is located in the constellation of Fornax (the Furnace), is seen in Figure 17-14.

A cluster is said to be either poor or rich, depending on how many galaxies it contains. For example, the Milky Way Galaxy, the Andromeda Galaxy, and the Large and Small Magellanic clouds belong to a poor cluster called

Figure 17-14 *A cluster of galaxies* This cluster of galaxies, *called the Fornax cluster, is about 60 million-light years from Earth. Both elliptical and spiral galaxies are easily identified. An* enlarged view of the barred spiral near the edge of this picture *appears in the photograph on the first page of this chapter. (ROE/AAT Board)*

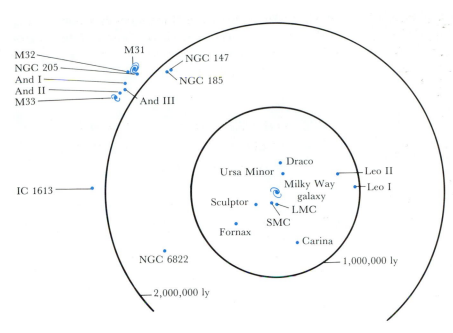

Figure 17-15 The Local Group *Our Galaxy belongs to a poor cluster consisting of about 30 galaxies, a dozen of which are dwarf ellipticals. The Andromeda Galaxy (M31) is the largest and most massive galaxy in the Local Group. The second largest is the Milky Way. M31 and the Milky Way Galaxy are each surrounded by a dozen satellite galaxies.*

the **Local Group.** The Local Group contains nearly two dozen galaxies, most of which are dwarf ellipticals. A map of the Local Group is shown in Figure 17-15.

The nearest fairly rich cluster is the Virgo cluster. It is a sprawling collection of over 1000 galaxies covering a 10° × 12° area of the sky. The distance to the Virgo cluster is about 50 million light-years, too far away for Cepheid variables to be seen from our Galaxy. Instead, the distance to the Virgo cluster has been determined by the apparent brightness of O and B supergiant stars and by the apparent brightness of globular clusters surrounding some of the galaxies. The overall diameter of the Virgo cluster is about 7 million light years.

The center of the Virgo cluster is dominated by three giant elliptical galaxies. Two of them appear in Figure 17-16. These enormous galaxies are 2 million light-years in diameter, 20 times as large as an ordinary elliptical or spiral. In other words, one giant elliptical is roughly the same size as the entire Local Group!

In addition to using the rich-versus-poor classification, astronomers often categorize clusters of galaxies as regular or irregular, depending on the overall shape of the cluster. The Virgo cluster, for example, is called irregular because of the asymmetrical way its galaxies are scattered about a sprawling region of the sky. In contrast, a regular cluster has a distinctly spherical appearance, with a marked concentration of galaxies at its center. This distribution is what would be expected for a system of objects whose energy has been equitably shared through gravitational interactions over the ages.

The nearest example of a rich, regular cluster is the Coma cluster, located 300 million light-years from us in the constellation of Coma Berenices. Despite the great distance to this cluster, more than 1000 bright galaxies within it are easily visible on photographic plates. Certainly there must be many thousands of dwarf ellipticals that are too faint to be detected from our distance. The

Figure 17-16 The center of the Virgo cluster *This fairly rich cluster lies about 50 million light-years from us. Only the center of this huge cluster appears in this view. Note the two giant elliptical galaxies (M84 and M86). (Royal Observatory, Edinburgh)*

total membership of the Coma cluster may be as many as 10,000 galaxies. The core of the Coma cluster is dominated by two giant ellipticals surrounded by many normal-sized galaxies (Figure 17-17).

A correlation exists between the regular-versus-irregular classification scheme and the dominant types of galaxies in a cluster. Rich, regular clusters, like the Coma cluster, contain mostly elliptical and S0 galaxies. Only 15% of the Coma cluster's galaxies are spirals and irregulars. Irregular clusters, such as the Virgo cluster and the Hercules cluster (Figure 17-18), have a more even mixture of galaxy types. For instance, of the 200 brightest galaxies in the Hercules cluster, 68% are spirals, 19% are ellipticals, and the rest are irregulars.

Clusters of galaxies are themselves grouped together in huge associations called **superclusters.** A typical supercluster contains dozens of individual clusters spread over a volume up to 100 million light years across. Patterns in the distribution of galaxies can be seen in maps such as Figure 17-19, which covers many hundreds of square degrees. Note the delicate, filamentary structure spread across the sky.

The arrangement of superclusters in space was clarified in the early 1980s when astronomers began discovering enormous **voids,** where exceptionally few galaxies are found. These voids are roughly spherical and measure 100

Figure 17-18   The Hercules cluster   *This irregular cluster, which is about 700 million light-years from Earth, contains a high proportion of spiral galaxies, often associated in pairs and small groups. (NOAO)*

million to 400 million light-years in diameter. Recent surveys reveal that galaxies are concentrated on the surfaces of these voids (Figure 17-20). The distribution of clusters of galaxies is therefore said to be "sudsy" because it resembles a collection of giant soap bubbles. Galaxies surround voids in the same way a soap film is concentrated on the surface of bubbles. Many astronomers suspect that this sudsy pattern contains important clues about conditions shortly after the Big Bang that led to the formation of clusters of galaxies.

17-3 Most of the matter in the universe has yet to be observed

A cluster of galaxies must be a gravitationally bound system. In other words, there must be enough matter in the cluster to produce gravity sufficient to prevent the galaxies from wandering away. Nevertheless, careful examination of a rich cluster, like the Coma cluster, typically reveals that the mass of the visually luminous matter is not at all sufficient to bind the cluster gravitationally. The observed line-of-sight speeds of the cluster galaxies, measured by Doppler shifts, are so large that more mass than has been observed is needed to keep them bound in orbit about the center of the cluster. This dilemma is

Figure 17-17   The center of the Coma cluster   *This rich, regular cluster is about 300 million light-years from Earth. Only the cluster's center, dominated by two giant galaxies, appears in this view. Regular clusters are composed mostly of elliptical and S0 galaxies and are common sources of X rays. (Courtesy of R. Schild)*

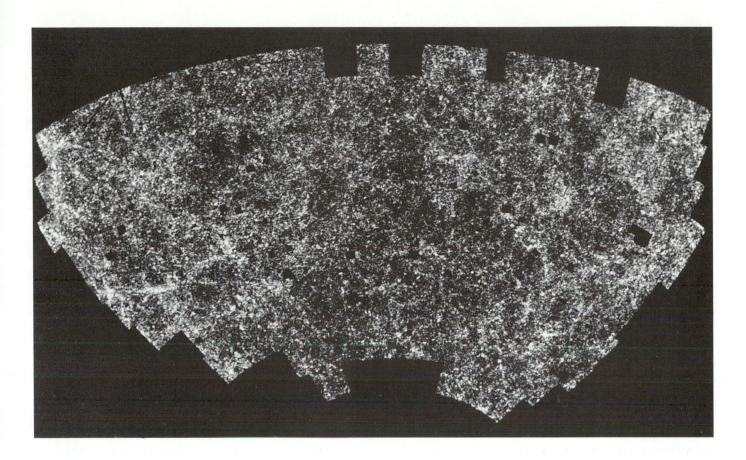

Figure 17-19 **Two million galaxies** *This map shows the distribution of roughly two million galaxies over about 10% of the sky. Each dot on the map is shaded according to the number of galaxies it contains. White dots indicate more than 20 galaxies, gray dots indicate between 1 and 19 galaxies, and the map is black where there are no galaxies. Note that the clusters are not smoothly distributed across the sky but rather form a lacy, filamentary structure. The small, bright patches are individual galaxy clusters. The larger, elongated bright areas are superclusters and filaments, which generally surround darker voids containing few galaxies. Statistical analysis of this map shows that galaxies clump together on distance scales up to about 150 million light-years. Over distances greater than this, the distribution of galaxies in the universe appears to be roughly uniform. (S. J. Maddox, W. J. Sutherland, G. P. Efstathiou, and J. Loveday; Oxford Astrophysics)*

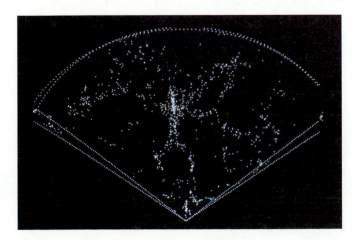

Figure 17-20 **A slice of the universe** *This map shows the locations of 1099 galaxies in a thin slice of the universe. To produce this map, a team of astronomers painstakingly measured the redshifts of galaxies out to a distance of nearly a billion light years from Earth in a strip of the sky 6° wide and 153° long. Note that most galaxies surround voids that are roughly circular. This distribution of galaxies is like a slice cut through the suds in a kitchen sink. (Courtesy of V. de Lapparent, M. Geller, and J. Huchra)*

called the **dark-matter problem.** A lot of nonluminous matter must be scattered about each of the clusters, or else the galaxies would long ago have wandered away in random directions and the clusters would no longer exist today. Analyses demonstrate that the total mass needed to bind a typical rich cluster is 10 times greater than the mass of material that shows up on visual photographs.

Some of this mystery has been solved recently by X-ray astronomers. Satellite observations of rich clusters have revealed that X rays pour from the space between galaxies in rich clusters. This flow is evidence of substantial amounts of hot intergalactic gas at temperatures between 10 and 100 million kelvin. Analyses show that the mass of this hot intergalactic gas is typically as great as the combined mass of all the visible galaxies in the cluster.

Unfortunately, this discovery of hot intergalactic gas in rich clusters solves only part of the missing-mass problem. Most astronomers agree that a great deal of matter still remains to be discovered in rich clusters. One popular speculation is that these clusters may contain a lot of undetected dim stars. These faint stars could be located in extended halos surrounding individual galaxies and scattered throughout the spaces between the galaxies of a cluster.

Evidence to support the idea of extended halos comes from the rotation curves of galaxies. Many galaxies have rotation curves similar to that of our Milky Way Galaxy (recall Figure 16-13). These rotation curves remain remarkably flat out to surprisingly great distances from the galaxy's center. For example, Figure 17-21 shows the rotation curves of four spiral galaxies. In all cases, the orbital speed is fairly constant out to distances beyond which the galaxies' stars and nebulae are so dim and widely scattered that reliable measurements are not possible. Nevertheless, we have still not detected the true edge of these and many similar galaxies. In the outer portions of a galaxy we should see a decline in orbital speed, in accordance with Kepler's third law. Because this decline has not been observed, astronomers conclude that there must be a considerable amount of dark matter extending well beyond the visible portion of a galaxy's disk.

Many proposals have been made to account for this dark matter. Some of the more exotic suggestions include black holes and various massive particles left over from the creation of the universe. The more mundane suggestions include dim stars and Jupiterlike planets. The nature of this unseen matter is one of the greatest mysteries in modern astronomy.

17-4 Colliding galaxies produce starbursts, spiral arms, and other spectacular phenomena

The galaxies in a cluster are all in orbit about their common center of mass. Occasionally two galaxies pass close enough to each other to collide. When they do collide, their stars pass by each other. There is so much space between the stars that the probability of two stars crashing into each other is extremely small. However, the galaxies' huge clouds of interstellar gas and dust cannot interpenetrate, and so they slam into each other, producing strong shock waves. The stars keep right on going, but the colliding interstellar clouds are stopped in their tracks. In this way, two colliding galaxies may be stripped of their interstellar gas and dust. The violence of the collision heats the gas stripped from these galaxies to extremely high temperatures. This process may be a major source of the hot intergalactic gas often observed in rich, regular clusters.

In a less violent collision or a near miss between two galaxies, the compressed interstellar gas may have enough time to cool sufficiently to allow many protostars to form. Such collisions can thus stimulate prolific star formation, which may account for the **starburst galaxies** that blaze with the light of numerous newborn stars. These galaxies are characterized by bright centers surrounded by clouds

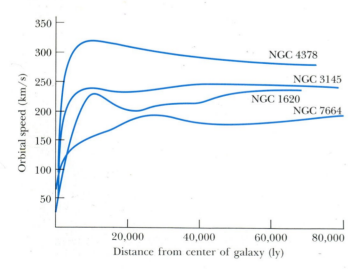

Figure 17-21 *The rotation curve of four spiral galaxies* This graph shows the orbital speed of material in the disks of four spiral galaxies. Many galaxies have flat rotation curves, indicating the presence of extended halos of dark matter. (Adapted from V. Rubin and K. Ford)

Figure 17-22 A starburst galaxy *Prolific star formation is occurring at the center of this galaxy, called M82, located about 120 million light-years from Earth. This activity was probably triggered by a tidal interaction with neighboring galaxies. Note the turbulent appearance of the interstellar gas and dust around the galaxy's center. (Lick Observatory)*

of warm interstellar dust, indicating a recent, vigorous episode of star birth (Figure 17-22). The warm dust is so abundant that starburst galaxies are among the most luminous objects in the universe at infrared wavelengths.

The starburst galaxy M82 in Figure 17-22 is one member of a nearby cluster that includes the beautiful spiral galaxy M81 and a fainter companion called NGC 3077 (Figure 17-23*a*). A recent radio survey of that region of the sky revealed enormous streams of hydrogen gas connecting the three galaxies (Figure 17-23*b*). The loops and twists in these streamers suggests that the three galaxies have had several close encounters over the ages. Incidentally, a similar stream of hydrogen gas connects our Galaxy with its nearest neighbor, the Large Magellanic Cloud.

Gravitational interaction between colliding galaxies can also hurl thousands of stars out into intergalactic space along huge, arching streams. These shapes have been dramatically illustrated in computer simulations like the one shown in Figure 17-24. Note the remarkable

a

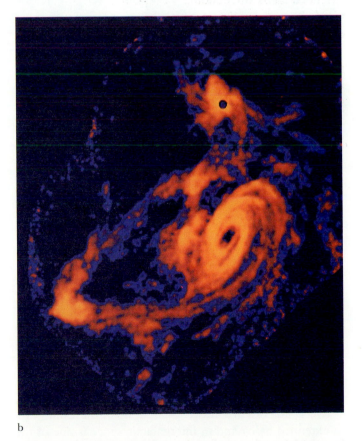

b

Figure 17-23 The M81–M82–NGC 3077 cluster *The starburst galaxy M82 is in a nearby cluster whose three members are connected by streamers of hydrogen gas. (a) This wide-angle photograph shows the three galaxies at visual wavelengths.*

(b) This mosaic of 13 fields observed with radio telescopes of the Very Large Array shows the streams of hydrogen gas that connect the three galaxies. (Palomar Sky Survey; M. S. Yun and P. Ho, Center for Astrophysics)

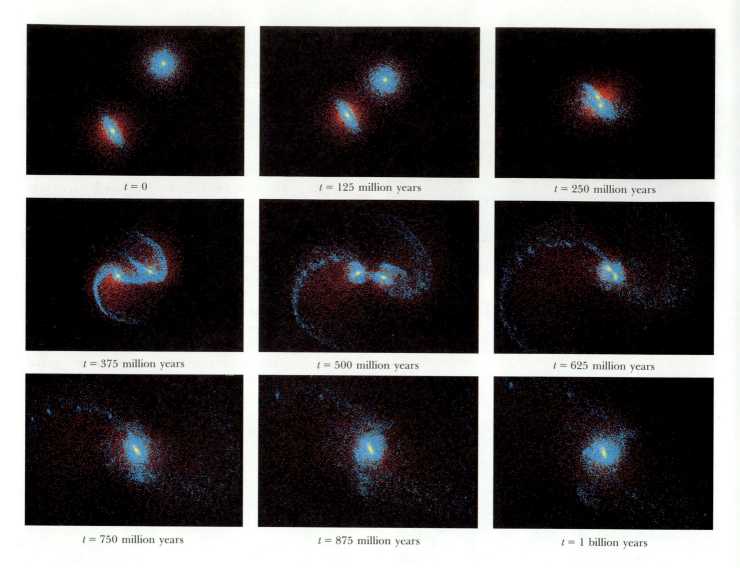

$t = 0$ $t = 125$ million years $t = 250$ million years

$t = 375$ million years $t = 500$ million years $t = 625$ million years

$t = 750$ million years $t = 875$ million years $t = 1$ billion years

Figure 17-24 A simulated collision between two galaxies
These frames from a computer simulation show the collision and merging of two disk-shaped galaxies. Stars in the disk of each galaxy are colored blue, while stars in their central bulges are *yellow. Red indicates dark matter that surrounds each galaxy. The pictures display progress at 125-million-year intervals. Compare the frame at* t = 625 *million years with the photograph of NGC 2623 in Figure 17-25. (Courtesy of J. Barnes)*

similarity between this simulation and the colliding galaxies shown in Figure 17-25.

In a rich cluster there must be many near misses between galaxies. If galaxies are surrounded by extended halos of dim stars, these near misses could strip the galaxies of their outlying stars. In this way, a loosely dispersed sea of dim stars might come to populate the space between galaxies in a cluster. Searching for these dim stars in extended halos and in intergalactic space is one of the projects assigned to the Hubble Space Telescope.

During any collision between galaxies, some of the stars are flung far and wide, scattering material into inter-

galactic space. However, other stars lose energy and momentum. As these stars slow down, the galaxies may merge. Several dramatic examples of **galaxy mergers** have been discovered recently (Figure 17-26). Astronomers also speak of **galactic cannibalism,** which occurs when a large galaxy captures and devours a smaller one. Cannibalism differs from mergers in that the dining galaxy is bigger than its dinner, whereas merging galaxies are about the same size.

Many astronomers suspect that galactic cannibalism explains why giant ellipticals are so huge. As we have seen, giant galaxies typically occupy the centers of rich

Figure 17-25   A colliding pair of galaxies with "antennae"
Many pairs of colliding galaxies exhibit long "antennae" of stars
ejected by the collision. This particular system, called NGC
2623, is also a significant source of radio radiation. Super-
computer simulations, like the one shown in Figure 17-24,
give important insights into possible histories of such systems.
(Palomar Observatory)

clusters. In many cases, smaller galaxies are located around these giants (examine Figure 17-9). As they pass through the extended halo of a giant elliptical, these smaller galaxies slow down and are eventually devoured by the larger galaxy.

Figure 17-27 shows a computer simulation in which a large, disk-shaped galaxy devours a small satellite galaxy. The large galaxy consists, by mass, of 90% stars (in blue) and 10% gas (in white). It is surrounded by a halo of dark matter having a mass about 3.3 times that of the disk. The satellite galaxy, which has a tenth of the mass of the large galaxy, contains only stars (in orange). Initially the satellite is in circular orbit about the large galaxy. Note that spiral arms appear in the large galaxy as the collision proceeds. Two billion years elapse as the satellite spirals in toward the core of the large galaxy. Although much material is stripped from the satellite, most of its stars plunge into the nucleus of the large galaxy.

Close encounters between galaxies provide a third way of forming spiral arms, in addition to density waves and self-propagating star formation. Computer simulations clearly demonstrate that spiral arms can be created during a collision that draws out long streamers of stars and gas. For instance, the spiral arms of M51 (examine Figure 17-2) were produced when a second galaxy pulled material out of the disk of M51. The disruptive galaxy is now located at the end of one of the spiral arms created by the collision. Some astronomers argue that the spiral arms of our Milky Way Galaxy were similarly produced by a close encounter with the Large Magellanic Cloud.

As we shall see in the next chapter, some astronomers suspect very massive black holes may be located at the centers of certain peculiar galaxies that are unusually powerful sources of energy. A cannibalistic collision of the

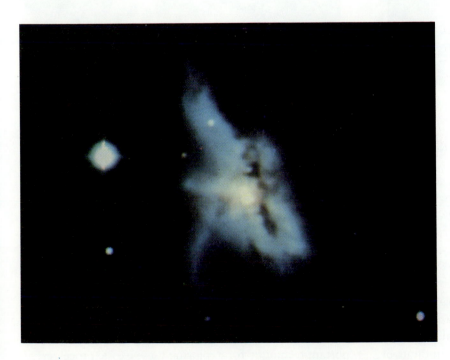

Figure 17-26   Merging galaxies   This contorted object in the constellation of Ophiuchus consists of two spiral galaxies in the process of merging. The collision between the two galaxies has triggered an immense burst of star formation. (Courtesy of W. C. Keel)

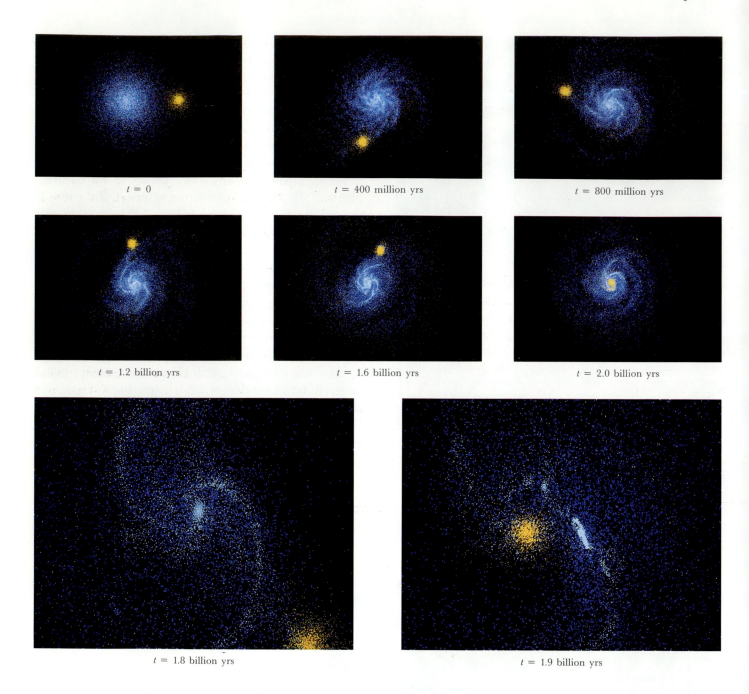

$t = 0$

$t = 400$ million yrs

$t = 800$ million yrs

$t = 1.2$ billion yrs

$t = 1.6$ billion yrs

$t = 2.0$ billion yrs

$t = 1.8$ billion yrs

$t = 1.9$ billion yrs

Figure 17-27 A simulated galactic cannibalism This simulation, performed at the Pittsburgh Supercomputing Center, shows a small galaxy (stars in orange) being devoured by a larger, disk-shaped galaxy (stars in blue, gas in white). The upper six pictures display progress at 400-million-year intervals. Note how spiral arms are induced in the disk galaxy as a result of its interaction with the satellite galaxy. The lower two pictures, which cover the interval from 1.8 to 1.9 billion years, are close-up views of the satellite galaxy plunging into the core of the disk galaxy. As the satellite galaxy sweeps through the inner regions of the disk galaxy, a significant amount of gas becomes concentrated along one of the spiral arms. Vigorous star formation would be expected in these gas clouds. (Courtesy of L. Hernquist)

type shown in Figure 17-27 would "feed" the black hole at the core of such a galaxy. Astrophysicists calculate that the in-falling matter would release enormous amounts of energy, thereby explaining the galaxy's power output. Computer simulations are becoming an increasingly important technique for exploring these exotic processes.

17-5 Galaxies formed billions of years ago from the gravitational contraction of huge clouds of primordial gas

Astronomers can probe the past to gain important clues about galactic evolution simply by looking deep into space. The more distant a galaxy is, the longer its light takes to reach us. Consequently, as we examine galaxies at increasing distances from Earth, we are actually looking farther and farther back in time, seeing galaxies at increasingly earlier stages of their lives.

By observing remote galaxies, astronomers have discovered that galaxies were bluer and brighter in the past than they are today. These changes in color and brightness suggest that a newly formed galaxy has an abundance of young, bright, hot, massive stars. As the galaxy ages, these O and B stars become red supergiants and eventually die off. The galaxy therefore gradually becomes somewhat redder and dimmer.

From studies of various types of galaxies, astronomers now know that ellipticals formed nearly all their stars in one vigorous burst of activity that lasted for only about a billion years. In contrast, spiral galaxies have been forming stars steadily for at least the past 10 billion years, although at a gradually decreasing rate. Indeed, there is still plenty of interstellar hydrogen in the disks of spiral galaxies like our Milky Way to fuel star formation today. Figure 17-28 compares the rates at which spiral and elliptical galaxies form stars.

It seems reasonable to suppose that galaxies formed from huge clouds of primordial hydrogen and helium roughly 10 to 15 billion years ago. This birth date for both spirals and ellipticals is deduced from the ages of the most ancient, metal-poor stars that both types of galaxies contain. Under the action of gravity, these pregalactic clouds of gas started to contract and form protogalaxies studded with the first generation of stars.

The rate of star formation in a protogalaxy determines whether it becomes a spiral or an elliptical. If the rate is low, then the gas has plenty of time to settle in a flattened disk. A flattened disk is the natural consequence of the overall rotation that the original gas cloud possessed. Star formation continues in this disk, because it contains an ample supply of hydrogen, and thus a spiral galaxy is created. But if the stellar birthrate is high, then virtually all of the gas is used up in the creation of stars before a disk can form. In this case, an elliptical galaxy is created. These contrasting scenarios are depicted in Figure 17-29.

Galactic evolution is a difficult and often controversial subject in which many questions remain. For instance, our discussion of the birth of a galaxy began with a cloud of precisely the right mass and size to guarantee that it would contract to become a galaxy. But where did this cloud come from? And what happened in the early universe to cause the primordial hydrogen and helium to gather in clouds destined to evolve into galaxies instead of becoming objects a million times bigger or smaller? Even more troublesome is the issue of the unseen mass. The observable stars, gas, and dust in a galaxy account for only about 10% of its mass. We have little idea what the remaining 90% looks like, how it's distributed in space, or what it's made of. With this level of ignorance, some astronomers argue that any discussion of galactic evolution is woefully inadequate.

17-6 The redshifts of remote galaxies are proportional to their distances from Earth

Whenever an astronomer finds an object in the sky that can be seen or photographed, the natural inclination is to attach a spectrograph to a telescope and record the spectrum. As long ago as 1914, V. M. Slipher, working at the Lowell Observatory in Arizona, began taking spectra of "spiral nebulae." He was surprised to discover that, of the 15 spiral nebulae he studied, the spectral lines of 11 of them were shifted toward the red end of the spectrum, indicating substantial recessional velocities for

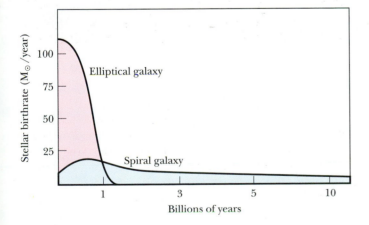

Figure 17-28 The stellar birth rate in galaxies *Most of the stars in an elliptical galaxy are created in a brief burst of star formation when the galaxy is very young. In spiral galaxies, stars form at a more leisurely pace that extends over billions of years. (Adapted from J. Silk)*

Figure 17-29 *The creation of spiral and elliptical galaxies* A galaxy begins as a huge cloud of primordial gas that collapses gravitationally. (a) If the rate of star birth is low, then much of the gas collapses to form a disk and a spiral galaxy is created. (b) If the rate of star birth is high, then the gas is converted into stars before a disk can form, resulting in an elliptical galaxy.

those objects. This marked dominance of redshifts was presented by Curtis in the Shapley–Curtis debate as evidence that these spiral nebulae could not be ordinary nebulae in our Milky Way Galaxy.

During the 1920s Edwin Hubble and Milton Humason photographed the spectra of many galaxies with the 100-inch telescope on Mount Wilson. Five representative elliptical galaxies and their spectra are shown in Figure 17-30. As indicated by this illustration, there seems to be a direct correlation between the distance to a galaxy and the size of its redshift. In other words, nearby galaxies are moving away from us slowly, and more distant galaxies are rushing away from us much more rapidly. This recessional motion pervades the universe and is called the **Hubble flow.**

Using various techniques, Hubble estimated the distances to a number of galaxies. Using the Doppler effect (recall Figure 5-19), Hubble calculated the speed at which each galaxy is receding from us. When he plotted the data on a graph of distance versus speed, he found that the points lie nearly along a straight line. Figure 17-31 is a modern version of Hubble's graph.

This relationship between the distances to galaxies and their redshifts is one of the most important astronomical

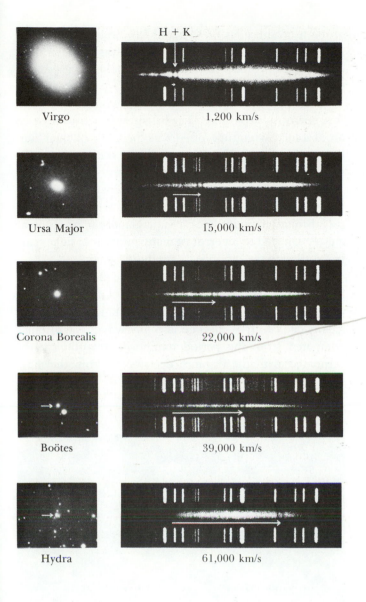

Virgo   H + K   1,200 km/s

Ursa Major   15,000 km/s

Corona Borealis   22,000 km/s

Boötes   39,000 km/s

Hydra   61,000 km/s

Figure 17-30   Five galaxies and their spectra *The photographs of these five elliptical galaxies all have the same magnification. The photographs are labeled according to the constellation in which each galaxy is located. The spectrum of each galaxy is the hazy band between the comparison spectra. In all five cases, the so-called H and K lines of calcium are seen. The recessional velocity (calculated from the Doppler shifts of the H and K lines) is given below each spectrum. Note that the fainter—and thus more distant—a galaxy is, the greater is its redshift. (The Observatories of the Carnegie Institution of Washington)*

discoveries of the twentieth century. As we shall see in Chapter 19, this relationship tells us that we are living in an expanding universe. In 1929 Hubble published this discovery, which is now known as the **Hubble law.**

The Hubble law is most easily stated as the formula

$$v = H_0 r$$

where v is the recessional velocity, r is the distance, and H_0 is a constant, commonly called the **Hubble constant.** This is the formula for the straight line displayed in Figure 17-31; the Hubble constant tells us the slope of this line.

The exact value of the Hubble constant is a topic of heated debate among astronomers. Various teams of astronomers who have used different techniques for estimating the distances to remote galaxies have arrived at differing values for H_0. Nevertheless, the most recent work seems to be converging on a value of about

$$H_0 = 80 \text{ km/s/Mpc}$$

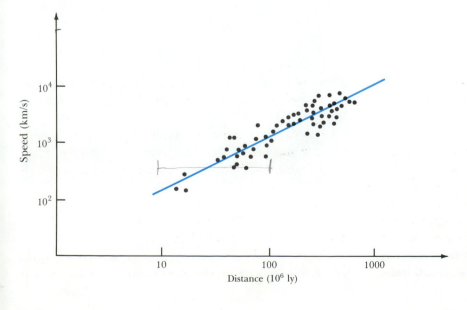

Figure 17-31 The Hubble law *The distances and recessional velocities of 60 Sc spiral galaxies are plotted on this graph. The straight line is the "best fit" for the data. This linear relationship between distance and speed is called the Hubble law. (Adapted from Sandage and Tammann)*

(say "80 kilometers per second per megaparsec"). In other words, for each million parsecs to a galaxy, the galaxy's speed away from us is larger by 80 km/s. For example, a galaxy located 100 million parsecs from Earth should be rushing away from us with a speed of 8000 km/s.

The Hubble constant is one of the most important numbers in all physical science. It expresses the rate at which the universe is expanding and thus tells us the age of the universe. Naturally, astronomers are very interested in an accurate determination of H_0.

In order to determine the Hubble constant, an astronomer must measure the redshifts and distances to many galaxies. Although redshift measurements can be quite precise, it is very difficult to gauge the distances to remote galaxies accurately. Indeed, conflicting measurements of distance are the main reason that the value of H_0 is so controversial.

Recall from our discussion in Chapter 12 that we can always find the distance to an object if we know both its apparent and absolute magnitudes. Astronomers use the term **standard candle** to denote any object whose absolute magnitude is known. Cepheid variables, the brightest supergiants, globular clusters, and supernovae are all useful as standard candles, because astronomers know something about their absolute magnitudes. To determine the distance to a galaxy, an astronomer must measure the apparent magnitude of one or more of these standard candles in that galaxy. As soon as both the apparent and absolute magnitudes are known, the distance to the galaxy can be easily calculated.

The distances to nearby galaxies can be determined by fairly reliable methods. For example, Cepheid variables can be seen out to 20 Mly from Earth. The distances to galaxies in this nearby volume of space can thus be determined from the period–luminosity law.

Beyond 20 Mly, even the brightest Cepheid variables, which have absolute magnitudes of about −6, fade from view. Astronomers then turn to more luminous stars. The brightest red supergiants have absolute magnitudes of −8; the brightest blue supergiants have absolute magnitudes of −9. These two types of stars can be seen out to distances of 50 million and 80 million light-years, respectively. Out to these limits, you can determine the distances to galaxies from the apparent magnitudes of these luminous supergiants.

Beyond 80 Mly, individual stars are no longer discernible. Astronomers therefore turn to entire clusters and nebulae. The brightest globular clusters, which have a total absolute magnitude of about −10, can be seen out to 130 Mly from Earth. The brightest H II regions have absolute magnitudes of −12 and can be detected out to 300 Mly. From the measured apparent brightness of these clusters and nebulae, distances to remote galaxies can be estimated.

Finally, to get beyond 300 Mly, astronomers must wait for supernova explosions. The brightest supernovae reach an absolute magnitude of −19 at the peak of their outbursts (Figure 17-32). These brilliant outbursts can be seen out to distances of 8 billion light-years from Earth.

Astronomers go to great lengths to check the reliability of their standard candles. After all, a tiny mistake in the absolute magnitude of a supergiant star or globular cluster can lead to an error of many millions of light years in calculating the distance to a remote galaxy.

The major obstacle in determining the Hubble constant is that the farther we look into space, the fewer standard candles we have. For example, the distance to a nearby galaxy can be cross-checked in many ways. The distance computed from the period–luminosity relation can be compared with the distance determined from the mag-

Figure 17-32 A supernova in M61 *In 1961 a supernova was discovered in this spiral galaxy, which is a member of the Virgo cluster. Supernovae can be seen in extremely remote galaxies and are important standard candles used to determine the distances to these faraway galaxies. (Lick Observatory)*

nitudes of the most luminous supergiants. Then these results can be compared with the magnitudes of the galaxy's globular clusters and the angular sizes of its H II regions. After all these steps, the results can then be averaged to obtain a distance to the galaxy in which astronomers can have some confidence.

As we turn to more distant galaxies, however, we can see fewer and fewer standard candles. For example, the nearest rich, regular cluster (the Coma cluster in Figure 17-17) is so far away that it is impossible to see individual stars. With fewer standard candles, fewer cross-checks can be made. The distance to these remote galaxies thus becomes less certain. Unfortunately, remote galaxies are precisely the objects whose distances we must determine to find the value of the Hubble constant. Uncertainty in determining distances is the cause of our uncertainty in the value of H_0.

Recently two astronomers discovered an important correlation, called the **Tully–Fisher relation,** between the width of the hydrogen 21-cm line and the absolute magnitude of spiral galaxies. This relation may be the best distance indicator currently available. That such a relationship should exist can be seen as follows. Radiation from the approaching side of a galaxy is blueshifted while that from the galaxy's receding side is redshifted. As a result, the 21-cm line is broadened by an amount directly related to how fast a galaxy is rotating, which is related to the galaxy's mass by Kepler's third law. Furthermore, the more massive a galaxy is, the more stars it contains, and so the brighter it is. Consequently, the width of a galaxy's 21-cm line is directly related to the galaxy's luminosity. Since line widths can be measured quite accurately, astronomers can use the Tully–Fisher relation to determine the luminosities of spiral galaxies.

Astronomers are hopeful that techniques like the Tully–Fisher relation along with instruments like the Hubble Space Telescope (HST) and the Keck Telescope will solve many of the problems of determining H_0. For instance, after its optical and mechanical defects are corrected, the HST is expected to produce exceptionally sharp views of galaxies, which will allow astronomers to identify standard candles out to distances well beyond the reach of earlier generations of telescopes. Advances such as these will contribute significantly to our understanding of the structure and evolution of galaxies and the universe.

Summary

- The Hubble classification system groups galaxies into four major categories: spirals, barred spirals, ellipticals, and irregulars.

  Spiral galaxies and barred spiral galaxies are sites of active star formation.

  Elliptical galaxies are virtually devoid of interstellar gas and dust; no star formation is occurring in these galaxies.

- Galaxies are grouped in clusters rather than scattered randomly through the universe.

  A rich cluster contains hundreds or even thousands of galaxies; a poor cluster may contain only a few dozen galaxies.

  A regular cluster has a nearly spherical shape with a central concentration of galaxies; in an irregular cluster, the distribution of galaxies is asymmetrical.

  Our Galaxy is a member of a poor, irregular cluster called the Local Group.

  Rich, regular clusters contain mostly elliptical and S0 galaxies; irregular clusters contain more spiral and irregular galaxies.

  Giant elliptical galaxies are often found near the centers of rich clusters.

- The observable mass of a cluster of galaxies is not large enough to account for the observed motions of the galaxies; a large amount of unobserved mass must be present between the galaxies. This situation is referred to as the dark-matter problem.

  Hot intergalactic gases emit X rays in rich clusters. Extended halos of dim stars probably surround all galaxies.

- When two galaxies collide, their stars pass each other, but their interstellar media collide violently, either stripping the gas and dust from the galaxies or triggering prolific star formation.

The gravitational effects of a galactic collision can throw stars out of their galaxies into intergalactic space.

Galactic mergers may occur; a large galaxy in a rich cluster may grow steadily through galactic cannibalism, perhaps producing a giant elliptical galaxy.

- Galaxies probably formed roughly 10 to 15 billion years ago by the gravitational contraction of huge clouds of hydrogen and helium.

The rate of star formation in a protogalaxy determines whether it will become a spiral galaxy or an elliptical galaxy. A low rate of star birth results in a spiral galaxy because the interstellar gas has time enough to collapse and form a disk. A high rate of star birth turns the gas into stars before a disk can form, resulting in an elliptical galaxy.

- There is a simple linear relationship between the distance from the Earth to a galaxy and the redshift of that galaxy (which is a measure of the speed with which it is receding from us); this relationship is the Hubble law, $v = H_0 r$.

Standard candles, such as Cepheid variables, the brightest supergiants, globular clusters, H II regions, and supernovae in a galaxy, are used in estimating intergalactic distances. Because of difficulties in measuring the distances to galaxies, the value of the Hubble constant H_0 is not known with certainty; this leads to uncertainties about the rate at which the universe is expanding and about the age of the universe.

Review questions

1 What was the Shapley–Curtis debate all about? Was a winner declared at the end of the debate? Whose ideas turned out to be correct?

2 How did Edwin Hubble prove that the "Andromeda Nebula" is not a nebula in our Milky Way Galaxy?

3 What is the Hubble classification scheme? Which category includes the biggest galaxies? Into which category do the smallest galaxies fall? Which type of galaxy is the most common?

4 In which types of galaxies are new stars most likely

forming? Describe the observational evidence that supports your answer.

5 How is it possible that galaxies in our Local Group still remain to be discovered? In what part of the sky would these galaxies be located? What sorts of observations might reveal these galaxies?

6 Are there any galaxies besides our own that can be seen with the naked eye? If so, which ones can you name?

7 What is the difference between a rich cluster of galaxies and a poor one? What is the difference between a regular cluster of galaxies and an irregular one?

8 How can a collision between galaxies produce a starburst galaxy?

9 Why do astronomers believe that there must be considerable quantities of dark matter in clusters of galaxies?

10 Explain why the dark matter in galaxy clusters could not be neutral hydrogen.

11 Outline a process that could have caused a protogalaxy to evolve into an elliptical galaxy. What would have had to occur differently for the protogalaxy to have become a spiral galaxy?

12 What is the Hubble law?

13 Some galaxies in the Local Group exhibit blueshifted spectral lines. Why aren't these blueshifts violations of the Hubble law? Explain.

14 What is a standard candle? Why are standard candles important to astronomers who try to measure the Hubble constant?

15 Why are there differing values of H_0?

16 What kinds of stars would you expect to find populating space between galaxies in a cluster?

Advanced questions

* 17 Suppose you were to take a spectrum of a distant galaxy and find that its redshift corresponds to a speed of 22,000 km/s. How far away is the galaxy?

* 18 A cluster of galaxies in the southern constellation of Pavo (the Peacock) is located 100 Mpc from Earth. How fast is this cluster receding from us?

19 How might you determine what fraction of a galaxy's redshift is caused by the galaxy's orbital motion about the center of mass of its cluster?

Discussion questions

20 Discuss the advantages and disadvantages of using the various standard candles to determine extragalactic distances.

21 Discuss whether the various Hubble types of galaxies actually represent some sort of evolutionary sequence.

22 Discuss the sorts of phenomena that can occur when galaxies collide. Do you think that such collisions can change the Hubble type of a galaxy?

For further reading

de Vaucouleurs, G. "The Distance Scale of the Universe." *Sky & Telescope,* December 1983 • This article gives interesting historical insights into the trials and tribulations of determining distances to galaxies.

———. "M31's Spiral Shape." *Sky & Telescope,* December 1987 • This article gives a fascinating historical account of how astronomers came to recognize the spiral structure of the Andromeda Galaxy.

Field, G. "The Hidden Mass in Galaxies." *Mercury,* May/June 1982 • This well-written article describes evidence for dark matter in the universe.

Gorenstein, P., and Tucker, W. "Rich Clusters of Galaxies." *Scientific American,* November 1978 • This article summarizes the main properties of rich clusters of galaxies.

Hartley, K. "Elliptical Galaxies Forged by Collision." *Astronomy,* May 1989 • This brief article summarizes arguments for the idea that collision and merging of spiral galaxies may give birth to elliptical galaxies.

Hodge, P. *Galaxies,* Harvard University Press, 1986 • Written in a friendly, informal style, this superb book covers galaxies and galactic evolution and closely related topics.

Keel, W. "Crashing Galaxies, Cosmic Fireworks." *Sky & Telescope,* January 1989 • This article discusses the many galactic forms and phenomena that result from collisions between galaxies.

Osterbrock, D., Brashear, R., and Gwinn, J. "Young Edwin Hubble." *Mercury,* January/February 1990 • This excellent article chronicles the adolescence and early adulthood of one of the greatest astronomers of all time.

Rubin, V. "Dark Matter in Spiral Galaxies." *Scientific American,* June 1983 • This fine article on dark matter gives many insights into how astronomers practice their profession.

Silk, J. "Formation of the Galaxies." *Sky & Telescope,* December 1986 • Taking clues from the many properties of galaxies, this article presents an overview of how galaxies probably form.

Smith, R. *The Expanding Universe: Astronomy's Great Debate.* Cambridge University Press, 1982 • This book gives a detailed, meticulously documented history of the birth and development of extragalactic astronomy.

Tully, R. "Unscrambling the Local Supercluster." *Sky & Telescope,* June 1982 • This article explains how astronomers discover and try to understand large-scale structures like the "Local Supercluster," that is composed of the Local Group and other nearby clusters of galaxies.

Wray, J. *The Color Atlas of Galaxies.* Cambridge University Press, 1988 • This gorgeous book contains a magnificent collection of color photographs of more than 600 galaxies.

The core of a radio galaxy   *This artist's rendition shows a scenario that many astronomers believe is responsible for double radio sources. A supermassive black hole at the center of a galaxy is surrounded by an accretion disk. In the inner regions of the accretion disk, matter crowding toward the hole is diverted outward along two oppositely directed beams. These beams deposit energy into two huge, radio-emitting lobes located on either side of the galaxy. (Astronomy)*

18

Quasars and Active Galaxies

Since the 1960s, astronomers have discovered thousands of objects in the sky whose large redshifts indicate that they are extremely far away, typically more than a billion light-years from Earth. To be observable at all at such huge distances, these so-called quasars and active galaxies must be exceptionally luminous. Indeed, observations indicate that a typical quasar emits the energy output of 100 galaxies from a volume roughly the size of our solar system. Some quasars and the nuclei of active galaxies also exhibit puzzling features such as jets that propel matter outward at relativistic speeds. The preponderance of evidence leads many astronomers to believe that supermassive black holes are responsible for the energy emitted by these luminous objects.

The development of radio astronomy in the late 1940s ranks among the most important scientific accomplishments of the twentieth century. Until then, everything known about the distant universe had to be gleaned from visual observations. Radio telescopes provided a view of the universe in a wavelength range far beyond that of visible light. Many surprising discoveries emerged from this new ability to examine the previously invisible universe.

The first radio telescope was built in 1936 by an amateur astronomer, Grote Reber, in his backyard in Illinois. By 1944 Reber had detected strong radio emissions from Sagittarius, Cassiopeia, and Cygnus. Two of these sources, nicknamed Sgr A and Cas A, happen to be in our Galaxy—they are the galactic nucleus and a supernova remnant. In 1951 Walter Baade and Rudolph Minkowski used the 200-inch optical telescope on Palomar Mountain to discover a strange-looking galaxy at the position of the third source, called Cygnus A (Cyg A). Figure 18-1 is a photograph of the optical counterpart of Cyg A.

The peculiar galaxy associated with Cygnus A is very dim. Nevertheless, Baade and Minkowski managed to photograph its spectrum, which shows a number of emission lines, all shifted by 5.7% toward the red end of the spectrum. A redshift of 5.7% corresponds to a speed of 17,000 km/s. According to the Hubble law, this speed corresponds to a distance of 700 million light-years!

The enormous distance to Cygnus A astounded astronomers, because Cyg A is one of the brightest radio sources in the sky. Although Cyg A is barely visible through the 200-inch telescope at Palomar, its radio waves can be picked up by amateur astronomers with backyard equipment. The energy output in radio waves from Cyg A must therefore be enormous. In fact, Cyg A shines with a radio luminosity 10^7 times as bright as that of an ordinary galaxy such as M31 in Andromeda. The object corresponding to Cyg A must obviously be something quite extraordinary.

18-1 Quasars look like stars but have huge redshifts

During the late 1950s and early 1960s, radio astronomers were busy making long lists of all the radio sources they were finding across the sky. One of the most famous lists, titled the *Third Cambridge Catalogue* (the first two catalogues produced by the British team were filled with inaccuracies) was published in 1959. It lists 471 radio sources. Even today astronomers often refer to these sources by their "3C numbers." With the discovery of the extraordinary luminosity of Cyg A (designated 3C 405, because it is the 405th source on the Cambridge list), astronomers were eager to learn whether any other sources in the 3C catalog had similarly extraordinary properties.

One interesting case is 3C 48. In 1960 Allan Sandage used the 200-inch Palomar telescope to discover a "star" at the location of this radio source (Figure 18-2). Because ordinary stars are not strong sources of radio emission, 3C 48 must be something unusual. Indeed, its spectrum showed a series of emission lines that no one could identify. Although 3C 48 was clearly an oddball, many astronomers thought it was just another strange star in our Galaxy.

Another such "star," called 3C 273, was discovered in 1962. It was found to have a luminous "jet" protruding

Figure 18-1 Cygnus A (also called 3C 405) This strange-looking galaxy was discovered at the location of the radio source Cygnus A (Cyg A). This galaxy has a redshift corresponding to a recessional speed of 6% of the speed of light. According to the Hubble law, this redshift corresponds to a distance of about 700 million light-years from Earth. Because Cyg A is one of the brightest radio sources in the sky, the energy output of this remote galaxy must be enormous. (Palomar Observatory)

Figure 18-2 The quasar 3C 48 *For several years astronomers believed erroneously that this object is simply a peculiar nearby star that happens to emit radio waves. Actually, the redshift of this starlike object is so great that, according to the Hubble law, it must be roughly 4 billion light-years away. (Palomar Observatory)*

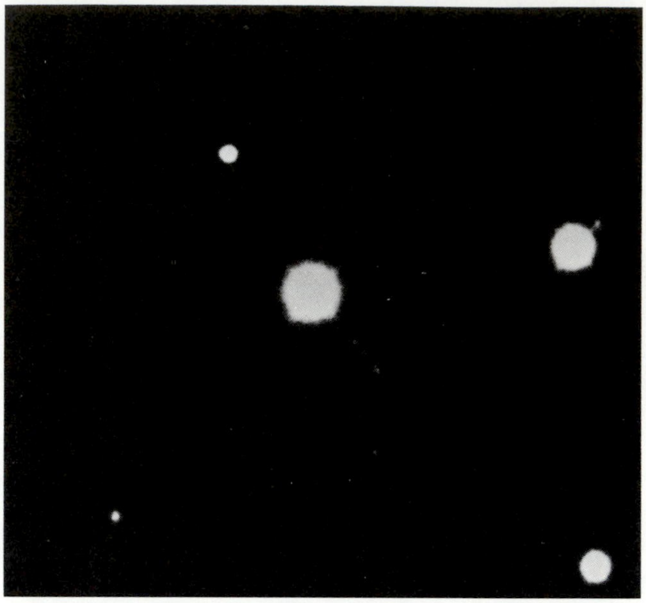

Figure 18-3 The quasar 3C 273 *This greatly enlarged view shows the starlike object associated with the radio source 3C 273. Note the luminous jet to one side of this "star." By 1963 astronomers discovered that the redshift of this "star" is so great that its distance, according to the Hubble law, is nearly 2 billion light-years from Earth. (NOAO)*

from one side, as shown in Figure 18-3. And, as was the case with 3C 48, this object contains a series of bright spectral lines that no one could identify.

The reason astronomers had difficulty identifying the emission lines in the spectra of 3C 48 and 3C 273 was that they had assumed these starlike objects were peculiar stars nearby in our Galaxy. After all, they certainly look like stars.

In 1963 a breakthrough finally occurred when Maarten Schmidt at the California Institute of Technology found that four of the brightest spectral lines of 3C 273 are positioned relative to one another just like four familiar spectral lines of hydrogen. However, these emission lines of 3C 273 are at much longer wavelengths than the usual positions of the hydrogen lines. In other words, the light from 3C 273 has a substantial redshift.

Stellar spectra exhibit comparatively small Doppler shifts, because a star in our Galaxy cannot be moving extremely fast relative to the Sun without soon escaping from the Galaxy. Schmidt thus conjectured that 3C 273 might not be a nearby star after all. Pursuing this hunch, he promptly identified all four spectral lines as being

hydrogen lines that have suffered an enormous redshift corresponding to a speed of almost 15% of the speed of light. According to the Hubble law, this huge redshift implies the incredible distance to 3C 273 of roughly 2 billion light-years.

Because of their strong radio emission and starlike appearances, 3C 48 and 3C 273 were dubbed **quasi-stellar radio sources.** This term was soon shortened to quasars.

Figure 18-4 shows the spectrum of 3C 273. Instead of using photography to record a spectrum, many observatories now use a charged-coupled device (CCD) at the focus of a spectrograph. As discussed in Chapter 5 (review Figure 5-12), the output of this device is a graph of intensity versus wavelength on which emission lines appear as peaks. The graph in Figure 18-4 was obtained in this way. The four hydrogen lines are identified.

The spectral lines of 3C 273 are brighter than the intensity of the background radiation at other wavelengths. The background is called the **continuum.** The emission lines are caused by excited atoms that are emitting radiation at specific wavelengths. Spectra of ordinary galaxies have dark absorption lines (recall Figure 17-30). Most

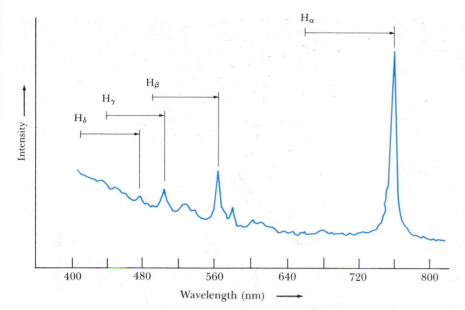

Figure 18-4 The spectrum of 3C 273 Four bright emission lines caused by hydrogen dominate the spectrum of 3C 273. The arrows indicate how far these spectral lines are redshifted from their usual wavelengths.

quasars and many peculiar galaxies exhibit strong emission lines in their spectra, a sign that something unusual is going on.

Inspired by Schmidt's success, astronomers next identified the spectral lines of 3C 48 as having suffered a redshift corresponding to a velocity of nearly one-third the speed of light. Therefore 3C 48 must be nearly twice as far away as 3C 273, or about 4 billion light-years from Earth according to the Hubble law.

Incidentally, these distances to quasars assume that the Hubble constant (H_0) is 80 km/s/Mpc. As discussed in the previous chapter, the value of the Hubble constant is not known with certainty. A higher value of H_0 would mean that the quasars are nearer to us than the distances given in this chapter; conversely, a lower value would mean they are farther away.

Thousands of quasars have been discovered since the pioneering days of the early 1960s. All quasars look like stars, and all have enormous redshifts. A few quasars have redshifts corresponding to speeds greater than 90% of the speed of light. Figure 18-5 shows the spectrum of a quasar whose redshift corresponds to a speed of 92% of the speed of light. This is such an enormous redshift that the Lyman alpha line is shifted all the way from its usual

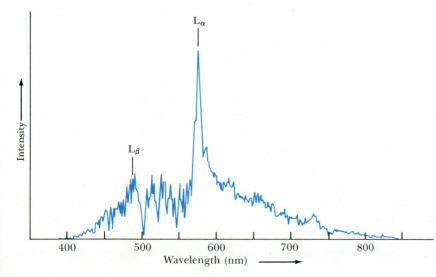

Figure 18-5 The spectrum of a high-redshift quasar The light from this quasar, known as PKS 2000-330, is so highly redshifted that spectral lines normally in the far-ultraviolet (L_α and L_β) can be seen at visible wavelengths. Note the large number of deep absorption lines on the short-wavelength side of L_α. These lines, collectively called the "Lyman- alpha forest," are probably caused by remote clouds of gas along our line of sight to the quasar. Hydrogen in these clouds absorbs photons from the quasar at wavelengths less redshifted than the quasar's L_α line.

wavelength of 121.6 nm in the far-ultraviolet into the middle of the visible spectrum. From the Hubble law, it follows that the distances to these high-redshift quasars is typically in the range of 10 to 13 billion light-years—their light has taken 10 to 13 billion years to reach us. When we look at these quasars, we are seeing objects as they existed when the universe was very young.

18-2 A quasar emits a huge amount of energy from a small volume

Galaxies are big and bright. A typical large galaxy, like our own Milky Way, contains several hundred billion stars and shines with a luminosity of 10 billion Suns. The most gigantic and most luminous galaxies, such as the giant ellipticals, are only 10 times brighter, shining with the brilliance of 100 billion Suns. Beyond 8 billion light-years from Earth, even the brightest galaxies are too faint to be easily detected. Most ordinary galaxies are too dim to be detected at half that distance. The fact that quasars can be seen at distances where galaxies are not visible means that quasars are more luminous than galaxies. Indeed, a typical quasar is 100 times brighter than a typical galaxy like our Milky Way.

In the mid-1960s, several astronomers discovered that some of the newly identified quasars had been photographed inadvertently in the past. For example, 3C 273 was found on numerous photographs, including one taken in 1887. By carefully examining the images of quasars on these old photographs, astronomers found that quasars fluctuate in brightness, occasionally flaring up. See, for example, the data from old photographs

of another quasar, 3C 279, plotted in Figure 18-6. Note the prominent outbursts that occurred around 1937 and 1943. During these outbursts, the luminosity of 3C 279 increased by a factor of at least 25. Because of the enormous distance to this quasar, it must have been shining with a brilliance at least 10,000 times as great as that of the entire Milky Way.

Fluctuations in the brightness of quasars allow astronomers to place strict limits on the sizes of quasars, because an object cannot be observed to vary in brightness faster than the time it takes light to travel across that object. For example, an object that is 1 ly in diameter cannot vary in brightness with a period of less than 1 year.

To understand this limitation, imagine an object that measures 1 ly across, as shown in Figure 18-7. Suppose the entire object emits a brief flash of light. Photons from that part of the object nearest the Earth arrive at our telescopes first. Photons from the middle of the object arrive at Earth six months later. And finally, light from the far side of the object arrives a year after the first photons. Although the object emitted a sudden flash of light, we observe a gradual change in brightness that lasts a full year. In other words, the flash is stretched out over an interval equal to the difference in time that light takes to travel between the nearest and most remote observable regions of the object.

The brightness of many quasars varies over intervals of only a few weeks or months. Some quasars actually fluctuate from night to night. Recent X-ray data from the Einstein Observatory reveal large variations in as little as three hours.

This rapid flickering means that quasars must be small. The energy-emitting region of a typical quasar—the "powerhouse" that blazes with the luminosity of 100 galaxies—is less than 1 light-day in diameter. If quasars

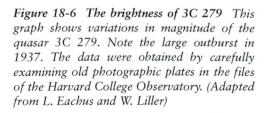

Figure 18-6 The brightness of 3C 279 This graph shows variations in magnitude of the quasar 3C 279. Note the large outburst in 1937. The data were obtained by carefully examining old photographic plates in the files of the Harvard College Observatory. (Adapted from L. Eachus and W. Liller)

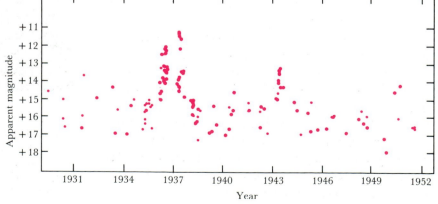

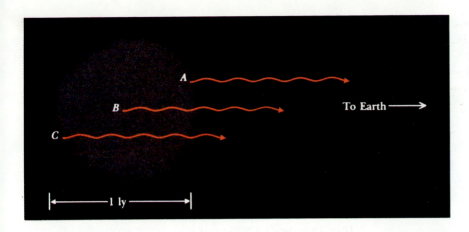

Figure 18-7 *A limit to the speed of variations in brightness* *The rapidity with which the brightness of an object can vary is limited by the time that light takes to travel across the object. Even if the object, shown here to be 1 ly in size, emits a sudden flash of light, photons from point A arrive at Earth one year before photons from point C. Thus, as seen from Earth, the sudden flash is observed as a gradual change in brightness over a full year.*

are indeed at the huge distances indicated by their redshifts, they must be producing the luminosity of 100 galaxies in a volume roughly the same size as our solar system!

18-3 Active galaxies bridge the gap in energy output between ordinary galaxies and quasars

In the 1960s, the gap in energy output between ordinary galaxies and quasars seemed so huge that some astronomers preferred to question the interpretation of quasar redshifts rather than accept the existence of such highly luminous objects. In recent years, however, astronomers have discovered various kinds of peculiar galaxies whose luminosities fall between those of ordinary galaxies and quasars. Some of these strange galaxies have unusually bright, starlike nuclei, others have strong emission lines in their spectra, and still others are highly variable. Some have jets and beams of radiation emanating from their cores, and most of these objects are more luminous than ordinary galaxies. All are called **active galaxies.**

The first active galaxies were discovered in 1943 during a survey of spiral galaxies by Carl Seyfert at the Mount Wilson Observatory. Called **Seyfert galaxies,** these luminous objects have bright, starlike nuclei and strong emission lines in their spectra. For example, NGC 4151 (Figure 18-8) has an extremely rich spectrum with many prominent emission lines. Some of these emission lines are produced by iron atoms with a dozen or more electrons stripped away, indicating that NGC 4151 contains some extremely hot gas. Seyfert galaxies also exhibit variability in brightness. For instance, sometimes the magnitude of NGC 4151 changes over a few days.

Another example of a Seyfert galaxy is NGC 1068, shown in Figure 18-9. At infrared wavelengths this galaxy shines with the brilliance of 10^{11} Suns. This extraordinary luminosity has been observed to vary by as much as 7×10^9 Suns over only a few weeks. In other words, the infrared power output of the nucleus of NGC 1068 rises and falls by an amount nearly equal to the total luminosity of our entire Galaxy.

Figure 18-8 *The Seyfert galaxy NGC 4151* *This is one of the best-studied Seyfert galaxies. Because of its bright, starlike nucleus and emission-line spectrum, NGC 4151 might be mistaken for a quasar if it were very far away. According to the Hubble law, the redshift of this galaxy indicates a distance of 40 million light-years from Earth. (Palomar Observatory)*

Figure 18-9 The Seyfert galaxy NGC 1068   This Seyfert galaxy, also called M77 or 3C 71, is renowned for its extraordinary infrared luminosity, which varies over intervals as short as a few weeks. Note how bright the inner spiral arms are compared with the outer spiral arms. This galaxy is about the same distance from Earth as NGC 4151 in Figure 18-8. (Courtesy of R. Schild)

Many more Seyfert galaxies have been discovered in recent years. Approximately 10% of the most luminous galaxies in the sky are Seyfert galaxies. Some of the brightest Seyfert galaxies shine as brightly as faint quasars, which leads many astronomers to suspect that the nuclei of Seyfert galaxies are, in fact, low-luminosity quasars.

Some Seyfert galaxies exhibit vestiges of violent, explosive phenomena in their nuclei. For instance, NGC 1275 in Figure 18-10 has filaments of gas tens of thousands of light-years long protruding from its nucleus in all directions. Spectroscopic studies indicate that this gas is being blasted away from the galaxy's nucleus at 3000 km/s. The nucleus of this galaxy is a strong source of radio waves and X rays (Figure 18-11). In 1977 Vera Rubin and her colleagues at the Carnegie Institution of Washington reported observations demonstrating that NGC 1275 actually consists of two galaxies. As we saw in the previous chapter, a collision or close encounter between two galaxies can result in the ejection of matter into intergalactic space.

The high-speed ejection of matter from active galaxies is seen best at nonvisible wavelengths. For instance, Figure 18-12a is a photograph of the peculiar galaxy NGC 5128 in the southern constellation of Centaurus. An unusual broad dust lane stretches across the galaxy.

Figure 18-10 The active galaxy NGC 1275 (also called 3C 84) This Seyfert galaxy, located in the Perseus cluster, is a strong source of X rays and radio radiation. According to the Hubble law, the galaxy's redshift indicates a distance of nearly 300 million light-years from Earth. Note the streamers of gas. Spectroscopic observations confirm significant mass ejection from the galaxy's center. (NOAO)

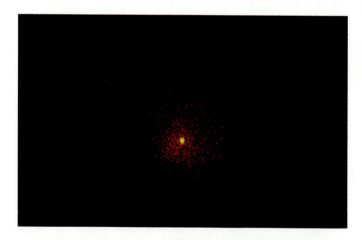

Figure 18-11 An X-ray image of NGC 1275 This picture from the Einstein Observatory shows the X-ray appearance of the Seyfert galaxy NGC 1275. Most of the X-ray emission comes from a point source at the galaxy's nucleus. (Harvard-Smithsonian Center for Astrophysics)

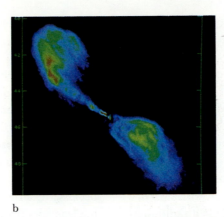

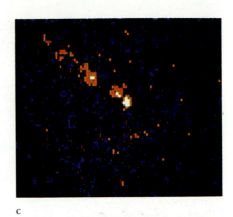

a

b

c

Figure 18-12   The peculiar galaxy NGC 5128 (also called Centaurus A)   This extraordinary galaxy is located in the constellation of Centaurus, roughly 13 million light-years from Earth. These three views show visible, radio, and X-ray images of this galaxy, all to the same scale. (a) This photograph at visible wavelengths shows a dust lane across the face of the galaxy. (b) This radio image shows that vast quantities of radio radiation pour from extended regions of the sky on either side of the dust lane. (c) This X-ray picture from the Einstein Observatory shows that NGC 5128 has a bright X-ray nucleus. An X-ray jet protrudes from the galaxy's nucleus along a direction perpendicular to the galaxy's dust lane. (NOAO; D. A. Chartee and J. O. Burns; Harvard-Smithsonian Center for Astrophysics)

The galaxy NGC 5128 is one of the brightest sources of radio waves in the sky. It was one of the first sources discovered when radio telescopes were erected in Australia. As a radio source, it is called Centaurus A, because of its location in the sky. In part, the brightness of Centaurus A at radio wavelengths comes from its proximity to Earth, only 13 million light-years away. As shown in Figure 18-12*b*, radio waves pour from two regions, called **radio lobes,** on either side of the galaxy's dust lane. Farther from the dust lane is a second set of radio lobes that spans a volume 2 Mly across. Recent X-ray pictures of NGC 5128 reveal an X-ray jet (Figure 18-12*c*) sticking out of the galaxy's nucleus. This jet, which is perpendicular to the galaxy's dust lane, is aimed toward one of the radio lobes. These observations suggest that particles and energy stream out of the galaxy's nucleus toward the radio lobes.

By 1970 radio astronomers had discovered dozens of objects similar to Centaurus A that are now called **double radio sources.** An active galaxy, usually resembling a giant elliptical, is often found between the two radio lobes. For instance, the visible galaxy associated with Cygnus A (recall Figure 18-1) is located between two radio lobes, shown in Figure 18-13.

All double radio sources seem to have some sort of central "engine" that squirts electrons and magnetic field outward along two oppositely directed jets at speeds very near the speed of light. After traveling many thousands or even millions of light-years, this ejected material slows down, allowing the electrons and the magnetic field to produce the radio radiation that we detect. A specific type of radio emission called synchrotron radiation occurs whenever electrons move in a spiral within a magnetic

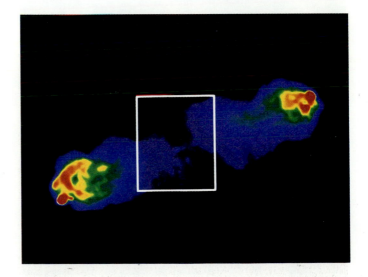

Figure 18-13 A radio image of Cygnus A   This color-coded radio picture was produced at the Very Large Array. Most of the radio emission from Cygnus A comes from the radio lobes located on either side of the peculiar galaxy seen in Figure 18-1. These two radio lobes are each about 160,000 light-years from the optical galaxy, and each contains a brilliant, condensed region of radio emission. The rectangle indicates the area covered by Figure 18-1. (VLA, NRAO)

field. The radio waves that come from the lobes of a double radio source have all the characteristics of synchrotron radiation.

The idea that a double radio source involves powerful jets of particles traveling near the speed of light is supported by the existence of **head–tail sources,** so named because each such source appears to have a "head" of concentrated radio emission, with a weaker "tail" trailing behind it. A good example is the active elliptical galaxy NGC 1265 in the Perseus cluster of galaxies. NGC 1265 is known to be moving at a high speed (2500 km/s) relative to the cluster as a whole. Figure 18-14 is a radio map of NGC 1265. Note that the radio emission has a distinctly windswept appearance. Just as smoke pouring from a steam locomotive trails a rapidly moving train, particles ejected along two jets from this galaxy are deflected by the galaxy's passage through the sparse intergalactic medium.

At radio wavelengths, the double radio sources are among the brightest objects in the universe. The energy contained in the radio lobes of a typical double radio source roughly equals the energy released by 10 billion supernova explosions.

A final class of active galaxies is the **N galaxies,** so named because they have bright nuclei. Certain extreme examples of these galaxies are the **BL Lacertae objects,** named after their prototype, BL Lacertae (BL Lac), in the constellation of Lacerta (the Lizard). BL Lacertae objects are also sometimes called **blazars.**

BL Lacertae (Figure 18-15) was first discovered in 1929, when it was mistaken for a variable star, largely because its brightness varies by a factor of 15 in only a few months. BL Lac's most intriguing characteristic is a totally featureless spectrum that exhibits neither absorption nor emission lines.

Careful examination of BL Lac revealed some "fuzz" around its bright, starlike core. By blocking out the light from the bright center of BL Lac, astronomers have been able to obtain a spectrum of this fuzz. This spectrum contains many spectral lines and strongly resembles the spectrum of an elliptical galaxy (Figure 18-16). In other words, a BL Lacertae object is an elliptical galaxy with a

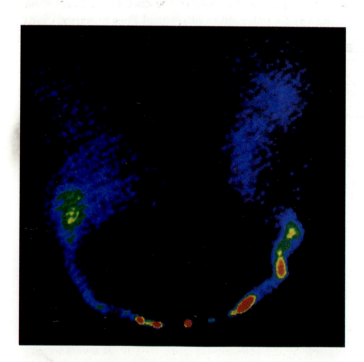

Figure 18-14 **The head–tail source NGC 1265** *The active elliptical galaxy NGC 1265 would probably be an ordinary double radio source except that the galaxy is moving at a high speed through the intergalactic medium. Because of this motion, the two jets trail the galaxy, giving this radio source a distinctly windswept appearance. (NRAO)*

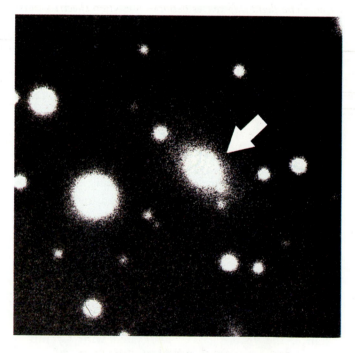

Figure 18-15 **BL Lacertae** *This superb photograph shows "fuzz" around BL Lacertae. BL Lacertae objects appear to be giant elliptical galaxies with bright, starlike nuclei, much as Seyfert galaxies are spiral galaxies with quasarlike nuclei. BL Lacertae objects contain much less gas and dust than do Seyfert galaxies. (Courtesy of T. D. Kinman; NOAO)*

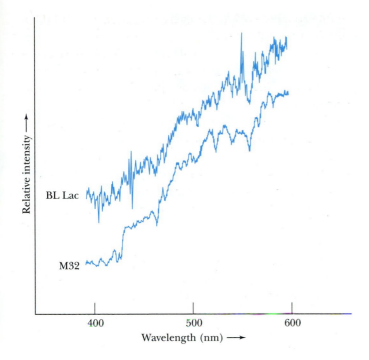

Figure 18-16 The spectrum of BL Lacertae The spectrum of the "fuzz" surrounding BL Lacertae (BL Lac) is shown, along with the spectrum of M32, a small elliptical galaxy in the Local Group. The slight differences between these two spectra can be explained by assuming that BL Lac is a giant elliptical galaxy. The spectrum of BL Lac is redshifted by an amount corresponding to a distance of 1 billion light-years from Earth. (Adapted from J. S. Miller, H. B. French, and S. A. Hawley)

bright starlike center, much as a Seyfert galaxy is a spiral galaxy with a quasarlike center.

These discoveries have prompted many astronomers to suspect that quasars are the superluminous centers of very distant, very active galaxies. Painstaking observations have in fact revealed faint galaxylike fuzz around several quasars. Of course, quasars are extremely far away and are thus difficult to observe. Therefore clues about the engine that powers a quasar might come from studying relatively nearby active galaxies.

18-4 Supermassive black holes may be located at the centers of some galaxies

How do quasars and active galaxies produce such enormous amounts of energy from such small volumes? As long ago as 1968, the British astronomer Donald

Lynden-Bell suggested that an extremely massive black hole could be the engine that powers a quasar or active galaxy. Astronomers today often use the term supermassive black hole when referring to a black hole that may contain millions or even billions of solar masses. At the center of a quasar or active galaxy, nature may be drawing upon the tremendous energy tied up in the gravitational field of a supermassive black hole.

As we saw in Chapter 15, finding black holes is a difficult business. At best, we can see only the effects of the hole's gravity and try to rule out non-black-hole explanations of the data. This general approach has been applied to several nearby galaxies, including M31 and M32.

The Andromeda Galaxy (M31) is the largest, most massive galaxy in the Local Group. At a distance of only 2.2 million light-years from Earth, M31 is so close to us that details in its core as small as 1 pc across can be resolved under the best seeing conditions. A wide-field view of M31 is seen in Figure 17-3. A photograph of the core of M31 is shown in Figure 18-17.

In the mid-1980s several astronomers made careful spectroscopic observations of the core of M31. By measuring the Doppler shifts of spectral lines at various locations in the core, they could determine the orbital speeds of stars about the galaxy's nucleus. The results show that stars within the central 50 ly of the galaxy's nucleus are orbiting the nucleus at exceptionally high speeds, which suggests that a massive object is located at the galaxy's center. Without the gravity of such an object to keep the stars in their high-speed orbits, they would have escaped from the galaxy's core long ago. From such observations, astronomers estimate that the mass of the central object is about 50 million solar masses. That much matter confined to such a small volume strongly suggests the existence of a supermassive black hole.

Located near M31 is a small elliptical galaxy called M32 (Figure 18-18a). High-resolution spectroscopy of this galaxy indicates that stars close to the center of M32 are orbiting the galaxy's nucleus at unusually high speeds, which could be explained by the presence of a supermassive black hole there. Furthermore, a picture taken by the Hubble Space Telescope (Figure 18-18b) shows a remarkable concentration of stars at the core of M32. The density of stars there is more than a hundred million times greater than the density of stars in the Sun's neighborhood. This strong concentration of stars also suggests the presence of a supermassive black hole whose powerful gravity causes the stars to crowd around it.

Figure 18-17 The central bulge of M31 *The central bulge of the Andromeda Galaxy can be seen with the naked eye on a clear, moonless night. A photograph of the entire galaxy is shown in Figure 17-3. Spectroscopic observations suggest that a supermassive black hole is located at the center of this large, nearby galaxy. (NOAO)*

Astronomers have uncovered evidence of supermassive black holes in other, more remote galaxies. John Kormendy used a 3.6-m telescope on Mauna Kea to examine the core of M104 spectroscopically (Figure 18-19). Once again high-speed gas orbiting the galaxy's nucleus was found. These observations suggest that the center of this galaxy is dominated by a supermassive black hole containing a billion solar masses. Similar spectroscopic observations of the edge-on S0 galaxy NGC 3151 reveal a 10^9-$M_\odot$ supermassive black hole at that galaxy's core also.

a

b

Figure 18-18 The elliptical galaxy M32 (a) *This small galaxy is a satellite of M31, a portion of which is seen at the left of this wide-angle photograph. Both galaxies are about 2.2 million light-years from Earth.* (b) *This high-resolution image from the* Hubble Space Telescope *shows the center of M32. Note the concentration of stars at the nucleus of the galaxy. The area covered in this view is 175 light-years on a side. (Palomar Observatory; NASA, ESA)*

Figure 18-19 The Sombrero Galaxy (also called M104) This spiral galaxy in Virgo is nearly edge-on to our Earth-based view. Spectroscopic observations suggest that a billion-solar-mass black hole is located at the galaxy's center. (ESO)

Recent observations with the Hubble Space Telescope (HST) point to the possibility of a supermassive black hole at the center of the giant elliptical galaxy M87. Located some 52 million light years from Earth, M87 is an active galaxy that has long been recognized as quite unusual. In 1918 Heber Curtis at the Lick Observatory reported that the center of M87 has "a curious straight ray . . . apparently connected with the nucleus by a thin line of matter." Figure 18-20*a* is a long exposure of M87 and several neighboring galaxies; M87's nucleus and jet are buried in the galaxy's glare. Figure 18-20*b* is a short exposure that reveals the bright core of M87 and a jet that is about 6000 light-years long. M87 is also a powerful source of radio waves and X rays.

In the early 1990s the center of M87 was photographed with exceptional clarity by the HST (Figure

a

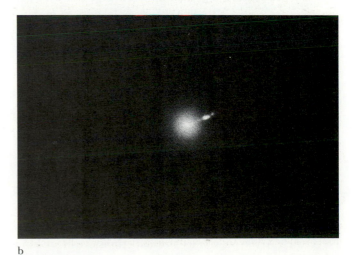

b

Figure 18-20 The giant elliptical galaxy M87 The giant elliptical galaxy M87 is located near the center of the sprawling, rich Virgo Cluster, which is about 50 million light-years from Earth. (a) This long exposure shows the extent of M87 and some smaller, neighboring galaxies. The numerous fuzzy spots that surround M87 are globular clusters; each contains about a million stars. (b) This short exposure shows the bright core of M87 and the jet. (Anglo-Australian Observatory; Lick Observatory)

Figure 18-21   The center of M87   This picture from the Hubble Space Telescope shows the starlike nucleus of M87 and its jet that extends outward some 6000 light-years. M87's extraordinarily bright nucleus may be the result of a supermassive black hole whose gravity causes an enormous number of stars to crowd around it. (NASA and ESA)

18-21). The HST image shows an exceptionally bright, starlike nucleus. To produce this fiery glow, stars must be packed so tightly at the center of M87 that their density is at least three hundred times greater than that normally found at the centers of giant ellipticals. This dense clustering of stars suggests that a black hole with a mass of nearly 3 billion Suns may be located at the center of M87.

18-5   A supermassive black hole may be the "central engine" that powers an active galaxy or a quasar

There is no conclusive evidence that supermassive black holes exist at the centers of galaxies. All of the observations discussed above give only circumstantial indications. Nevertheless, the possibility that supermassive black holes might be quite commonplace has inspired astrophysicists to formulate mechanisms that could give rise to quasars and active galactic nuclei. Various realistic scenarios about how to tap the gravitational energy of an extremely

massive black hole have been worked out. In essence, these schemes are scaled-up versions of the explanation of Cygnus X-1 discussed in Chapter 15. A large black hole is needed to produce a large energy output.

Imagine a billion-solar-mass black hole sitting at the center of a galaxy. Because the centers of galaxies are congested places, we would expect this supermassive black hole to be surrounded by a huge rotating accretion disk of matter captured by the hole's gravity. According to Kepler's third law, the inner regions of this accretion disk would orbit the hole more rapidly than would the outer parts. Thus, the rapidly spinning inner regions would be constantly rubbing against the more slowly moving gases in the outer regions, and the resulting friction would heat up the gases as they spiral toward the hole.

Because of the constant inward crowding of hot gases, the pressures surrounding the hole would be tremendous. Of course, some of the inflowing material might be swallowed by the hole, but the pressures in the accretion disk would be so great that most of the hot gases would never really get near the hole. Furthermore, the inner edge of the accretion disk could be orbiting the black hole rapidly enough to prevent matter from plunging into the hole, thus adding to the congestion surrounding the hole. In an attempt to relieve this congestion, matter would be violently ejected along the perpendicular to the accretion disk, that being the direction along which the hot gases experience the least resistance. The result would be two oppositely directed beams of relativistic particles. The overall scenario is sketched in Figure 18-22, and an artist's rendition is seen on the opening page of this chapter.

How is this ejected matter confined in narrow jets? Recent studies demonstrate that there is a natural "self-focusing" effect whenever any high-speed matter penetrates a medium instead of squirting out into empty space. For instance, consider the behavior of water squirting out of a nozzle on an ordinary garden hose. As water squirts out of the nozzle, the unconfined stream broadens and the spray fans out through a wide angle. If the nozzle is placed in a swimming pool, however, the stream of water does not fan out as much in the water as it did in the air. Similarly, as the two jets of hot gas leave the vicinity of a black hole, they must blast their way through some gas that is still crowding inward. Passage through this material causes the jets to become extremely narrow, concentrated beams. This self-focusing effect helps explain not only the double radio sources but also the jets and beams we see protruding from active galaxies and some quasars.

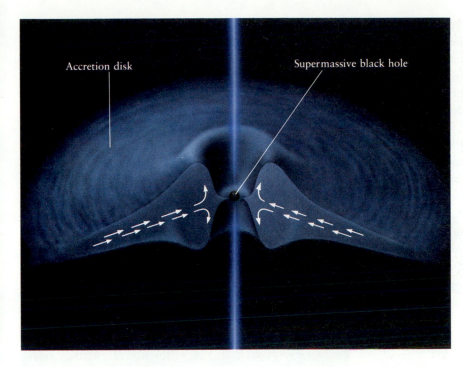

Figure 18-22 A supermassive black hole as the "central engine" The energy output of an active galaxy or a quasar may involve an extremely massive black hole that captures matter from its surroundings. In the scenario depicted here, the inflow of material through an accretion disk is redirected to produce two powerful jets of particles traveling near the speed of light. (Also see artist's rendition on the first page of this chapter.)

The model of a supermassive black hole surrounded by an accretion disk with oppositely directed high-speed jets has received considerable attention in recent years, because it apparently offers a single explanation for a wide variety of quasars and active galaxies. Specifically, the main difference between double radio sources, quasars, and blazars may be the angle at which the central engine is viewed. As Figure 18-23 shows, an observer sees a double radio source when the accretion disk is viewed nearly edge-on, so that the jets are nearly in the plane of the sky. At a steeper angle, the observer sees a quasar. If one of the jets is aimed almost directly at Earth, a blazar or N galaxy is seen.

In 1992 the Hubble Space Telescope took a picture that some astronomers interpret as showing an accretion disk around a supermassive black hole at the center of the

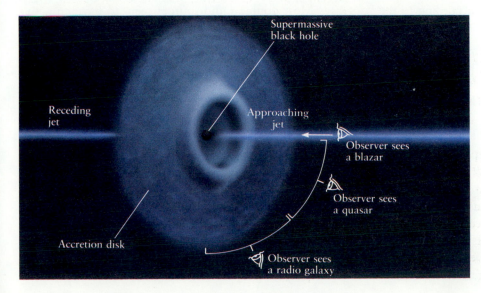

Figure 18-23   The orientation of the "central engine" and its jets   Double radio sources, quasars, and blazars may be the same type of object viewed from different directions. If one of the jets is aimed almost directly at the Earth, we see a BL Lacertae object or an N galaxy. If the jet is somewhat tilted to our line of sight, we see a quasar. If the jets are nearly perpendicular to our line of sight, we see a double radio source.

spiral galaxy M51. An overall view of this galaxy, sometimes called the Whirlpool Galaxy, appears in Figure 17-2. Figure 18-24 is HST's image of the galaxy's nucleus. The dark, horizontal bar in the picture is our edge-on view of a ring of dust and gas about 100 light-years in diameter. This bar is identified as the accretion disk, because ground-based radio and optical observations show a double-lobed structure that is bisected by the bar. The fainter, inclined bar in the HST image is a mystery.

The possible existence of supermassive black holes in quasars and active galaxies is one of the most exciting topics in modern astronomy. In the 1990s, instruments like the Hubble Space Telescope and the Keck Telescope will continue to probe the nuclei of galaxies with unprecedented resolution. These observations will undoubtedly give us a much better understanding of the powerful engines at the cores of active galaxies and quasars.

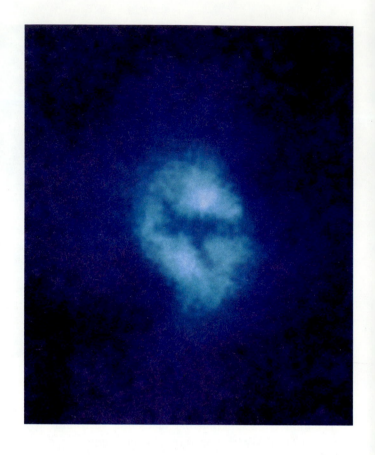

Figure 18-24 The nucleus of M51 This image, taken by the HST in 1992, shows a dark X silhouetted across the nucleus of the spiral galaxy M51. The dark, horizontal bar may be an edge-on view of a huge dusty ring that surrounds a supermassive black hole. The fainter, inclined bar is puzzling. (NASA, ESA)

Summary

- A quasar, or quasi-stellar object, is an object that looks like a star but has a huge redshift. This corresponds to an extreme distance from the Earth according to the Hubble law.

  To be seen from Earth, a quasar must be very luminous, typically about 100 times brighter than an ordinary galaxy.

  Relatively rapid fluctuations in the brightnesses of quasars indicate that they cannot be much larger than the diameter of our solar system.

- An active galaxy is an extremely luminous galaxy that has one or more unusual features: an unusually bright, starlike nucleus; strong emission lines in its spectrum; rapid variations in luminosity; or jets or beams of radiation emanating from its core.

  An active galaxy with a bright, starlike nucleus and strong emission lines in its spectrum is categorized as a Seyfert galaxy.

  Most double radio sources seem to have an active galaxy located between the two radio lobes that distinguish this type of radio source.

  A head–tail radio source seems to show evidence of relativistic particle jets emerging from an active galaxy.

  BL Lacertae objects, blazars, and N galaxies have bright nuclei whose cores show relatively rapid variations in luminosity.

  Quasars are probably very distant active galaxies.

- Spectroscopic observations of certain galaxies seem to indicate the presence of huge concentrations of matter (perhaps supermassive black holes) at their centers.

- The strong energy emission from quasars, active galaxies, and double radio sources may be produced as matter falls toward a supermassive black hole at the center of the object.

   Matter spiraling in toward a supermassive black hole would be channeled into two oppositely directed beams that propel particles and energy into intergalactic space.

   The orientation of these beams relative to our line of sight may be the factor that determines whether we see a quasar, a blazar, or a double radio source.

Review questions

1 Suppose you suspected a certain object in the sky to be a quasar. What sort of observations might you perform to find out if it were indeed a quasar?

2 Explain why astronomers do not use any of the standard candles described in Chapter 17 to determine the distances to quasars.

3 Explain how the rate of variability of a source of light can be used to place an upper limit on the size of the source.

4 What is an active galaxy? How many different kinds of active galaxies can you name? How do they differ from one another?

5 Why do astronomers believe that the energy-producing region of a quasar is very small?

6 How is synchrotron radiation produced?

7 Why does it seem reasonable to suppose that quasars are extremely distant active galaxies?

8 What is a double radio source?

9 How does SS433 (discussed in Chapter 14) compare with a typical double radio source?

10 What is a supermassive black hole? What observational evidence suggests that supermassive black holes might be located at the centers of certain galaxies?

11 Why do many astronomers believe that the "engine" at the center of a quasar is a supermassive black hole surrounded by an accretion disk?

12 How might the orientation of the jets emanating from the center of a galaxy be related to the type of active galaxy that we observe?

Advanced questions

13 In the 1960s it was suggested that quasars might be compact objects ejected at high speeds from the centers of nearby ordinary galaxies. Why does the absence of blue-shifted quasars disprove this hypothesis?

14 Why do you suppose there are no quasars relatively near our galaxy?

15 When quasars were first discovered, many astronomers were optimistic that these extremely luminous objects could be used to probe distant regions of the universe. For example, it was hoped that quasars would provide high-redshift data from which the Hubble constant could be accurately determined. Why do you suppose these hopes have not been realized?

16 Just before this book went to press, the scientists responsible for the Hubble Space Telescope image of the jet in M87 (see Figure 18-21) announced that they plan to use the Hubble Space Telescope to make detailed spectroscopic observations of the nucleus of M87. Consult recent issues of such magazines as *Sky & Telescope* and *Science News* to see if these observation have been made. What were the results and the conclusions? Do they support or refute the suspicion that a supermassive black hole is located at the center of M87?

17 Just before this book went to press, scientists using the Hubble Space Telescope reported their discovery of a bar across the nucleus of M31 similar to the barlike structures they discovered in M51 (see Figure 18-24). Consult recent issues of such magazines as *Sky and Telescope* and *Science News* to see if a photograph of this bar has been published. Do the observations support or refute the suspicion that a supermassive black hole is located at the center of M31?

Discussion questions

18 Speculate on the possibility that quasars, double radio sources, giant elliptical galaxies, and so on form some sort of evolutionary sequence.

19 Some quasars show several sets of absorption lines whose redshifts are less than the redshift of the quasars' emission lines. For example, the quasar PKS 0237–23 has five sets of absorption lines, all with redshifts somewhat less than the redshift of the quasar's emission lines. Propose an explanation for these sets of absorption lines.

For further reading

Blandford, R., and others. "Cosmic Jets." *Scientific American*, May 1982 • This article, which describes jets emanating from the centers of active galactic nuclei, includes an explanation of superluminal (i.e., faster than light) motions sometimes observed in the structure of such jets.

Burns, J., and Price, R. "Centaurus A: The Nearest Active Galaxy." *Scientific American*, November 1983 • By examining the nearby active galaxy NGC 5128, the authors piece together a rather comprehensive model of active galactic nuclei and double radio sources.

McCarthy, P. "Measuring Distances to Remote Galaxies and Quasars." *Mercury*, January-February 1988 • This brief article presents an exceptionally clear summary of how the distances to remote objects depends on such factors as the Hubble constant.

Mood, J. "Star Hopping to a Quasar." *Astronomy*, May 1987 • If you have access to a fairly large telescope and want to hunt for the brightest-appearing quasar in the sky, you might consult this article on 3C 273.

Preston, R. *First Light*. Atlantic Monthly Press, 1987 • This book, which takes the reader behind the scenes at the Palomar Observatory, includes an exciting and well-written section on Maarten Schmidt's pioneering research on quasars.

Shipman, H. *Black Holes, Quasars, and the Universe*. 2nd ed. Houghton Mifflin, 1980 • This clear, cogent text has an excellent section on quasars.

Verschuur, G. *The Invisible Universe Revealed*. Springer Verlag, 1987 • This fascinating and well-written book has four excellent chapters summarizing our knowledge of radio galaxies and quasars.

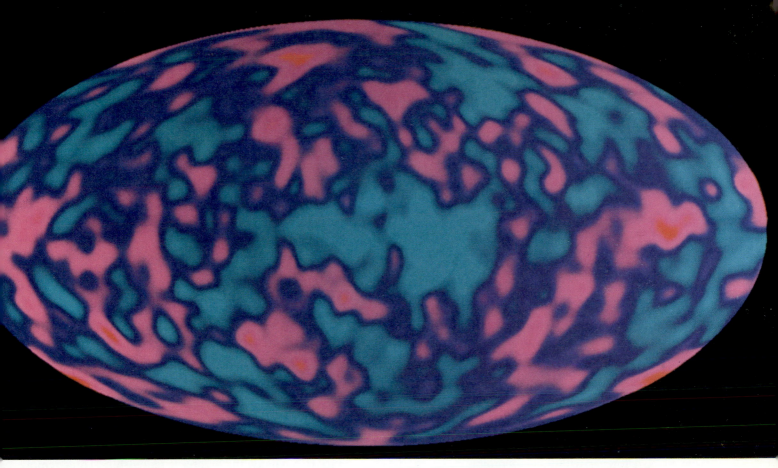

Structure of the early universe This microwave map of the sky, produced from data taken by NASA's Cosmic Background Explorer (COBE), shows temperature variations in the cosmic microwave background. Pink regions are about 0.0003 K warmer than the average temperature of 2.73 K; blue regions are about 0.0003 K cooler than the average. These tiny temperature fluctuations date back to the earliest moments of the universe and may be directly related to the large-scale structure of the universe today. (NASA Goddard Space Flight Center)

19

Cosmology and the Creation of the Universe

We live in an expanding universe. This universal expansion—which is actually the expansion of space—began with an explosion at the beginning of time called the Big Bang. The universe now is full of microwave photons, ghostly relics of the primordial fireball that filled all space shortly after the Big Bang. The smoothness of this microwave background, coupled with the fact that space is almost perfectly flat, suggests that the newborn universe suddenly "inflated" to many times its original size. Looking far into the future, we discover that the universe will either expand forever or collapse in a Big Crunch, depending on the density of matter throughout space. To understand why the universe turned out the way it did, we explore the idea that all the fundamental forces had the same strength immediately after the Big Bang. We then see how the various familiar forces and particles that comprise ordinary matter "froze out" of the universe as it expanded and cooled.

As foolish as it may seem, one of the most profound questions you can ask is, "Why is the sky dark at night?" This question apparently haunted Johannes Kepler as long ago as 1610, and it was popularized in the early 1800s by the German amateur astronomer Heinrich Olbers.

To appreciate the problem, we must begin by assuming that the universe is infinite and that stars are scattered more or less randomly across this infinite expanse of space. Isaac Newton argued that no other assumption makes sense. If the universe were not infinite or if stars were grouped in only one part of the universe, then the gravitational forces between the stars would soon cause all this matter to fall together into a compact blob. Obviously, this has not happened. Thus, as classical Newtonian mechanics would have it, we must be living in a universe that is both infinite and static. Only then does a star feel a uniform gravitational pull from every part of the sky, from all the other stars in the universe. According to this model, the universe is infinitely old and will continue to exist forever without undergoing major changes in structure.

Imagine looking out into space in such a static, infinite universe. Because space goes on forever, with stars scattered throughout it, your line of sight must eventually hit a star. No matter where you look in the sky, you should see a star. The entire sky should thus be as bright as an average star. Even at night, the entire sky should be blazing like the surface of the Sun. That this is not so is the dilemma called **Olbers's paradox**.

19-1 We live in an expanding universe

Olbers's paradox tells us that there is something wrong with Newton's idea of an infinite, static universe. According to the classical, Newtonian picture of reality, space is laid out in all directions like a great, flat sheet of inflexible, rectangular graph paper. This rigid, three-dimensional space stretches on and on, totally independent of stars or galaxies or anything else. Similarly, a Newtonian clock ticks steadily and monotonously forever, never slow-ing down or speeding up.

Albert Einstein demonstrated that this view of space and time is wrong. In his special theory of relativity, he proved that measurements with clocks and rulers depend on the motion of the observer. As explained in Chapter

15, Einstein's general theory of relativity tells us that the shape of space is profoundly influenced by the masses occupying that space. Specifically, matter tells space how to curve, and curved space tells matter how to move.

Shortly after formulating the general theory of relativity in 1915, Albert Einstein applied his ideas to the structure of the universe. The prevailing view was that the universe is infinite and static. To Einstein's dismay, his calculations could not produce a truly static universe. His equations instead indicated that the universe must be either expanding or contracting. The Newtonian view was so strong that Einstein doubted the implications of his equations, and so he missed the opportunity to propose that we live in an expanding universe.

Edwin Hubble is usually credited with discovering that we live in an expanding universe. As we saw in Chapter 17, Hubble discovered the simple linear relationship between the distances to remote galaxies and the redshifts of those galaxies' spectral lines. This relationship, now called the Hubble law, states that the greater the distance to a galaxy, the greater is its redshift. Thus, remote galaxies appear to be moving away from us with speeds proportional to their distances from us (review Figure 17-31). Since the galaxies are getting farther and farther apart as time goes on, astronomers say that the universe is expanding.

What does it mean to say that "the universe is expanding?" According to general relativity, space is not rigid. The amount of space between widely separated locations can change over time. A good analogy is that of a person blowing up a balloon, as sketched in Figure 19-1. Small coins, each representing a galaxy, are glued onto the surface of the balloon. As the balloon expands, the amount of space between the coins gets larger and larger. In the same way, as the universe expands, the amount of space between widely separated galaxies grows larger and larger. The expansion of the universe *is* the expansion of space.

The expanding balloon analogy illustrates several important characteristics of the expanding universe. Imagine sitting on one of the coins in Figure 19-1. As the balloon expands, you see the other coins moving away from you. Specifically, you observe that nearby coins move away from you slowly, whereas more distant coins move away more rapidly. You find this relationship, which is just like the Hubble law, no matter which coin you decide to call home. Your coin is not at the center of the balloon, of course. Indeed, the surface of the balloon does not have a center: You could walk around the surface forever without finding the center. And the surface of the balloon

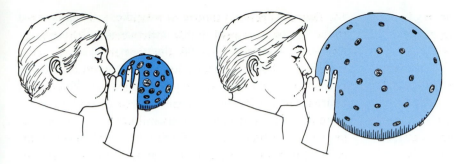

Figure 19-1 *The expanding balloon analogy The expanding universe can be compared to the expanding surface of an inflating balloon. All the coins on the balloon recede from one another as the balloon expands, just as all the galaxies recede from each other as the universe expands. (Adapted from C. Misner, K. Thorne, and J. Wheeler)*

has no edge: You could explore every inch of the surface and never find an edge.

Just as the surface of the balloon has neither center nor edge, our universe has no center or edge. No matter what galaxy you call home, all the other galaxies are receding from you. No one is ever at the "center" of the universe. Questions like "What is beyond the edge of the universe?" or "What is the universe expanding into?" are as meaningless as asking "What is north of the Earth's north pole?"

The fact that space is continually expanding explains how photons from remote galaxies are redshifted. Imagine a photon coming toward us from a distant galaxy. As the photon travels through space, space is expanding, and so the photon's wavelength becomes stretched. When the photon reaches our eyes, we see a drawn-out wavelength, which is the redshift. The longer the photon's journey, the more its wavelength will have been stretched. Thus, photons from distant galaxies have larger redshifts than those of photons from nearby galaxies, which is expressed as the Hubble law.

A redshift caused by the expansion of the universe is properly called a **cosmological redshift** to distinguish it from a Doppler shift. Doppler shifts are caused by an object's *motion through* space, whereas a cosmological redshift is caused by the *expansion of* space.

It is important to realize that the expansion of space occurs primarily in the voids that separate clusters of galaxies. Just as the coins in Figure 19-1 do not expand as the balloon inflates, galaxies themselves do not expand. Einstein and others have established that an object held together by its own gravity, like a galaxy, is always contained within a patch of nonexpanding space. A galaxy's gravitational field produces this nonexpanding patch, which is indistinguishable from the flat, rigid space of Newton. Thus, the Earth and your body are not getting bigger and bigger. Only the distance between widely separated galaxies increases with time.

19-2 The expanding universe probably emerged from an explosive event called the Big Bang

The universe has been expanding for billions of years, so there must have been a time in the ancient past when all the matter in the universe was concentrated in a state of infinite density. Presumably, some sort of colossal explosion initiated the expansion of the universe. This explosion, commonly called the Big Bang, marks the creation of the universe.

To calculate the time elapsed since the Big Bang, imagine watching a movie of any two galaxies separated by a distance r and receding from each other with a velocity v. Now run the film backward, and observe the two galaxies approaching each other as time runs in reverse. We can calculate the time T_0 it will take for the galaxies to collide by using the simple equation

$$T_0 = r/v$$

Employing the Hubble law, $v = H_0 r$, to replace the velocity v in this equation and using the facts that 1 Mpc equals 3.09×10^{19} km and one year contains 3.16×10^7 seconds, we get

$$\frac{1}{H_0} = \frac{1}{80 \text{ km/s/Mpc}} = 12 \text{ billion years}$$

Actually, the true age of the universe is probably about 15 billion years, which is roughly the age of the oldest stars.

The finite age of the universe offers a resolution of Olbers's paradox. The reason the entire sky is not as

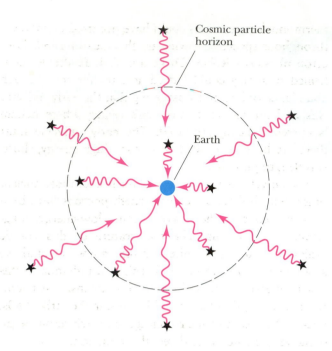

Figure 19-2   The observable universe *The radius of the cosmic particle horizon is equal to the distance that light has traveled since the Big Bang. Because the Big Bang occurred about 15 billion years ago, the cosmic particle horizon is about 15 billion light-years away. We cannot see stars beyond the cosmic particle horizon because their light has not had enough time to reach us.*

bright as the surface of the Sun is that we cannot see any stars that are more than 15 billion light-years away. The universe may indeed be infinite, with galaxies scattered throughout its limitless expanse. But the universe is less than 15 billion years old, so the light from stars more than 15 billion light-years away has just not had enough time to get here.

You can think of the Earth as being at the center of an enormous sphere with a radius of roughly 15 billion light-years (Figure 19-2). The surface of this sphere is called the **cosmic particle horizon.** Our entire **observable universe** is located inside this sphere. We cannot see anything beyond the cosmic particle horizon, because the travel time for light from these incredibly remote objects is greater than the age of the universe. Throughout the observable universe, the galaxies are distributed sparsely enough that most of our lines of sight do not hit any stars, which explains why the night sky is dark.

The concept of a Big Bang origin of the universe is a straightforward, logical consequence of having an expanding universe. If you can just imagine far enough into the past, you can arrive at a time nearly 15 billion years ago when the density of matter throughout the universe

was infinite. The entire universe was like the singularity at the center of a black hole. For this reason, a better name for the Big Bang is the **cosmic singularity.**

There are many misconceptions about the Big Bang. For one thing, it was not at all like an exploding bomb. When a bomb explodes, pieces of debris fly off *into space* from a central location. If you could trace all the pieces back to their origin, you could find out exactly where the bomb had been. You cannot do this procedure with the Big Bang, however, because the cosmic singularity was not a point somewhere in space. When the density of matter throughout the universe was infinite, the cosmic singularity filled *all* space. The Big Bang therefore occurred everywhere throughout the entire universe. Indeed, we can think of the Big Bang as an explosion of space at the beginning of time.

Comparing the Big Bang to a black hole singularity helps us appreciate certain aspects of the creation of the universe. As we saw in Chapter 15, matter is crushed to infinite density at the center of a black hole. This location, called the singularity, is characterized by infinite curvature where space and time are all tangled up. Without a clear background of space and time, such concepts as *past, future, here,* and *now* cease to have any meaning.

At the moment of the Big Bang, a state of infinite density filled the universe. Space and time throughout the universe were completely jumbled up in a state of infinite curvature like that at the center of a black hole. Thus, we cannot use the laws of physics to tell us what happened at the moment of the Big Bang. And we certainly cannot use science to tell us what existed *before* the Big Bang. These things are fundamentally unknowable. The phrases "*before* the Big Bang" or "at the *moment* of the Big Bang" are meaningless, because time itself did not really exist.

A very short time after the Big Bang, space and time did begin to exist in the way we think of them today. This interval, called the **Planck time,** equals 10^{-43} second. From the moment of the Big Bang, at time $t = 0$, to the Planck time 10^{-43} second later, all known science fails us. We do not know how space, time, and matter behaved under these extreme circumstances.

The depiction of the Big Bang as a space–time singularity rests on the validity of the general theory of relativity. That theory is, after all, the best description of gravity we have, and it clearly predicts the existence of singularities at the centers of black holes as well as throughout all space at the beginning of time. Because of these singularities, however, many physicists believe that general relativity may not give a correct picture of conditions earlier than the Planck time. These scientists look forward to the development of improved theories, sometimes called "quantized gravity" or "supergrand unified field theories," to shed new light on the nature of the Big Bang and perhaps do away with singularities. In that case, we might be able to figure out what existed before the Big Bang. We shall have more to say about these speculative developments near the end of this chapter.

19-3 The microwave radiation that fills all space is evidence for a hot Big Bang

One of the major successes of modern astronomy involves discoveries about the origin of the heavy elements. We know today that all the heavy elements are created in the fiery infernos at the centers of stars. As astronomers began understanding the details of thermonuclear synthesis in the 1960s, a new problem arose: There is too much helium around. For example, the Sun consists of about 74% hydrogen and 25% helium, leaving only 1% for all the remaining heavier elements combined. This 1% can be understood as material produced at the centers of ancient stars that long ago cast these heavy elements into space. Some freshly made helium certainly accompanied these heavy elements, but it was not nearly enough helium to account for one-quarter of the Sun's mass.

Shortly after World War II, Ralph Alpher and George Gamow proposed that the universe immediately following the Big Bang must have been so incredibly hot that thermonuclear reactions could have occurred everywhere throughout space. Following up this idea in 1960, Princeton physicists Robert Dicke and P. J. E. Peebles confirmed that they could indeed account for today's high abundance of helium by assuming that the early universe was at least as hot as the Sun's center, where helium is currently being produced. The early universe must therefore have been filled with many high-energy, short-wavelength photons.

The universe has expanded so much since those ancient times that all those short-wavelength photons have been so stretched that they have become low-energy, long-wavelength photons. The temperature of this cosmic radiation field is now quite low, only a few kelvin above absolute zero. According to Wien's law, radiation at this low temperature should have its peak intensity at microwave wavelengths of a few millimeters. In the early 1960s, Dicke, Peebles, and their colleagues at Princeton began designing an antenna to detect this radiation.

Meanwhile, just a few miles from Princeton University, Arno Penzias and Robert Wilson of Bell Telephone Laboratories were working on a new horn antenna designed to relay telephone calls to Earth-orbiting communications satellites (Figure 19-3). Penzias and Wilson were deeply puzzled when, no matter where they pointed their antenna in the sky, they detected a faint background noise. Thanks to a colleague, they soon learned of the work of Dicke and Peebles and came to realize that they had detected the cooled-down cosmic background radiation left over from the hot Big Bang.

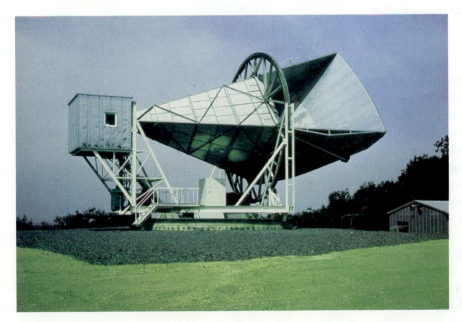

Figure 19-3 The Bell Labs horn antenna *This horn antenna at Holmdel, New Jersey, was used by Arno Penzias and Robert Wilson in 1965 to detect the cosmic microwave background. (Bell Labs)*

Figure 19-4 The Cosmic Background Explorer (COBE) *This satellite, launched in 1989, is measuring the spectrum and angular distribution of the cosmic microwave background over a wavelength range of 1 μm to 1 cm. COBE (pronounced CO-bee) is looking for deviations from a perfect blackbody spectrum and for temperature variations across the sky. (Courtesy of J. Mather; NASA)*

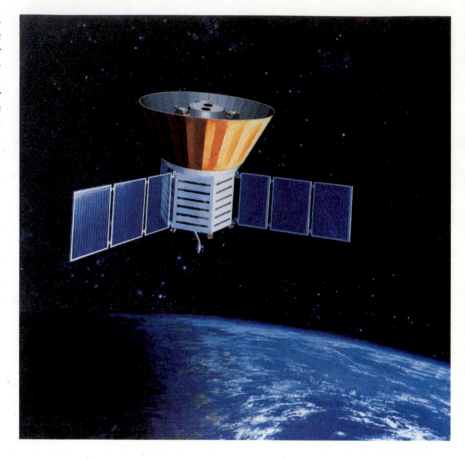

Since those pioneering days, scientists have made many measurements of the intensity of this background radiation at a variety of wavelengths. The most accurate measurements come from the Cosmic Background Explorer satellite (COBE), which was placed in orbit about the Earth in 1989 (Figure 19-4). Data from COBE's spectrometer shown in Figure 19-5 demonstrate that this ancient radiation has the spectrum of a blackbody with a temperature of 2.73 K. This radiation field, which fills all of space, is called the **cosmic microwave background.**

An important feature of the microwave background is that its intensity is almost perfectly isotropic (meaning "the same in all directions"). In other words, we detect nearly the same intensity from all parts of the sky. However, extremely accurate measurements first made from high-flying airplanes and more recently from COBE reveal a very slight variation in temperature across the sky. The microwave background is slightly warmer than average in the direction of the constellation of Leo and slightly cooler in the opposite direction (toward Aquarius). Between the warm spot in Leo and the cool spot in

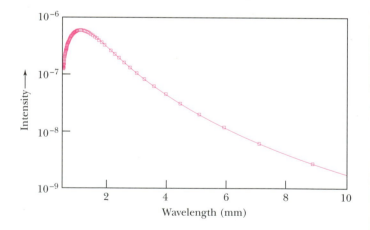

Figure 19-5 The spectrum of the cosmic microwave background *The little squares on this graph are COBE's measurements of the brightness of the cosmic microwave background plotted against wavelength. The data fall along a blackbody curve for 2.73 K to a remarkably high degree of accuracy. The peak of the curve occurs at a wavelength of 1.1 mm, in accordance with Wien's law. (Courtesy of E. Cheng; NASA COBE Science Team)*

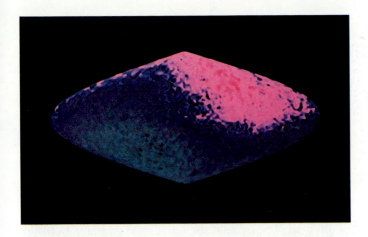

Figure 19-6 The microwave sky This map of the microwave sky was produced from data taken by instruments on board COBE. The galactic center is in the middle of the map, and the plane of the Milky Way runs horizontally across the map. Color indicates temperature; magenta is warm and blue is cool. The temperature variation across the sky is caused by the Earth's motion through the microwave background. (NASA)

Aquarius, the background temperature across the sky declines in a smooth fashion. A map of the microwave sky showing this anisotropy is seen in Figure 19-6.

This temperature variation can be explained as a result of the Earth's overall motion through the cosmos. If we were at rest with respect to the microwave background, the radiation would be truly isotropic. Because we are moving through this radiation field, however, we see

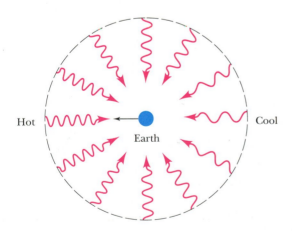

Figure 19-7 Our motion through the microwave background Because of the Doppler effect, the microwave background is slightly warmer in that part of the sky toward which we are moving. Recent measurements indicate that our Galaxy, along with the rest of the Local Group, is moving in the general direction of the Centaurus Cluster.

a Doppler shift. Specifically, we see shorter-than-average wavelengths in the direction toward which we are moving, as sketched in Figure 19-7. A decrease in wavelength corresponds to an increase in photon energy and thus to an increase in temperature. The observed temperature excess corresponds to a speed of 390 km/s.

Conversely, we see longer-than-average wavelengths in that part of the sky from which we are receding. An increase in wavelength corresponds to a decline in photon energy and hence to a decline in temperature. We are thus traveling from Aquarius toward Leo at a speed of 390 km/s. Taking into account the known velocity of the Sun around the center of our Galaxy, we find that the entire Milky Way Galaxy is moving at 600 km/s in the general direction of the Centaurus Cluster, possibly because of the gravitational pull of an enormous mass, dubbed the **Great Attractor**, in that direction.

19-4 The future of the universe is determined by the average density of matter in it

During the 1920s, general relativity theory was applied to cosmology by Alexandre Friedmann in Russia, Georges Lemaître in France, Willem de Sitter in the Netherlands, and, of course, Einstein himself. The resulting picture of the structure and evolution of the universe, called **relativistic cosmology,** is in good agreement with our intuitive notion that gravity should be slowing the cosmological expansion.

To envision the effect of gravity, consider a cannonball shot upward from the surface of the Earth. If the cannonball's speed is less than the escape speed (about 11 km/s), the ball will fall back to Earth. If the cannonball's speed equals the escape velocity, it will just barely escape falling back to Earth. And if the ball's speed exceeds the escape velocity, it can easily leave the Earth and never fall back or stop despite the relentless pull of gravity.

The universe obeys a similar set of rules. If the average density of matter throughout space is low, the gravity associated with this matter is weak and the expansion of the universe will continue forever. Even infinitely far into the future, galaxies will continue to rush away from each other. In such circumstances, we say that the universe is **unbounded,** or open.

Conversely, if the average density of matter across space is high, then the resulting gravity will be strong enough to eventually halt the expansion of the universe.

The universe will reach a maximum size and then begin contracting as gravity starts to pull the galaxies back toward each other. In this case, we say that the universe is **bounded,** or closed.

Separating these two scenarios is the situation in which we say that the universe is **marginally bounded.** In this case, the average density of matter across space is called the **critical density** (ρ_c), so that the galaxies just barely manage to keep moving away from each other. This situation is analogous to the cannonball leaving the Earth with a speed exactly equal to the escape velocity.

Estimates of the critical density depend on the Hubble constant. Using a Hubble constant of 80 km/s/Mpc, we find that $\rho_c = 2.4 \times 10^{-26}$ kg/m³, which is equivalent to a density of about 14 hydrogen atoms per cubic meter of space.

The average density of luminous matter that we see in the sky seems to be about 5×10^{-28} kg/m³, which is about 1/50 of the critical density. As discussed in Chapter 17, the dark-matter problem suggests that we are seeing only a small fraction of all the matter in the universe. Most of the mass of the universe seems to be dark and remains to be discovered. Some physicists argue that this dark matter is probably not composed of particles like protons, neutrons, or electrons; otherwise, we would have already detected it. The dark matter may therefore be vastly different from anything we have ever encountered. Because of these uncertainties, present-day observations and estimates of the density are not accurate enough to tell us whether the universe is bounded or unbounded.

19-5 The rate of deceleration of the universe is related to the redshifts of extremely distant galaxies

Another way of determining the future of the universe is to measure the rate at which cosmological expansion is slowing down. Because of their mutual gravitational attraction, galaxies have not been flying away from each other with a constant velocity. Gravity has caused the speed of separation between galaxies to decrease gradually since the Big Bang. Thus, the expansion rate of the universe has been decreasing.

The deceleration of the universe shows up as a deviation from the straight-line relationship predicted by the Hubble law. Suppose that you were to measure the red-shifts of galaxies several billion light-years from Earth. The light from these galaxies has taken billions of years to get to your telescope, so your measurements reveal how fast the universe was expanding billions of years ago. Because the universe was expanding faster in the past than it is today, your data will deviate slightly from the straight-line Hubble law.

The graphs in Figure 19-8 display the relationship between the deceleration of the universe and the Hubble law extended to great distances. Astronomers denote the amount of deceleration with the **deceleration parameter** (q_0). Appropriately, $q_0 = 0$ corresponds to no deceleration at all. This is possible if the universe is completely empty and thus there is no gravity to slow down the expansion. As sketched in Figure 19-8b, a $q_0 = 0$ universe expands forever at a constant rate.

The case $q_0 = \frac{1}{2}$ corresponds to a marginally bounded universe. Such a universe just barely manages to expand forever, because it contains matter at the critical density ρ_c. If q_0 is between 0 and $\frac{1}{2}$, the universe is unbounded and will continue to expand forever. Such a universe contains matter at less than the critical density. If q_0 is greater than $\frac{1}{2}$, the universe is bounded and is filled with matter of a density greater than ρ_c. This universe is doomed to collapse upon itself and ultimately end in a **Big Crunch** in the extremely distant future.

In principle it should be possible to determine q_0 by measuring the redshifts and distances of many remote galaxies and then plotting the data on a Hubble diagram like the one in Figure 19-8a. If the data points fall above the $q_0 = \frac{1}{2}$ line, the universe is bounded. If the data points fall between the $q_0 = 0$ and $q_0 = \frac{1}{2}$ lines, the universe is unbounded.

Unfortunately, such observations are extremely difficult to make. The deceleration of the universal expansion is so slight that galaxies nearer than a billion light-years are of no help in determining q_0. And beyond a billion light-years, uncertainties cloud determinations of distance. For example, Figure 19-9 shows data obtained by Allan Sandage with the 200-inch telescope at the Palomar Observatory. Note how the data points are scattered about the $q_0 = \frac{1}{2}$ line. This scatter, a result of observational uncertainties, prevents us from determining conclusively whether the universe is bounded or unbounded. Nevertheless, because the data lie close to the $q_0 = \frac{1}{2}$ line, many astronomers think that the universe is not far from being marginally bounded.

A troublesome dilemma arises when we calculate the age of the universe with different values of q_0. If $q_0 = 0$,

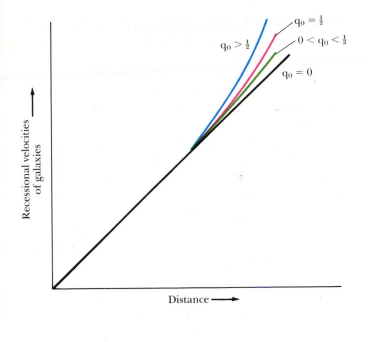

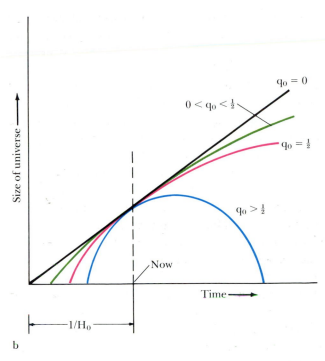

a

b

Figure 19-8 Deceleration and the Hubble diagram *These two graphs compare (a) the appearance of the Hubble diagram and (b) the evolution of the universe. The case $q_0 = 0$ is an empty universe. The case $q_0 = \frac{1}{2}$ is a marginally bounded universe. If* $0 < q_0 < \frac{1}{2}$, *then the universe is unbounded and will expand forever. If $q_0 > \frac{1}{2}$, the universe is bounded and will someday collapse.*

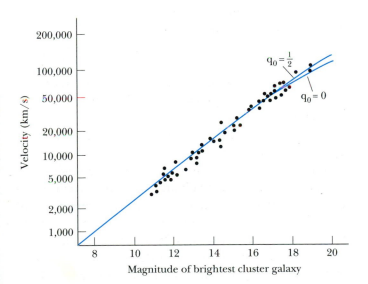

Figure 19-9 The Hubble diagram *This graph shows the Hubble diagram extended to include extremely remote galaxies. The magnitude of the brightest galaxy in a cluster is directly correlated with the distance to the cluster. If the data points fall between the curves marked $q_0 = 0$ and $q_0 = \frac{1}{2}$, then the universe is unbounded. If the data points fall above the curve marked $q_0 = \frac{1}{2}$, then the universe is bounded. (Adapted from A. Sandage)*

then the universe is empty and, as we saw earlier, its age is $1/H_0$. For a Hubble constant of 80 km/s/Mpc, that turns out to be 12 billion years. If the universe is marginally bounded, then the age of the universe equals $\frac{2}{3}(1/H_0)$, or about 8 billion years. If the universe is unbounded, it has an age of 8 to 12 billion years. If the universe is bounded, then its age is less than 8 billion years.

All of these estimates for the age of the universe are uncomfortably low. It is impossible for the universe to be younger than the stars it contains, some of which seem to be 15 billion years old. Either astronomers' understanding of the ages of stars is not correct, or a straightforward relativistic cosmology involving only two constants (H_0 and q_0) does not adequately describe the universe.

When Einstein first applied his general theory of relativity to the structure of the universe, he was frustrated by the inability of his calculations to produce a static universe, in accordance with the prevailing view of the time. In desperation, he added a term called the **cosmological constant** (Λ) to his equations to ensure that his calculations would predict a static universe. This cosmological constant represents a pressure that balances gravitational attraction in the static universe and prevents it from collapsing. "It was the greatest blunder of my life," he later lamented, because he missed the opportunity of proposing that the universe is, in fact, not static.

The cosmological models discussed above, which give a maximum age of the universe of 12 billion years, assume that $\Lambda = 0$. Inspired by such discrepancies as the ages of the oldest stars, some astronomers have resurrected the idea of a nonzero cosmological constant. The presence of such a term in the equations of general relativity greatly complicates the behavior of the universe. One possibility is that the universe started with an explosive Big Bang and vigorous expansion, but then slowed down and entered a quasi-stationary state during which it did not expand very much. This quasi-stationary state ended when the repulsive pressure associated with a nonzero Λ took over and propelled the universe to the expansion rate we see today. The age of this "loitering universe" could be much greater than the ages of the oldest stars, yet still give us a Hubble constant of 80 km/s/Mpc.

19-6 The shape of the universe is related to the rate of deceleration and to the average density of matter

There is another way of investigating the fate of the universe. Our present understanding of the universe is based on Einstein's general theory of relativity, which explains that gravity curves the fabric of space. The gravity of all the matter in the universe should thus give space an overall curvature. By measuring the curvature of space, we should be able to discover whether the universe is bounded or unbounded, and thus we might discover its ultimate fate.

To see what astronomers mean by the curvature of the universe, imagine shining two powerful laser beams out into space. Suppose that we can align these two beams so that they are perfectly parallel as they leave the Earth.

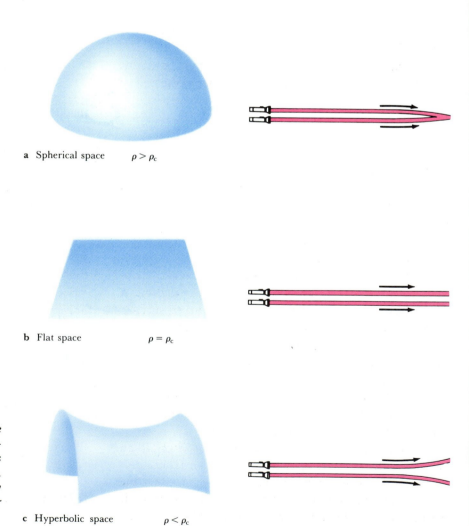

a Spherical space $\rho > \rho_c$

b Flat space $\rho = \rho_c$

c Hyperbolic space $\rho < \rho_c$

Figure 19-10 The geometry of the universe
The "shape" of space is determined by the matter contained in the universe. The curvature is either **(a)** *positive,* **(b)** *zero, or* **(c)** *negative, depending on whether the average density throughout space is greater than, equal to, or less than the critical density.*

Table 19-1 The shape and fate of the universe ($\Lambda = 0$)

Geometry of space	Curvature of space	Average density throughout space	Deceleration parameter (q_0)	Type of universe	Ultimate future of the universe
Spherical	Positive	Greater than critical density	Greater than $\frac{1}{2}$	Closed	Eventual collapse
Flat	Zero	Exactly equal to critical density	Exactly equal to $\frac{1}{2}$	Flat	Perpetual expansion (just barely)
Hyperbolic	Negative	Less than critical density	Between 0 and $\frac{1}{2}$	Open	Perpetual expansion

Finally, suppose that nothing gets in the way of these two beams, so that we can follow them for billions of light-years across the universe, across the space whose curvature we wish to detect.

With this arrangement, there are only three possibilities. First, we might find that our two beams of light remain perfectly parallel, even after traversing billions of light-years. In this case, we would say that space is not curved: The universe has *zero curvature* and space is *flat*.

Alternatively, we might find that our two beams of light gradually get closer and closer together as they move across the universe, so that they eventually intersect at some enormous distance from Earth. In this case, space would not be flat. Recall that lines of longitude on the Earth's surface are parallel at the equator but intersect at the poles. Thus, in this case, the shape of the universe would be analogous to the surface of a sphere. We would then say that space is *spherical* and the universe has *positive curvature*.

The third and final possibility is that the two parallel beams of light would gradually diverge, becoming farther and farther apart as they move across the universe. In this case, we say that the universe has *negative curvature*. A saddle is a good example of a negatively curved surface. Parallel lines drawn on a saddle always diverge. Mathematicians say that saddle-shaped surfaces are hyperbolic. Thus, in a negatively curved universe, we would describe space as *hyperbolic*.

These three situations are sketched in Figure 19-10. To illustrate these cases, three surfaces are drawn: a sphere, a plane, and a saddle. Of course, real space is three-dimensional, but it is much easier to visualize an analogous two-dimensional surface. Thus, as you examine the drawings in Figure 19-10, remember that the real universe has one more dimension. For example, if the universe is hyperbolic, the geometry of space is the three-dimensional analogue of the two-dimensional surface of a saddle.

Each of the three possible geometries corresponds to a different behavior and fate of the universe. Flat space corresponds to the marginally bound case ($q_0 = \frac{1}{2}$), in which the galaxies just barely manage to keep receding from each other. This flat space scenario divides the positive curvature cases from the negative curvature cases. If the density across space is greater than the critical density, then $q_0 > \frac{1}{2}$, and space is positively curved. Conversely, if the density across space is less than the critical density, then $0 < q_0 < \frac{1}{2}$, and space is negatively curved. These relationships, which are valid only if $\Lambda = 0$, are summarized in Table 19-1.

Note that both the flat and the hyperbolic universes are infinite. They extend forever in all directions, so that they have neither an edge nor a center. In contrast, the spherical universe is finite. However, it also lacks a center or an edge. A good analogy is the Earth. Even though the Earth has a finite surface area (511 million square kilometers), you could walk forever around it without ever finding the center or an edge. Likewise, relativistic cosmology strictly rules out any possibility of a center or edge to the universe.

19-7 The flatness of the universe and the isotropy of the microwave background suggest that a period of vigorous inflation followed the Big Bang

Ever since Edwin Hubble discovered that the universe is expanding, astronomers have struggled to determine the deceleration parameter. During the 1960s and 1970s, various teams of astronomers reported various values for q_0, some slightly larger than $q_0 = \frac{1}{2}$ and some smaller. Because $q_0 = \frac{1}{2}$ is the special case of a flat universe separating a bounded cosmological model from an unbounded model, the predicted fate of the universe swung back and forth. According to some data, the universe seems to be just barely open and infinite, whereas other data indicate that the universe is just barely closed and ultimately doomed to collapse.

Motivated by the fact that q_0 may be nearly equal to $\frac{1}{2}$, physicists began looking for special conditions associated with a flat universe in which the average density of matter (ρ) is exactly equal to the critical density, ρ_c. Specifically, suppose the average density of matter during the Big Bang was slightly larger (or smaller) than the critical density. How would this deviation grow (or decrease) as the universe evolved?

We saw that the earliest understandable moment in the universe was the Planck time, about 10^{-43} second after the Big Bang. Between $t = 0$ and $t = 10^{-43}$ second, the universe was so dense and particles were interacting so violently that no known theory can properly describe what happened then. However, immediately after the Planck time, the fate of the universe became very sensitive to the density of matter. Calculations demonstrate that the slightest deviation from the precise critical density would have mushroomed very rapidly, doubling itself every 10^{-35} second. If the density were slightly less than ρ_c, the universe would soon have become wide open and virtually empty. If, on the other hand, the density were slightly greater than ρ_c, the universe would soon have become tightly closed and so packed with matter that the entire cosmos would have rapidly collapsed in a Big Crunch. In other words, immediately after the Big Bang, the fate of the universe hung in the balance, like a pencil teetering on its point, so that the tiniest deviation from the precise equality $\rho = \rho_c$ would have rapidly propelled the universe away from the special case of $q_0 = \frac{1}{2}$.

Observations reveal that q_0 is today approximately $\frac{1}{2}$. Consequently, the density of the universe immediately after the Big Bang must have been equal to the critical density to an incredibly high degree of precision. Calculations demonstrate that, in order for q_0 to be roughly $\frac{1}{2}$ today, ρ right after the Big Bang must have been equal to ρ_c to more than 50 decimal places!

What could have happened immediately after the Planck time to ensure that $\rho = \rho_c$ to such an astounding degree of accuracy? Because $\rho = \rho_c$ means that space is flat, this enigma is called the **flatness problem.**

A second enigma, closely related to the flatness problem, is the isotropy of the cosmic microwave background. We saw that the microwave background is so incredibly uniform across the sky that sensitive temperature measurements can reveal our motion through this radiation field. Subtracting the effects of our motion, we find that the temperature of the microwave background is the same in all parts of the sky to an accuracy of 1 part in 10,000.

To appreciate this so-called **isotropy problem**, think about microwave radiation coming at us from two opposite parts of the sky. This radiation left over from the Big Bang has been traveling toward us for nearly 15 billion years. The total distance between opposite sides of the observable universe is roughly 30 billion light-years, so these widely separated regions have absolutely no connection with each other. Why, then, do these unrelated parts of the universe have the same temperature?

In the early 1980s, Alan Guth at Stanford University offered a remarkable solution to the problems of the flatness of the universe and the isotropy of the microwave background. Guth analyzed the suggestion that the universe had experienced a brief period of extremely rapid expansion shortly after the Planck time (Figure 19-11). During this **inflationary epoch**, as it is called, the universe ballooned outward in all directions to become many billions of times its original size. This ballooning occurs because, very briefly, the cosmological constant in Guth's equations becomes extremely large and provides a powerful repulsive pressure that literally blows up the universe.

Inflation placed much of the material that was originally near our location far beyond the edge of the observable universe today. Thus, the cosmic particle horizon is now expanding *into* space containing matter and radiation that was once in close contact with our location.

Inflation accounts for the flatness of the observable universe. Even though the newborn universe might have begun with a highly curved shape, inflation expanded the universe so much that the observable portion we can see today looks quite flat. The observable universe, which extends outward from Earth roughly 15 billion light-years in all directions, is such a tiny fraction of the inflated universe that any overall curvature is virtually undetectable. To see why, think about a small portion of the Earth's surface, such as your backyard. For all practical purposes, it is impossible to detect the Earth's curvature over such a small area, so your backyard looks very flat. Like your backyard, the segment of space we can observe looks very flat.

Inflation also accounts for the isotropy of the microwave background. When we examine microwaves that

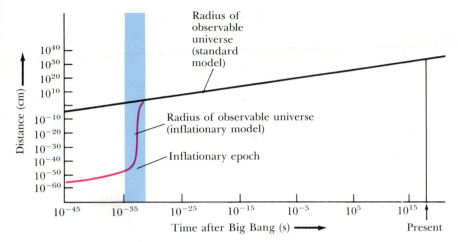

are from opposite parts of the sky, we are seeing radiation from regions of the universe that were originally in such intimate contact that they completely shared all physical properties. Before the onset of inflation, these regions therefore had the same temperature. Inflation then separated the once-intimate regions by a distance greater than the distance that light takes to travel across the observable universe. Since inflation, widely separated parts of the sky have simply cooled down to the 2.73 K temperature we observe today.

Finally, it is important to note that the concept of inflation does not violate Einstein's dictum that nothing can travel faster than the speed of light. Remember that the expansion of the universe is the expansion of space and does not involve the motion of objects through space. Inflation was a brief moment during which the distances between particles suddenly increased by an enormous amount. This expansion was accomplished entirely by a sudden vigorous expansion of space. Particles did not move through space; space inflated.

19-8 Cosmic background radiation dominated the universe during the first million years

Everything in the universe falls into one of two categories: matter or energy. The matter is contained in such objects as stars, planets, and galaxies, all of which are composed of particles like electrons, protons, and neutrons. The radiant energy in the universe consists of photons. There are, of course, many starlight photons traveling across space, but the vast majority of photons in the universe are members of the cosmic microwave background.

Matter prevails over radiation today because the energy carried by microwave photons is so small. Using Einstein's famous equation $E = mc^2$, which expresses a fundamental relationship between mass and energy, astronomers can compare the matter density of the universe with the energy density of the microwave radiation that fills all space. The average density of matter throughout space today is much larger than the comparable energy density associated with the microwave background. Astronomers therefore say that we live in a **matter-dominated universe.**

Although the universe is dominated by matter, the number of photons in the microwave background is very large. From the physics of blackbody radiation it can be demonstrated that there are today 550 million photons in every cubic meter of space. In contrast, if all the visible matter in the universe were uniformly spread throughout space, there would be roughly one hydrogen atom in every three cubic meters of space. In other words, photons outnumber atoms by roughly a billion to one. This radiation field no longer has much "clout," though, because its photons have been redshifted to long wavelengths and low energies after nearly 15 billion years of being stretched by the expansion of the universe.

Although matter dominates the universe today, this was not always so. To see why, think back toward the Big Bang. Because the universe was more compressed, the density of matter was greater. The photons in the background radiation were also crowded more closely together then, but an additional effect must be considered. In those times, the photons were less redshifted and thus had shorter wavelengths and higher energy than they do today. Because of this added energy, there was a time in the ancient past when radiation held sway over matter.

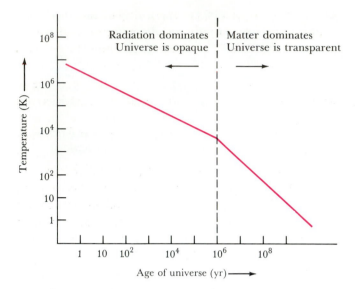

Figure 19-12 **The declining temperature of the universe** As the universe expands, the photons in the radiation background become increasingly redshifted and the temperature of the radiation field falls. Roughly 1 million years after the Big Bang, when the temperature fell below 3000 K, hydrogen atoms formed and the universe became transparent.

During that time, the energy density of the radiation field was greater than the average density of matter. Astronomers call this state a **radiation-dominated universe.**

The transition from a radiation-dominated universe to a matter-dominated universe occurred about 1 million years after the Big Bang. Since that time, the wavelengths of photons have been stretched by a factor of 1000. Today these microwave photons typically have wavelengths of about 1 mm, but when the universe was one million years old they had wavelengths of about 0.001 mm.

We can use Wien's law to calculate the temperature of the cosmic background radiation at the time of this transition from a radiation-dominated universe to a matter-

dominated one. Just as a peak wavelength λ_{max} of 1 mm corresponds to a blackbody temperature of 3 K, a peak wavelength of 0.001 mm corresponds to 3000 K. The temperature history of the universe is graphed in Figure 19-12.

The nature of the universe changed in a fundamental way when the temperature of the radiation field fell to 3000 K. To understand this change, recall that hydrogen is by far the most abundant element in the universe. Of course, a hydrogen atom consists of a single proton orbited by a single electron. It takes only a little energy to knock the proton and electron apart. In fact, a radiation field warmer than about 3000 K easily ionizes hydrogen. Thus, hydrogen atoms could not exist earlier than about 1 million years after the Big Bang. As sketched in Figure 19-13a, the background photons prior to $t = 1$ million years had energies great enough to prevent electrons and protons from getting together to form hydrogen atoms. Only since $t = 1$ million years have these photons been redshifted enough to permit hydrogen atoms to exist (see Figure 19-13b).

Prior to $t = 1$ million years, the universe was completely filled with a shimmering expanse of high-energy photons colliding vigorously with protons and electrons. This state of matter, called a plasma, is opaque, just as the glowing gases inside a neon advertising sign or fluorescent light bulb are opaque. The term **primordial fireball** describes the universe during this time.

After $t = 1$ million years, the photons no longer had enough energy to keep the protons and electrons apart. As soon as the temperature of the radiation field fell below 3000 K, protons and electrons everywhere began combining to form hydrogen atoms. Because hydrogen is transparent, the universe suddenly became transparent! The photons that just a few seconds earlier had been vigorously colliding with charged particles could now stream

Figure 19-13 **The era of recombination** (a) Before recombination, photons in the cosmic background prevented protons and electrons from forming hydrogen atoms. (b) As soon as hydrogen atoms could survive, the universe became transparent. The transition from opaque to transparent occurred roughly 1 million years after the Big Bang.

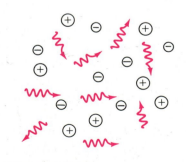

a Before recombination

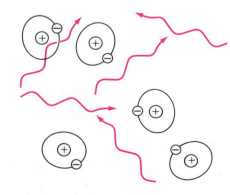

b After recombination

unimpeded across space. Today we see these same photons in the microwave background.

This dramatic moment, when the universe went from being opaque to being transparent, is referred to as the **era of recombination.** Because the universe was opaque prior to $t = 1$ million years, we cannot see any farther into the past than the era of recombination. The microwave background contains the most ancient photons we shall ever be able to observe. This microwave background is today only a ghostly relic of its former dazzling splendor.

19-9 During the first second, most of the matter and antimatter in the universe annihilated each other

When the universe was young, it was dominated by a hot radiation field that filled all space. We have also seen that the photons of this radiation field are a billion times more plentiful than ordinary particles of matter. Where did all these photons and particles come from? To search for clues, we must delve still further into the past and examine conditions within a few seconds of the Big Bang.

As we move back in time toward the Big Bang, we find that the energy of the photons in the universe increases. If we go far enough into the past, we come to the time when these photons possessed enough energy to create matter according to Einstein's equation $E = mc^2$. This equation explains that matter and energy can be converted one into the other. To make a particle of mass m, you need an amount of energy E that is at least as great as mc^2, where c is the speed of light.

The creation of matter from energy is routinely observed in laboratory experiments involving high-energy gamma rays. When a highly energetic gamma ray photon collides with a second photon, both vanish and matter appears in their place, as sketched in Figure 19-14. This process, called **pair production,** always creates one ordinary particle and one antiparticle. For instance, if one of the particles is an electron, the other is an antielectron (Figure 19-15).

There is nothing mysterious about antimatter. An antiparticle is like an ordinary particle except that it has an opposite electric charge. For example, an antielectron (e^+) has a positive charge, whereas an ordinary electron (e^-) has a negative charge. If a particle and its antiparticle collide, they annihilate each other and their mass is converted back into high-energy photons. Particles and antiparticles are always created or destroyed in equal numbers, thus ensuring that the total electric charge in the universe remains constant. Physicists often refer to this balance between matter and antimatter as a "symmetry."

Close to the time of the Big Bang, photon energy was high enough to initiate pair creation throughout space. When the universe was 1 second old, the temperature of the radiation field was about 6 billion kelvin. This was so hot that colliding photons produced electrons and antielectrons. Thus at times earlier than $t = 1$ second, the universe contained vast numbers of these particles. When the universe was 0.0001 second old, the temperature was 10 trillion kelvin. That was hot enough to create protons and antiprotons or neutrons and antineutrons. Thus at times earlier than $t = 0.0001$ second, the universe was teeming with these particles also. In other words, shortly after the Big Bang all space was chock full of particles and antiparticles immersed in an inconceivably hot bath of high-energy photons.

Now let us think forward in time from these early moments when the universe was filled with particles and antiparticles. As the universe expanded, all the gamma ray

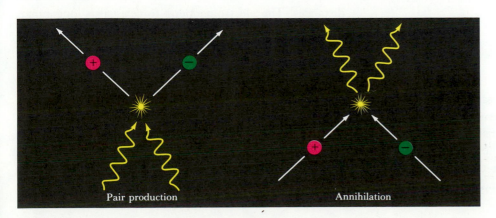

Pair production   Annihilation

Figure 19-14 Pair production and annihilation *A particle and an antiparticle can be created during the collision of high-energy photons. Conversely, a particle and an antiparticle can annihilate each other by giving up energy in the form of gamma rays.*

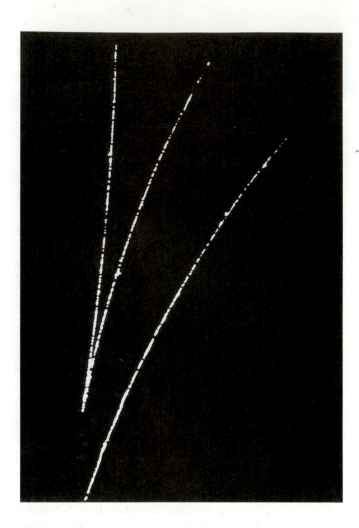

Figure 19-15 *The creation of matter from energy* *This photograph shows the conversion of a gamma ray into matter inside a bubble chamber, a device filled with liquid hydrogen that is designed to make the path of a charged particle visible as a long row of tiny bubbles. The path of the gamma ray is not visible because photons are electrically neutral. Near the bottom of the photograph, the energy carried by the gamma ray is converted into an electron and an antielectron. Because of a magnetic field surrounding the bubble chamber, the electron is deflected to the right while the antielectron veers toward the left. The path of a stray electron is also seen at the right. (Courtesy of Lawrence Berkeley Laboratory)*

photons became increasingly redshifted. The temperature soon declined to a level at which the gamma rays were no longer energetic enough to create pairs of particles and antiparticles. Collisions between existing particles and antiparticles did add photons to the cosmic radiation background, but collisions between photons no longer replenished the supply of particles and antiparticles. The first second of the universe was thus dominated by the

wholesale annihilation of vast amounts matter and antimatter.

In laboratory experiments, physicists observe a strict equality in numbers of particles and antiparticles. Particles and antiparticles are always created or destroyed in equal numbers. If this symmetry is truly valid, then for every proton there should have been an antiproton. For every electron there should have been an antielectron. This poses a dilemma because by the time the universe was 1 second old, every particle should have been annihilated by an antiparticle, leaving no matter at all in the universe.

Obviously this did not happen. Physicists therefore say that a "symmetry breaking" occurred during the earliest moments of the universe, which caused the number of particles to be slightly greater than the number of antiparticles. For every billion antiprotons, a billion plus one protons must have existed. For every billion antielectrons, a billion plus one electrons existed. This particular slight excess is suggested by the fact that there are roughly a billion photons in the microwave background today for each proton and neutron in the universe.

19-10 *Speculative theories suggest that all four forces had the same strength immediately after the Big Bang*

All the behavior and interaction of everything in the universe can be explained as the result of just *four* physical forces: gravity, electromagnetism, and the strong and weak nuclear forces. These four forces differ sharply from one other.

We are all familiar with the force of gravity, the long-range force that dominates the universe over astronomical distances. The electromagnetic force is also long range, because its influence extends to infinity, as gravity's does. Just as the force of gravity holds the Moon in orbit about the Earth, the electromagnetic force holds electrons in orbit about the nuclei in atoms. We do not generally observe the long-distance effects of the electromagnetic force, however, because in most cases there is a negative electric charge for every positive charge and a south magnetic pole for every north magnetic pole. Over large volumes of space, the net effects of electromagnetism effectively cancel each other. A similar canceling does not occur with gravity because there is no "negative mass."

The strong and the weak nuclear forces are both said to be short range because their influence extends only over distances less than about 10^{-15} m. The **strong nuclear force** holds protons and neutrons together inside the nuclei of atoms. Without this force, nuclei would disintegrate because of the electromagnetic repulsion of the positively charged protons. Thus, the strong nuclear force overpowers the electromagnetic force inside nuclei. The **weak nuclear force** is so weak that it does not hold anything together. Instead, it is at work in certain kinds of radioactive decay, such as the transformation of a neutron into a proton.

Numerous experiments in nuclear physics strongly suggest that protons and neutrons are composed of more basic particles called **quarks,** the most common varieties being "up" quarks and "down" quarks. A proton is composed of two up quarks and one down quark, whereas a neutron is made of two down quarks and one up quark.

In the 1970s the concept of quarks gave rise to a more fundamental description of the strong and weak nuclear forces. In its most basic form, the strong nuclear force is the force that holds quarks together. Similarly, the weak nuclear force is at work whenever a quark changes from one variety to another. For example, when a neutron decays into a proton, one of the neutron's down quarks changes into an up quark.

A highly successful description of forces, called **quantum field theory,** explains that two particles exert a force on each other by exchanging a third particle. The exchanged particle is the carrier of the force. For instance, the gravitational force occurs when particles exchange **gravitons,** and the weak nuclear force exists when particles exchange **weakons.** The carrier of the electromagnetic force is the photon, and quarks stick together by exchanging **gluons.** Thus we may summarize the four forces as indicated in Table 19-2.

To examine details of the forces, scientists use particle accelerators that hurl high-speed electrons and protons at targets. In such experiments, physicists find that the different forces begin to look the same as the particles' speeds approach the speed of light. In fact, during experiments at the CERN accelerator in Europe in the 1980s, particles were slammed together with such violence that the electromagnetic force and the weak nuclear force had equal strength—they were "unified."

In recent years physicists have labored to produce a comprehensive theory that would completely describe these forces. Some of the most promising attempts predict that all four forces would have the same strength if only we could slam particles together with energies trillions of times greater than that in the CERN accelerator. Physicists have no hope of building accelerators that powerful. However, the universe was so hot and particles were moving with such high speeds immediately after the Big Bang that all four forces were unified into a single "superforce." The earliest moments of the universe thus provide a laboratory wherein scientists can explore some of the most elegant ideas in physics. Figure 19-16 shows the age and temperature of the universe when the various forces become unified.

Many of the ideas connecting particle physics with cosmology are very new and still quite speculative. Nevertheless, Figure 19-17 attempts to summarize these ideas. During the Planck time (from $t = 0$ to $t = 10^{-43}$ second), particles collided with such high energies that all four forces were unified. By the end of the Planck time, however, the energy of particles in the universe had declined to the extent that gravity was no longer unified with the

Table 19-2 The four forces

Force	Relative strength	Particles exchanged	Particles acted upon	Range	Example
Strong	1	Gluons	Quarks	10^{-15} m	Holding nuclei together
Electromagnetic	1/137	Photons	Charged particles	Infinite	Holding atoms together
Weak	1/10,000	Weakons	Quarks, electrons, neutrinos	$<10^{-16}$ m	Radioactive decay
Gravity	6×10^{-39}	Gravitons	Everything	Infinite	Holding the solar system together

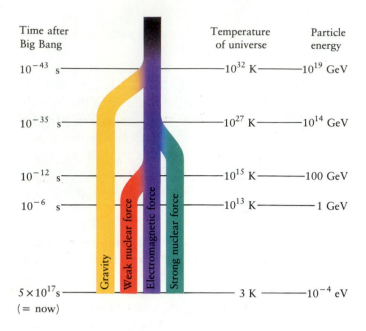

Time after
Big Bang

10^{-43} s

10^{-35} s

10^{-12} s

10^{-6} s

5×10^{17}s
(= now)

Temperature
of universe

10^{32} K

10^{27} K

10^{15} K

10^{13} K

3 K

Particle
energy

10^{19} GeV

10^{14} GeV

100 GeV

1 GeV

10^{-4} eV

Gravity

Weak nuclear force

Electromagnetic force

Strong nuclear force

Figure 19-16 **Unification of the four forces** *The strength of the four physical forces depends on the speed or energy with which these particles interact. As shown in this diagram, the higher the temperature of the universe, the more the forces resemble each other. Also included here is the age of the universe at which the various forces were equal.*

other three forces. We can therefore say that, at $t = 10^{-43}$ second, gravity "froze out" of the otherwise unified hot soup that filled all space. The temperature of the universe was 10^{32} K when gravity emerged as a separate force.

By $t = 10^{-35}$ second, the temperature of the universe had fallen to 10^{27} K and the energy of particles had declined to the extent that the strong nuclear force was no longer unified with the electromagnetic and weak nuclear forces. Thus, at $t = 10^{-35}$ second, the strong nuclear force made its appearance, freezing out of an otherwise unified hot soup. Calculations suggest that the inflationary epoch lasted from $t = 10^{-35}$ second to about $t = 10^{-24}$ second, during which time the universe increased its size enormously, perhaps by a factor of 10^{50}.

At $t = 10^{-12}$ second, when the temperature of the universe had dropped to 10^{15} K, a final "freeze out" separated the electromagnetic force from the weak nuclear force. From that moment on, all four forces interacted with particles essentially as they do today.

The next significant event occurred at $t = 10^{-6}$ second, when the universe's temperature was 10^{13} K. Prior to this moment, particles collided so violently that individual protons and neutrons could not exist, because they were constantly being fragmented into quarks. After this moment, appropriately called **confinement**, quarks could finally stick together to form individual protons and neutrons.

As noted on Figure 19-17, all the primordial helium was produced by $t = 3$ minutes, and the universe became transparent to photons at $t = 1$ million years. These photons are today observed as the cosmic microwave background, the ghostly reminder of the first few moments of the universe.

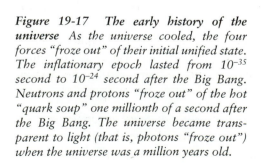

Figure 19-17 **The early history of the universe** *As the universe cooled, the four forces "froze out" of their initial unified state. The inflationary epoch lasted from 10^{-35} second to 10^{-24} second after the Big Bang. Neutrons and protons "froze out" of the hot "quark soup" one millionth of a second after the Big Bang. The universe became transparent to light (that is, photons "froze out") when the universe was a million years old.*

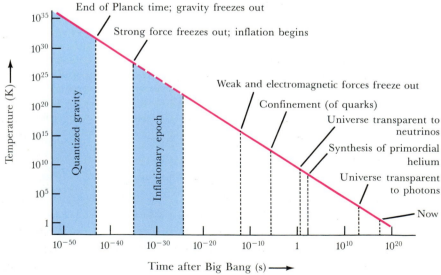

End of Planck time; gravity freezes out

Strong force freezes out; inflation begins

Weak and electromagnetic forces freeze out

Confinement (of quarks)

Universe transparent to neutrinos

Synthesis of primordial helium

Universe transparent to photons

Now

Temperature (K)

10^{35}

10^{30}

10^{25}

10^{20}

10^{15}

10^{10}

10^{5}

1

Quantized gravity

Inflationary epoch

10^{-50} 10^{-40} 10^{-30} 10^{-20} 10^{-10} 1 10^{10} 10^{20}

Time after Big Bang (s) ⟶

Summary

- The universe began as an infinitely dense cosmic singularity that expanded explosively in the event called the Big Bang, which can be described as an explosion of space at the beginning of time.

  The Hubble law describes the ongoing expansion of the universe; the space between widely separated galaxies is growing ever larger.

  The observable universe extends about 15 billion light-years in every direction from the Earth. We cannot see objects beyond the cosmic particle horizon at the distance of 15 billion light-years because light from these objects has not had enough time to reach us.

  It is meaningless to speak of an edge or center to the universe.

- A Hubble constant of 80 km/s/Mpc implies that the age of the universe $(1/H_0)$ is less than 12 billion years. This conclusion is distressing because the oldest stars seem of have ages of about 15 billion years.

  By adding a cosmological constant (Λ) to the equations that describe relativistic cosmology, it is possible to calculate models of the universe that are at least 15 billion years old, but have the same Hubble constant we observe today.

- Before the Planck time (about 10^{-43} second after calculated time of the Big Bang), the universe was so dense that known laws of physics do not properly describe the behavior of space, time, and matter.

- The cosmic microwave background radiation is the greatly redshifted remnant of the very hot universe that existed about 1 million years after the Big Bang.

  During the first million years of the universe, matter and energy formed an opaque plasma called the primordial fireball.

  By 1 million years after the Big Bang, the expansion of the universe had caused the temperature of the primordial fireball to fall below 3000 K so that protons and electrons could combine to form hydrogen atoms; this event is called the era of recombination.

  The universe became transparent during the era of recombination, meaning that the microwave background radiation contains the oldest photons in the universe.

  The universe today is matter dominated, but during the first million years, before the background radiation became greatly redshifted, the universe was radiation dominated.

- The average density of matter in the universe determines the curvature of space and the ultimate fate of the universe.

  If the average density of matter in the universe is greater than the critical density ρ_c, then space is spherical (with positive curvature), the deceleration parameter q_0 has a value greater than $\frac{1}{2}$, and the universe is closed (bounded) and will ultimately collapse.

  If the average density of matter in the universe is less than ρ_c, then space is hyperbolic (with negative curvature), q_0 has a value less than $\frac{1}{2}$, and the universe is open (unbounded) and will continue to expand forever.

  If the average density of matter in the universe is exactly equal to ρ_c, then space is flat (with zero curvature), q_0 is exactly equal to $\frac{1}{2}$, the universe is marginally bounded, and expansion will just barely continue forever.

- That the universe is nearly flat and that the cosmic microwave background is almost perfectly isotropic may be explained as the result of a brief period of very rapid expansion (the inflationary epoch).

  During the inflationary period, much of the material originally near our location moved far beyond the limits of our observable universe. The observable universe today is thus expanding into space containing matter and radiation that was in close contact with our matter and radiation during the first instant after the Big Bang.

- Four basic forces—gravity, electromagnetism, the strong nuclear force, and the weak nuclear force—explain all of the interactions observed in the universe.

Theories that attempt to explain all four forces in terms of a single consistent set of physical laws suggest that all four forces were identical just after the Big Bang.

At the end of the Planck time, gravity "froze out" to become a distinctive force. A short time later, the strong nuclear force became a distinctive force. A final "freeze-out" separated the electromagnetic force from the weak nuclear force.

Review questions

1 What does it mean when astronomers say that we live in an expanding universe?

2 What is Olbers's paradox, and how can it be resolved?

3 Explain the difference between a Doppler shift and a cosmological redshift. Why is it incorrect to think of the redshifts of remote galaxies and quasars as being a result of the Doppler effect?

4 How does modern cosmology preclude the possibility of a center or an edge to the universe?

5 In what ways are the fate of the universe, the geometry of the universe, and the average density of the universe related?

6 Suppose that the universe will expand forever. What will eventually become of the microwave background radiation?

7 What does it mean to say that the universe is matter dominated? When was the universe radiation dominated?

8 What is the difference between an electron and an antielectron?

9 Where did most of the photons in the cosmic microwave background come from?

10 Describe an example of each of the four basic forces in the physical universe.

11 What is the observational evidence for (**a**) the Big Bang, (**b**) the inflationary epoch, and (**c**) the confinement of quarks?

Advanced questions

12 Verify by direct calculation that $1/H_0$ is 12 billion years if H_0 is 80 km/s/Mpc.

13 Before the cosmic microwave background was discovered, it seemed possible that we might be living in a "steady state" universe whose overall properties do not change with time. The steady-state model, like the Big Bang model, assumes an expanding universe, but it does not assume a "creation event." Instead, in the steady-state theory matter is assumed to be created continuously everywhere in space to ensure that the average density of the universe remains constant. Explain why the cosmic microwave background is a major blow to the steady-state theory.

14 As this book was going to press, California astronomers Wendy Freedman and Barry Madore published a recalibration of standard candles that yielded a Hubble constant of 85 km/s/Mpc. Consult recent issues of *Sky & Telescope*, *Astronomy*, and *Science News* to see if any other determinations of H_0 have been published since this 1992 result. Do they agree with the Freedman–Madore value? Are there any new ideas in the literature you have researched to account for the discrepancy between $1/H_0$ and the ages of the oldest stars?

Discussion questions

15 Suppose we were living in a radiation-dominated universe. How would such a universe be different from what we now observe?

16 Discuss the theological implications of the idea that we cannot use science to tell us what existed before the Big Bang.

17 Do you think that there can be "other universes," regions of space and time that are not connected to our universe? Should astronomers be concerned with such possibilities? Why or why not?

For further reading

Barrow, J., and Silk, J. *The Left Hand of Creation: Origin and Evolution of the Universe.* Basic Books, 1983 • This masterful book tells the exciting story of how space, time, and matter may have been created.

Davies, P. *The Forces of Nature.* 2nd ed. Cambridge University Press, 1986 • This is perhaps one of the best introductions to particle physics and grand unified field theory.

———. *Superforce*. Simon & Schuster, 1984 • This lucid and entertaining treatise examines the quest for a grand unified theory of physical reality.

Dicus, D., and others. "The Future of the Universe." *Scientific American*, March 1983 • This fascinating article, which speculates about the future of the universe through the year 10^{100}, includes descriptions of proton decay and the evaporation of black holes.

Disney, M. *The Hidden Universe*. Macmillan, 1984 • This book on dark matter in the universe illustrates how astronomers develop and investigate new ideas.

Dressler, A. "The Large-Scale Streaming of Galaxies." *Scientific American*, September 1987 • This article describes the strategy that astronomers used to piece together observations that led to the discovery of the Great Attractor.

Gribbin, J. *In Search of the Big Bang*. Bantam, 1986 • This very readable introduction to modern cosmology describes the history, observations, and ideas leading to our modern understanding of the creation and structure of the universe.

Guth, A., and Steinhardt, P. "The Inflationary Universe." *Scientific American*, May 1984 • This interesting article explains why the universe probably experienced a sudden but brief period of vigorous expansion very early in its history.

Harrison, E. *Darkness at Night*. Harvard University Press, 1987 • This definitive book on Olbers's paradox uses eloquent, nontechnical language to explain why the night sky is dark.

Kanipe, J. "Beyond the Big Bang." *Astronomy*, April 1992 • This article eloquently describes some of the issues that astronomers today are exploring as they try to understand details of the creation of the universe.

Monda, R. "Shedding Light on Dark Matter." *Astronomy*, February 1992 • This well-written article surveys the current state of our understanding of dark matter and includes speculations about what that dark matter may be made of.

Silk, J. *The Big Bang*. W. H. Freeman & Company, 1989 • This revised and updated edition of a fascinating book describes many of the processes that made the universe the way it is.

The prevalence of life This painting, entitled DNA Embraces the Planets, *artistically expresses the suspicion of many scientists that carbon-based life may be a common phenomenon in the universe. Other scientists argue, however, that we may be unique and no intelligent alien civilizations exist. In either case, humanity has the clear mandate to preserve and protect the abundance of life forms with which we share our planet. (Courtesy of J. Lomberg)*

Afterword:
The Search for Extraterrestrial Life

The heavens inspire us to contemplate a variety of profound topics, including the creation of the universe, the nature of the stars, and the formation of the Earth. Of all the fascinating subjects we might explore, perhaps none is as compelling as the question of extraterrestrial life. Are we alone? Does life exist elsewhere in the universe? What are the chances that we might someday make contact with an alien civilization?

There are no firm answers to such questions. Earth is the only planet on which life is known to exist, and our probes to other worlds have so far failed to detect any life forms on them. This lack of data does not, however, negate the possibility of extraterrestrial biology.

As you have seen throughout this book, one of the great lessons of modern astronomy is that our circumstances are quite ordinary. Contrary to the beliefs of our ancestors, we do not occupy a special location like the "center of the universe." Over the past four centuries it has become clear that we inhabit one of nine planets orbiting an unremarkable star—just one of billions in an undistinguished galaxy. Is it possible that we are also biologically commonplace? The answer to this question is sought by scientists involved in SETI, the search for extraterrestrial intelligence.

Although the possibility of an alternative form of biochemistry cannot be ruled out, scientists at present confine their search to life as we now know it. All terrestrial life is based on the unique properties of the carbon atom. Carbon is an extremely versatile element whose atoms are capable of forming chemical bonds that result in especially long and complex molecules. Among these carbon-based compounds, called **organic molecules,** are the molecules of which living organisms are made.

Organic molecules can be linked together to form elaborate structures such as chains, lattices, and fibers. Some of these structures are capable of complex, self-regulating chemical reactions. Furthermore, the primary constituents of organic molecules—carbon, hydrogen, nitrogen, oxygen, sulfur, and phosphorus—are among the most abundant elements in the universe. Indeed, the versatility and abundance of carbon suggest that extraterrestrial biology may also be based on organic chemistry.

Organic molecules are scattered abundantly throughout the Galaxy. In interstellar clouds, carbon atoms have combined with other elements to produce an impressive variety of organic compounds. Since the 1960s radio astronomers have detected telltale microwave emission lines from interstellar clouds that help identify dozens of these carbon-based chemicals. Examples include ethyl alcohol (CH_3CH_2OH), formaldehyde (H_2CO), methyl cyanoacetylene (CH_3C_3N), and acetaldehyde (CH_3CHO).

Further evidence of extraterrestrial organic molecules comes from newly fallen meteorites called carbonaceous chondrites (Figure A-1), which are often found to contain a variety of organic substances. As noted in Chapter 10, carbonaceous chondrites are ancient meteorites dating from the formation of the solar system. So it seems reasonable to conclude that, even from their earliest days, the planets have been continually bombarded with organic compounds.

Interstellar space is not the only source of organic material. In a classic experiment performed in 1952, American chemists Stanley Miller and Harold Urey demonstrated that simple chemicals can combine to form prebiological compounds under supposedly primitive Earthlike conditions. In a closed container, they subjected a mixture of hydrogen, ammonia, methane, and water vapor to an electric arc (to simulate lightning bolts) for a week. At the end of this period, the inside of the container had become coated with a reddish-brown substance rich in compounds essential to life.

Scientists today tend to believe that Earth's primordial atmosphere was probably composed of carbon dioxide, nitrogen, and water vapor outgassed from volcanoes along with some hydrogen. Modern versions of the Miller–Urey experiment (Figure A–2) using these common gases have also produced a wide variety of organic compounds.

It is important to emphasize that scientists have not created life in a test tube. Biologists have yet to figure out, among other things, how these organic molecules gathered into cell-like arrangements and managed to develop systems for self-replication. Nevertheless, since so many chemical components of life are so easily synthesized under conditions that simulate the primordial Earth, it seems reasonable to suppose that life could have originated in chemical processes involving materials from the Earth or from space.

A widespread abundance of organic precursors does not guarantee that life is commonplace throughout the universe. If a planet's environment is hostile, life may either never get started or quickly become extinct. It seems quite possible, however, that there are planets orbiting other stars. Perhaps conditions on some of these worlds are sufficiently suitable for life as we know it.

The development of life on Earth seems to suggest that extraterrestrial life, including intelligent species, might evolve on habitable planets, if there were sufficient time and hospitable conditions. How might we ascertain

Figure A-1 *A carbonaceous chondrite* Carbonaceous chondrites are ancient meteorites that date back to the formation of the solar system. Chemical analysis of newly fallen specimens disclose that they are rich in organic molecules, many of which are the chemical building blocks of life. This sample is a piece of a large carbonaceous chondrite that fell in Mexico in 1969. (From the collection of R. A. Oriti)

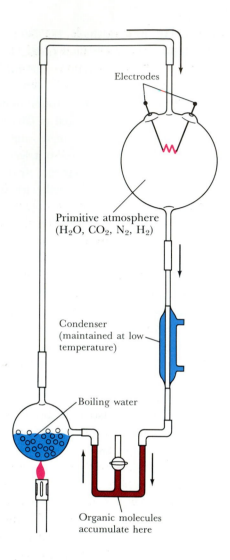

Electrodes

Primitive atmosphere
(H_2O, CO_2, N_2, H_2)

Condenser
(maintained at low
temperature)

Boiling water

Organic molecules
accumulate here

*Figure A-2   The Miller–Urey experiment updated   Modern
versions of this classic experiment prove that numerous organic
compounds important to life can be synthesized from gases that
were present in Earth's primordial atmosphere. This experiment
supports the hypothesis that life on Earth arose as a result of
ordinary chemical reactions.*

whether such worlds exist, given the tremendous distances that indubitably separate us from them? Many astronomers hope to learn about extraterrestrial civilizations by detecting radio transmissions from them. As we have seen, radio waves can travel immense distances without being significantly altered by the gas and dust through which they pass. Because of this ability to penetrate the interstellar medium, radio waves are a logical choice for interstellar communication.

Over the past several decades, astronomers have proposed various ways to search for alien radio transmission, and several limited searches have been undertaken. One of the first searches occurred in 1960, when Frank Drake used a radio telescope at the National Radio Astronomy Observatory in West Virginia to "listen" to two Sunlike stars, τ Ceti and ε Eridani, without success. About 40 similar unsuccessful searches have taken place since then using radio telescopes in both the United States and the former Soviet Union. For instance, in 1973 astronomers "listened" to 600 nearby solar-type stars for an hour each, but no unusual signals were detected. Since 1983 a radio telescope belonging to Harvard University has been used along with a sophisticated computer program to scan a wide range of frequencies over a large portion of the sky. So far nothing has turned up.

Should we be discouraged by this lack of success? What are the chances that a radio astronomer might someday detect radio signals from an extraterrestrial civilization? The first person to tackle such issues was Frank Drake. Drake proposed that the number of technologically advanced civilizations in the Galaxy (designated by the letter N) could be estimated with the equation:

$$N = R_* f_p n_e f_l f_i f_c L$$

where

R_* = rate at which solar-type stars form in the Galaxy

f_p = fraction of stars that have planets

n_e = number of planets per solar system suitable for life

f_l = fraction of those habitable planets on which life actually arises

f_i = fraction of those life forms that evolve into intelligent spedies

f_c = fraction of those species that develop adequate technology and then choose to send messages out into space

and

L = lifetime of that technologically advanced civilization

The Drake equation is enlightening because it expresses the number of extraterrestrial civilizations as a product of terms, some of which can be estimated from what we know about stars and stellar evolution. For instance, the first two terms, R_* and f_p, can be determined by observation. In estimating R_*, we should probably exclude massive stars (those larger than about l.5 $M_\odot$), because they have main sequence lifetimes shorter than the time it took for intelligent life to develop here on Earth. Life on Earth originated some 3.5 to 4.0 billion years ago. If that is typical of the time needed to evolve higher life forms, then a massive star probably becomes a red giant or a supernova before intelligent creatures can appear on any of its planets.

Although low-mass stars have much longer lifetimes, they, too, seem unsuited for life because they are so cool. Only planets very near a low-mass star would be sufficiently warm for life as we know it. However, a planet that close can become tidally coupled to the star, with one side continuously facing the star while the other is in perpetual frigid darkness.

This leaves us with main sequence stars like the Sun, those with spectral types between F5 and M0. Based on statistical studies of star formation in the Milky Way, some astronomers estimate that roughly one of these Sunlike stars forms in the Galaxy each year, thus yielding a value for R_* of 1 per year.

We learned in Chapter 6 that the planets in our solar system formed as a natural consequence of the birth of the Sun, and we have seen evidence suggesting that similar processes of planetary formation may be commonplace around single stars. Yet no planet outside our solar system has been discovered so far. Nevertheless, many astronomers give f_p a value of 1, meaning they believe it likely that most Sunlike stars have planets.

Unfortunately, the rest of the terms in the Drake equation are very uncertain. Let's play with some hypothetical values. The chances that a planetary system has an Earth-like world are not known. Were we to consider our own solar system as representative, we could put n_e at 1. Let's be more conservative, however, and suppose that 1 in 10 solar-type stars is orbited by a habitable planet, making $n_e = 0.1$. From what we know about the evolution of life on Earth, we might assume that, given appropriate conditions, the development of life is a certainty, which would make $f_l = 1$. This is, of course, an area of intense interest to biologists. For the sake of argument, we might also assume that evolution naturally leads to the development of intelligence (a conjecture that is hotly debated) and also make $f_i = 1$. It's anyone's guess as to whether these intelligent extraterrestrial beings would attempt communication with other civilizations in the Galaxy, but were we to assume they would, f_c would be put at 1 also.

The last variable, L, the longevity of civilization, is the most uncertain of all and certainly cannot be subjected to testing! Looking at our own example, we see a planet whose atmosphere and oceans are increasingly polluted by creatures that possess nuclear weapons. If we are typical, perhaps L is as short as 100 years. Putting all these numbers together, we arrive at

$$N = 1 \times 1 \times 0.1 \times 1 \times 1 \times 1 \times 100 = 10$$

In other words, out of the hundreds of billions of stars in the Galaxy, we would estimate that there are only ten technologically advanced civilizations from which we might receive communication.

A wide range of values has been proposed for the terms in the Drake equation, and these various guesses produce vastly different estimates of N. Some scientists argue that there is exactly one advanced civilization in the Galaxy and that we are it. Others speculate that there may be hundreds or thousands of planets inhabited by intelligent creatures.

If extraterrestrial beings were purposefully sending messages into space, it seems reasonable that they might choose a frequency that is fairly free of emissions from extraneous sources. SETI pioneer Bernard Oliver has pointed out that a range of relatively noise-free frequencies exists in the neighborhood of the microwave emission lines of hydrogen and hydroxyl radical (OH) (Figure A-3). This region of the microwave spectrum is called the "water hole," a humorous reference to the H and OH lines being so close together. Or perhaps they would choose to transmit at a wavelength of 21 cm, because astronomers studying the distribution of hydrogen around the Galaxy would already have their radio telescopes tuned to that wavelength (recall Figure 16-7).

Even if there are only a few alien civilizations scattered across the Galaxy, we have the technology to detect radio transmissions from them. The most ambitious plan ever proposed was Project Cyclops, which would have consisted of 1000 to 2500 radio antennae, each 30 m in diameter (Figure A-4). This colossal array would have been so sensitive that it could have detected signals from virtually anywhere in the Galaxy. Unfortunately, that project was not funded.

On a more modest scale, NASA has funded a SETI program called the Microwave Observing Project, which began in October 1992. NASA scientists are using radio telescopes along with sophisticated electronic equipment and powerful computers to conduct both a targeted search and an all-sky survey.

Figure A-3 **The "water hole"** *The so-called water hole is a range of radio frequencies— from about 1000 to 10,000 megahertz (Mhz)—that happens to have relatively little cosmic noise. Some scientists suggest that this noise-free region would be well suited for interstellar communication. (Adapted from C. Sagan and F. Drake)*

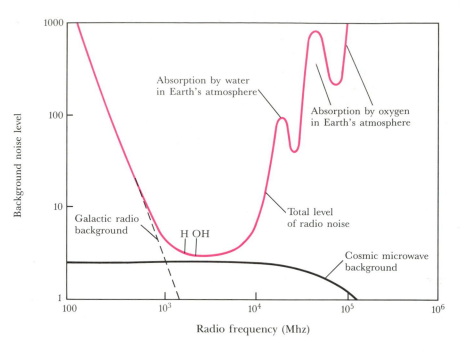

Figure A-4 **Project Cyclops** *Project Cyclops was the grandest plan ever seriously considered for the purpose of searching for extraterrestrial civilizations. It would have consisted of 1000 to 2500 radio antennae spread over an area 16 km in diameter. (NASA)*

The targeted search will examine some 800 nearby solar-type stars over a frequency range that covers the long-wavelength side of the "water hole." The largest radio telescopes are being used in order to achieve the highest possible sensitivity. Specialized equipment for processing digital signals enables NASA scientists to study tens of millions of individual frequency channels simultaneously. This equipment can automatically detect continuous waves or pulses, whether they remain constant in frequency or drift slowly because of some relative motion between the transmitter and the receiver.

The all-sky survey uses several dish-shaped antennae of NASA's Deep Space Network (Figure A-5) to extend the search over the entire sky. Some sensitivity is sacrificed, but the survey does cover the entire "water hole." Digital signal processing equipment automatically examines ten million frequency channels in an effort to detect continuous carrier wave signals from fixed positions in the sky.

Both search modes are conducted simultaneously so that signals can be confirmed independently. New technology developed for the Microwave Observing Project permits signal recognition during the observations and the ability to search for a wide variety of signals. The project is almost completely automated in order to sift through incoming data at a rate far beyond human capabilities.

The detection of a message from an alien civilization would be one of the greatest events in human history. Such a message could dramatically change the course of civilization through the sharing of scientific information or an awakening of social or humanistic enlightenment. In only a few years our technology, industry, and social structure might advance the equivalent of centuries into the future. Such changes would touch every person on Earth. Mindful of these profound implications, scientists push ahead with the search for extraterrestrial communication.

Figure A-5 **A deep-space tracking antenna** *This dish-shaped antenna, located in the Mojave Desert in California, was originally built by NASA to track interplanetary spacecraft. It is now being used in an all-sky survey, along with an antenna at the Arecibo Observatory in Puerto Rico, to search for extraterrestrial intelligence. In 1996 an antenna in Canberra, Australia, will join the network. (Jet Propulsion Laboratory)*

For further reading

Baugher, J. *On Civilized Stars: The Search for Intelligent Life in Outer Space.* Prentice-Hall, 1985 • This nontechnical book, written by a physicist, speculates on extraterrestrial intelligence and the possibility of our communicating with it.

Beatty, J. K. "The New, Improved SETI." *Sky & Telescope,* May 1983 • This article discusses new directions in SETI research that emerged in the early 1980s.

Davies, P. *The Cosmic Blueprint.* Simon & Schuster, 1988 • This extraordinary book examines connections between physics and biology in order to understand the process of life more fully.

Dawkins, R. *The Blind Watchmaker.* Norton, 1987 • This well written book surveys our modern understanding of the process of evolution.

Feinberg, G., and Shapiro, R. *Life beyond Earth: The Intelligent Earthling's Guide to Life in the Universe.* • This excellent introduction to exobiology includes some fascinating speculation about possible forms of life.

Finney, B., and Jones, E. *Interstellar Migration and the Human Experience.* University of California Press, 1985 • These proceedings of a conference of astronomers, anthropologists, and philosophers draw upon large-scale human migrations on Earth to speculate about galactic colonization.

Goldsmith, D., ed. *The Quest for Extraterrestrial Life: A Book of Readings.* University Science Books 1980 • This fascinating compilation includes articles by Carl Sagan, Frank Drake, Fred Hoyle, and Bernard Oliver, to name just a few.

Goldsmith, D., and Owen, T. *The Search for Life in the Universe.* Benjamin/Cummings, 1980 • This book sets the stage with long sections on astronomy, biology, and planetary evolution before dealing with the search for extraterrestrial life.

Gould, S. *Wonderful Life.* Scribner, 1990 • This fascinating book explains how a recent reinterpretation of fossils from the "Cambrian explosion"—a sudden proliferation of life forms that occurred about 600 million years ago—is changing the way we think about evolution.

Hart, M., and Zuckerman, B., eds. *Extraterrestrials—Where Are They?* Pergamon Press, 1982 • This thought-provoking volume examines the implications of our failure to observe extraterrestrials.

Horowitz, N. *To Utopia and Back: The Search for Life in the Solar System.* W. H. Freeman and Company, 1986 • This concise, nontechnical introduction to the processes and origin of life pays particular attention to the results from the *Viking* landers on Mars.

Kutter, G. S. *The Universe and Life.* Jones and Bartlett, 1987 • The first half of this well-written textbook covers astronomy while the second half is devoted to biology and evolution.

Marx, G., ed. *Bioastronomy—The Next Steps.* Kluwer, 1988 • Although this book consists of the proceedings of a technical conference on SETI, many of its papers can be easily understood by someone who has read this textbook.

Papagiannis, M. D. "Bioastronomy: The Search for Extraterrestrial Life." *Sky & Telescope,* June 1984 • This article eloquently but briefly examines various issues involved in SETI research.

Regis, E. *Extraterrestrials: Science and Alien Intelligence.* Cambridge University Press, 1985 • This engrossing collection of essays explores philosophical and moral issues surrounding the search for alien intelligence.

Rood, R. and Trefil, J. *Are We Alone?* Scribner, 1981 • This book takes an informal, friendly, occasionally skeptical look at modern ideas about extraterrestrial life.

Sagan, C., and Drake, F. "The Search for Extraterrestrial Intelligence." *Scientific American,* May 1975 • This classic article explores the possibility and methodology of communicating with extraterrestrials.

Schorn, R. A. "Extraterrestrial Beings Don't Exist." *Sky & Telescope,* September 1981 • This one-page article looks at the possibility that intelligent life might be extremely rare.

Shklovskii, I. S., and Sagan, C. *Intelligent Life in the Universe,* Holden-Day, 1966 • This somewhat dated but nevertheless classic book was one of the first to examine the question of extraterrestrial life in light of our modern understanding of astronomy and biology.

Wilson, A. C. "The Molecular Basis of Evolution." *Scientific American,* October 1985 • This interesting article surveys recent developments in our understanding of evolution at the molecular level.

Appendixes

The terrestrial worlds *This montage of photographs taken by various spacecraft shows the terrestrial planets and the seven largest moons of the solar system at the same scale. (Prepared for NASA by S. P. Meszaros)*

1 The planets: Orbital data

Planet	Semimajor axis (AU)	Semimajor axis (10⁶ km)	Sidereal period (yr)	Sidereal period (d)	Synodic period (d)	Mean orbital speed (km/s)	Orbital eccentricity	Inclination of orbit to ecliptic (°)
Mercury	0.3871	57.9	0.2408	87.97	115.88	47.9	0.206	7.00
Venus	0.7233	108.2	0.6152	224.70	583.92	35.0	0.007	3.39
Earth	1.0000	149.6	1.0000	365.26	——	29.8	0.017	0.00
Mars	1.5237	227.9	1.8809	686.98	779.94	24.1	0.093	1.85
(Ceres)	2.7656	413.7	4.603		466.6	17.9	0.097	10.61
Jupiter	5.2028	778.3	11.862		398.9	13.1	0.048	1.31
Saturn	9.5388	1427.0	29.458		378.1	9.6	0.056	2.49
Uranus	19.1914	2871.0	84.01		369.7	6.8	0.046	0.77
Neptune	30.0611	4497.1	164.79		367.5	5.4	0.010	1.77
Pluto	39.5294	5913.5	248.54		366.7	4.7	0.248	17.15

$1 M_\odot = 3.33 \times 10^5 M_\oplus$ (earth masses)
$= 333{,}000 M_\oplus$

2 The planets: Physical data

Planet	Equatorial diameter (km)	Equatorial diameter (Earth = 1)	Mass (Earth = 1)	Mean density (kg/m³)	Rotation period* (d)	Inclination of equator to orbit (°)	Surface gravity (Earth = 1)	Albedo	Brightest visual magnitude	Escape velocity (km/s)
Mercury	4,878	0.38	0.055	5430	58.65	2(?)	0.39	0.106	−1.9	4.3
Venus	12,102	0.95	0.815	5250	−243.01	177.3	0.88	0.65	−4.4	10.4
Earth	12,756	1.00	1.000	5520	0.997	23.4	1.00	0.37	——	11.2
Mars	6,786	0.53	0.107	3950	1.026	25.2	0.38	0.15	−2.0	5.0
Jupiter	142,984	11.21	317.94	1330	0.410	3.1	2.34	0.52	−2.7	59.6
Saturn	120,536	9.45	95.18	690	0.426	26.7	0.93	0.47	+0.7	35.5
Uranus	51,118	4.01	14.53	1290	−0.746	97.9	0.79	0.50	+5.5	21.3
Neptune	49,528	3.88	17.14	1640	0.800	29.6	1.12	0.5	+7.8	23.3
Pluto	2,300	0.18	0.002	2030	−6.387	122.5	0.04	0.5	+15.1	1.1

*Negative values indicate retrograde rotation.

Moons of Saturn and Uranus *This montage of photographs taken by the Voyager spacecraft shows the medium-sized satellites of Saturn and Uranus along with our Moon, all at the same scale. (Prepared for NASA by S. P. Meszaros)*

3 Satellites of the planets

Planet	Satellite	Discoverer(s)	Mean distance from planet (km)	Sidereal period (d)	Orbital eccentricity	Diameter of satellite* (km)	Approximate magnitude at opposition
Earth	Moon	——	384,400	27.322	0.05	3476	−13
Mars	Phobos	Hall (1877)	9,380	0.319	0.01	28 × 23 × 20	+11
	Deimos	Hall (1877)	23,460	1.263	0.00	16 × 12 × 10	12
Jupiter	Metis	Synnott (1979)	127,960	0.295	0.00	(40)	+18
	Adrastea	Jewitt et al. (1979)	128,980	0.298	0(?)	24 × 16 × 20	19
	Amalthea	Barnard (1892)	181,300	0.498	0.00	270 × 200 × 155	14
	Thebe	Synnott (1979)	221,900	0.675	0.01	(100)	16
	Io	Galileo (1610)	421,600	1.769	0.00	3630	5
	Europa	Galileo (1610)	670,900	3.551	0.01	3138	5
	Ganymede	Galileo (1610)	1,070,000	7.155	0.00	5262	5
	Callisto	Galileo (1610)	1,883,000	16.689	0.01	4800	6
	Leda	Kowal (1974)	11,094,000	238.72	0.15	(16)	20
	Himalia	Perrine (1904)	11,480,000	250.57	0.16	(180)	15
	Lysithea	Nicholson (1938)	11,720,000	259.22	0.11	(40)	18
	Elara	Perrine (1905)	11,737,000	259.65	0.21	(80)	17
	Ananke	Nicholson (1951)	21,200,000	631	0.17	(30)	19
	Carme	Nicholson (1938)	22,600,000	692	0.21	(44)	18
	Pasiphae	Melotte (1908)	23,500,000	735	0.38	(70)	17
	Sinope	Nicholson (1914)	23,700,000	758	0.28	(40)	18
Saturn	Pan	Showalter (1990)	133,570	0.573	0.00	20	+19
	Atlas	Terrile (1980)	137,640	0.602	0(?)	40 × 30 × 30	18
	Prometheus	Collins et al. (1980)	139,350	0.613	0.00	140 × 80 × 100	16
	Pandora	Collins et al. (1980)	141,700	0.629	0.00	110 × 70 × 100	16
	Epimethius	Walker (1966)	151,422	0.694	0.01	140 × 100 × 100	16
	Janus	Dolfus (1966)	151,472	0.695	0.01	220 × 160 × 200	14
	Mimas	Herschel (1789)	185,520	0.942	0.02	390	13
	Enceladus	Herschel (1789)	238,020	1.370	0.00	500	12
	Tethys	Cassini (1684)	294,660	1.888	0.00	1050	10
	Telesto	Smith et al. (1980)	294,660	1.888	0(?)	(24)	19
	Calypso	Smith et al. (1980)	294,660	1.888	0(?)	30 × 20 × 25	19
	Dione	Cassini (1684)	377,400	2.737	0.00	1120	10
	Helene	Laques et al. (1980)	377,400	2.737	0.01	40 × 30 × 30	18
	Rhea	Cassini (1672)	527,040	4.518	0.00	1530	10

(continued)

3 Satellites of the planets *(continued)*

Planet	Satellite	Discoverer(s)	Mean distance from planet (km)	Sidereal period (d)	Orbital eccentricity	Diameter of satellite* (km)	Approximate magnitude at opposition
	Titan	Huygens (1655)	1,221,850	15.945	0.03	5150	8
	Hyperion	Bond (1848)	1,481,000	21.277	0.10	410 × 260 × 220	14
	Iapetus	Cassini (1671)	3,561,000	79.331	0.03	1440	11
	Phoebe	Pickering (1898)	12,952,000	550.48	0.16	220	16
Uranus	Cordelia	*Voyager 2* (1986)	49,750	0.335	0(?)	(30)	+24
	Ophelia	*Voyager 2* (1986)	53,760	0.376	0(?)	(30)	24
	Bianca	*Voyager 2* (1986)	59,160	0.435	0(?)	(40)	23
	Cressida	*Voyager 2* (1986)	61,770	0.464	0(?)	(70)	22
	Desdemona	*Voyager 2* (1986)	62,660	0.474	0(?)	(60)	22
	Juliet	*Voyager 2* (1986)	64,360	0.493	0(?)	(80)	22
	Portia	*Voyager 2* (1986)	66,100	0.513	0(?)	(110)	21
	Rosalind	*Voyager 2* (1986)	69,930	0.558	0(?)	(60)	22
	Belinda	*Voyager 2* (1986)	75,260	0.624	0(?)	(70)	22
	Puck	*Voyager 2* (1986)	86,010	0.762	0(?)	150	20
	Miranda	Kuiper (1948)	129,780	1.414	0.00	470	16
	Ariel	Lassell (1851)	191,240	2.520	0.00	1160	14
	Umbriel	Lassell (1851)	265,970	4.144	0.00	1170	15
	Titania	Herschel (1787)	435,840	8.706	0.00	1580	14
	Oberon	Herschel (1787)	582,600	13.463	0.00	1520	14
Neptune	Naiad	*Voyager 2* (1989)	48,230	0.296	0(?)	(50)	+25
	Thalassa	*Voyager 2* (1989)	50,070	0.312	0(?)	60	24
	Despoina	*Voyager 2* (1989)	52,530	0.333	0(?)	80	23
	Galatea	*Voyager 2* (1989)	61,950	0.429	0(?)	180	23
	Larissa	*Voyager 2* (1989)	73,550	0.554	0(?)	150	21
	Proteus	*Voyager 2* (1989)	117,640	1,121	0(?)	415	20
	Triton	Lassell (1846)	354,800	5.877	0.00	2700	14
	Nereid	Kuiper (1949)	5,513,400	359.16	0.75	(340)	19
Pluto	Charon	Christy (1978)	19,640	6.387	0.00	1190	+17

*A diameter given in parentheses is estimated from the amount of sunlight it reflects.

4 The nearest stars*

Name	Parallax (arc sec)	Distance (ly)	Spectral type†	Radial velocity (km/s)	Proper motion (arc sec/yr)	Apparent visual magnitude	Luminosity (Sun = 1.0)
Sun			G2 V			−26.7	1.0
Proxima Centauri	0.772	4.2	M5e	−16	3.85	11.05	0.00006
α Centauri A	0.750	4.3	G2 V	−22	3.68	−0.01	1.6
α Centauri B			K0 V			1.33	0.45
Barnard's star	0.545	5.9	M5 V	−108	10.31	9.54	0.00045
Wolf 359	0.421	7.6	M8e	+13	4.70	13.53	0.00002
BD +36°2147	0.397	8.1	M2 V	−84	4.78	7.50	0.0055
Luyten 726-8A	0.387	8.4	M6e	+29	3.36	12.52	0.00006
Luyten 726-8B (UV Ceti)			M6e	+32		13.02	0.00004
Sirius A	0.377	8.6	A1 V	−8	1.33	−1.46	23.5
Sirius B			wd			8.3	0.003
Ross 154	0.345	9.4	M5e	−4	0.72	10.45	0.00048
Ross 248	0.314	10.3	M6e	−81	1.60	12.29	0.00011
ε Eridani	0.303	10.7	K2 V	+16	0.98	3.73	0.30
Ross 128	0.298	10.8	M5	−13	1.38	11.10	0.00036
61 Cygni A	0.294	11.2	K5 V	−64	5.22	5.22	0.082
61 Cygni B			K7 V			6.03	0.039
ε Indi	0.291	11.2	K5 V	−40	4.70	4.68	0.14
BD +43°44A	0.290	11.2	M1 V	+13	2.90	8.08	0.0061
BD +43°44B			M6 V	+20		11.06	0.00039
Luyten 789-6	0.290	11.2	M7e	−60	3.26	12.18	0.00014
Procyon A	0.285	11.4	F5 IV–V	−3	1.25	0.37	7.65
Procyon B			wd			10.7	0.00055
BD +59°1915A	0.282	11.5	M4	0	2.29	8.90	0.0030
BD +59°1915B			M5	+10	2.27	9.69	0.0015
CD −36°1668	0.279	11.7	M2 V	+10	6.90	7.35	0.0013
G51-15	0.278	11.7			1.27	14.81	0.00001
τ Ceti	0.277	11.9	G8 V	−16	1.92	3.50	0.45
BD +5°1668	0.266	12.2	M5	+26	3.77	9.82	0.0015
Luyten 725-32 (YZ Ceti)	0.261	12.4	M5e	+28	1.32	12.04	0.0002
CD −39°14192	0.260	12.5	M0 V	+21	3.46	6.66	0.028
Kapteyn's star	0.256	12.7	M0 V	+245	8.72	8.84	0.0039
Kruger 60 A	0.253	12.8	M3	−26	0.86	9.85	0.0016
Kruger 60 B			M5e			11.3	0.0004

*In the case of double stars, like Sirius A and B or Procyon A and B, both members have the same parallax, distance, etc. These redundant values are not repeated for the second (dimmer) member.

†"wd" stands for white dwarf; "e" means that the star's spectrum contains emission lines.

5 The visually brightest stars*

Star	Name	Apparent visual magnitude	Spectral type	Absolute magnitude	Distance (ly)	Radial velocity (km/s)	Proper motion (arc sec/yr)
α CMa A	Sirius	−1.46	A1 V	+1.4	9	−8	1.324
α Car	Canopus	−0.72	F0 I	−8.5	1200	+21	0.025
α Boo	Arcturus	−0.04	K2 III	−0.2	36	−5	2.284
α Cen A	Rigel Kents	0.00	G2 V	+4.4	4	−25	3.676
α Lyr	Vega	0.03	A0 V	+0.5	26	−14	0.345
α Aur	Capella	0.08	G8 III	−0.5	42	+30	0.435
β Ori A	Rigel	0.12	B8 Ia	−7.1	910	+21	0.001
α CMi A	Procyon	0.38	F5 IV	+2.6	11	−3	1.250
α Eri	Achernar	0.46	B3 IV	−1.6	85	+19	0.098
α Ori	Betelgeuse	0.50	M2 Iab	−5.6	310	+21	0.028
β Cen	Hadar	0.61	B1 II	−5.1	460	−12	0.035
α Aql	Altair	0.77	A7 IV–V	+2.2	17	−26	0.658
α Tau A	Aldebaran	0.85	K5 III	−0.3	68	+54	0.202
α Sco A	Antares	0.96	M1 Ib	−4.7	330	−3	0.029
α Vir	Spica	0.98	B1 V	−3.5	260	+1	0.054
β Gem	Pollux	1.14	K0 III	+0.2	36	+3	0.625
α PsA	Fomalhaut	1.16	A3 V	+2.0	22	+7	0.367
α Cyg	Deneb	1.25	A2 Ia	−7.5	1800	−5	0.003
β Cru	Mimosa	1.25	B0.5 III	−5.0	420	+20	0.049
α Leo A	Regulus	1.35	B7 V	−0.6	85	+4	0.248

*Acrux, the brightest star in Crux (the Southern Cross), appears to the naked eye as a star of magnitude +0.87, which suggests that it should be in this table. A small telescope, however, reveals that Acrux is actually a double star whose blue-white components have visual magnitudes of 1.4 and 1.9, and so they are dimmer than any of the stars listed here.

Glossary

absolute magnitude   The apparent magnitude that a star would have at a distance of 10 parsecs. (page 227)

absolute zero   A temperature of −273°C (or 0 K) where all molecular motion stops; the lowest possible temperature. (pages 75–76)

absorption line   A dark line at a specific wavelength in a continuous spectrum. (pages 83–84)

absorption line spectrum   Dark lines superimposed on a continuous spectrum. (pages 83–84)

acceleration   A change in velocity. (page 46)

accretion   The gradual accumulation of matter in one location, usually from the action of gravity. (page 102)

accretion disk   A disk of matter orbiting a star or black hole. (page 240)

active galactic nucleus   The center of a galaxy that is emitting exceptionally large amounts of energy; the center of a Seyfert galaxy or quasar. (page 345ff)

active galaxy   A very luminous galaxy, often containing an active galactic nucleus. (page 345)

aerosol   A system of tiny particles or droplets dispersed in a gas. (page 181)

AGB star   An asymptotic giant branch star; a red supergiant. (page 266)

albedo   The fraction of sunlight that a planet, asteroid, or satellite reflects.

amino acid   A class of chemical compounds that are the building blocks of proteins. (page 198)

angle   The opening between two straight lines that meet at a point. (page 7)

Angstrom (Å)   A unit of length equal to 10^{-8} cm.

angular diameter (angular size)   The angle subtended by the diameter of an object. (page 7)

angular measure   A method of describing the size of an angle. (page 7)

angular momentum   A measure of the momentum associated with rotation. (page 101)

angular resolution   The angular size of the smallest detail that can be distinguished with a telescope. (page 62)

annihilation   The process by which the mass of a particle and an antiparticle is converted into energy. (page 371)

annular eclipse   An eclipse of the Sun in which the Moon is too distant to cover the Sun completely so that a ring of sunlight is seen around the Moon at mid-eclipse. (pages 30–31)

anorthosite   A light-colored rock found throughout the lunar highlands and in certain very old mountains on Earth. (pages 170–171)

anticyclone   A large atmospheric system characterized by high air pressure at its center and circular wind motion, which is clockwise in Earth's Northern Hemisphere. (page 141)

antimatter   Matter consisting of antiparticles such as antiprotons, antielectrons (positrons), and antineutrons. (page 371)

aperture   The diameter of an opening; the diameter of the primary lens or mirror of a telescope.

aphelion The point in its orbit where a planet is farthest from the Sun. (page 41)

apogee The point in its orbit where a satellite or the Moon is farthest from the Earth.

Apollo asteroid An asteroid whose orbit brings it closer to the Sun than to the Earth.

apparent magnitude A measure of the brightness of light from a star or other object as seen from Earth. (page 227)

A ring The outer portion of Saturn's main ring system. (page 144)

asteroid (minor planet) One of tens of thousands of small, rocky planetlike objects in orbit about the Sun. (pages 96, 190)

asteroid belt A $1\frac{1}{2}$-AU-wide region between the orbits of Mars and Jupiter in which most asteroids are found. (pages 190–191)

asthenosphere A warm, plastic layer of the mantle beneath the lithosphere of the Earth. (page 121)

astronomical unit (AU) The semimajor axis of the Earth's orbit; the average distance between the Earth and the Sun. (page 9)

astronomy The branch of science dealing with objects and phenomena that lie beyond the Earth's atmosphere.

astrophysics That part of astronomy dealing with the physics of astronomical objects and phenomena.

asymptotic giant branch star A star in the extreme upper right corner of the Hertzsprung–Russell diagram; a red supergiant. (page 266)

atom The smallest particle of an element that has the properties that characterize that element. (page 75)

atomic number The number of protons in the nucleus of an atom. (pages 81, 85)

aurora Light radiated by atoms and ions in the Earth's upper atmosphere, mostly in the polar regions. (pages 133, 217)

autumnal equinox The intersection of the ecliptic and the celestial equator where the Sun crosses the equator from north to south. (page 19)

average density The mass of an object divided by its volume. (page 93)

Balmer lines Emission or absorption lines in the

hydrogen spectrum involving electron transitions between the second and higher energy levels. (page 85)

Balmer series All of the Balmer lines. (page 85)

Barnard object Any one of several dark nebulae discovered by E. E. Barnard. (page 245)

barred spiral galaxy A spiral galaxy in which the spiral arms begin from the ends of a "bar" running through the nucleus rather than from the nucleus itself. (pages 320–321)

belt asteroid An asteroid whose orbit lies largely in the asteroid belt. (page 190)

belts (on Jupiter) Dark, reddish bands in Jupiter's cloud cover. (page 139)

Big Bang An explosion throughout all space roughly 15 billion years ago from which the universe emerged. (page 360)

Big Crunch The gravitational collapse of the universe; the ultimate fate of a bounded universe. (page 364)

binary asteroid Two asteroids revolving about each other. (page 193)

binary star Two stars revolving about each other; a double star. (pages 234–241)

bipolar sunspot group A sunspot that has roughly equal areas covered by north and south magnetic polarity. (page 217)

blackbody A hypothetical perfect radiator that absorbs and reemits all radiation falling upon it. (page 76)

blackbody curve The curve obtained when the intensity of radiation from a blackbody at a particular temperature is plotted against wavelength. (pages 77–78)

blackbody radiation Electromagnetic radiation emitted by a blackbody. (page 77)

black hole An object whose gravity is so strong that the escape velocity exceeds the speed of light. (pages 271, 288ff)

blazar A BL Lacertae object. (page 348)

BL Lacertae object A type of active galaxy. (pages 348–349)

blueshift A shift toward shorter wavelengths; the Doppler shift of light from an approaching source. (pages 87–88)

Bode's law A numerical sequence that gives the

approximate distances of the planets from the Sun in astronomical units. (page 189)

Bohr atom A model of the atom, described by Niels Bohr, in which electrons revolve about the nucleus in circular orbits. (page 86)

bounded universe A universe that expands, reaches a maximum size, and then contracts. (page 364)

breccia A rock formed by the sudden amalgamation of various rock fragments under pressure. (page 171)

B ring The brightest portion of Saturn's main ring system. (page 144)

burster A nonperiodic X-ray source that emits powerful bursts of X rays. (page 285)

caldera The crater at the summit of a volcano. (page 123)

Callisto One of the four Galilean satellites. (pages 178–180)

carbonaceous chondrite A class of extremely ancient, carbon-rich meteorites. (page 197)

carbon burning The thermonuclear fusion of carbon nuclei. (page 269)

Cassegrain focus An optical arrangement in a reflecting telescope in which light rays are reflected by a secondary mirror to a focus behind the primary mirror. (page 60)

Cassini division A prominent gap in Saturn's rings discovered in 1675 by G. D. Cassini. (page 144)

celestial equator A great circle on the celestial sphere 90° from the celestial poles. (page 18)

celestial mechanics The branch of astronomy dealing with the motions and gravitational interactions of objects in the solar system.

celestial poles Points about which the celestial sphere appears to rotate. (page 18)

celestial sphere A sphere of very large radius centered on the observer; the apparent sphere of the sky. (page 18)

Celsius temperature scale *See* temperature (Celsius). (pages 75–76)

center of mass That point in an isolated system that moves at a constant velocity in accordance with Newton's first law. (page 235)

central bulge A spherical distribution of stars that surrounds the nucleus of a spiral galaxy like the Milky Way. (page 304)

Cepheid variable One of two types of yellow, supergiant pulsating stars. (pages 261–262)

Cerenkov radiation Radiation produced by particles traveling though a substance faster than light can. (page 273)

Ceres The largest asteroid and the first to be discovered. (pages 189–191)

Chandrasekhar limit The maximum mass of a white dwarf. (page 268)

charge-coupled device (CCD) A type of solid-state silicon wafer designed to detect photons. (page 63)

chemical differentiation The separation of a planet's internal structure into layers of differing density. (page 130)

chromatic aberration An optical defect whereby different colors of light passing through a lens are focused at different locations. (pages 57–58)

chromosphere The middle layer in the solar atmosphere, between the photosphere and the corona. (pages 210–212)

circumstellar shell A shell of gas that surrounds a star. (page 260)

close binary A binary star whose members are separated by a few stellar diameters. (page 239)

Clouds of Magellan Two nearby galaxies visible to the naked eye from southern latitudes. (pages 323–324)

cluster of galaxies A collection of galaxies containing a few to several thousand member-galaxies. (pages 324–326)

cocoon nebula Gas and dust that usually surround a protostar. (page 247)

collision-ejection theory A theory according to which the Moon may have been created by the impact of a planet-sized object that struck the Earth. (pages 106–107)

color index The difference in the magnitudes of a star measured in two separate wavelength bands. (page 229)

coma (of a comet) The diffuse gaseous component of the head of a comet. (pages 198–199)

comet A small body of ice and dust in orbit about the Sun. While passing near the Sun, a comet's vaporized ices give rise to a coma and tail. (pages 96, 198–202)

condensation temperature The temperature at which a substance turns from gas to solid at low pressure. (page 100)

conduction The transfer of heat by passing energy directly from atom to atom. (page 207)

configuration (of a planet) A particular geometric arrangement of the Earth, a planet, and the Sun. (page 38)

confinement The moment shortly after the Big Bang when quarks bound together to form particles like protons and neutrons. (page 374)

conic section The curve of intersection between a circular cone and a plane. This curve can be a circle, ellipse, parabola, or hyperbola. (pages 46–47)

conjunction The alignment of two bodies in the solar system with the Earth, so that they appear in the same part of the sky as seen from Earth. (page 38)

conservation of angular momentum The law of physics stating that the total amount of angular momentum in an isolated system remains constant. (page 278)

constellation A configuration of stars often named after an object, person, or animal. (pages 14–15)

contact binary A close binary system in which both stars fill or overflow their Roche lobes. (page 240)

continental drift The gradual movement of the continents over the surface of the Earth from plate tectonics. (page 119)

continuous spectrum A spectrum of light over a range of wavelengths without any spectral lines. (pages 83–84)

continuum An unbroken spectrum spanning a range of wavelengths. (page 342)

convection The transfer of energy by moving currents of fluid or gas containing that energy. (pages 121, 207)

convective zone (**convective envelope**) A layer in a star where energy is transported outward by means of convection. (page 208)

core (of a planet) The central portion of a planet. (page 131)

core helium burning Helium burning at the center of a star. (page 257)

core hydrogen burning Hydrogen burning at the center of a star. (page 254)

corona The Sun's outer atmosphere. (pages 212–214)

coronagraph A specially designed telescope with a baffle that blocks out the solar disk so that the corona can be photographed. (page 213)

coronal hole A dark region of the Sun's inner corona, as seen at X-ray wavelengths. (page 214)

cosmic microwave background Photons from every part of the sky with a blackbody spectrum at 2.73 K; the cooled-off radiation from the primordial fireball that filled all space. (page 362)

cosmic particle horizon A sphere, centered on the Earth, whose radius equals the distance traveled by light since the time of the Big Bang. (page 360)

cosmic singularity The Big Bang. (page 360)

cosmological constant (Λ) A number sometimes inserted in the equations of general relativity that represents a pressure that opposes gravity throughout the universe. (pages 365–366)

cosmological model A specific theory about the organization and evolution of the universe.

cosmological redshift An increase in wavelength caused by the expansion of the universe. (page 359)

cosmology The study of the organization and evolution of the universe. (page 358ff)

coudé focus A reflecting telescope in which a series of mirrors direct light to a remote focus away from the moving parts of the telescope. (pages 60–61)

crater A circular depression on a planet or satellite caused by the impact of a meteoroid. (pages 166, 168–169, 194)

crescent moon One of the phases of the Moon in which it appears less than half full. (pages 21–24)

C ring The faint, inner portion of Saturn's main ring system; the crepe ring. (page 144)

critical density The average density throughout the universe at which space is flat and galaxies just barely continue receding from each other infinitely far into the future. (pages 364)

critical surface An imaginary figure-8-shaped surface surrounding the stars in a binary that delineates the gravitation domain of each star; Roche lobes. (page 240)

current sheet A flattened distribution of electrically charged particles surrounding Jupiter in the plane of its magnetic equator. (page 154)

cyclone A large atmospheric system characterized by

low air pressure at its center and circular wind motion, counterclockwise in Earth's Northern Hemisphere. (page 141)

dark-matter problem The contention that most of the matter in the universe is underluminous and possibly quite different from ordinary matter. (page 328)

dark nebula A cloud of interstellar gas and dust that obscures the light of more distant stars. (pages 245–246)

deceleration parameter (q_0) A number that specifies how fast the expansion of the universe is slowing down. (pages 364–365)

deferent A stationary circle in the Ptolemaic system along which another circle (an epicycle) moves carrying a planet, the Sun, or the Moon. (page 37)

degeneracy The condition in which all the allowed states for particles (electrons or neutrons) have been filled, thereby causing the substance to behave differently from an ordinary gas. (page 257)

degenerate-electron pressure A powerful pressure produced by degenerate electrons. (page 257)

degenerate-neutron pressure A powerful pressure produced by degenerate neutrons. (page 277)

degree A unit of angular measure or of temperature measure. (pages 7, 75)

density The ratio of the mass of an object to its volume. (page 93)

density wave A region of compression in a spiral galaxy. (page 310)

density-wave theory An explanation of spiral arms in galaxies proposed by C. C. Lin and F. Shu. (pages 310–311)

detached binary A close binary system in which the surfaces of both stars are inside their Roche lobes. (page 240)

deuterium An isotope of hydrogen whose nuclei each contain one proton and one neutron; heavy hydrogen.

differential rotation The rotation of a nonrigid object in which parts adjacent to each other do not always stay close together. (pages 139, 216, 308)

differentiation (chemical) The separation of different kinds of material in different layers inside a planet. (page 130)

diffraction grating A system of closely spaced slits or

reflecting strips (usually on a piece of glass) used to produce a spectrum.

direct motion The gradual, eastward apparent motion of a planet against the background stars, as seen from Earth. (page 37)

disk (of a galaxy) A flattened assemblage of stars, gas, and dust in a spiral galaxy like the Milky Way. (page 303)

diurnal Daily. (page 14)

diurnal motion Motion in one day. (page 14)

Doppler effect The apparent change in wavelength of radiation due to relative motion between the source and the observer along the line of sight. (pages 87–88)

double-line spectroscopic binary A spectroscopic binary whose spectrum exhibits spectral lines of both stars. (page 237)

double radio source An extragalactic radio source characterized by two large regions of radio emission, often located on either side of an active galaxy. (page 347)

D ring An extremely faint ring system inside Saturn's C ring that extends down to Saturn's cloud tops. (page 148)

dust tail (of a comet) That part of a comet's tail caused by dust particles escaping from its nucleus. (page 201)

dwarf elliptical galaxy A small elliptical galaxy with far fewer stars than a typical galaxy. (page 322)

dynamo effect The generation of a magnetic field by circulating electric charges.

eccentricity A number between 0 and 1 that indicates how flattened an ellipse is (the eccentricity of a circle is zero). (page 93)

eclipse The cutting off of part or all of the light from one celestial object by another. (pages 26–32)

eclipse path The track of the tip of the Moon's shadow along the Earth's surface during a total or annular solar eclipse. (page 30)

eclipse season A period during the year when a solar or lunar eclipse is possible.

eclipsing binary star A double star system in which the stars periodically pass in front of each other, as seen from Earth. (pages 237–239)

ecliptic The apparent annual path of the Sun on the celestial sphere. (pages 18–19)

electromagnetic radiation   Radiation consisting of oscillating electric and magnetic fields such as gamma rays, X rays, visible light, ultraviolet and infrared radiation, radio waves, and microwaves. (pages 55–56)

electromagnetic spectrum   The entire array or family of electromagnetic radiation. (page 56)

electromagnetic theory   A comprehensive description of electricity and magnetism first formulated by J. C. Maxwell. (page 55)

electron   A negatively charged subatomic particle usually found in orbit about the nucleus of an atom. (page 85)

electron volt (eV)   The energy acquired by an electron accelerated through an electric potential of one volt. (page 78)

element   A substance that cannot be decomposed by chemical means into simpler substances. (pages 80–81)

ellipse   A conic section obtained by cutting completely through a circular cone with a plane. (pages 40, 46)

elliptical galaxy   A galaxy with an elliptical shape and no conspicuous interstellar material. (pages 321–323)

elongation   The angular distance between a planet and the Sun. (page 38)

emission line   A bright spectral line. (page 83)

emission line spectrum   A spectrum that contains emission lines. (pages 83–84)

emission nebula   A glowing gaseous nebula whose light comes from fluorescence caused by a nearby star. (page 248)

energy   The ability to do work.

energy flux   The amount of energy emitted from each square meter of an object's surface per second. (page 76)

energy level (in an atom)   A particular amount of energy possessed by an atom above the atom's least energetic state. (page 87)

Enke division   A thin gap in Saturn's A ring, possibly first seen by J. F. Enke in 1838. (page 146)

epicycle   In the Ptolemaic system, a moving circle about which planets revolve. (page 37)

equations of stellar structure   A set of relationships that describe the interactions of matter, energy, and gravity inside a star. (page 208)

equinox   One of the intersections of the ecliptic and the celestial equator. (page 19)

era of recombination   The moment, roughly one million years after the Big Bang, when the universe became transparent. (pages 370–371)

ergosphere   The region of space immediately outside the event horizon of a rotating black hole where it is impossible to remain at rest. (page 293)

E ring   One of two extremely faint ring systems located well beyond the outer edge of Saturn's main ring system. (page 148)

escape velocity   The speed needed by one object to achieve a parabolic orbit away from a second object and thereby permanently move away from the second object.

Europa   One of the Galilean satellites. (pages 177–178)

event horizon   The location around a black hole where the escape velocity equals the speed of light; the surface of a black hole. (pages 291–292)

evolutionary track   On the Hertzspring–Russell diagram, the path followed by a point representing an evolving star. (pages 246–247)

excitation   The process of imparting energy to an atom or ion.

exponent   A number placed above and after another number to denote the power to which the latter is to be raised, as in 10^4. (page 8)

extragalactic   Beyond the Milky Way Galaxy.

eyepiece   A magnifying lens used to view the image produced at the focus of a telescope. (page 57)

faculae   Bright patches often found near sunspots in the solar photosphere. (page 216)

Fahrenheit scale   *See* temperature (Fahrenheit). (pages 75–76)

first quarter moon   A phase of the waxing moon when Earth-based observers see half of the Moon's illuminated hemisphere. (pages 21–24)

flare   A sudden, temporary outburst of light from an extended region of the solar surface. (page 216)

flatness problem   The quandary associated with the fact that space throughout the universe seems to be essentially flat. (page 368)

flocculent spiral galaxy   A spiral galaxy whose spiral

arms are broad, fuzzy, and poorly demarcated. (page 309)

focal length   The distance from a lens or mirror where converging light rays meet. (pages 57, 60)

focus (pl., **foci**)   The point where light rays converged by a lens or mirror meet. (pages 57, 60)

force   That which can change the momentum of an object. (pages 46–47)

frequency   The number of waves that cross a given point per unit time; the number of vibrations per unit time.

F ring   A thin ring just beyond the outer edge of Saturn's main ring system. (pages 146–148)

full moon   A phase of the Moon during which its full daylight hemisphere can be seen from Earth. (pages 22–23)

galactic cannibalism   A collision between two galaxies of unequal mass and size in which the smaller galaxy seems to be absorbed into the larger galaxy. (page 330)

galactic cluster   A loose association of young stars in the disk of the galaxy. (pages 249–250)

galactic equator   The intersection of the principal plane of the Milky Way with the celestial sphere.

galactic nucleus   The center of a galaxy; the center of the Milky Way Galaxy. (page 303)

galaxy   A large assemblage of stars; the galaxy of which the Sun and nearby stars are members. (page 320ff)

galaxy merger   The collision and subsequent coalescence of two galaxies. (page 330)

Galilean satellite   Any one of the four large moons of Jupiter. (page 173ff)

gamma rays   The most energetic form of electromagnetic radiation. (page 55)

Ganymede   One of the Galilean satellites. (pages 178–180)

general theory of relativity   A description of gravity formulated by Einstein explaining that gravity affects the geometry of space and the flow of time. (pages 49–51, 289–290)

geo-   A prefix referring to the Earth.

geocentric   Centered on the Earth.

geomagnetic   The Earth's magnetic field.

giant (star)   A star whose diameter is roughly 10 to 100 times that of the Sun. (pages 232–233)

giant elliptical galaxy   A very large, extremely massive elliptical galaxy, usually located near the center of a rich cluster of galaxies. (page 322)

giant molecular cloud   A large cloud of cool gas in a galaxy. (page 251)

gibbous moon   A phase of the Moon in which more than half, but not all, of the Moon's daylight hemisphere is visible from Earth. (pages 22–24)

globular cluster   A large spherical cluster of stars usually found in the outlying regions of a galaxy. (page 258)

globule   A small, dense, dark nebula.

gluon   A particle that is exchanged between quarks. (page 373)

grand-design spiral   A spiral galaxy whose spiral arms are thin, graceful, and well defined. (page 309)

grand unified theory   A theory that attempts to explain the physical forces.

granulation   The rice-grain-like structure of the solar photosphere. (pages 210–211)

granule   A lightly colored feature about 1000 kilometers in diameter in the solar photosphere. (pages 210–211)

grating   An optical device consisting of closely spaced lines ruled on a piece of glass that is used to disperse light into a spectrum. (page 82)

gravitation (**gravity**)   The tendency of matter to attract matter. (page 47)

gravitational lens   The distortion of the appearance of an object by a source of gravity. (pages 293–296)

gravitational radiation (**gravitational waves**)   Ripples in the overall geometry of space produced by moving objects. (page 292)

graviton   A particle that carries the gravitational force. (page 373)

Great Attractor   A hypothetical enormous mass in the general direction of the constellation Centaurus toward which the Local Group and most nearby clusters of galaxies seem to be traveling. (page 363)

Great Dark Spot   A large, dark oval in Neptune's southern hemisphere. (pages 158–159)

greatest elongation   The largest possible angle between the Sun and an inferior planet. (pages 38–39)

Great Red Spot   A large, red-orange oval in Jupiter's southern hemisphere. (pages 138–140)

greenhouse effect   The trapping of infrared radiation near a planet's surface by the planet's atmosphere. (pages 114–115)

G ring   One of two extremely faint ring systems located well beyond the outer edge of Saturn's main ring system. (page 148)

ground state   The lowest energy level of an atom. (page 87)

halo (of a galaxy)   A spherical distribution of globular clusters that surrounds a galaxy. (pages 303–304)

harmonic law   Kepler's third law. (page 41)

head–tail source   A radio galaxy whose radio emission is deflected away from the galaxy. (page 348)

helio-   A prefix referring to the Sun.

heliocentric   Centered on the Sun. (page 38)

helioseismology   The study of vibrations of the solar surface. (page 220)

helium burning   The thermonuclear fusion of helium to produce carbon and oxygen. (pages 256–257)

helium flash   The nearly explosive ignition of helium burning in the dense core of a red giant star. (page 257)

helium-shell flash   The explosive ignition of helium burning in a thin shell surrounding the core of a low-mass star. (page 266)

Hertzsprung–Russell (H–R) diagram   A plot of the absolute magnitude or luminosity of stars against their surface temperatures. (pages 232–233)

heterogeneous accretion theory   A theory of the formation of the inner planets involving planetesimals of varying chemical composition. (page 103)

highlands (lunar)   Heavily cratered regions of the lunar surface. (page 169)

high-pressure system   A region in a planet's atmosphere of higher-than-average atmospheric pressure. (page 140)

homogeneous accretion theory   A theory of the formation of the inner planets involving planetesimals of essentially the same chemical composition. (page 103)

H I region   A region of neutral hydrogen in interstellar space.

horizontal branch   A group of post-helium-flash stars on the Hertzsprung–Russell diagram of a typical globular cluster near the main sequence. (page 259)

hot-spot volcanism   The creation of volcanoes on a planet's surface caused by a reservoir of hot magma in the planet's mantle. (page 122)

H II region   A region of ionized hydrogen in interstellar space. (page 248)

Hubble classification   A system of classifying galaxies according to their appearance into one of four broad categories: spirals, barred spirals, ellipticals, and irregulars. (page 320)

Hubble constant   The constant of proportionality in the relation between the velocities of remote galaxies and their distances. (page 335)

Hubble flow   The recession of the galaxies caused by the expansion of the universe. (page 334)

Hubble law   The relationship which states that the redshifts of remote galaxies are directly proportional to their distances from Earth. (page 335)

hydrocarbon   A chemical composed of hydrogen and carbon. (page 181)

hydrogen burning   The thermonuclear fusion of hydrogen nuclei to produce helium. (page 207)

hydrogen envelope   A large, tenuous sphere of hydrogen gas surrounding the head of a comet. (page 199)

hydrostatic equilibrium   A balance between the weight of a layer in a star and the pressure that supports it. (page 207)

hyperbola   A conic section formed by cutting a circular cone with a plane at an angle steeper than the sides of the cone. (pages 46–47)

hypothesis   A tentative theory, advanced to explain certain facts or phenomena, which can be tested. (page 2)

image   The optical representation of an object produced by the focusing of light rays by lenses or mirrors.

impact breccia   A rock consisting of various fragments cemented together by the impact of a meteoroid. (page 171)

impact crater   A crater on the surface of a planet or

moon produced by the impact of an asteroid or meteoroid. (pages 193–194)

inertia The property of matter that requires a force to act on it to change its state of motion.

inferior conjunction The configuration when Mercury or Venus is between the Sun and the Earth. (pages 38–39)

inflationary epoch A brief period shortly after the Big Bang during which the scale of the universe increased very rapidly. (page 368)

infrared radiation Electromagnetic radiation of a wavelength longer than visible light yet shorter than radio waves. (page 56)

instability strip A region on the Hertzsprung–Russell diagram occupied by pulsating stars. (page 261)

interferometry A method of increasing resolving power by combining electromagnetic radiation obtained along two or more paths. (page 65)

interstellar dust Microscopic solid grains of various compounds in interstellar space. (page 301)

interstellar gas Sparse gas in interstellar space. (page 250)

interstellar medium Interstellar gas and dust. (page 250)

inverse-square law A statement of the fact that the apparent brightness of a light source is inversely proportional to the square of its distance from an observer. (pages 237–228)

Io One of the Galilean satellites. (pages 173–177)

ion An atom that has become electrically charged due to the addition or loss of one or more electrons. (page 87)

ionization The process by which an atom loses electrons. (page 87)

ion tail (of a comet) The relatively straight tail of a comet produced by the solar wind acting on ions. (page 201)

iron meteorite A meteorite composed of iron with a small admixture of nickel; also called an iron. (pages 196–197)

irregular galaxy An asymmetrical galaxy having neither spiral arms nor an elliptical shape. (page 323)

isotope Any of several forms for the same chemical element whose nuclei all have the same number of protons but different numbers of neutrons. (page 85)

isotropic The same in all directions. (page 362)

isotropy problem The quandary associated with the fact that the temperature of the cosmic microwave background is essentially the same in all directions. (page 368)

Jovian planet Any of the four largest planets: Jupiter, Saturn, Uranus, and Neptune. (page 96)

kelvin *See* temperature (Kelvin). (pages 75–76)

Kepler's laws Three statements, discovered by Johannes Kepler, that describe the motions of the planets. (pages 40–42)

kiloparsec (kpc) One thousand parsecs; about 3260 light-years. (page 10)

Kirchhoff's laws Three statements formulated by Gustav Kirchhoff describing spectra and spectral analysis. (page 83)

Kirkwood gaps Gaps in the spacing of asteroid orbits discovered by Daniel Kirkwood. (page 191)

last quarter moon A phase of the waning moon when Earth-based observers see half of the Moon's illuminated hemisphere. (pages 22–24)

law of cosmic censorship The assertion that a singularity must be surrounded by an event horizon. (page 292)

law of equal areas Kepler's second law. (page 41)

laws of physics Basic principles that govern the behavior of physical reality. (page 3)

light Electromagnetic radiation. (page 54)

light curve A graph that displays variations in the brightness of a star or other astronomical object. (page 238)

light-gathering power A measure of how much light a telescope intercepts and brings to a focus. (pages 61–62)

light-year The distance light travels in a vacuum in one year. (page 10)

limb (of Sun or Moon) The apparent edge of the Sun or Moon as seen in the sky.

limb darkening The phenomenon whereby the Sun is darker near its limb than near the center of its disk. (page 211)

limiting magnitude   The faintest magnitude that can be observed with a given telescope under certain conditions.

line of nodes   A line connecting the nodes of an orbit; the line along which the plane of the Moon's orbit intersects the plane of the ecliptic. (page 27)

liquid metallic hydrogen   A metal-like form of hydrogen that can be produced under extreme pressure. (pages 153–154)

lithosphere   The solid, upper layer of the Earth; essentially the Earth's crust. (page 121)

LMC   The Large Magellanic Cloud, a companion galaxy to the Milky Way. (page 323)

Local Group   The cluster of galaxies of which our own galaxy is a member. (page 325)

low-pressure system   A region in a planet's atmosphere of lower-than-average atmospheric pressure. (pages 140–141)

luminosity   The rate at which electromagnetic radiation is emitted from a star or other object. (pages 206, 227)

luminosity class   The classification of a star of a given spectral type according to its luminosity. (pages 233–234)

lunar   Referring to the Moon.

lunar eclipse   An eclipse of the Moon. (pages 26–29)

Lyman series   A series of spectral lines of hydrogen produced by electron transitions to and from the lowest energy state of the hydrogen atom. (pages 86–87)

magnetic dynamo model (of the solar cycle)   A theory that explains phenomena of the solar cycle as a result of periodic winding and unwinding of the Sun's magnetic field in the solar atmosphere. (pages 218–219)

magnetic field   A region of space near a magnetized body within which magnetic forces can be detected.

magnetogram   A picture of the solar atmosphere that displays the Sun's magnetic field. (pages 217–218)

magnetosphere   The region around a planet occupied by its magnetic field. (pages 132, 154)

magnification (magnifying power)   The number of times larger in angular diameter an object appears through a telescope than with the naked eye. (page 57)

magnitude   A measure of the amount of light received from a star or other luminous object. (page 227)

magnitude scale   The system of denoting magnitudes. (page 227)

main sequence   A grouping of stars on the Hertzsprung–Russell diagram extending diagonally across the graph from the hottest, brightest stars to the dimmest, coolest stars. (page 232)

main sequence star   A star whose surface temperature and luminosity place it on the main sequence on the Hertzsprung–Russell diagram. (page 232)

major axis (of an ellipse)   The longest diameter of an ellipse. (page 40)

mantle (of a planet)   That portion of a terrestrial planet located between its crust and core. (page 131)

mare (pl., **maria**)   Latin for "sea;" a large, relatively crater-free plain on the Moon. (page 168)

mare basalt   Dark, solidified lava that covers the lunar maria. (pages 170–171)

marginally bounded universe   A universe that just barely manages to expand forever. (page 364)

mass   A measure of the total amount of material in an object. (page 46)

mass loss   The phenomenon whereby stars shed their outer layers. (page 260)

mass–luminosity relation   The relationship between the masses and luminosities of main sequence stars. (pages 235–236)

mass transfer   The movement of gas from one star in a close binary to the other. (page 239)

matter-dominated universe   A universe in which the density of matter exceeds the energy density of the radiation field that fills all space. (page 369)

mechanics   The branch of physics dealing with the behavior and motions of objects acted upon by forces.

megaparsec (Mpc)   One million parsecs. (page 10)

mesosphere   The layer in the Earth's atmosphere above the stratosphere. (page 113)

metal-poor star   A star whose abundance of heavy elements is significantly less than that of the Sun. (page 260)

metal-rich star   A star whose abundance of heavy elements is comparable to that of the Sun. (page 259)

meteor   The luminous phenomenon seen when a

meteoroid enters the Earth's atmosphere; a "shooting star." (page 195)

meteorite A fragment of a meteoroid that has survived passage through the Earth's atmosphere. (page 195)

meteoroid A small rock in interplanetary space. (pages 193, 195)

meteor shower Many meteors that seem to radiate from a common point in the sky. (page 202)

micrometer (μm) A unit of length equal to 10^{-6} meter. (page 56)

microwaves Short-wavelength radio waves. (page 56)

Milky Way Galaxy Our galaxy. (page 300ff)

minor axis (of an ellipse) The smallest diameter of an ellipse.

minor planet *See* asteroid.

minute of arc One-sixtieth of a degree. (page 7)

model A hypothesis that has withstood observational or experimental tests. (page 3)

model (of the Sun or a star) A delineation of internal characteristics of the Sun or a star. (page 207)

molecule A combination of two or more atoms.

momentum A measure of the inertia of an object; an object's mass multiplied by its velocity.

monochromatic Of one wavelength or color.

nanometer (nm) A unit of length equal to 10^{-9} meter. (page 55)

nebula (pl., nebulae) A cloud of interstellar gas and dust. (page 5)

N galaxy A galaxy with a bright, starlike nucleus. (page 348)

neon burning The thermonuclear fusion of neon nuclei. (page 269)

neutrino A subatomic particle with no electric charge and little or no mass, yet which is important in many nuclear reactions. (pages 206, 271)

neutrino oscillation A hypothetical phenomenon whereby a neutrino changes from one type to another. (page 210)

neutron A subatomic particle with no electric charge and with a mass nearly equal to that of the proton. (page 85)

neutronization The combining of protons and electrons to produce neutrons. (page 270)

neutron star A very compact, dense star composed almost entirely of neutrons. (pages 271, 277)

New General Catalogue (NGC) A catalogue of star clusters, nebulae, and galaxies first published in 1888. (page 318)

new moon A phase of the Moon when it is nearest the Sun in the sky. (pages 21, 23)

Newtonian mechanics A branch of physics based on Newton's laws. (page 3)

Newtonian reflector An optical arrangement in a reflecting telescope in which a small mirror reflects converging light rays to a focus on one side of the telescope tube. (page 60)

Newton's laws The laws of mechanics and gravitation formulated by Isaac Newton. (pages 44–47)

node The intersection of an orbit with a reference plane such as the plane of the celestial equator or the ecliptic. (page 27)

no-hair theorem The assertion that all of the properties of a black hole are specified by only three quantities: its mass, it electric charge, and its angular momentum. (page 293)

nonthermal radiation Radiation emitted by charged particles moving through a magnetic field; synchrotron radiation.

northern lights Light radiated by atoms and ions in the Earth's upper atmosphere, mostly in the northern polar regions; *aurora borealis*. (page 133)

nova A star that experiences a sudden outburst of radiant energy, temporarily increasing its luminosity by a factor of between 10^4 and 10^6. (page 284)

nuclear Referring to the nucleus of an atom.

nuclear density The density of matter in the nucleus of an atom; about 10^{17} kg/m^3. (page 270)

nucleus (of an atom) The massive part of an atom, composed of protons and neutrons, about which electrons revolve. (pages 84–85)

nucleus (of a comet) A collection of ices and dust that constitute the solid part of a comet. (page 199)

nucleus (of a galaxy) The concentration of stars and dust at the center of a galaxy. (page 303)

OBAFGKM   The sequence of stellar spectral classification. (page 231)

OB association   An association of very young, massive stars predominantly of spectral types O and B. (page 248)

objective   The principal lens or mirror of a telescope. (page 57)

oblateness   A measure of how much a flattened sphere (or spheroid) differs from a perfect sphere. (page 152)

observable universe   All space that is nearer to us than the distance traveled by light since the time of the Big Bang. (page 360)

occultation   The eclipsing of an astronomical object by the Moon or a planet.

oceanic rift   The boundary between two tectonic plates where an upwelling of magma produces seafloor spreading. (page 121)

Olber's paradox   An argument based on Newtonian cosmology that the night sky should be as bright as the Sun's surface. (page 358)

Oort cloud   A region surrounding the solar system where comets spend most of their time. (page 201)

open cluster   A loose association of young stars in the disk of the galaxy; a galactic cluster. (page 249)

opposition   The configuration of a planet when it is at an elongation of 180° and thus appears opposite the Sun in the sky. (pages 38–39)

optics   The branch of physics dealing with the behavior and properties of light. (page 53ff)

orbit   The path of an object that is moving about a second object or point.

organic molecule   A carbon-based compound. (page 379)

outgassing   Volcanic processes by which gases escape from a planet's crust into its atmosphere. (page 127)

oxygen burning   The thermonuclear fusion of oxygen nuclei. (page 269)

pair production   The creation of a particle and an antiparticle from energetic photons. (page 371)

Pallas   The second asteroid to be discovered. (page 186)

parabola   A conic section formed by cutting a circular cone at an angle parallel to the sides of the cone. (pages 46–47)

parallax   The apparent displacement of an object caused by the motion of the observer. (page 225)

parsec (pc)   A unit of distance; 3.26 light-years. (page 10)

partial eclipse   A lunar or solar eclipse in which the eclipsed object does not appear completely covered. (page 29)

Paschen series   A series of spectral lines of hydrogen produced by electron transitions between the third and higher energy levels of the hydrogen atom. (pages 86–87)

Pauli exclusion principle   A principle of quantum mechanics that says that two identical particles cannot have the same position and momentum. (page 257)

penumbra   The portion of a shadow in which only part of the light source is covered by an opaque body. (page 28)

penumbral eclipse   A lunar eclipse in which the Moon passes only through the Earth's penumbra. (page 28)

perigee   The point in its orbit where a satellite or the Moon is nearest the Earth.

perihelion   The point in its orbit where a planet is nearest the Sun. (page 41)

period   The interval of time between successive repetitions of a periodic phenomenon.

periodic table   A listing of the chemical elements according to their properties, invented by D. Mendeleev. (pages 80–81)

period–luminosity relation   A relationship between the period and average luminosity of a pulsating star. (page 262)

permafrost   Permanently frozen subsoil. (page 130)

perturbation   A small, disturbing effect.

phases (of the Moon)   The appearance of the Moon as it orbits the Earth. (pages 21–24)

photodisintegration   The breakup of nuclei in the core of a massive star by energetic gamma rays. (page 270)

photometry   The measurement of light intensities. (page 229)

photon   A discrete unit of electromagnetic energy. (page 78)

photosphere   The region in the solar atmosphere from which most of the visible light escapes into space. (pages 210–212)

on observation, experimentation, and the formulation of hypotheses that are testable. (page 2)

seafloor spreading The process whereby magma upwelling along rifts in the ocean floor causes adjacent segments of the Earth's crust to separate. (pages 120–121)

second of arc One-sixtieth of a minute of arc. (page 7)

seismic waves Vibrations traveling through a terrestrial planet, usually associated with earthquakelike phenomena. (page 131)

seismograph A device used to record and measure seismic waves, such as those produced by earthquakes. (page 131)

seismology The study of earthquakes and related phenomena. (page 131)

self-propagating star formation The process whereby the birth of stars in one part of a galaxy stimulates star formation in a neighboring region of that galaxy. (page 309)

semidetached binary A close binary system in which one star fills or is overflowing its Roche lobe. (page 240)

semimajor axis Half of the major axis of an ellipse. (page 40)

Seyfert galaxy A spiral galaxy with a bright nucleus whose spectrum exhibits emission lines. (pages 345–346)

shell helium burning Helium burning that occurs in a thin shell surrounding the core of a star. (page 266)

shell hydrogen burning Hydrogen burning that occurs in a thin shell surrounding the core of a star. (page 255)

shepherd satellite A small satellite whose gravity is responsible for maintaining a sharply defined ring of matter around a planet such as Saturn or Uranus. (page 148)

shock wave An abrupt, localized region of compressed gas caused by an object traveling through the gas at a speed greater than the speed of sound. (page 132)

shooting star *See* meteor.

sidereal month The period of the Moon's revolution about the Earth with respect to the stars. (pages 23–24)

sidereal period The orbital period of one object about another with respect to the stars. (page 39)

single-line spectroscopic binary A spectroscopic binary whose spectrum exhibits the spectral lines of only one of its two stars. (page 237)

singularity A place of infinite curvature of space and time; the center of a black hole. (page 291)

SMC The Small Magellanic Cloud, a companion galaxy to the Milky Way. (pages 323–324)

solar activity Phenomena that occur in the solar atmosphere, such as sunspots, plages, flares, and so forth.

solar corona The Sun's outer atmosphere. (pages 212–213)

solar cycle A 22-year cycle during which the Sun's magnetic field reverses its polarity. (page 218)

solar eclipse An eclipse of the Sun. (pages 26–31)

solar flare A violent eruption on the Sun's surface. (pages 132, 216)

solar nebula The cloud of gas and dust from which the Sun and solar system formed. (pages 100–101)

solar seismology The study of the Sun's interior from observations of vibrations of its surface. (pages 219–220)

solar system The Sun, planets, their satellites, asteroids, comets, and related objects that orbit the Sun. (pages 4, 94–95)

solar wind A radial flow of particles (mostly electrons and protons) from the Sun. (pages 105, 212, 214)

solstice Either of two points along the ecliptic at which the Sun reaches its maximum distance north or south of the celestial equator. (pages 19-20)

southern lights Light radiated by atoms and ions in the Earth's upper atmosphere, mostly in the southern polar regions; *aurora australis*. (page 133)

special theory of relativity A description of mechanics and electromagnetic theory formulated by Einstein according to which measurements of distance, time, and mass are affected by the observer's motion. (page 289)

spectral analysis The identification of chemicals by the appearance of their spectra. (page 80)

spectral line A dark or bright line at a specific wavelength in a spectrum. (page 79)

spectral type A classification of stars according to the appearance of their spectra. (page 231)

spectrogram The photograph of a spectrum. (page 82)

spectrograph A device for photographing a spectrum. (pages 81–82)

spectroscope   A device for directly viewing a spectrum. (page 81)

spectroscopic binary star   A double star whose binary nature can be deduced from the periodic Doppler shifting of lines in its spectrum. (page 237)

spectroscopy   The study of spectra.

spectrum   The result of dispersing a beam of electromagnetic radiation so that different wavelengths are separated in space. (page 55)

speed   The rate at which an object moves.

spherical aberration   An optical defect whereby different portions of a lens or mirror have slightly different focal lengths. (page 63)

spicule   A narrow jet of rising gas in the solar chromosphere. (pages 211–212)

spin   A small amount of angular momentum possessed by certain particles such as electrons, protons, and neutrons. (pages 304–305)

spiral arms   Lanes of interstellar gas, dust, and young stars that wind outward in a plane from the central regions of a galaxy. (pages 251–252, 304)

spiral galaxy   A flattened, rotating galaxy with pinwheel-like spiral arms winding outward from the galaxy's nucleus. (page 220)

standard candle   An object whose known luminosity can be used to deduce the distance to a galaxy. (page 336)

star   A self-luminous sphere of gas.

starburst galaxy   A galaxy where there is an exceptionally high rate of star formation. (pages 328–329)

Stefan–Boltzmann law   A relationship between the temperature of a blackbody and the rate at which it radiates energy. (page 76)

Stefan's law   *See* Stefan–Boltzmann law. (page 76)

stellar evolution   The changes in size, luminosity, temperature, and so forth, that occur as a star ages. (page 245)

stellar model   The result of theoretical calculations that give details of physical conditions inside a star. (page 208)

stony-iron meteorite   A meteorite composed of roughly equal amounts of rock and iron. (page 196)

stony meteorite   A meteorite composed of rock with very little iron; also called a stone. (pages 195–196)

stratosphere   A layer in the Earth's atmosphere directly above the troposphere. (page 113)

strong nuclear force   The force that binds protons and neutrons together in nuclei. (page 373)

subduction zone   A location where colliding tectonic plates cause the Earth's crust to be pulled down into the mantle. (page 121)

subgiant   A star whose luminosity is between that of main sequence stars and normal giants of the same spectral type. (page 233)

summer solstice   The point on the ecliptic where the Sun is farthest north of the celestial equator. (pages 19–20)

Sun   The star about which the Earth and other planets revolve. (page 205ff)

sunspot   A temporary cool region in the solar photosphere. (pages 214–215)

sunspot cycle   The semiregular 11-year period with which the number of sunspots fluctuates. (page 216)

sunspot maximum   The time during the solar cycle when there are exceptionally many sunspots on the Sun. (page 216)

sunspot minimum   The time during the solar cycle when there are exceptionally few sunspots on the Sun. (page 216)

supercluster   A collection of many clusters of galaxies. (page 326)

supergiant   A star of very high luminosity. (page 233)

supergranule   A large convective cell in the Sun's photosphere. (page 212)

superior conjunction   The configuration of a planet being behind the Sun. (pages 38–39)

supermassive black hole   A black hole whose mass exceeds a thousand solar masses. (pages 313, 349–354)

supernova   A stellar outburst during which a star suddenly increases its brightness roughly a millionfold. (pages 5, 198, 271)

supernova remnant   A nebula left over after the detonation of a supernova. (pages 275–276)

synchrotron radiation   The radiation emitted by

charged particles moving through a magnetic field; nonthermal radiation. (page 313)

synodic month The period of revolution of the Moon with respect to the Sun; the length of one cycle of lunar phases. (pages 23–24)

synodic period The interval between successive occurrences of the same configuration of a planet. (page 39)

tail (of a comet) Gas and dust particles from a comet's nucleus that have been swept away from the comet's head by the radiation pressure of sunlight and the solar wind. (pages 199–201)

telescope An instrument for viewing remote objects. (page 53ff)

temperature (Celsius) Temperature measured on a scale where water freezes at 0° and boils at 100°. (pages 75–76)

temperature (Fahrenheit) Temperature measured on a scale where water freezes at 32° and boils at 212°. (pages 75–76)

temperature (Kelvin) Absolute temperature measured in Celsius degree intervals. (pages 75–76)

terminator The line dividing day and night on the surface of the Moon or a planet; the line of sunset or sunrise.

terra (pl., **terrae**) Cratered region of the Moon; lunar highland. (page 169)

terrestrial planet Any of the planets Mercury, Venus, Earth, or Mars, and sometimes also the Galilean satellites and Pluto. (page 96)

theory A hypothesis that has been demonstrated to describe a range of phenomena accurately. (page 3)

thermal energy The energy associated with the motions of atoms or molecules in a substance.

thermal equilibrium A balance between the input and outflow of heat in a system. (page 207)

thermal radiation The radiation naturally emitted by any object not at absolute zero. (pages 76–78)

thermonuclear fusion A reaction in which the nuclei of atoms are fused together at a high temperature. (pages 206–207)

thermonuclear reaction A reaction resulting from a high-speed collision of nuclear particles moving rapidly because they are at a high temperature. (page 207)

thermosphere A layer in the Earth's atmosphere above the mesosphere. (page 113)

tidal force A gravitational force whose strength and/or direction varies over a body and thus tends to deform the body. (page 175)

total eclipse A solar eclipse during which the Sun is completely hidden by the Moon, or a lunar eclipse during which the Moon is completely immersed in the Earth's umbra. (pages 28–31)

transit The passage of a small object in front of a larger object. (page 164)

triple point The pressure and temperature at which a substance can simultaneously exist as a solid, a liquid, and a gas. (page 181)

Trojan asteroid One of several asteroids that share Jupiter's orbit about the Sun. (page 192)

troposphere The lowest level in the Earth's atmosphere. (page 113)

T Tauri stars Young variable stars associated with interstellar matter that show erratic changes in luminosity. (page 249)

T Tauri wind A flow of particles away from a T Tauri star. (page 105)

Tully–Fisher relation A correlation between the width of the 21-cm line of a spiral galaxy and its absolute magnitude. (page 337)

turbulence Random motions in a gas or liquid.

turnoff point The location of the brightest main sequence stars on the Hertzsprung–Russell diagram of a globular cluster. (page 259)

UBV system A system of stellar magnitudes involving measurements of starlight intensity in the ultraviolet, blue, and visible spectral regions. (page 229)

ultraviolet radiation Electromagnetic radiation of wavelengths shorter than those of visible light but longer than those of X rays. (page 56)

umbra The central, completely dark portion of a shadow. (page 28)

unbounded universe A universe that expands forever. (page 363)

universal constant of gravitation The constant of proportionality in Newton's law of gravitation. (page 47)

universal law of gravitation   Newton's law of gravitation, which describes how the gravitational force between two bodies depends on their masses and separation. (page 47)

universe   All space along with all the matter and radiation in space.

Van Allen belts   Two doughnut-shaped regions around the Earth where many charged particles (protons and electrons) are trapped by the Earth's magnetic field. (page 132)

variable star   A star whose luminosity varies. (page 261)

velocity   A quantity that specifies both direction and speed.

vernal equinox   The point on the ecliptic where the Sun crosses the celestial equator from south to north. (page 19)

visual binary star   A double star in which the two components can be resolved through a telescope. (page 235)

VLBI   Very-long-baseline interferometry; a method of connecting widely separated radio telescopes to make observations of very high resolution. (page 65)

void   A huge, roughly spherical region where exceptionally few galaxies are found. (page 329)

volatile element   An element that boils at a relatively low temperature. (page 171)

vortex   A whirling mass of liquid or gas; a whirlpool.

wavelength   The distance between two successive peaks in a wave. (pages 55)

waning   An adjective that means "decreasing," as in the waning crescent moon or the waning gibbous moon. (page 22)

waxing   An adjective that means "increasing," as in the waxing crescent moon or the waxing gibbous moon. (page 21)

weak nuclear force   A nuclear interaction involved in certain kinds of radioactive decay. (page 373)

weakon   A particle that is responsible for the weak nuclear force. (page 373)

weight   The force with which a body presses down on the surface of the Earth. (page 46)

white dwarf   A low-mass star that has exhausted all its thermonuclear fuel and contracted to a size roughly equal to the size of the Earth. (pages 5, 233, 268–269)

Widmanstätten patterns   Crystalline structure seen in certain types of meteorites. (pages 196–197)

Wien's law   A relationship between the temperature of a blackbody and the wavelength at which it emits the greatest intensity of radiation. (page 77)

winter solstice   The point on the ecliptic where the Sun reaches its greatest distance south of the celestial equator. (pages 19–20)

X ray   Electromagnetic radiation whose wavelength is between that of ultraviolet light and gamma rays. (page 56)

year   The period of revolution of the Earth about the Sun.

Zeeman effect   A splitting or broadening of spectral lines because of a magnetic field. (pages 216–217)

zenith   The point on the celestial sphere opposite to the direction of gravity. (page 20)

zero-age main sequence (ZAMS)   The main sequence of young stars that have just begun to burn hydrogen at their cores. (page 258)

zodiac   A band of twelve constellations around the sky centered on the ecliptic. (page 21)

zones (on Jupiter)   Light-colored band in Jupiter's cloud cover. (page 139)

Answers to Selected Exercises

Chapter 1

8 (a) km = kilometer (b) cm = centimeter (c) s = second
   (d) km/s = kilometers per second (e) mph = miles per hour
   (f) m = meter (g) m/s = meters per second (h) hr = hour
   (i) yr = year (j) g = gram (k) kg = kilogram
9 $1° = 3600$ seconds of arc
10 (a) 10^7 (b) 4×10^5 (c) 6×10^{-2} (d) 1.7×10^{10}
11 $8\frac{1}{3}$ minutes
12 (a) 8.6 years (b) 8.16×10^{13} km
13 8700 km
14 29 million Suns

Chapter 2

15 At or between latitudes of 23.4°N and 23.4°S
16 At the South Pole; maximum elevation = 23.4°
    on December 21
17 Due east

Chapter 3

15 5.2 AU^2 in 1991; 26 AU^2 in five years
16 a = 100 AU; maximum distance = 200 AU
17 2.8 years
19 It would be weaker by a factor of 100.

Chapter 4

13 $\frac{1}{25}$
14 The light-gathering power of the Palomar telescope is one
    million times greater than that of the human eye.
16 (a) 222× (b) 100× (c) 36×

Chapter 5

10 $7\frac{1}{2}$ trips around the Earth
11 238 nm
12 6520 K
13 $\lambda_{max} = 250$ nm; sixteen times brighter than the Sun
14 The cooler star would be one-sixteenth as bright as the Sun.
15 13 km/s toward the Earth

16 21 km/s away from the Earth
17 86,000 km/s = $\frac{2}{7}$ the speed of light

Chapter 6

15 Assuming an average density of 1500 kg/m^3, the planet's
    diameter would be about $1\frac{1}{2}$ times the Earth's diameter.

Chapter 7

16 Core = 17%, mantle = 82%, crust = 1%
17 About 220 million years ago

Chapter 11

15 1400 kg/m^3
17 4.9%; the Sun's chemical composition will be 69%
    hydrogen and 30% helium
18 Photosphere: 500 nm = visible light; chromosphere:
    58 nm = ultraviolet light; corona: 1.93 nm = X rays

Chapter 12

18 4.31 parsecs
21 $10 \, M_\odot$; $10^{-3} \, L_\odot$

Chapter 13

19 2000
20 200 times longer

Chapter 14

22 About 6500 years ago

Chapter 16

18 $22\frac{1}{2}$ times
20 Once every 50,000 years

Chapter 17

17 About 30 Mpc
18 8000 km/s

Illustration Credits

Front cover Courtesy of Leon Golub, Smithsonian Astrophysical Observatory and IBM Research.

Contents p. vii: Copyright © 1984 Anglo-Australian Telescope Board; p. viii: W. J. Kaufmann; p. ix: NASA; p. x: National Optical Astronomy Observatories; p. xi: NASA; p. xii: National Optical Astronomy Observatories.

Chapter 1 p. 1: Photography by David F. Malin of the Anglo-Australian Observatory from original negatives by the U.K. 1.2-m Schmidt telescope, copyright © 1987 Royal Observatory, Edinburgh; Fig. 1-1: Dennis. L. Mammana; Figs. 1-2, 1-3: NASA; Fig. 1-4: National Optical Astronomy Observatories; Fig. 1-5: Copyright © 1981 Anglo-Australian Telescope Board; Fig. 1-6: Lick Observatory photograph; Fig. 1-7: Copyright © R. J. Dufour, Rice University; Fig. 1-8: Palomar Observatory photograph; Fig. 1-10: Scientific American Books, NASA, and the Anglo-Australian Observatory; Fig. 1-12: Lockheed Missiles & Space Company, Inc.

Chapter 2 p. 13: Copyright © 1980 Anglo-Australian Telescope Board; Fig. 2-1: British crown copyright, reproduced with permission of the Controller of the Britannic Majesty's Stationary Office; Fig. 2-2: (left) Robert C. Mitchell, Central Washington University, (right) drawing from Elijah Burritt's *Atlas,* Janus Publications, Wichita Omnisphere Earth-Space Center, 220 S. Main, Wichita, Kansas; Fig. 2-13: NASA; Fig. 2-15: Lick Observatory photographs; Fig. 2-23: Mike Harms; Fig. 2-24: William R. Dellinges; Fig. 2-26: NASA; Fig. 2-27: Dennis di Cicco.

Chapter 3 p. 35: NASA; Fig. 3-8: New Mexico State University Observatory; Fig. 3-10: Clifford Holmes; Figs. 3-11, 3-12: Yerkes Observatory; Fig. 3-14: Copyright © 1986 Jack B. Marling, Lumicon; Fig. 3-15: Lick Observatory photographs.

Chapter 4 p. 53: California Association for Research in Astronomy; Fig. 4-8: Yerkes Observatory; Fig. 4-12: National Optical Astronomy Observatories; Fig. 4-14: Multiple-Mirror Telescope Observatory; Fig. 4-15: California Institute of Technology; Fig. 4-17: Smithsonian Institution Astrophysical Observatory; Fig. 4-18: Patrick Seitzer, NOAO; Figs. 4-19, 4-20: National Radio Astronomy Observatory; Fig. 4-21: (a) NASA, (b) Copyright © 1982 Associated Universities, Inc., under contract with the National Science Foundation (VLA observations by Imke de Pater, J. R. Dickel); Figs. 4-22, 4-23: NASA: Fig. 4-24: (a) George R. Carruthers, NRL; (b) NASA, (c) Robert C. Mitchell, Central Washington University; Figs. 4-25, 4-26: NASA; Fig. 4-27: Space Telescope Science Institute, NASA, and ESA; Fig. 4-28: TRW, Inc.; Fig. 4-29: NASA; Fig. 4-30: (a) Max Planck Institut für Radioastronomie, (b) Jet Propulsion Laboratory, (c) Griffith Observatory, (d) Royal Observatory, Edinburgh, (e) Elihu A. Boldt, NASA.

Chapter 5 p. 74: Bausch and Lomb; Fig. 5-5: National Optical Astronomy Observatories; Fig. 5-8: The Observatories of the Carnegie Institution of Washington; Fig. 5-10: Palomar Observatory photograph; Fig. 5-15: The Observatories of the Carnegie Institution of Washington; Fig 5-20: National Optical Astronomy Observatories.

Chapter 6 p. 92: Photography by David F. Malin of the Anglo-Australian Observatory from original negatives by the U.K. 1.2-m Schmidt telescope, copyright © 1979 Royal Observatory, Edinburgh; Figs. 6-3, 6-4, 6-5, 6-6, 6-7: NASA; Fig. 6-8: Copyright © 1980 Anglo-Australian Telescope Board; Fig. 6-9: Copyright © 1984 Anglo-Australian Telescope Board; Fig. 6-11: University of Arizona and Jet Propulsion Laboratory; Fig. 6-15: Copyright © 1982 Anglo-Australian Telescope Board; Figs. 6-16, 6-17: NASA; Fig. 6-18: Willy Benz, Harvard-Smithsonian Center for Astrophysics.

Chapter 7 p. 111: NASA and ESA; Figs. 7-1, 7-3: NASA; Fig. 7-7: Stephen M. Larson; Fig. 7-8: NASA and USGS; Figs. 7-10, 7-11: NASA; Fig. 7-13: W. Kaufmann; Fig. 7-15: Marie Tharp and Bruce Heezen copyright © 1977; Figs. 7-18, 7-19, 7-20, 7-21, 7-22, 7-23, 7-24: NASA; Fig. 7-25: Stephen P. Meszaros, NASA; Fig. 7-26: A. Post, Project Office Glaciology, U.S. Geological Survey; Fig. 7-27: Carle M. Pieters and the U.S.S.R.

Academy of Sciences; Fig. 7-28: NASA; Figs. 7-29, 7-30: NASA and USGS; Fig. 7-33: Courtesy of Syun-Ichi Akasofu, Geophysical Institute, University of Alaska, Fairbanks.

Chapter 8 p. 137: Stephen P. Meszaros, NASA; Fig. 8-1: Stephen P. Larson; Figs. 8-2, 8-3, 8-6, 8-7, 8-8, 8-11, 8-12: NASA; Fig. 8-13: Lowell Observatory; Figs. 8-14, 8-15, 8-16, 8-17, 8-18, 8-19, 8-20, 8-21, 8-23, 8-24, 8-26: NASA; Figs. 8-29, 8-30: New Mexico State University Observatory; Figs. 8-32, 8-33, 8-34, 8-35, 8-36, 8-37: NASA.

Chapter 9 p. 163: NASA; Fig. 9-1: Yerkes Observatory; Fig. 9-2: New Mexico State University Observatory; Fig. 9-3: (left) NASA, (right) Lick Observatory photograph; Fig. 9-4: NASA; Fig 9-5: The Observatories of the Carnegie Institution of Washington; Fig. 9-8: Lick Observatory photograph; Figs. 9-9, 9-10, 9-12, 9-13, 9-14, 9-15, 9-16, 9-17, 9-18, 9-19, 9-20, 9-21, 9-22, 9-23, 9-24, 9-25, 9-26, 9-27, 9-28, 9-30, 9-31: NASA; Fig. 9-32: NASA and ESA.

Chapter 10 p. 188: Copyright © 1985 Anglo-Australian Telescope Board; Figs. 10-1, 10-5: Yerkes Observatory; Fig. 10-6: NASA; Fig. 10-7: Meteor Crater Enterprises, Arizona; Fig. 10-8: Walter Alvarez; Figs. 10-9, 10-10, 10-11, 10-12, 10-13, 10-14: Ronald A. Oriti; Fig. 10-15: John A. Wood; Fig. 10-16: Hans Vehrenberg; Fig. 10-17: Max Planck Institut für Aeronomie; Fig. 10-19: Johns Hopkins University and the Naval Research Laboratory; Figs. 10-20, 10-21: Lick Observatory photographs; Fig. 10-23: Palomar Observatory photograph; Fig. 10-24: New Mexico State University Observatory; Fig. 10-25: TASS, Sovfoto.

Chapter 11 p. 205: Leon Golub, Center for Astrophysics, and Serge Koutchmy, Centre National de la Recherche Scientifique; Fig. 11-1: Celestron International; Fig. 11-4: Raymond Davis, Brookhaven National Laboratory; Figs. 11-5, 11-6, 11-7: National Optical Astronomy Observatories; Fig. 11-9: Roland and Margorie Christen, taken with an Astro-Physics 130-mm StarFire EDT refractor; Fig. 11-10: NASA and the High Altitude Observatory; Fig. 11-11: NASA and Harvard College Observatory; Fig. 11-12: National Optical Astronomy Observatories; Fig. 11-13: The Observatories of the Carnegie Institution of Washington; Fig. 11-15: National Optical Astronomy Observatories; Fig. 11-16: Naval Research Laboratory; Figs. 11-17, 11-18: National Optical Astronomy Observatories; Fig. 11-21: National Solar Observatory; Fig. 11-22: Kenneth G. Libbrecht, Big Bear Solar Observatory.

Chapter 12 p. 224: Copyright © 1977 Anglo-Australian Telescope Board; Fig. 12-8: Courtesy of Nancy Houk, Nelson Irvine, and David Rosenbush; Fig. 12-13: Yerkes Observatory; Fig. 12-17: Lick Observatory photograph.

Chapter 13 p. 244: Copyright © 1981 Anglo-Australian Telescope Board; Fig. 13-1: Photography by David F. Malin of the Anglo-Australian Observatory from original negatives by the U.K. 1.2-m Schmidt telescope, copyright © 1979 Royal Observatory, Edinburgh; Fig. 13-2: Copyright © 1980 Anglo-

Australian Telescope Board; Fig. 13-4: National Optical Astronomy Observatories; Fig. 13-5: Copyright © 1980 Anglo-Australian Telescope Board; Fig. 13-6: Copyright © 1981 Anglo-Australian Telescope Board; Fig. 13-7: Photography by David F. Malin of the Anglo-Australian Observatory from original negatives by the U.K. 1.2-m Schmidt telescope, copyright © 1985 Royal Observatory, Edinburgh; Fig. 13-8: (a) Courtesy of Ronald J. Maddalena, Mark Morris, J. Moscowitz, and P. Thaddeus; Fig. 13-9: Copyright © 1980 Anglo-Australian Telescope Board; Fig. 13-10: Copyright © 1981 Anglo-Australian Telescope Board; Fig. 13-12: (a) Copyright © 1981 Anglo-Australian Telescope Board, (b) Copyright © 1983 Anglo-Australian Telescope Board; Fig. 13-13: Courtesy of Hans Vehrenberg; Fig. 13-14: Copyright © 1984 Anglo-Australian Telescope Board; Fig. 13-17: U.S. Naval Observatory; Fig. 13-20: Lick Observatory photograph; Fig. 13-21: Copyright © 1980 Anglo-Australian Telescope Board; Fig. 13-22: Copyright © 1979 Anglo-Australian Telescope Board.

Chapter 14 p. 265: Lick Observatory photograph; Fig. 14-2: Copyright © 1979 Anglo-Australian Telescope Board; Fig. 14-3: U.S. Naval Observatory; Fig. 14-5: Courtesy of R. B. Minton; Fig. 14-8: European Southern Observatory; Fig. 14-9: Copyright © 1987 Anglo-Australian Telescope Board; Fig. 14-10: Courtesy of K. S. Luttrell; Fig. 14-11: NASA, ESA; Fig. 14-12: The Observatories of the Carnegie Institution of Washington; Fig. 14-14: Palomar Observatory photograph; Fig. 14-15: Photography by David A. Malin of the Anglo-Australian Observatory from original negatives by the U.K. 1.2-m Schmidt telescope, copyright © 1979 Royal Observatory, Edinburgh; Fig. 14-16: S. S. Murray, Harvard-Smithsonian Center for Astrophysics, (b) NRAO/AUI (VLA observations by A. Angerhofer, R. A. Perley, B. Balick, D. K. Milne); Fig. 14-18: Palomar Observatory photograph; Fig. 14-20: Lick Observatory photographs; Fig. 14-21: NASA; Fig. 14-25: NRAO/AUI (VLA observations by R. M. Hjellming and K. J. Johnson); Fig. 14-26: Lick Observatory photographs.

Chapter 15 p. 288: Courtesy of D. Norton, Science Graphics; Fig. 15-6: National Radio Astronomy Observatory, operated by Associated Universities, Inc., under contract with the National Science Foundation (VLA observations by P. E. Greenfield, D. H. Roberts, B. F. Burke); Fig. 15-7: NASA, ESA; Fig. 15-8: Computations by E. Falco (MIT), M. Kurtz, R. Schild, M. Schneps, copyright © 1985 Smithsonian Astrophysical Observatory; Fig. 15-9: National Optical Astronomy Observatories; Fig. 15-10: National Radio Astronomy Observatory, operated by Associated Universities, Inc., under contract with the National Science Foundation (VLA observations by J. N. Hewitt and E. L. Turner); Fig. 15-11: Courtesy of J. Kristian, The Observatories of the Carnegie Institution of Washington.

Chapter 16 p. 300: Courtesy of Dennis di Cicco; Fig. 16-1: Steward Observatory, University of Arizona; Fig. 16-3: Harvard Observatory; Fig. 16-4: NOAO; Fig. 16-6: U.S. Naval Observatory; Fig. 16-9: Courtesy of G. Westerhout; Fig. 16-10:

(a) Copyright © 1986 Anglo-Australian Telescope Board; (b) National Radio Astronomy Observatory, operated by Associated Universities, Inc. under contract with the National Science Foundation; Fig. 16-14: Courtesy of P. Seiden, D. Elmegreen, B. Elmegreen, and A. Mobarak, IBM; Fig. 16-17: (a) Courtesy of Dennis di Cicco, (b and c) NASA; Fig. 16-18: Copyright © 1983 Anglo-Australian Telescope Board; Fig. 16-19: (a) National Radio Astronomy Observatory, operated by Associated Universities, Inc., under contract with the National Science Foundation (VLA observations by F. Yusef-Zadeh, M. R. Morris, D. R. Chance), (b) K-Y Lo, N. Killeen, NCSA.

Chapter 17 p. 317: European Southern Observatory; Fig. 17-1: Courtesy of Lund Humphries; Fig. 17-2: National Optical Astronomy Observatories; Figs. 17-3 and 17-4: Palomar Observatory, copyright © California Institute of Technology; Fig. 17-5: (Sa) Courtesy of Rudolph Schild, Harvard-Smithsonian Center for Astrophysics, (Sb, Sc) Courtesy of Philip E. Seiden, IBM; Fig. 17-6: (Sa) NOAO, (Sb, Sc) Courtesy of Rudolph Schild, Harvard-Smithsonian Center for Astrophysics; Fig. 17-7: (SBa, SBc) Courtesy of Philip E. Seiden, IBM, (SBb) Courtesy of Rudolph Schild, Harvard-Smithsonian Center for Astrophysics; Fig. 17-8: Yerkes Observatory; Fig. 17-9: Courtesy of Rudolph Schild, Harvard-Smithsonian Center for Astrophysics; Fig. 17-10: Copyright © 1987 Anglo-Australian Telescope Board; Figs. 17-12, 17-13, 17-14: Photography by David A. Malin of the Anglo-Australian Observatory from original negatives by the U.K. 1.2-m Schmidt telescope, copyright © 1984 Royal Observatory, Edinburgh; Fig. 17-16: Photography by David A. Malin of the Anglo-Australian Observatory from original negatives by the U.K. 1.2-m Schmidt telescope, copyright © 1987 Royal Observatory, Edinburgh; Fig. 17-17: Courtesy of Rudolph Schild, Harvard-Smithsonian Center for Astrophysics; Fig. 17-18: National Optical Astronomy Observatories; Fig. 17-19: S. J. Maddox, W. J. Sutherland, G. P. Efstathiou, and J. Loveday, Oxford Astrophysics; Fig. 17-20: V. de Lapparent, M. Geller, and J. Huchra; Fig. 17-22: Lick Observatory photograph; Fig. 17-23: (a) M. Yun and P. Ho, Harvard-Smithsonian Center for Astrophysics, (b) Copyright © 1960 National Geographic Society—Palomar Sky Survey, reproduced by permission of the California Institute of Technology; Fig. 17-24: Joshua Barnes, Canadian Institute for

Theoretical Astrophysics; Fig. 17-25: Palomar Observatory photograph; Fig. 17-26: Courtesy of W. C. Keel, University of Alabama; Fig. 17-27; Courtesy of Lars Hernquist, Institute for Advanced Study with simulations performed at the Pittsburgh Supercomputing Center; Fig. 17-30: The Observatories of the Carnegie Institution of Washington; Fig. 17-32: Lick Observatory photograph.

Chapter 18 p. 340: Courtesy of *Astronomy* Magazine, Kalmbach Publishing Co.; Figs. 18-1, 18-2: Palomar Observatory photographs; Fig. 18-3: National Optical Astronomy Observatories; Fig. 18-8: Palomar Observatory photograph; Fig. 18-9: Courtesy of Rudolph Schild, Harvard-Smithsonian Center for Astrophysics; Fig. 18-10: National Optical Astronomy Observatories; Fig. 18-11: J. E. Grindlay, Harvard-Smithsonian Center for Astrophysics; Fig. 18-12: (a) National Optical Astronomy Observatories, (b) David A. Chartee, University of Illinois, and Jack O. Burns, New Mexico State University, (c) E. J. Schreier, Harvard-Smithsonian Center for Astrophysics; Fig. 18-13: National Radio Astronomy Observatory, operated by Associated Universities, Inc. under contract with the National Science Foundation (VLA observations by P. A. Scheuer, R. A. Laing, R. A. Perley); Fig. 18-14: National Radio Astronomy Observatory, operated by Associated Universities, Inc. under contract with the National Science Foundation (VLA observations by C. P. O'Dea, F. N. Owen); Fig. 18-15: Courtesy of T. D. Kinman, NOAO; Fig. 18-17: National Optical Astronomy Observatories; Fig. 18-18: (a) Palomar Observatory, copyright © California Institute of Technology, (b) NASA, ESA; Fig. 18-19: National Optical Astronomy Observatories; Fig. 18-20: (a) Copyright © 1987 Anglo-Australian Telescope Board, (b) Lick Observatory photograph; Figs. 18-21, 18-24: NASA, ESA.

Chapter 19 p. 357: NASA Goddard Space Flight Center; Fig. 19-3: Bell Telephone Laboratories; Fig. 19-4: Courtesy of John Mather, NASA; Fig. 9-6: NASA; Fig. 19-15: Lawrence Berkeley Laboratory;

Afterword p. 378: Courtesy of Jon Lomberg; Fig. A-1: From the collection of Ronald A. Oriti, Santa Rosa Junior College; Fig. A-4: NASA Fig. A-5: Jet Propulsion Laboratory.

Appendixes pp. 385, 386: Stephen P. Meszaros, NASA.

Index

Star Charts

The following set of star charts, one for each month of the year, are from the *Griffith Observer* magazine. To use these charts, first select the chart that best corresponds to the date and time of your observations. Hold the chart vertically and turn it so that the direction you are facing shows at the bottom.

THE NIGHT SKY IN JANUARY

Chart time (Local Standard Time):

10 pm...First of January
9 pm...Middle of January
8 pm...Last of January

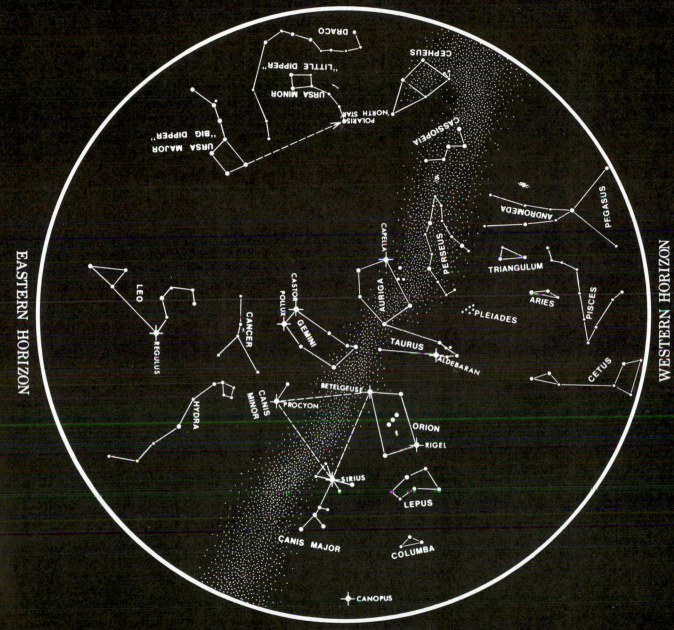

SOUTHERN HORIZON

THE NIGHT SKY IN FEBRUARY

Chart time (Local Standard Time):

10 pm...First of February
9 pm...Middle of February
8 pm...Last of February

THE NIGHT SKY IN MARCH

Chart time (Local Standard Time):

10 pm...First of March
9 pm...Middle of March
8 pm...Last of March

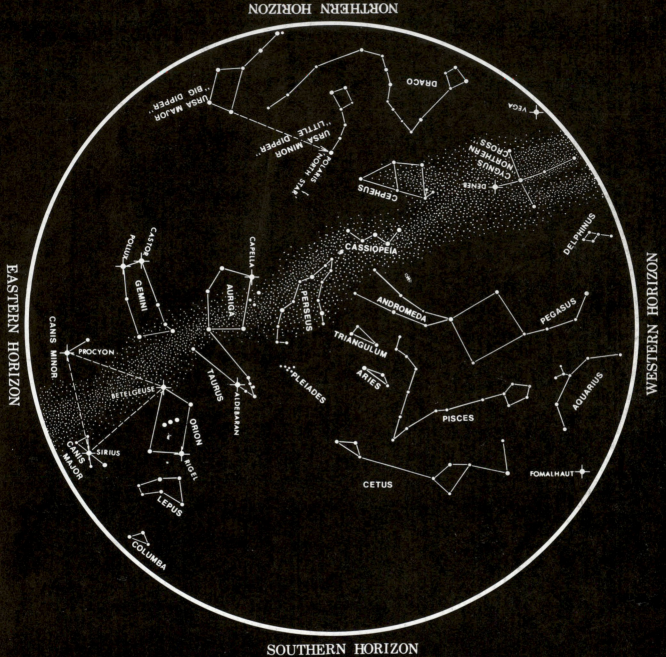

SOUTHERN HORIZON

THE NIGHT SKY IN DECEMBER

Chart time (Local Standard Time):

10 pm...First of December
9 pm...Middle of December
8 pm...Last of December

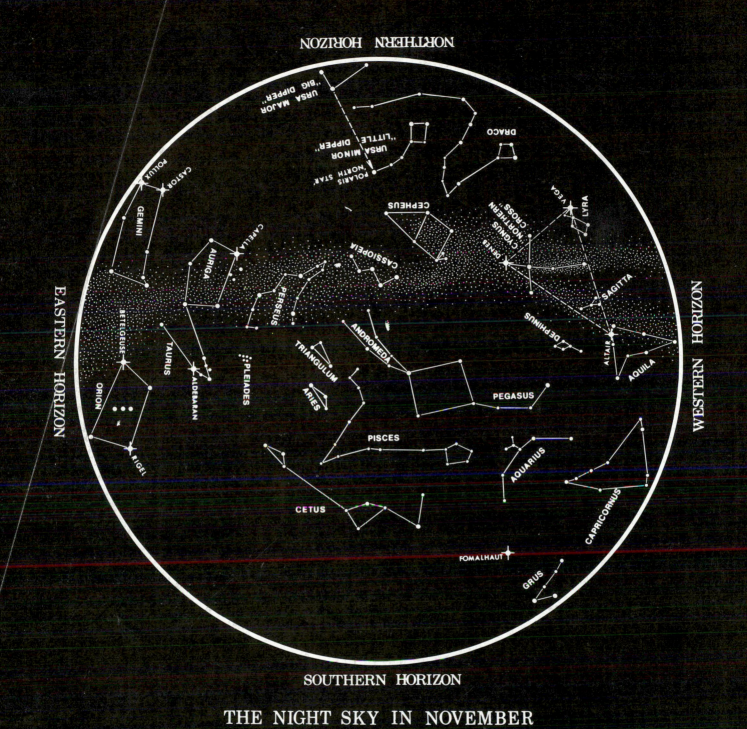

THE NIGHT SKY IN NOVEMBER

Chart time (Local Standard Time):

10 pm...First of November
9 pm...Middle of November
8 pm...Last of November

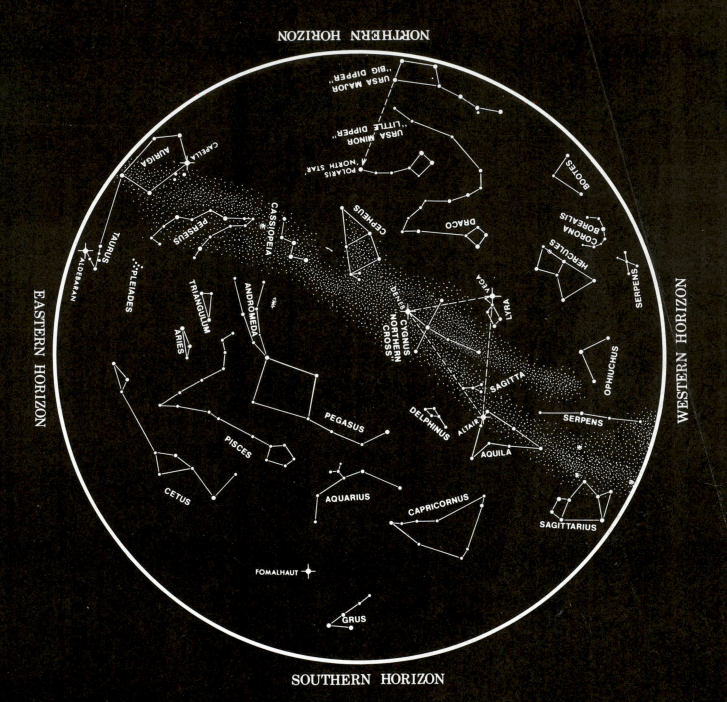

THE NIGHT SKY IN OCTOBER

Chart time (Daylight Savings Time):

11 pm...First of October
10 pm...Middle of October
9 pm...Last of October

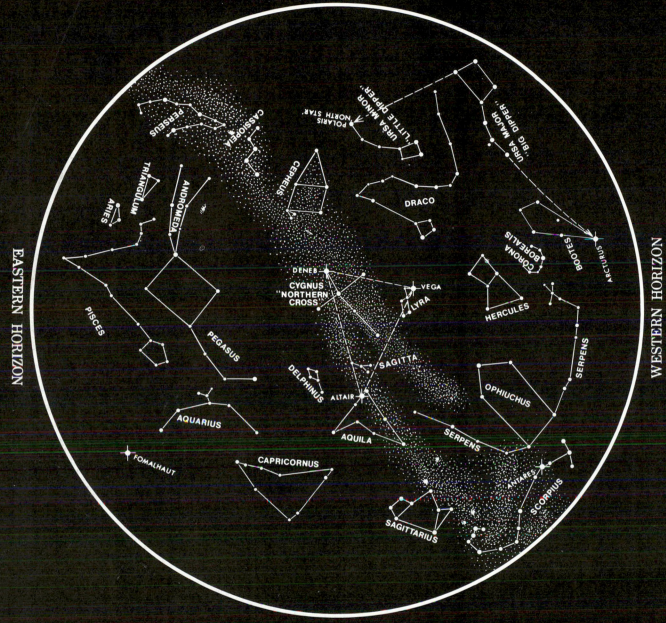

THE NIGHT SKY IN SEPTEMBER

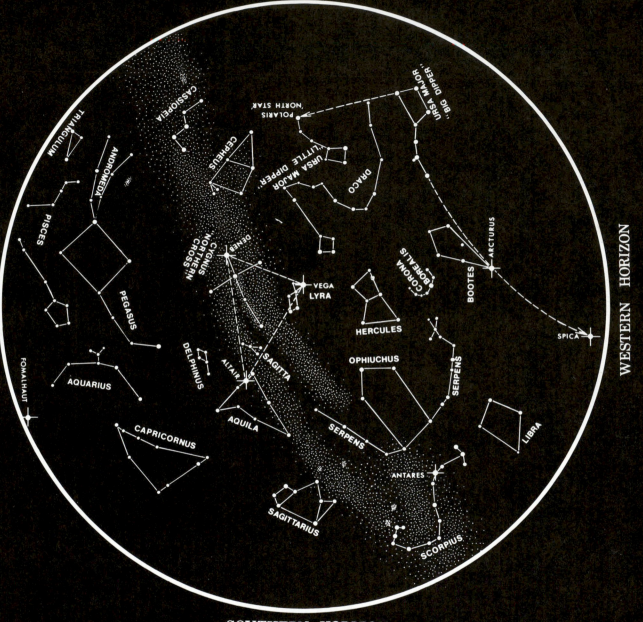

SOUTHERN HORIZON

THE NIGHT SKY IN AUGUST

Chart time (Daylight Savings Time):

11 pm...First of August
10 pm...Middle of August
9 pm...Last of August

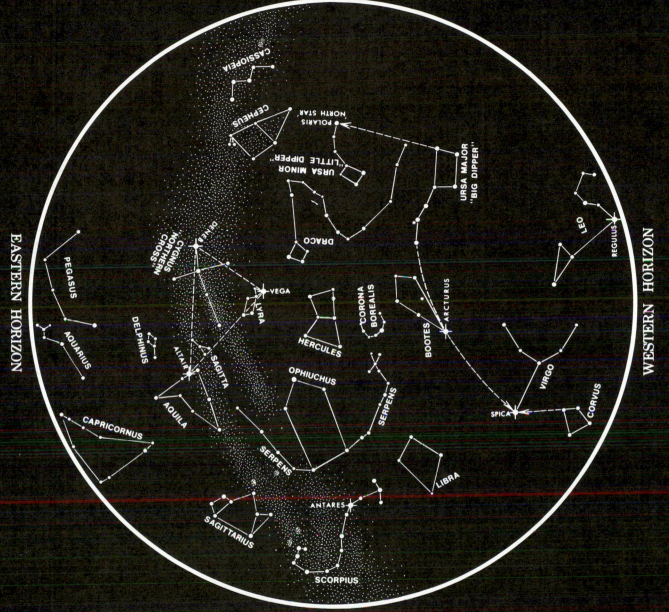

SOUTHERN HORIZON

THE NIGHT SKY IN JULY

Chart time (Daylight Savings Time):

11 pm...First of July

10 pm...Middle of July

9 pm...Last of July

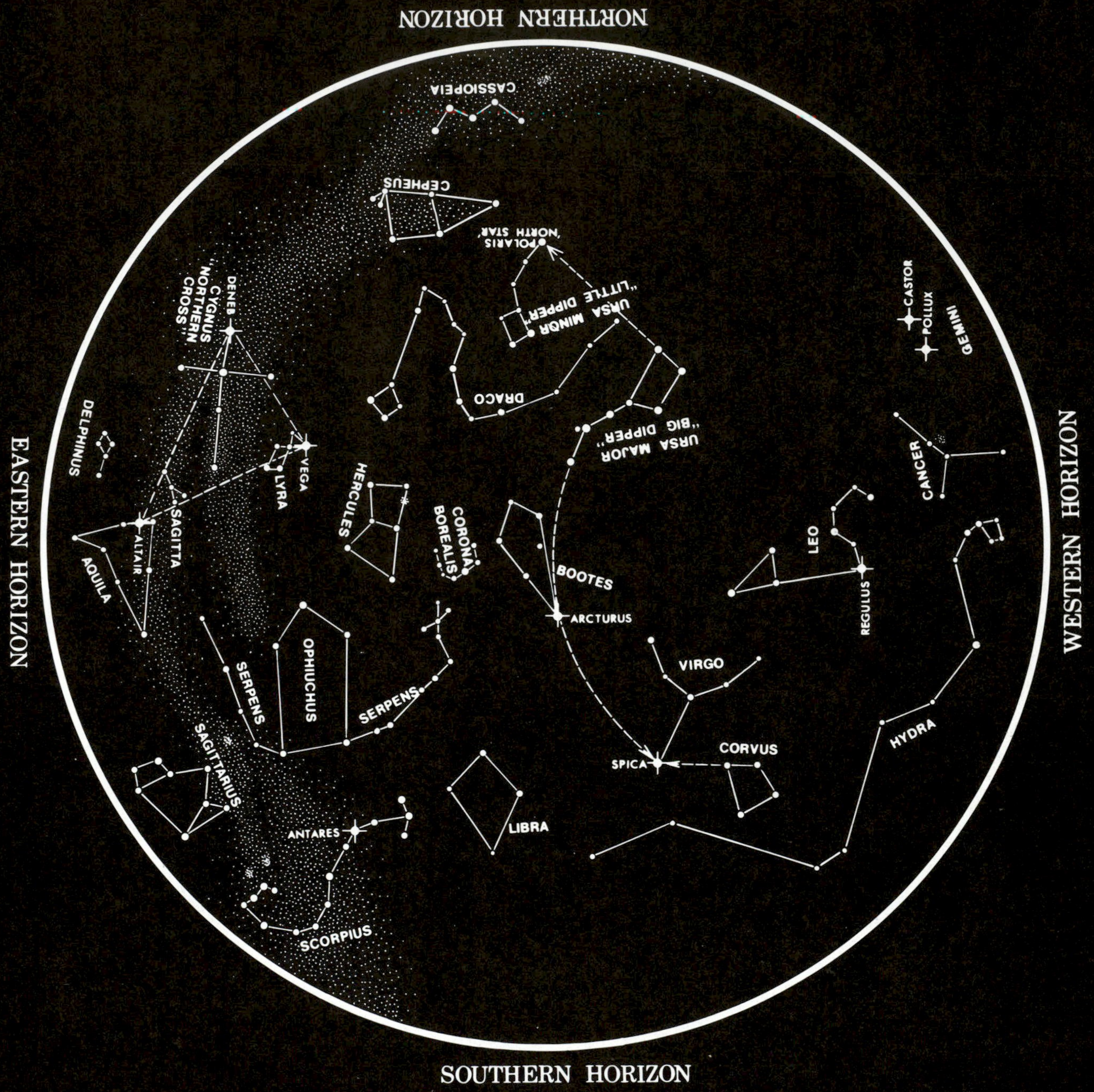

NORTHERN HORIZON

EASTERN HORIZON

WESTERN HORIZON

SOUTHERN HORIZON

THE NIGHT SKY IN JUNE

Chart time (Daylight Savings Time):

11 pm...First of June

10 pm...Middle of June

9 pm...Last of June

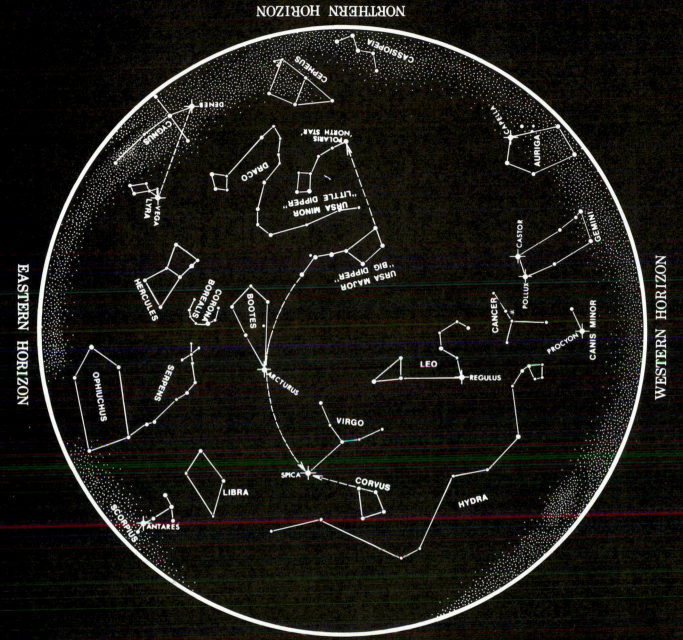

SOUTHERN HORIZON

THE NIGHT SKY IN MAY

Chart time (Daylight Savings Time):

11 pm...First of May
10 pm...Middle of May
9 pm...Last of May

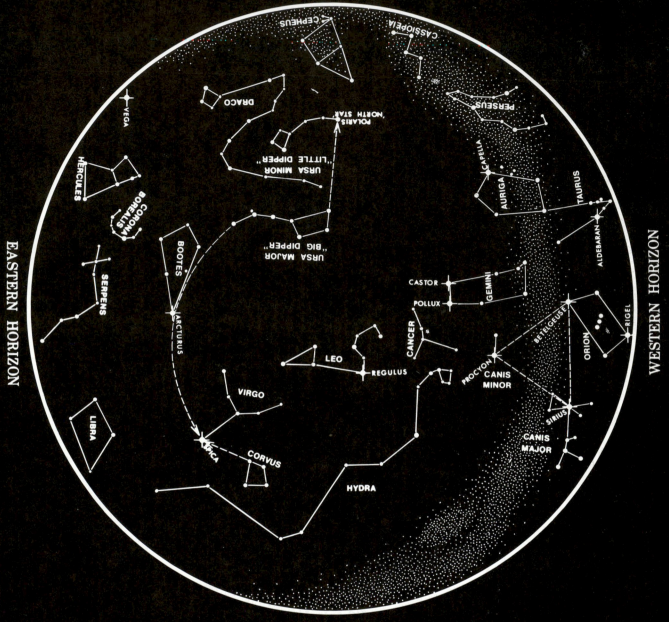

NORTHERN HORIZON

EASTERN HORIZON

WESTERN HORIZON

SOUTHERN HORIZON

THE NIGHT SKY IN APRIL

Chart time (Daylight Savings Time):

11 pm...First of April
10 pm...Middle of April
9 pm...Last of April